# Advances in High-speed Rail Technology

More information about this series at http://www.springer.com/series/13506

Youtong Fang · Yuehong (Helen) Zhang
Editors

# China's High-Speed Rail Technology

## An International Perspective

ZHEJIANG UNIVERSITY PRESS
浙江大学出版社

Springer

*Editors*
Youtong Fang
Zhejiang University
Hangzhou
China

Yuehong (Helen) Zhang
Zhejiang University
Hangzhou
China

ISSN 2363-5010 ISSN 2363-5029 (electronic)
Advances in High-speed Rail Technology
ISBN 978-981-10-5609-3 ISBN 978-981-10-5610-9 (eBook)
https://doi.org/10.1007/978-981-10-5610-9

Jointly published with Zhejiang University Press, Hangzhou

The print edition is not for sale in China Mainland. Customers from the China Mainland please order the print book from: Zhejiang University Press.

Library of Congress Control Number: 2017946043



Printed on acid-free paper

This Springer imprint is published by Springer Nature
The registered company is Springer Nature Singapore Pte Ltd.
The registered company address is: 152 Beach Road, #21-01/04 Gateway East, Singapore 189721, Singapore

# Foreword

The steam locomotive was first developed by Richard Trevithick in 1804, but several significant advances were made before its use on the Stockton and Darlington Railway in 1825, closely followed by the first true intercity railway in the world, between Liverpool and Manchester in 1830. George Stephenson and his son Robert were intimately associated with these early railways, so perhaps unsurprisingly, but not quite accurately, George Stephenson is known worldwide as the father of the railways. The railway rapidly became the most important mass transportation system because of its convenience, speed, efficiency, and considerable economic advantage over other existing but much slower modes. Although many improvements took place in the succeeding century, by the early 1960s, the railway generally was in decline in competition with, principally, the automobile. But in 1964, the opening of Japanese Tokaido Shinkansen marked the entry into the world of high-speed rail operating at high capacity and the rebirth of the railway. Modern high-speed rail technology assembles the world's advanced techniques in material science, control science, advanced manufacturing, electrical science and computing, which has made a big difference in safety, comfort, and speed compared with traditional railway technology, and has significantly changed people's travel habits and view of space and time.

The introduction of the railway to China was slow and tentative. A merchant called Durand from Britain built the 500 m of track in Beijing in 1865 as a demonstration. However, it was soon demolished by the Qing government as it was said to "have scared the dragon" and was not good for *Fengshui*. After several other abortive attempts, the first railway between Beijing and Jingzhang, about 200 km, was opened in 1909. However, even a century later in the early 2000s, there was still great gap between China and other developed countries in terms of per capita railway mileage and overall technical level of railways. Although a high share of goods and many people were transported by rail, capacity was a major constraint. For example, at times such as the Spring Festival, the railway supply can hardly match the enormous demand. To solve the problems, Chinese enterprises have set out to introduce high-speed rail technology from Japan, Germany, France, and Canada under the organization of the Ministry of Railways since 2004. And on the basis of that, they have developed their own high-speed rail system of research, design, manufacture, construction, operation, and maintenance. In 2008, the Beijing-Tianjin intercity line opened. After that, Wuhan-Guangzhou, Beijing-Shanghai, and some other 350 km/h high-speed railway opened successively. Nowadays, both the scale and the speed indicators of China's high-speed rail have reached an advanced level in world. The new high-speed line length of 20,000 km now exceeds the total length of high-speed line in the rest of the world combined and is due to top 38,000 km by 2025. In total, China now has the second largest railway network in the world and the railway is firmly in place as the preferred mode of transport over longer distances in China. The achievements of the last 20 years or so have been phenomenal and deserve to be understood and appreciated by wide audiences throughout the world. It is my hope that this book will go some way to achieving this.

Professor Youtong Fang is an expert in the field of railway traction power, as well as my old friend in this industry. The permanent magnet traction motor he designed for the 350 km/h high-speed railway has been adopted, and the energy-saving effect is remarkable according to the test results of the line. At the same time, Prof. Fang participated in the organization of China's high-speed rail technology innovation. In 2010, Helen Zhang, the Chief Editor of Journal of Zhejiang University-SCIENCE A: Applied Physics & Engineering (JZUS-A), who has been concerned about the progress of high-speed rail technology in China, invited Prof. Fang as a guest editor, soliciting the first special issue about Chinese high-speed rail. Right during this period, the Yongwentai railway accident occurred on July 23, 2011, which caused various international comments about the high-speed rail technology of China. In December of the same year, under Prof. Fang' s organization, JZUS-A published the first special issue in English on China's high-speed railway technology, followed by another three special issues on high-speed rail of China, which aroused considerable interest throughout the world.

In 2014, Prof. Fang invited me to visit the China CSR, Qingdao Sifang, China Tonghao Group, Zhejiang University, and Beijing Jiaotong University. Also, we took the high-speed train from Hangzhou to Beijing. By sharing viewpoints with Chinese scholars, and from my perspective of studying railway for many years, I wrote an article called "Background of recent developments of passenger railways in China, the UK and other European countries," in which I commented that the high-speed train system has tremendous advantages in increasing the efficiency and convenience of transport without adding to carbon generation. It consists of a brief introduction to the history and comparisons of railways in the UK and China, a description of rail speedup in the last few decades in the UK, and notes of current high-speed trains. Similar brief details are given of high-speed train in Europe. Brief mention is made of comparative railway safety. The development of high-speed rail in China is discussed, and the UK high-speed development plan is briefly introduced. Later during Prof. Fang's visit to Britain, he talked to me about his plan, to publish a book which is a compilation of recent studies on high-speed rail technology in China from an international perspective, after getting the permission from JZUS-A and authors. And the book will contribute to the international counterparts. He asked if I would like to write the preface for the book, and I readily agreed.

This book involves not only a comparison of railway development in China and abroad, such as Europe, the UK, and Japan, but also the strategic thinking of the development of high-speed rail of China. Furthermore, this book introduces the research on the some key technologies, covering various aspects of infrastructure, vehicles, signals, materials, dynamics, traction power supply, transportation organization, and so on. Although it cannot cover everything, this book gives an introduction of the research and technical work of Chinese scholars, which is very helpful for foreign scholars and industry to understand the new technology of Chinese high-speed rail. I hope that the publication of this book is able to attract more foreign scholars to participate in China's high-speed rail innovation, especially at the moment when China's high-speed rail is moving toward international prominence. Furthermore, I hope that it will serve to spread knowledge of China's recent remarkable railway progress and achievements.

Imperial College London Roderick A. Smith
January 2017

# Preface

In 1865, the British merchant Durand constructed China's first railway in Beijing, starting the research on railway technology in China. In 2002, the "China Star" high-speed train achieved a top speed of 321 km/h in a test run. In just a few years, research in this field is on a fast track and parallel study in the most developed countries. A new era started in 2008, as symbolized by the first commercial high-speed railway from Beijing to Tianjing. In about seven years, 19,000 km of high-speed railway has been built in China, longer than the total length in the rest of the world, demonstrating top-rank capability in both research and manufacturing. There is no doubt that this advanced high-speed rail technology could not have been achieved without the research of the overseas pioneers. At the same time, what China now does in high-speed rail promotes development throughout the world. In other words, academic exchange of high-speed rail technology between China and other countries is of great importance to both China and the rest of the world.

In 2010, Mrs. Helen Zhang, Chief Editor of *Journal of Zhejiang University-SCIENCE A: Applied Physics & Engineering* (*JZUS-A*), invited Prof. Youtong Fang, Director of the High-Speed Railway Research Center in Zhejiang University, to be the guest editor in chief for the first special issue of *JZUS-A* to summarize cutting-edge high-speed railway technology in China. Up to now, three special issues have been published on this topic, and these have aroused worldwide interest.

Study of the high-speed train itself is at the core of this research, and recent research in China has focused on the theory and practice of wheel–rail, fluid–structure, and bow–net interactions. Wei-hua Zhang, Xue-song Jin et al. developed high-speed rail wheel–rail dynamics, Yun-min Chen, Xue-cheng Bian et al. carried out systematic study on subgrade settlement, Xue-ming Shao, Yao Zheng et al. developed the computational fluid mechanics method for the high-speed train, Weihua Zhang et al. analyzed the dynamic performance of the bow–net system, Liang Meng et al. developed the catenary contact wire with higher strength and higher conductivity, Limin Jia, Qinfen Lu, Xiaoyan Huang, Ping Tan, Gang Chen et al. did excellent research on transportation organization, traction power, communication signals, and train intellectualization. The international perspective is also highlighted by the comprehensive review on the development of the China high-speed railway by Prof. Roderick A. Smith, the former Chief Scientific Advisor to Department of Transport, Fellow of the Royal Academy of Engineering, and Satoru Sone, Professor of Tokyo University, and Associate Director of the Japan Railway Technology Association.

The three special issues summarized not only in-depth research, but also the development route, especially charting the course of innovation that China has followed in high-speed rail technology. If we could collect all this work in one book, it would allow our international counterparts to better understand the research of Chinese scholars, and further promote the academic exchange of ideas and development. To this end, all the papers come from *JZUS-A*, and were reprinted in this book with permission from *JZUS-A* and Springer, as well as all authors.

I would like to thank the copy editors of the book, Ms. Hanfeng Lin and Ms. Xinxin Zhang. This book could not have been published without their hard work. I am also really grateful to every author—this book is a product of your intelligence and work. I also need to thank Prof. Roderick A. Smith for writing the foreword.

Finally, the publication of the book has also benefited from the support of the Natural Science Foundation of China (51637009, U1434202, 51477149, 51577166, 51507152), the "One belt one road" development open research alliance of science and technology, and the China Academy of West Region Development.

Qiushi Garden
January 2017

Youtong Fang

# Contents

# Editors and Contributors

## About the Editors

**Youtong Fang** is a Professor at Zhejiang University. He is Chairman of the High-Speed Rail Research Center in Zhejiang University, Deputy Director of the National Intelligent Train Research Center, on the committee of the China High-Speed Rail Innovation Plan, and expert of the National 863 Program (China High Technology Development Program) in modern transportation and advanced carrying technology. He is also director of three projects of the National Natural Science Foundation of China (NSFC) and more than ten National 863 projects and Technology Support Programs. Recent work has been on electrical machines and drives. His research interests include permanent magnet (PM) machines and drives for traction applications.

**Yuehong (Helen) Zhang** is Chief Editor of the *Journal of Zhejiang University-SCIENCE A (Applied Physics and Engineering)*. Since 2010, under her planning, *JZUS-A* has published three special issues on China's High-Speed Railway Technology. She is also a Board Member of PILA/CrossRef (2014–2017), a Council Member of ALPSP (2011–2016), and Vice President of the Society of China University Journals. She has often been invited to give speeches at international publishing seminars, including the 4th World Conference on Research Integrity (www.wcri2015.org). She received a research grant from the Committee on Publication Ethics (COPE) in 2011 and has published many papers in international journals, including several short papers in *Nature*. In 2016, she published two books: *Against Plagiarism: A Guide for Editors and Authors*, published by Springer, and *Chinese Cultural Kaleidoscope*, published by Zhejiang University Press.

## Contributors

**Guo-huan Bao** Department of Materials Science and Engineering, Zhejiang University, Hangzhou, China

**Xue-cheng Bian** Department of Civil Engineering, MOE Key Laboratory of Soft Soils and Geoenvironmental Engineering, Zhejiang University, Hangzhou, China

**Wen-ping Cao** Newcastle University, Newcastle upon Type, UK

**Chang Chao** Department of Civil Engineering, MOE Key Laboratory of Soft Soils and Geoenvironmental Engineering, Zhejiang University, Hangzhou, China

**Da-wei Chen** National Engineering Laboratory for System Integration of High-Speed Train (South), CSR Qingdao Sifang Co., Ltd, Qingdao, China

**Jin-miao Chen** Department of Civil Engineering, Zhejiang University, Hangzhou, China

**Ren-peng Chen** Department of Civil Engineering, Zhejiang University, Hangzhou, China

**Yi Chen** Department of Materials Science and Engineering, Zhejiang University, Hangzhou, China

**Yun-min Chen** Department of Civil Engineering, MOE Key Laboratory of Soft Soils and Geoenvironmental Engineering, Zhejiang University, Hangzhou, China

**Jian Chu** State Key Laboratory of Industrial Control Technology & Institute of Cyber-Systems and Control, Zhejiang University, Hangzhou, China

**Jian Deng** School of Aeronautics and Astronautics, Zhejiang University, Hangzhou, China

**An-ping Dong** China Railway Construction Electrification Bureau Group Co., Ltd., Beijing, China

**Yuan-feng Duan** College of Civil Engineering and Architecture, Zhejiang University, Hangzhou, China

**Ke-qing Fan** School of Information Engineering, Wuyi University, Jiangmen, China

**You-tong Fang** College of Electrical Engineering, Zhejiang University, Hangzhou, China

**Wei-wei Fei** Research Institute of Railway & Urban Mass Transit, Tongji University, Shanghai, China

**Qing-hua Guan** State Key Laboratory of Traction Power, Southwest Jiaotong University, Chengdu, China

**Guang-xu Han** State Key Laboratory of Traction Power, Southwest Jiaotong University, Chengdu, China

**Jian Han** State Key Laboratory of Traction Power, Southwest Jiaotong University, Chengdu, China

**Bin He** State Key Laboratory of Traction Power, Southwest Jiaotong University, Chengdu, China

**Wei-ting He** State Key Laboratory of Industrial Control Technology & Institute of Cyber-Systems and Control, Zhejiang University, Hangzhou, China

**Xiao-yan Huang** College of Electrical Engineering, Zhejiang University, Hangzhou, China

**Zhang-wen Huang** College of Electrical Engineering, Zhejiang University, Hangzhou, China

**Bing Ji** School of Electrical and Electronic Engineering, Newcastle University, Newcastle upon Tyne, UK

**Li-min Jia** State Key Laboratory of Rail Traffic Control and Safety, Beijing Jiaotong University, Beijing, China

**Wan-feng Jin** Department of Civil Engineering, MOE Key Laboratory of Soft Soils and Geoenvironmental Engineering, Zhejiang University, Hangzhou, China

**Xue-song Jin** State Key Laboratory of Traction Power, Southwest Jiaotong University, Chengdu, China

**Kai Li** Department of Information Science and Electronic Engineering, Zhejiang University, Hangzhou, China

**Rui-ping Li** State Key Laboratory of Traction Power, Southwest Jiaotong University, Chengdu, China

**Tian Li** State Key Laboratory of Traction Power, Southwest Jiaotong University, Chengdu, China

**Xin-hua Li** School of Aeronautics and Astronautics, Zhejiang University, Hangzhou, China

**Zhi-hui Li** State Key Laboratory of Traction Power, Southwest Jiaotong University, Chengdu, China

**Jia Lin** State Key Laboratory of Industrial Control Technology & Institute of Cyber-Systems and Control, Zhejiang University, Hangzhou, China

**Liang Ling** State Key Laboratory of Traction Power, Southwest Jiaotong University, Chengdu, China

**Jia-bin Liu** Department of Materials Science and Engineering, Zhejiang University, Hangzhou, China

**Zheng Liu** School of Electrical and Electronic Engineering, Newcastle University, Newcastle upon Tyne, UK

**Qin-fen Lu** College of Electrical Engineering, Zhejiang University, Hangzhou, China

**Xiao-pei Lu** Department of Materials Science and Engineering, Zhejiang University, Hangzhou, China

**Ji-en Ma** College of Electrical Engineering, Zhejiang University, Hangzhou, China

**Liang Meng** Department of Materials Science and Engineering, Zhejiang University, Hangzhou, China

**Wen-ting Mo** IBM China Research Lab, Beijing, China

**Siu-wing Or** Department of Electrical Engineering, The Hong Kong Polytechnic University, Hong Kong, China

**Yong Qin** State Key Laboratory of Rail Traffic Control and Safety, Beijing Jiaotong University, Beijing, China

**Xue-ming Shao** Department of Mechanics, Zhejiang University, Hangzhou, China

**Roderick A. Smith** Department of Mechanical Engineering, Imperial College London, London, UK

**Satoru Sone** Kogakuin University, Tokyo, Japan

**Xue-guan Song** School of Mechanical Engineering, Dalian University of Technology, Dalian, China

**Chuan-ming Sun** CSR Qingdao Sifang Co. Ltd., Qingdao, China

**Ping Tan** State Key Laboratory of Industrial Control Technology & Institute of Cyber-Systems and Control, Zhejiang University, Hangzhou, China; School of Automation and Electrical Engineering, Zhejiang University of Science and Technology, Hangzhou, China

**Zheng Tan** Power Grid Technology Center, State Grid Jibei Electric Power Co. Ltd Research Institute, Beijing, China

**Zhi-feng Tang** College of Civil Engineering and Architecture, Zhejiang University, Hangzhou, China

**Chun Tian** Research Institute of Railway & Urban Mass Transit, Tongji University, Shanghai, China

**Jun Wan** Department of Mechanics, Zhejiang University, Hangzhou, China

**Bin Wang** College of Electrical Engineering, Zhejiang University, Hangzhou, China

**Han-lin Wang** Department of Civil Engineering, Zhejiang University, Hangzhou, China

**Heng-yu Wang** State Key Laboratory of Traction Power, Southwest Jiaotong University, Chengdu, China

**Li-tian Wang** China Railway Construction Electrification Bureau Group Co., Ltd., Beijing, China

**Li Wang** School of Traffic and Transportation, Beijing Jiaotong University, Beijing, China

**Rui-qian Wang** School of Urban Rail Transit, Changzhou University, Changzhou, China

**Xin Wang** College of Materials and Environmental Engineering, Hangzhou Dianzi University, Hangzhou, China

**Ze-feng Wen** State Key Laboratory of Traction Power, Southwest Jiaotong University, Chengdu, China

**Meng-ling Wu** Research Institute of Railway & Urban Mass Transit, Tongji University, Shanghai, China

**Xin-biao Xiao** State Key Laboratory of Traction Power, Southwest Jiaotong University, Chengdu, China

**Fang-fang Xie** School of Aeronautics and Astronautics, Zhejiang University, Hangzhou, China

**Hong-bing Xiong** Department of Mechanics, Zhejiang University, Hangzhou, China

**Jia-yang Xiong** State Key Laboratory of Traction Power, Southwest Jiaotong University, Chengdu, China

**Jie Xu** State Key Laboratory of Rail Traffic Control and Safety, Beijing Jiaotong University, Beijing, China

**Da-wei Yao** Department of Materials Science and Engineering, Zhejiang University, Hangzhou, China

**Li Yao** Danfoss (Tianjin) Ltd., Tianjin, China

**Jin Yu** National Engineering Laboratory for System Integration of High-Speed Train (South), CSR Qingdao Sifang Co., Ltd, Qingdao, China

**Meng-ge Yu** State Key Laboratory of Traction Power, Southwest Jiaotong University, Chengdu, China

**Wen-guang Yu** Department of Mechanics, Zhejiang University, Hangzhou, China

**Bin Zhang** State Key Laboratory of Fluid Power and Mechatronic Systems, Zhejiang University, Hangzhou, China

**He Zhang** Department of Electrical and Electronic Engineering, The University of Nottingham, Nottingham, UK

**Ji-ye Zhang** State Key Laboratory of Traction Power, Southwest Jiaotong University, Chengdu, China

**Jian-cheng Zhang** College of Electrical Engineering, Zhejiang University, Hangzhou, China

**Jian Zhang** College of Electrical Engineering, Zhejiang University, Hangzhou, China

**Jie Zhang** State Key Laboratory of Traction Power, Southwest Jiaotong University, Chengdu, China

**Ru Zhang** College of Civil Engineering and Architecture, Zhejiang University, Hangzhou, China

**Wei-hua Zhang** State Key Laboratory of Traction Power, Southwest Jiaotong University, Chengdu, China

**Guo-tang Zhao** State Key Laboratory of Traction Power, Southwest Jiaotong University, Chengdu 610031, China; China Railway Corporation, Beijing 100844, China

**Hong-ming Zhao** Zhejiang Insigma-Supcon Co., Ltd., Hangzhou, China

**Shu-min Zhao** College of Materials and Environmental Engineering, Hangzhou Dianzi University, Hangzhou, China

**Xin Zhao** State Key Laboratory of Traction Power, Southwest Jiaotong University, Chengdu, China

**Yang Zhao** College of Civil Engineering and Architecture, Zhejiang University, Hangzhou, China

**Yue Zhao** State Key Laboratory of Traction Power, Southwest Jiaotong University, Chengdu, China

**Yao Zheng** School of Aeronautics and Astronautics, Zhejiang University, Hangzhou, China

**Yong-jian Zhi** College of Electrical Engineering, Zhejiang University, Hangzhou, China

**Shuo-qiao Zhong** State Key Laboratory of Traction Power, Southwest Jiaotong University, Chengdu, China

**Jing Zhou** Department of Mechanical Engineering, Imperial College London, London, UK; College of Electrical Engineering, Zhejiang University, Hangzhou, China

**Li Zhou** State Key Laboratory of Traction Power, Southwest Jiaotong University, Chengdu, China

**Ning Zhou** State Key Laboratory of Traction Power, Southwest Jiaotong University, Chengdu, China

**Qiang Zhou** State Key Laboratory of Traction Power, Southwest Jiaotong University, Chengdu, China

**Min-hao Zhu** State Key Laboratory of Traction Power, Southwest Jiaotong University, Chengdu, China

**Yang-yong Zhu** Research Institute of Railway & Urban Mass Transit, Tongji University, Shanghai, China

# Copy editors:

Han-feng Lin, Xin-xin Zhang
Zhejiang University Press, Hangzhou, China

# Part I
# Overview of High-Speed Rail Technology Around the World

# Sustainability Development Strategy of China's High-Speed Rail

Ping Tan, Ji-en Ma, Jing Zhou and You-tong Fang

## Contents



## 1 Overview

The shortage of railway transport capability has restricted the development of China's economy. High-speed rail is vital for the development of the passenger railway, because of its huge transport capacity, safety, comfort, all-day operation, environ-

P. Tan
School of Automation and Electrical Engineering, Zhejiang University of Science and Technology, Hangzhou 310023, China

J. Ma
China Academy of West Region Development, Zhejiang University, Hangzhou 310027, China

J. Zhou · Y. Fang (✉)
College of Electrical Engineering, Zhejiang University, Hangzhou 310027, China
e-mail: youtong@zju.edu.cn

 
Y. Fang and Y.H. Zhang (eds.), *China's High-Speed Rail Technology*,
Advances in High-speed Rail Technology, https://doi.org/10.1007/978-981-10-5610-9_1

mentally friendly operation, and its sustainability. Since 2004, the former China Ministry of Railways has introduced and assimilated advanced technologies in other countries, used this to improve the technologies of the construction of the high-speed rail and train, and formed a nationwide cross-industry chain (Smith and Zhou 2014).

According to the China "Mid-long Term Railway Network Plan" (NDRC 2016), by 2020, the national railway operational mileage will reach 120,000 km. The double-track rate and electrochemical rate will reach 50 and 60%, respectively, and passenger transport and freight transport will be separate on main lines, of which the passenger dedicated line will reach 16,000 km. China's high-speed rail network covers a vast territory and experiences complex geographical, geological conditions, and climate. The rapid development and unique network conditions of China's high-speed rail make higher demands on the high-speed rail technology system, especially the high-speed train technology. These demands can be described as three main factors:

1. As high-speed train manufacture and system integration capability improve, so also must the corresponding design, optimization, experimental and evaluation techniques, as well as other supporting techniques.
2. The basic theories of the high-speed train must be developed based on research and must be suitable for the environmental conditions and operational requirements in China.
3. It is difficult for the existing railway research system to satisfy the needs of innovation, integrate domestic research resources, establish high-level research, and exploit and experiment with platform and industry–study–research–application in a combined industry alliance of the type needed to support systematic introduction and operation.

Against this background, in February 2008, the Ministry of Science & Technology and Ministry of Railways of China jointly published "China High Speed Train Independent Innovation Joint Action Plan." Under the support of this plan, on December 3, 2010, the new generation of high-speed train CRH380AL, having incorporated home-grown innovations, operated at 486.1 km/h high-speed rail test speed on the Beijing–Shanghai line pilot segment. All the train's performance indices fully satisfied the design requirements, which demonstrated that China's high-speed train technology had firmly established itself as the world's top level high-speed train technology. By the end of 2011, the Beijing–Shanghai high-speed rail was constructed and put into operation, representing, at that time, the world's longest mileage and highest technology standards with non-staged construction.

## 2 Innovation Achievements of China's High-Speed Rail Technology

The study and construction of high-speed rail in China have been going on for nearly 20 years. The first stage was from 1990 to 2007, during which there were five big accelerations in the development of the national rail, and the introduction and absorption of high-speed train technology from Germany, Japan, and France. The second stage is from 2008 until now, and this is the stage of independent innovation, the highlight of which is the launch of "China High Speed Train Independent Innovation Joint Action Plan." Supported by the Joint Action Plan, 25 key universities, 11 scientific research institutions, and 51 national laboratories and engineering technology research centers in China carried out a wide range of technical cooperation and exchange and quickly tackled the key technical problems, so as to ensure the development of the new-generation high-speed train (MST 2012).

## *2.1 Basic Theory Study of the High-Speed Train*

A high-speed train system dynamic theory was proposed, the study objective was extended from the single carriage to the whole train, and the research scope was extended from wheel–rail interface to wheel–rail, bow net, and fluid–structure interaction. Corresponding research was carried out through non-linear dynamic modeling, and bench and track testing.

On wheel–rail interaction, research was carried out on the wheel–rail profile, the matching of materials and hardness, and the matching relation between lines and high-speed trains, based on which the wheel–rail contact model under complicated conditions was established, and corresponding numerical methods were proposed and a high-speed train wheel–rail contact relation under system dynamics was put forward (Zhao et al. 2014).

On train fluid–structure, based on high-speed train numerical calculation software and a support platform, analysis was carried out on lateral wind dynamics and aerodynamic moment in train operation, as well as in safety analysis.

For the train bow net, the platform constructions, like the mechanism of the arc test rig and the pantograph vibration test rig, have been basically completed. The preliminary design of pantograph and catenary geometry and dynamic characteristics has been basically finished. The double-arc spacing calculation formula was proposed for the high-speed bow net system on the Beijing–Tianjin and Wuhan–Guangzhou Passenger Lines and ensured the stability of double arc at speeds under 380 km/h.

For train vibration and modal analysis, experiments were carried out on slab track at 350 km/h on the Beijing–Tianjin Intercity High Speed Line, the Wuhan–Guangzhou Passenger Line, and the Zhengzhou–Xi'an Passenger Line. These examined the system vibration phenomenon and characteristics caused by the periodical ratio relationship between wheel circumference, slab length, rail fixed length, and bridge span length, and this provided the basis for the design and structural optimization of the new-generation high-speed train. In addition, the general high-speed train comfort simulation platform was established through analysis of in-car comfort indices.

## 2.2 Design and Manufacturing Technologies of the High-Speed Train

During construction and improvement of a digital collaborative simulation platform, the technology architecture covering the design and experimental platform of the train set assembly, and the design of the train body and bogie are formed.

The development of the high-speed bogie digital design platform, the experimental platform, and the digital processing platform are completed; the batch production capability of the high-speed train bogie is now achieved. The system now uses a double-H welding bogie frame and bogie-integrated cast aluminum alloy transition corbel, a hollow car axle, and an aluminum alloy gear box structure, and this significantly reduces unsprung weight and wheel–rail dynamic interaction. A high flexibility air spring, a two-point air spring control system, and an adjustable length lateral rolling torsion bar device are adopted to improve the high-speed operation quality of the bogie. An elastic suspension traction drive structure is adopted to improve the comprehensive dynamic performance. Wheel-set guiding, an air spring, and a traction motor emergency system ensure the safety and reliability of the bogie.

According to the operational requirements of a high-speed train, aerodynamic performance evaluation of 20 design schemes of train head was carried out, to help finalize the design of the CRH380A series high-speed train head shape. The experiments on the Wuhan–Guangzhou, and Zhengzhou–Xi'an high-speed lines demonstrated that the total operational resistance of an eight-car train set at 350 km/h decreased 6.1%, the noise level decreased 7%, the tail lift force reduced 51.7%, and the lateral force reduced 6.1%, thus realizing the optimization design of the new generation of high-speed train, which provides important support for train head shape design.

The 1:8 model of the CRH380AL/BL series high-speed train and aerodynamic noise evaluation of key components were accomplished in a high-speed train aerodynamic acoustic wind tunnel. The adaptability of high-speed train aerodynamic noise measurement technology was studied and fully validated that the noise measured in the wind tunnel accurately reflects the train's aerodynamic noise in magnitude and frequency characteristics. The upgrading scheme for a pneumatic acoustic wind tunnel has successfully passed systemic proof, and it is estimated that the experimental speed in the wind tunnel will exceed 420 km/h.

The high-speed train aerodynamic numerical simulation platform is capable of modeling a 16-car full-size high-speed train and subgrade, bridge, tunnel, platform, and similar work conditions. Currently, vehicle appearance, new head shape series, and aerodynamic performance evaluation and optimization of the CRH2, CRH3, CRH380 series high-speed train, and higher speed test train under severe work conditions (cross wind, tunnel crossing) are carried out on this platform, which provides key systemic and theoretical support for resistance and noise reduction.

In car body modal optimization, there have been many new vibration reduction structures and materials, so as to achieve the required vibration reduction effects.

On the premise of a light car body, the first-order vertical bending frequency of a car body modal of vibration achieves a 10% increase and reaches 16.8 Hz; the first-order natural frequency of the floor increases by 22% and reaches 40.5 Hz; the first-order natural frequency of the end wall increases by 21% and reaches 48.2 Hz.

Concerning car body aerodynamic load, the multi-component, high-frequency, large-amplitude alternating aerodynamic load was studied, and airtight strength and tightness control standards for an operational speed of 350 km/h and above were established; new designs of cross section and bearing structure were adopted to optimize the pressure control method in a car; the air tightness of the car body increased from ±4000 to ±6000 Pa; the pressure variation rate is controlled below 2‰ atmosphere; and the maximum variation amplitude is controlled below 8‰ atmosphere pressure.

The design, simulation, and experimental platform of the traction drive system and brake system were established, and the demonstration base, pilot line, and production line for a high-speed train were constructed. This is capable of integrated manufacturing and mass production. The strength and reliability of bow net components were improved, and the dynamic current-receiving quality of the pantograph and catenary was increased (Ma et al. 2011).

The Chinese Train Control System (CTCS)-3 train operation control system set was completed, and a complete CTCS-3 train control system validation process was established. A sub-system special test, a laboratory integration test, an on-site installation, test and commissioning (ITC) test, an alignment test, and trial operation were completed. Problems such as electromagnetic interference, wireless communication interrupt, and Global System for Mobile Communications-Railway (GSM-R) network optimization were solved. This helped verify the reliability, safety, and stability of the system and improved the general technical level of the CTSC-3 train control system.

High-strength, high-conductivity contact wires with 37 kN tension force were developed and operated in the Beijing–Shanghai guideline, the strength of which reaches 570 MPa and the conductivity reaches 75% IACS. The dedicated "2 × 27.5 kV" switch cubicle realized large-scale domestic production.

The high-speed rail society–economy affected zone model was built, and a high-speed train passenger transport demand database was established, so as to accomplish high-speed rail basic database sharing and a service platform and the development of a high-speed rail passenger transport demand data service sub-system. The high-speed rail resource optimization system was established, and the "Rail Lines Video Query System" based on high-speed rail basic data sharing and service platform is able to provide data support and supportive decision-making for railway equipment management, scheduling command, rescue and relief work, and accident rescue.

The main pieces of equipment for a high-speed inspection car were developed, including a video device, a measurement device at the wiring interval, an automatic train control (ATC) determinator, a train wireless equipment measurement device, and a determination station; there was also an axle load transverse pressure measurement axis, an axle box acceleration measurement device; a track vertical displacement and vehicle shaking measurement device, line condition monitoring equipment, wheel load transverse pressure data processing equipment and video devices; stringing abrasion offset height measurement device, collector state monitoring device, pantograph observation device; electric power measurement station, data processing equipment, a power supply circuit measurement device, a train number ground measurement device, high-speed electric multiple unit (EMU) technology, the ground monitoring data analysis technology, a high-speed mobile comprehensive detection technology system including the open management and application technology of the high-speed railway infrastructure testing data.

## 3 Engineering Application of Domestic High-Speed Train

### *3.1 Opening and Operation of the Wuhan–Guangzhou High-Speed Train*

On December 26, 2009, the Wuhan–Guangzhou high-speed line was put into operation with a domestic high-speed train, which created the record of passing speed of 350 km/h in tunnels, 350 km/h double train in connection, and a double pantograph receiving current.

The Wuhan–Guangzhou high-speed line is the one under the world's most complicated working conditions. Its total length is 1068 km with 226 tunnels. The longest tunnel is Liuyanghe Tunnel at 10 km, and the inner and outside temperature difference of the tunnel is large. The percentage of bridge and tunnel is 67%. The Wuhan–Guangzhou high-speed rail features high operation speed, high passing speed, and high passing density. These unprecedented operating conditions pose first-ever challenges for a high-speed train.

The main technology innovations are as follows:

In terms of long-distance operation: (1) the reliability optimization of suspending items in view of high-speed complex air turbulence (including suspending below the train like a skirtboard, or a windshield); (2) with reference to complex aerodynamic effects, improvement of the car body air tightness, and optimization of the car body structural strength design; (3) aiming at continuous high-speed operation, optimization of the locate mode, location parameter, and secondary suspension parameters, and improvement of the coupling relationship between wheel and rail, to effectively control abrasion and wheel/rail interaction force; (4) in a complicated vibration environment, improvement of the structural strength design of the bogie and increase of the fatigue life by a factor of two.

In terms of train air tightness: Improve the air tightness of the whole train. When two trains pass each other with an operation speed of 350 km/h in a tunnel, it takes 252 s for the in-car pressure to reduce from 4000 to 1000 Pa, while the standard of pressure reduction is above 50 s; the pressure variation in-car is less than 1000 Pa, and the pressure variation rate in-car is less than 200 Pa/s.

In terms of vibration and noise reduction: Optimize the structural parameters of the bogie and systematically optimize the modal matching relation among bogie, car body, and line, for example, car sickness (0.5–1 Hz) and tiredness (8–10 Hz); track periodic turbulence: 6.5, 32, and 100 m.

In terms of wheel/rail matching: Continuously track the performance difference of various track profiles after abrasion with wheel tread, detecting the sensitive response region of the wheel/rail in high-speed conditions in an innovative manner, and achieve a series of solutions (rolling circle, contact point, rail grinding, and rotary wheel tread).

In optimization of aerodynamic performance and resistance: Operation resistance is reduced by 6%; currently, there are 66 trains operating on Wuhan–Guangzhou line, and the annual electricity saving is 120 million kWh.

## 3.2 *Opening of the Beijing–Shanghai High-Speed Line*

On April 27, 2010, the first sample car of the "Harmony" CRH380A high-speed train came off the assembly line successfully in China South Locomotive and Rolling Stock Co. Ltd. (SCR) Qingdao Sifang. The CRH380A high-speed train features 350 km/h operational speed and 380 km/h maximum speed and adopts a low resistance streamline head profile, high air tightness, and airtight car body; the advanced acoustic vibration reduction technology and innovations in strong green power traction system technologies not only ensure stable low-noise operation, but also realize low resistance, lightweight, regenerative braking, green power, and zero release. On December 3, 2010, the CRH380A high-speed train created an operation test speed of 486.1 km/h. The new generation of domestic high-speed train came into service and provides key technology elements for the Beijing–Shanghai high-speed line.

On June 30, 2011, the Beijing–Shanghai high-speed line completed construction and was put into operation. The total length of the Beijing–Shanghai high-speed line is 1318 km with a design speed of 350 km/h. This is the world's first high-speed line with such a long mileage and high-technology standards; the whole line fully adopts the technology and equipment of the domestic high-speed train, which demonstrates the highest level of China's high-speed rail and train.

## 3.3 *Domestic 400 km/h Comprehensive Inspection Car Came off the Assembly Line*

The comprehensive inspection car is important for periodic, comprehensive, and high-speed detection of a high-speed train with speeds above 200 km/h and has the comprehensive inspection ability of railway infrastructure such as rail, catenary, and communication signals.

In March 2011, the CRH380B-00 high-speed comprehensive inspection train, developed by the China Academy of Railway Sciences and Tangshan Railway Vehicle Co. Ltd., came off the assembly line and was put into operation. The designed maximum test speed of the train is 500 km/h, the synchronous detection speed reaches more than 350 km/h, and the maximum test speed reaches 400 km/h, and the train has the real-time detection ability of hundreds of parameters like high-speed rail wheel/rail dynamic and vehicle dynamic response, catenary, communication, and signals.

Relying on the comprehensive inspection car, the ground monitoring data analysis and processing center for comprehensive evaluation and decision support were established, expert analysis diagnosis, ground demonstration, data storage, and management systems were developed, and the open management platform for domestic high-speed rail infrastructure was built, all of which provide experimental verification for track irregularity standard management. Also developed are the Beijing–Shanghai high-speed line irregularity spectrum, a train/track system dynamic characteristic evaluation method based on the generalized energy method, a train/track system safety evaluation method using axle box acceleration rate, a train track dynamic simulation model, the relationship between contact line irregularity and bow net dynamic response, and a transponder message evaluation method for high-speed rail above 350 km/h. At the same time, the dynamic debugging and functional verification of the wireless transfer system and ground demonstration system were carried out.

## 4 Development Prospect of China's High-Speed Rail Technology

The construction of China's high-speed rail provides a convenient, economic, rapid, comfortable, and ecological mass transport tool for our people, and it also affects the regional economic pattern of China (Jin 2014). It has had a warm reception from our people, industry, investors, and the government. The next 10 years is an important period of China's high-speed rail construction and development. In 10-year time, the high-speed rail network covering China's mainland will be almost completed, and the high-speed rail will be the first choice for mid-long journeys. It will also promote the development of China's economy.

### *4.1 China's High-Speed Rail Network Plan in the Next 10 Years*

On June 29, 2016, the State Council Executive Meeting approved in principle the new "Mid-long Term Railway Network Planning," which requires the construction of a comprehensive transport system integrating road, navigation, and aviation, providing support for the development and upgrading of economy and society with the artery of transportation. It will involve construction of a high-speed railway network with "Eight Longitudinal" featuring the seaside and Beijing–Shanghai and "Eight Horizontal" featuring land bridge and riverside, and intercity railways as a supplement. It will form a 1- to 4-h transport circle between large and medium cities, and 0.5- and 4-h transport circle within urban clusters. At the same time, it will nurture and develop new economy formats of high-speed rail, promote the regional exchange and cooperation and resource optimization, accelerate the industrial gradient transfer, and drive the transformation and upgrading of the manufacturing industry and economy as a whole.

The detailed planning scheme includes the following:

First, construction of the "Eight Longitudinal and Eight Horizontal" high-speed rail main pathways (Fig. 1). The "Eight Longitudinal" pathways are as follows: the seaside path, Beijing–Shanghai path, Beijing–Hong Kong (Taiwan) path, Beijing–Harbin–Beijing–Hong Kong–Macao path, Hohhot–Nanjing path, Beijing–Kunming path, Baotou (Yinchuan)–Haikou path, and Lanzhou (Xi'an)–Guangzhou path; the "Eight Horizontal" pathways are as follows: the Suifenhe–Manzhouli path, Beijing–Lanzhou path, Qingdao–Yinchuan path, land bridge path, riverside path, Shanghai–Kunming path, Xiamen–Chongqing path, and Guangzhou–Kunming path.

Second, there will be expansion of the regional railway conjunction line. Based on the "Eight Longitudinal and Eight Horizontal" main pathways, it will plan and lay out high-speed rail regional connections, in order to further improve and increase the coverage of the high-speed rail network.

Third, there will be development of the intercity passenger line. While taking advantage of the high-speed rail and the conventional rail with intercity service, it will plan and construct the city cluster intercity passenger railway effectively linking medium-big cities and town centers, and supporting and leading the development of new urbanization.

The new projects of high-speed rail main paths adopt the 250 km/h and above standards in principle, in which 350 km/h standards were adopted for lines connecting large cities, with large population density, and a developed economy. The regional railway lines adopt 250 km/h and below standards. Intercity lines adopt 200 km/h and below standards.

According to the new railway network planning, up to 2020, a batch of major landmark projects will be completed and put into operation, in which high-speed rail has 3000 km in length, covering over 80% of big cities. By 2025, the railway network will reach 175,000 km, in which the high-speed railway will account for 38,000 km.

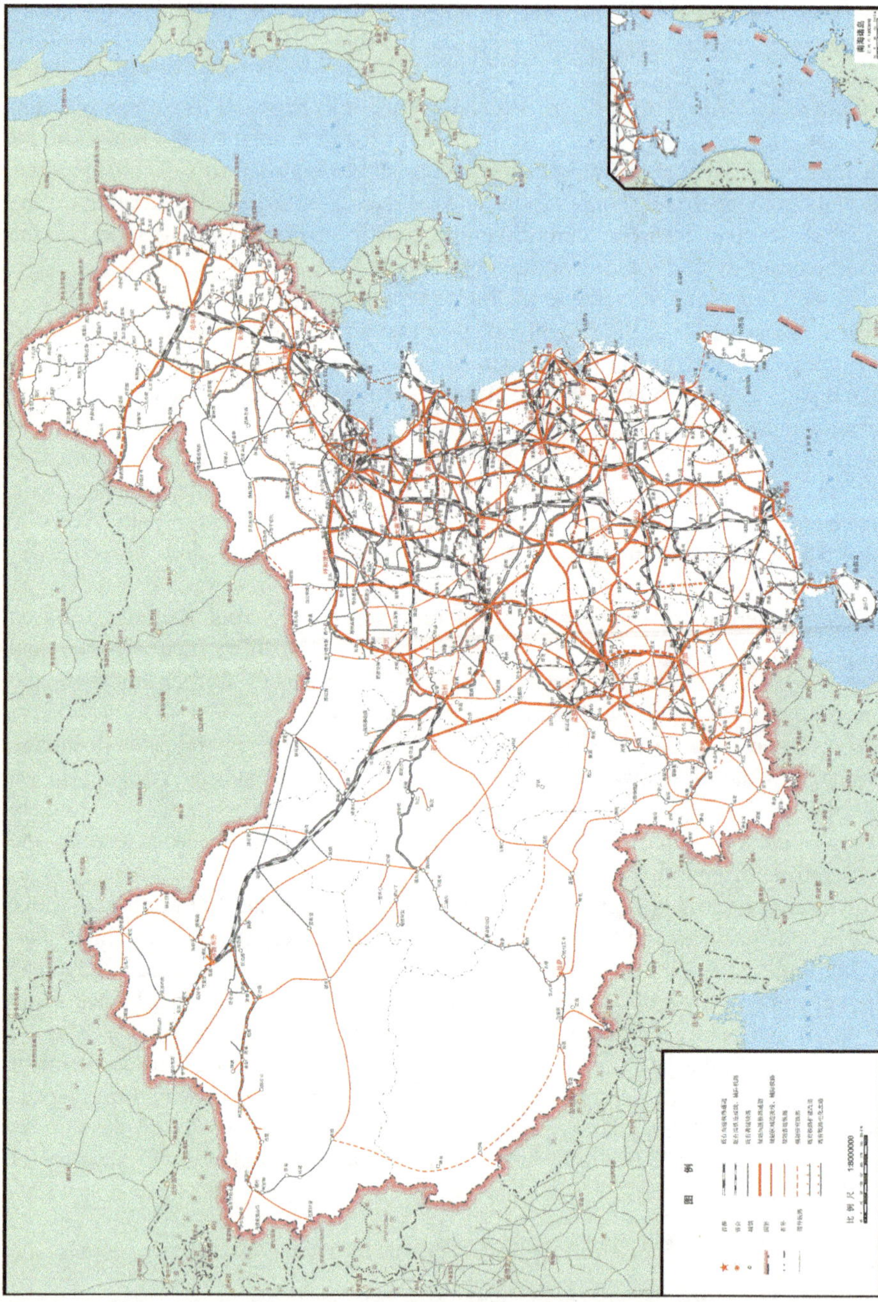

◄**Fig. 1** Mid-long-term railway network planning in China (http://www.gov.cn/xinwen/2016-07/20/content_5093165.htm). Translation of important places in "Eight Longitudinal and Eight Horizontal": 北京: Beijing, 上海: Shanghai, 香港: Hong Kong, 台湾: Taiwan, 哈尔滨: Harbin, 澳门: Macao, 呼和浩特: Hohhot, 南京: Nanjing, 昆明: Kunming, 包头: Baotou, 银川: Yinchuan, 海口: Haikou, 兰州: Lanzhou, 西安: Xi'an, 广州: Guangzhou, 绥芬河: Suifenhe, 满洲里: Manzhouli, 青岛: Qingdao, 厦门: Xiamen, 重庆: Chongqing. Translation of the legend: 首都: capital, 省会: provincial capital, 城镇: town, 国界: national boundaries, 省界: provincial boundaries, 国外铁路: railway in other countries, 既有高速铁路通道: existing high-speed rail, 既有高速连接线、城际铁路: existing high-speed rail connection inter-unban railway, 既有普速铁路: existing conventional rail, 规划高速铁路通道: planned high-speed rail, 规划区域连接线、城际铁路: planned regional connection inter-unban railway, 规划普速铁路: planned conventional rail, 规划研究铁路: planned research rail, 既有铁路扩能改造: expansion of existing railway, 既有铁路电化改造: electrification of existing railway

## 4.2 Study and Deployment of China's High-Speed Rail Research in the Next 5 Years

During the 13th Five-Year Plan, China launched the "Advanced Rail Transit" key program of the national key research and development plan. The main contents are given below.

### 4.2.1 Safety Assurance Technology of the Rail Transit System

There will be study of the sensing, evaluating, and alerting technique of the rail transit operational environment status, comprehensive monitoring and assurance technique of public right of way (ROW), decoupled and comprehensive safety assurance technique of the rail transit system, and the comprehensive safety assurance technique of the regional rail transit system, so as to form a rail transit safety-related holographic intelligent perception, fast identification, risk evaluation, early warning and emergency response, and constitutive security of carrying equipment. There will be construction of a comprehensive rail transit safety assurance technique platform including safety prediction evaluation theory and method, sets of safety standards techniques and regulations, and a technology support system. There will be construction of a rail transit safety assurance and emergency management integrated management platform. It should have the capability to lower rail transit safety accidents caused by technical issues by 50%, and effect a switch to active safety assurance.

### 4.2.2 High-Energy Efficiency of Traction Power Supply and Transmission Key Techniques of Rail Transit

There will be a revolution in traction drive technology. This will come about through the study of the following list of research topics: the study of virtual in-phase flexible power supply technology, the catenary system and power supply device with high flow characteristics, efficient converter device of rail transit train, contactless power supply techniques of rail transit, key techniques and equipment development of contactless power-supplied urban rail vehicle, key techniques and equipment development of interval power-supplied rail transit, master techniques regarding the train high-performance flexible power supply, high-conductivity wire materials, reducing the consumption of the converter, comprehensive usage of regenerative energy, traction converter technology based on new topological transformation, new materials and new structures, the realization of a virtual in-phase power supply, and high-efficiency traction converter technology and in-car equipment system. A comprehensive grasp of new power batteries, super capacitor energy storage application techniques, and car-ground integrated static dynamic contactless/wireless current collection technique will form safe, efficient, and economic rail transit hybrid energy storage and traction drive systems suitable for a multi-power supply mode and complicated application conditions (Huang et al. 2015).

### 4.2.3 Life Cycle Maintenance Technique of Rail Transit

This will come about through the study of rail transit integrated design–manufacture–operation technology concerning whole life cycle cost, rail transit train environmental-friendly technology, and the decoupled and efficiency improvement key technology in the rail transit energy consumption procedure, to form the whole life cycle design criteria integrating design, manufacturing, and operation. It will develop the integrated key technology concerning whole life cycle rail transit with environmental-friendly, comprehensive cost control, and efficiency improvement, realize the rail transit whole life cycle low cost, be environmental-friendly, and offer whole-scale efficiency improvement (Zhang et al. 2015).

### 4.2.4 Guided Transport System Mode Diversification and Equipment Study

Through the study of self-guided urban rail transit train system technology and equipment, and self-adjusted bogie key technology, the aim is to master the virtual rail guided transport system technology, to master the structural and parameter adapting and control techniques of wheel–rail bogie and line, and to create the supporting technology and equipment system of design, manufacture, evaluation, delivery, and operation of a guided transport system suitable for the town diversity of our country.

### 4.2.5 Key Technology for 400 km/h and Above High-Speed Passenger Transport Equipment

Through the study of train multi-effect coupled and smart control technology, the comprehensive comfort level control based on noise active control, the "gravity-resistance-driving force" multi-objective balanced energy saving technology, high safety factor walk system, structural fire proofing and electromagnetic compatibility technology, the key train technologies of various structural walk system, and transnational interconnected high-speed train equipment and operation maintenance system, the aim is to master the key technologies concerning system integrated, car body, bogie, traction and braking, power supply, train control, operational control, system operation and maintenance satisfying the needs of "One Belt One Road," and transnational interconnected adaptability and criteria system and accomplish the study of high-speed train and various structural trains with a speed level of 400 km/h and above. At the same time, we will systematically deepen and establish the key technology system regarding high-speed train multi-effect enhanced coupling and control, environmental-friendly enhancement, whole life cycle design and integration, reliability, availability, maintainability, and safety (RAMS) comprehensive performance improvement, to create the improved adaptability and enhanced technology of existing infrastructure and equipment, study the high-speed train system with operation speed of 400 km/h and above which is suitable for the high-speed rail infrastructure in China.

### 4.2.6 Railway Comprehensive Effectiveness and Service Level Improvement Under High-Speed Rail Network Conditions

Through the study of railway passenger freight service mode design and resource allocation under high-speed rail network, benefit and service level improvement technology of railway passenger freight transport, railway network operation assurance technology and operation, and service cooperative decision and support system, we will form the transportation technical standards supporting rail network comprehensive effectiveness and service level improvement and a new railway transportation engineering technology system, and realize and improve the "One Belt One Road" transnational transport and international competiveness.

### 4.2.7 Regional Rail Transport Co-transport and Service Technology

Through the study of regional transport comprehensive effectiveness improvement technology, regional rail transport safety assurance technology, and regional rail transport information service technology and system, we will form the regional rail transport multi-model co-transport, safety assurance, information service integrated technology, and system platform and satisfy the needs of regional rail transport comprehensive effectiveness and service level, to support the regional rail transport integrated transportation and service.

### 4.2.8 Space–Air–Train–Ground Integrated Rail Transport Safety and Control Technology

Through the study of rail transit dedicated static and dynamic hang platform system technologies, the space–air–train–ground–information integrated rail transit dedicated network technologies, rail transit system status information integration and processing techniques, vehicle mobile interconnection technology based on dedicated network, sparse low capacity road network train operation control system key technologies, and dynamic block system based on location information, we will form the dedicated static and dynamic air platform design, manufacture, operation and maintenance technology, form the dedicated space–air–train–ground integrated transportation and monitoring network which satisfies its interaction operation needs, form the multi-level, multi-granularity, high-dimensional holographic technology which satisfies the large-scale, all-weather, full coverage, all-around live monitoring needs of rail transit and its safe operation environment, form large-scale high-dimensional rail transit safety information monitoring and integration, analysis and application technology, and form high-property, low-cost, multi-functional new safety assurance mode of the wide-range sparse road network. At the same time, we will develop the key technologies concerning route control based on multiple information integration and location technique, multi-path information transmission and control, dynamic interval configuration braking and safety protection, develop, new train operation and control system with high maintainability featuring small-scale, low-density rail-side equipment and dynamic interval configuration, satisfying the needs of safety, efficient operation and sustainability of national defensive western and backcountry low-density transport network.

### 4.2.9 Rail Transit Freight Transportation Rapid Technology and Equipment Studies

Through the study of multi-mode freight transportation adapter system technology, key technologies and equipment study of 160 km/h freight train, road–rail convenient transport key technologies and equipment, and 250 km/h and above freight trains, we will form the rapid and convenient, multi-mode, high-speed standard regulation, and technique system, to satisfy the needs of national defense mobility and support the rapid, efficient, and low-cost railway-oriented comprehensive transport system.

### 4.2.10 Key Technology Study and Equipment Development of a Maglev Transport System

Through the study of key technologies and equipment study of the mid-speed maglev train, mid-speed train synchronous traction control techniques study, operation control technology of mid-speed maglev transport, independent technology integration demonstration model, and comprehensive evaluation of high-speed maglev transportation system, we will fully master the key technologies of high-efficiency, high-reliability suspension traction and operation control of mid-speed maglev, including hybrid suspension and synchronous traction, and construct a mid-speed Maglev test line. We will fully master the key technologies of high-speed maglev, break the limitation of intellectual property of other countries, to realize the independence of high-speed maglev, and be able to independently assemble a mid-long high-speed maglev transportation system.

Through the implementation of this project, the 400 km/h and above high-speed train and corresponding system will be delivered, and the life cycle operation cost will aim to be reduced by 20%.

## 5 Conclusions

The expansion of China's high-speed rail demands higher requirements for the study of high-speed rail technology. Research can be further carried out in the following areas (NSFC 2016):

1. Long-term service regression study of slab track;
2. High-speed rail freight transport technology;
3. High-speed rail driverless technology;
4. High-speed rail failure monitoring, diagnosis, and smart operation technology;
5. High-speed rail optimized operation;
6. Study and development of new-generation high-speed rail equipment.

Currently, the high-speed rail mileage in China has surpassed the total of all other countries. By 2025, the high-speed rail operation mileage will reach 38,000 km. Meanwhile, other countries are studying and investing in the construction of high-speed rail. It becomes increasingly important to improve the safety, comfort, economy, and mobility of the high-speed railway. Obviously, to further study the high-speed rail technology is not only necessary in itself, but also promotes the development of corresponding technology and industry.

## Appendix

**Highlight**
**Advantage of high-speed rail**

Traditional rail
The shortage of railway transport capability has restricted the development of China's economy

Modern high speed rail:
**Huge transport capacity**
**Safety**
**Comfort and fast**
**All-day operation**
**Environmentally-friendly operation**
**Sustainability**

**Innovation achievements of China's high-speed rail technology**

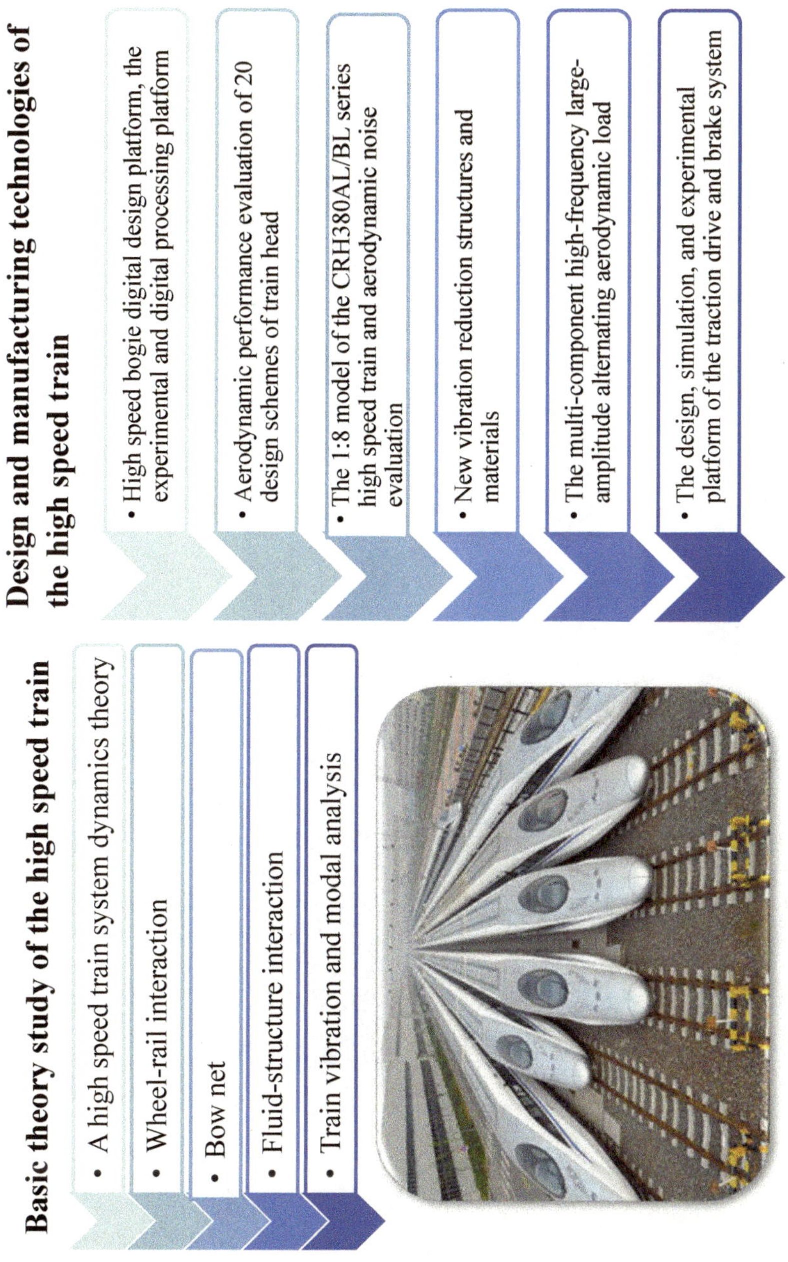

**Engineering application of domestic high-speed train**

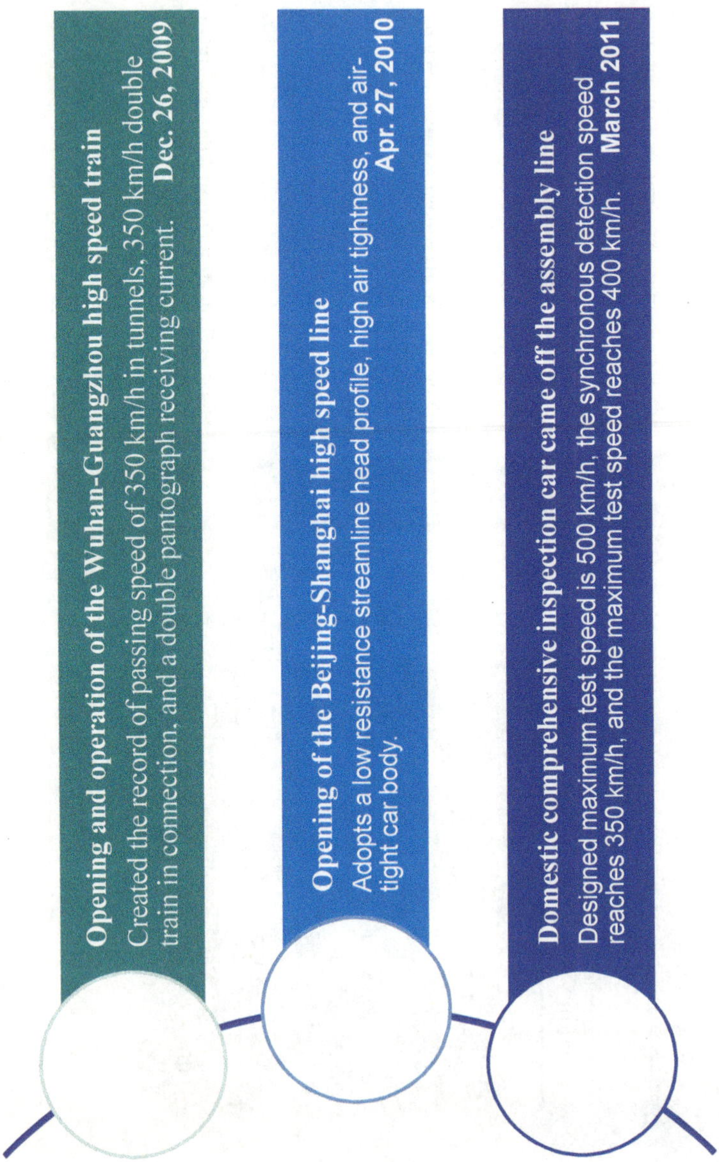

**Development prospect of China's high-speed rail technology**

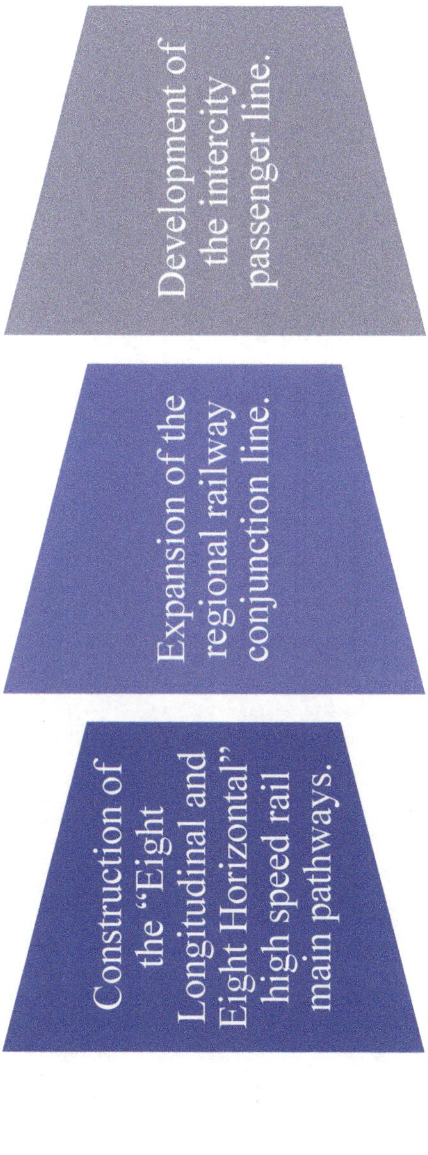


Up to 2020, high speed rail will reach 3000 km in length, covering over 80% of big cities. By 2025, the railway network will reach 175 thousand kilometers, in which the high speed railway will account for 38 thousand kilometers.

**Mid-long-term railway network planning in China**

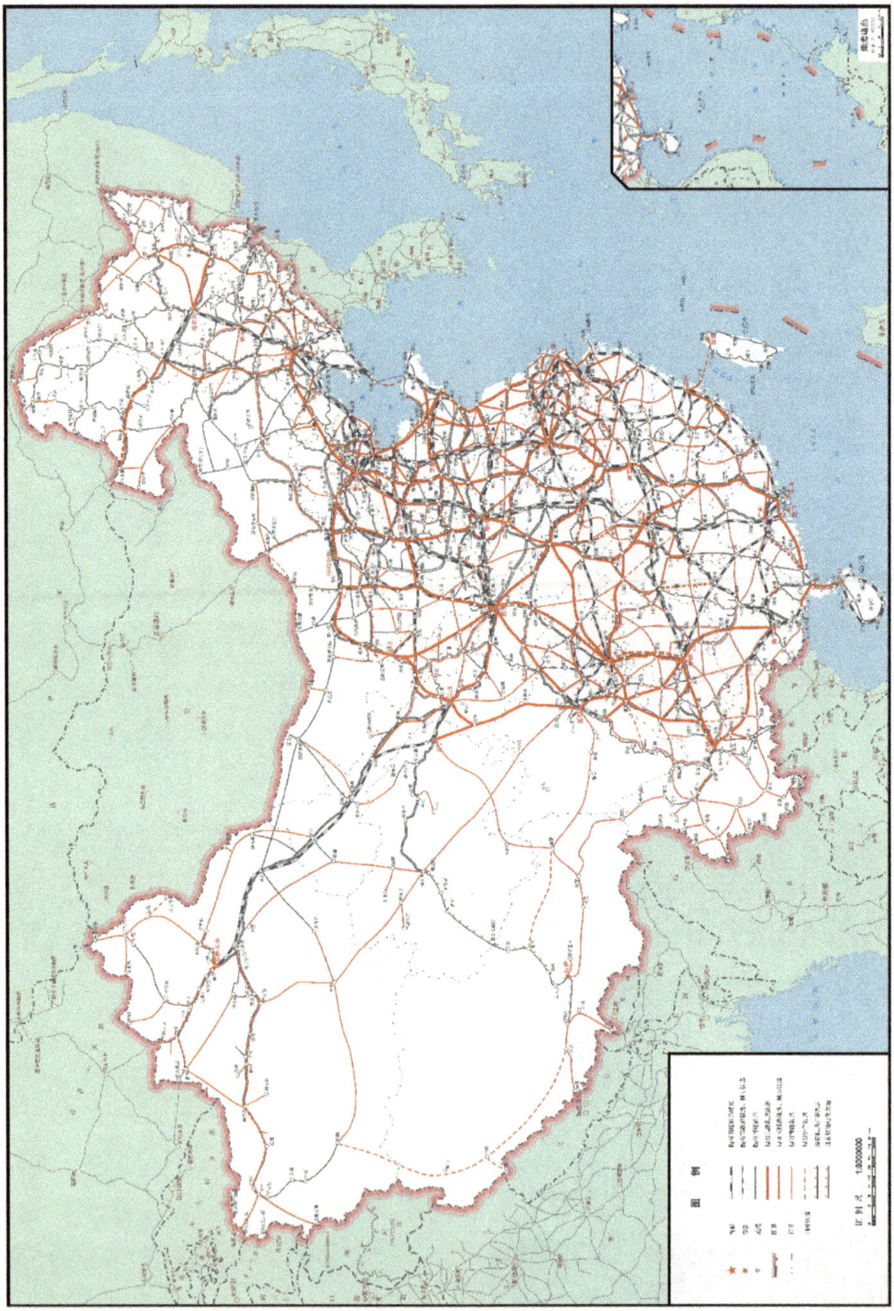

## Further research

Research can be further carried out in the following areas:

1. Long term service regression study of slab track;
2. High speed rail freight transport technology;
3. High speed rail driverless technology;
4. High speed rail failure monitoring, diagnosis and smart operation technology;
5. High speed rail optimized operation;
6. Study and development of new generation high speed rail equipment.

Obviously, to further study the high speed rail technology is not only necessary in itself, but also promotes the development of corresponding technology and industry.

## References

Huang, X. Y., Zhang, J. C., Sun, C. M., et al. (2015). A combined simulation of high-speed train permanent magnet traction system using dynamic reluctance mesh model and Simulink. *Journal of Zhejiang University-Science A (Applied Physics & Engineering), 16*(8), 607–615. doi:10.1631/jzus.A1400284.

Jin, X. S. (2014). Key problems faced in high-speed train operation. *Journal of Zhejiang University-Science A (Applied Physics & Engineering), 15*(12), 936–945. doi:10.1631/jzus.A1400338.

Ma, J. E., Zhang, B., Huang, X. Y., et al. (2011). Design and analysis of the hybrid excitation rail eddy brake system of high-speed trains. *Journal of Zhejiang University-Science A (Applied Physics & Engineering), 12*(12), 936–944. doi:10.1631/jzus.A11GT002.

MST (Ministry of Science and Technology of China). (2012). *The decade: Scientific development report on modern transportation*. Beijing, China (in Chinese): Science and Technology Literature Press.

NDRC (National Development and Reform Commission). (2016). *Mid-long term railway network plan*. China: NDRC. Available from http://wwwold.sdpc.gov.cn/zcfb/zcfbtz/201607/t20160720_811696.html. Accessed on November 23, 2016 (in Chinese).

NSFC (National Natural Science Foundation of China). (2016). *The national key research and development plan "advanced rail transit" key projects application guideline*. China: NSFC. Available from http://www.most.gov.cn/tztg/201605/t20160513_125542.htm. Accessed on November 23, 2016 (in Chinese).

Smith, R. A., Zhou, J. (2014). Background of recent developments of passenger railways in China, the UK and other European countries. *Journal of Zhejiang University-Science A (Applied Physics & Engineering), 15*(12), 925–935. doi:10.1631/jzus.A1400295.

Zhang, J., Ma, J. E., Huang, X. Y., et al. (2015). Optimal condition-based maintenance strategy under periodic inspections for traction motor insulations. *Journal of Zhejiang University-Science A (Applied Physics & Engineering), 16*(8), 597–606. doi:10.1631/jzus.A1400311.

Zhao, X., Wen, Z. F., Wang, H. Y., et al. (2014). Modeling of high-speed wheel-rail rolling contact on a corrugated rail and corrugation development. *Journal of Zhejiang University-Science A (Applied Physics & Engineering), 15*(12), 946–963. doi:10.1631/jzus.A1400191.

## Author Biographies

**Ping Tan** obtained his Ph.D. degree from the College of Control Science and Engineering at Zhejiang University, Hangzhou, China, in 2014. He has done research and development work on control systems in the SUPCON and INSIGMA Groups for more than ten years. He has been selected as one of the 151 Engineering Training Talents of Zhejiang Province and one of the 131 middle-aged and young talents of Hangzhou. He is the specialist of the Hangzhou Industrial and Information Technology Commission. He has won the first prize of Zhejiang Science and Technology Progress Award in 2006. He is the senior engineer of the Zhejiang University of Science and Technology. His research interests include distributed control systems, industrial Ethernet, fieldbus, vital computer, and signaling systems for the rail transit system.

**Ji-en Ma** Ph.D., is an associate professor in Zhejiang University, Hangzhou, China. She received a Ph.D. degree in mechatronics from Zhejiang University in 2009. Then, she did postdoctoral work in the College of Electrical Engineering, Zhejiang University, China. Her recent work is on electrical machines and drives. Her research interests include PM machines and drives for traction applications, and mechatronics machines such as the magneto fluid bearing.

**Jing Zhou** Ph.D., graduated from Xi'an Jiaotong University with a bachelor's degree in 2011 and then received a Ph.D. from Imperial College London in 2014. Then, she continued post-doctoral research in Zhejiang University, and her research mainly focuses on multi-field coupled analysis and wireless power transfer.

**You-tong Fang** is a professor at Zhejiang University. He is chairman of the High Speed Rail Research Center in Zhejiang University, deputy director of the National Intelligent Train Research Center, on the committee of the China High Speed Rail Innovation Plan, and expert of the National 863 Program (China High Technology Development Program) in modern transportation and advanced carrying technology. He is also director of three projects of the National Natural Science Foundation of China (NSFC) and more than ten National 863 projects and Technology Support Programs. His recent work has been on electrical machines and drives. His research interests include permanent magnet (PM) machines and drives for traction applications.

# KEY PROBLEMS FACED IN HIGH-SPEED TRAIN OPERATION

**Xue-song Jin**

**Contents**



## 1 Introduction

At present, the total length of China's high-speed railway is close to 12,000 km, more than the sum of the rest of the world's high-speed railway coverage. Compared with other national high-speed railways, in addition to the first-class quality of train and track line, the train running speed is higher, the train marshaling is longer (16 coaches), the direct operating miles are longer (e.g., Harbin to Shanghai is 2421 km and Beijing to Guangzhou is 2289 km), the track stiffness is larger and the track lines have a higher proportion of bridges. As China's high-speed railway network is completed in length and breadth, with a particular emphasis in the western areas, the direct operating mileage of trains will increase further, and the proportions of bridges and tunnels in the track lines will be further increased. Trains operating on long direct lines will be on continuous high-speed running, across different geographical areas, and in different running environments. Different regional geological conditions could influence the track's behavior in various ways, including changes to the track support stiffness. Climate differences offer a different wheel/rail running environment, and a different abrasion state of

X. Jin (✉)
State Key Laboratory of Traction Power, Southwest Jiaotong University, Chengdu 610031, China
e-mail: xsjin@home.swjtu.edu.cn

 
Y. Fang and Y.H. Zhang (eds.), *China's High-Speed Rail Technology*,
Advances in High-speed Rail Technology, https://doi.org/10.1007/978-981-10-5610-9_2

wheel/rail. In addition, the wheel/rail adhesive coefficient difference is larger along a long track line. Large temperature differences can change greatly the operational features of the train and the track structure, even to the extent of affecting the vehicle system damping noise reduction effect. All this directly affects train dynamic behavior and operational quality, which makes it difficult to maintain long-term comfort, high stability and safety. To achieve such goals requires high reliability of the train and track structure.

In high-speed operation, train wheels are excited by the irregularities of high-stiffness track (China's high-speed slab track) , and the carriages and the bogies are strongly affected by high-speed airflow. The train's dynamic behavior becomes more complicated. These track irregularities mainly consist of the normal random irregularity of wheel/rail running surfaces, the periodic sleeper support of the rail, the rail welding joint, and periodic bridge pillar supports of the track. These periodic supports form hard points along with the track in the vertical and lateral directions. The strong effect of periodic hard points at operational speeds of 200–350 km/h can be clearly seen in the measured acceleration and noise components in the frequency domain of the axle boxes, the bogie frames, and the coaches. Under conditions of high-speed and high-stiffness track excitation, the wheel-rail excitation energy is large with a wide frequency band, the excitation frequencies are very high, and the track's energy absorption is poor. The wheel/rail interaction can effectively transfer their energy into the bogies, the coaches and rail infrastructure, and it is easy to cause vehicle and track system resonance. The vibration frequencies increase as the train speed increases, and since most of the structural parts of train and track are in a rigid-flexible coupling condition, this can produce high-frequency structural noise.

High-speed trains need urgently to solve the following problems: (1) The quick wheel tread concave wear caused when the train runs at high speed, and the lateral oscillating of the train at the frequencies of 7–10 Hz. When the lateral oscillating amplitude is in excess of the prescribed threshold, the train monitoring system sends out an alarm signal and therefore the speed of the train is momentarily reduced (Cui et al. 2012); (2) The higher-order polygonal wear of wheel roundness leads to fierce vertical wheel/rail vibration, and therefore causes, at about 580 Hz, fierce vibration and abnormal noise in the carriage (8 dB beyond normal) (Cui et al. 2013; Han et al. 2014b), and leads to fatigue cracking of key vehicle parts (Tian et al. 2013); (3) The noise level is a little high inside and outside the train carriages and the noise at the ends of the carriage interior exceed the standard (Zhang et al. 2014); (4) The vehicle/track system structure parameters (mainly referring to the structural modal parameters) does not achieve the best matching condition and the vehicle structure vibration transfer characteristics are not clear. The train's ability to resist outside turbulence (track disturbance due to irregularities and airflow turbulence due to the high-speed running) is weak, and in high-speed runtime resonance or short time jitter readily occurs; (5) The track has high stiffness, with a strong propensity to transmit vibration and a low damping capacity. The vehicle dynamic response can clearly reflect the characteristics of the track, such as sleeper pitch, rail welding points, and the pier spacing of the bridges. Fatigue damage of the mortar layer and road base of the track develops, which leads to track performance degradation (Chen and Sun 2011); (6) The longitudinal dynamic behavior change of a long marshaling train and its influence on the lateral action of the train, and train safety assessment indicators, are all in need of further in-depth study. Relying only on existing theoretical analysis and experiment is insufficient for clear understanding and solution of the above problems. Railway engineers and researchers need to perfect a high-speed train-track coupling large system theory (Ling 2012), in order to help understand these problems.

This paper discusses the progress of the research on these problems, mainly from the point of view of high-speed train track coupling large system theoretical modeling, the wheel/rail relationship and the damping noise reduction theory and technology, including the articles published in this special issue.

## 2 Train-Track Coupling System Dynamic Model

Root causes of the issues listed in Sect. 1 are very much related to many factors of the train and the track coupling condition. But which are the major, minor, and secondary factors? A complete theory of modeling of train-track large-scale coupling system and the numerical analysis helps to determine the answer.

Even far in the past, there have been remarkable achievements in railway vehicle-track coupling dynamics modeling and its application. A series of researches have been carried out abroad. The theory of a dynamic simulation model started from the modeling of a single wheel in 1883 and this has continued from the multilayer track to the current mature single-vehicle/track coupling dynamics model (Ripke and Knothe 1995; Oscarsson and Dahlberg 1998). The model by Zhai et al. (1996) is the representative one in China. Its characteristic is that the ballasted track modeling considers five parameters nearly representing the dynamical behavior of the ballasted track. In a certain frequency range, it can characterize the effect of the ballast track characteristics on the vehicle-track coupling dynamics behavior. In these models, the vehicle modeling mainly depends on the multi-rigid-body system dynamics theory, where the components of the vehicle are treated to be rigid bodies, and they are connected with spring and damping elements. The dynamical behavior analysis of the vehicle is limited in the frequency range of 0–20 Hz (Knothe and Grassie 1993). These models are mainly used for railway vehicle stability, comfort, and safety assessment (Zhai et al. 2009). For a high-speed train, if the problems discussed in Sect. 1 are not included in the analysis, these models can also be used (Jin et al. 2013). However, in this case, the interaction between the vehicles of the train is ignored.

In recent years, the high-speed railway around the world has been developing rapidly. The scale of the network increases, the operational speed continuously increases as does the pass-through mileage. Clearly, problems have emerged. High-speed operation causes qualitative and quantitative changes in the dynamical behavior of train/track, and using the existing theory can neither explain nor solve these problems. High-speed train/track coupling large system theory needs to be developed and perfected (Zhang et al. 2007). Its development is mainly along the following paths:

1. The modeling considers the vertical, horizontal, and longitudinal dynamic behavior coupling of high-speed train and track. Ling et al. (2014) in this special issue models the connections between adjacent vehicles of the train and the effect of the connections on the vertical and horizontal behavior. The analysis of vehicle stability, comfort, and safety index has greatly improved, but we still lack understanding of the influence of the train longitudinal behavior, including acceleration, deceleration, vertical extrusion, collision, stretching, deformation and longitudinal wave, and structural characteristics, including train length, length of vehicles, and carriage longitudinal stiffness (Ling et al. 2014).
2. The modeling of the vehicle coupled with the track considers the effect of the rigid motion and the high-frequency deformation of the key structural parts, such as wheelsets (Zhong et al. 2013), bogie frame (Claus and Schiehlen 2002) and carriages (Zhou et al. 2009) of the vehicles, rails (Xiao et al. 2008), and sleepers and slabs (Xiao 2013) of the track. Here, in addition to the rail, the modeling of the sleeper and the wheelset mainly depends on the theory of beam model, and the modeling of the other main components is carried out by finite element methods, to determine their modals. Again according to the superposition principle, these models are put into the differential equations of the related parts of the vehicle and the track to find their solutions. Zhong et al. (2014) in this special issue considers the effect of wheels with 3–8 order polygonal wear on the wheel/rail rolling contact since the polygonal wear is much related to the wheelset 1–3 order bending resonances occurring when the train operates at high speeds. Ling et al. (2014) considers the influence of rail and track slab flexible deformation. However, because of the difficulty of the problem and computing limitations, the present rigid-flexible coupling modeling cannot consider the whole system of vehicle-track and is only local in scope. Thus, when at high speed, the local part models of the system, the overall modals, the local resonances, the system resonances as well as their relationship with wheel wear polygon, cannot be effectively identified and clearly understood.
3. The model of high-speed train considers the mutual effect of the train's behavior and airflow. Factors for this problem include train aerodynamic drag, pressure changes inside the carriage, train induced flow, two running trains meeting in opposite directions, airflow effect, ground surface effect, tunneling effect, aerodynamic noise, and structural vibration, etc. (Li et al. 2013; Yu et al. 2013; Yang et al. 2015). These factors have a close relationship with the geometry and characteristic dimensions of the train, the embankment, the tunnel, the vehicle dynamic performance, and the assembly process. The study makes great progress in theory and practical application on these problems, such as that now high-speed trains running under 350 km/h have better aerodynamic characteristics. Because of the complexity of these problems and the current limited calculation ability, in the analysis of the effect of high-speed airflow on the

behavior of the train, the modeling does not fully consider the dynamic characteristics of the vehicle structure, namely, the effect of the rigid motion and high-frequency deformation of the structural parts on the high-speed airflow (Yu et al. 2011). So far, we do not know what causes the local high-frequency vibration of the floor—is it the wall and roof of the carriage, the wheel-rail excitation or air turbulence? The components of vibration and noise in the frequency range of 250–800 Hz are dominant. What are the root causes of the phenomena? These are not clear and open questions. In current studies on these problems, the train structure and its local complex structures are greatly simplified, such as fluid-structure interaction of pantograph and catenary (Lee et al. 2007), windshield fluid-structure coupling between the carriages (Song et al. 2008), and bogie fluid-structure coupling (Moon et al. 2014).

An overall view will be clearer, and problems solved with further improvement of modeling of a high-speed train and track coupling system which can fully consider the structural rigid-flexible coupling and the fluid-structure coupling effect, combining this with the application of modern test technology.

## 3 Interaction of Wheel/Rail

Due to the long traveling distance and duration of high-speed trains running at high speed, structure materials of trains and track could change, and all kinds of hidden problems be gradually exposed, which could threaten safety. For China's high-speed railway, wheel/rail relationship problems mainly include: short pitch rail corrugation and wheel high-order polygonal wear that resulting in strong vibration and noise; rapid lateral concave wear on wheel tread and mismatching of low wear state of rail head that causing the lateral oscillation of the trains at high-speed, and therefore leading the train to slow down for a period. Fast wheel wear leads to shortening of the times between wheel repair intervals. Similarly, if the rail wear rate is much slower than that of the wheel, this will also lead to reductions in the intervals between essential repair of wheel and rail. Thus, wheel/rail material hardness matching remains to be further optimized.

Study of the relationship of wheel/rail includes the basic theory of wheel/rail in rolling contact or rolling contact mechanics and their application technologies. The application technologies of wheel/rail are divided into research on the matching of the geometric surfaces and materials of wheel/rail, wheel/rail adhesion and its control, wear and rolling contact fatigue damage, as well as wheel/rail noise and derailment. The study involves many subjects, such as system dynamics, materials science, tribology, solid mechanics, and calculation method, etc. (Jin and Shen 2001; Zhai et al. 2010).

The theory of wheel/rail rolling contact is the basic means of guiding wheel/rail application technology research. This is mainly through the five classic wheel-rail creep theoretical models which reflect the wheel/rail rolling contact mechanics behavior, and the wheel/rail 3D elastic-plastic theory of rolling contact, developed more recently. The five classic wheel/rail force models are, the 1D wheel/rail rolling contact model (Carter 1926), the 2D nonlinear wheel/rail creepage/force model (Vermeulen and Johnson 1964), the 3D linear creep theory model of wheel/rail (Kalker 1967) and the simplified theory model (Kalker 1982), and the 3D wheel-rail creepage/force model with consideration of the effect of wheel/rail small spin (Shen et al. 1983). These models were built mainly based on the Hertz contact theory hypothesis, expressed in analytical form, and have been widely used in railway vehicle-track coupling dynamics modeling and computational simulation. Their advantage is that they have fast speed in the wheel/rail force calculation. But these models do not consider the effect of train wheel rolling, wheel/rail transient behavior, wheel/rail environment boundary factors and changes in material properties. Thus, using them cannot effectively solve some problems in the service condition of wheel/rail, such as wheel high-order polygonal wear, rail short pitch corrugation, strong vibration noise, and wheel/rail rolling contact fatigue. It is urgent and necessary to develop a more complex wheel/rail rolling contact theory model to clearly understand and effectively solve them. This theoretical model should be the wheel/rail 3D elastic-plastic rolling contact theory model and the corresponding numerical method.

In the 1970s, Kalker (1990), according to the variational principle, used variational inequality to express the problem of elastic bodies in rolling contact with friction, and its corresponding numerical program, called "CONTACT", has been widely used in wheel/rail rolling contact behavior analysis. Using the theory it is possible to get information about the behavior of wheel/rail, whereas using the five classic models discussed above cannot (Jin and Shen 2008). The information includes the wheel/rail contact spot real shape, the stick-slip area distribution, the size and distribution of the tangential force, the normal force and the spin moment, the distribution of the friction work, and the wheel/rail stress distribution in the wheel/rail bodies. Compared to the above classic creep force models, however, its calculation speed is slow. The establishment of the theoretical model "CONTACT" does not depend on the assumptions of the Hertz contact theory, and thus it is called the non-Hertz rolling contact theory. Its numerical implementation process is based on the theory of elastic half space, and the numerical results of wheel/rail forces gained by using it are a little greater than the results of the finite element model. This is because the wheel/rail is regarded as two elastic half space objects in the "CONTACT" model, which exaggerates the actual wheel/rail contact rigidity. This model cannot consider the rolling speed change effect and wheel/rail transient behavior. Furthermore, it ignores the effects of the structural deformation of the wheel/rail and the wheel/rail surface state (roughness, the third medium), and the environmental conditions (temperature, humidity, and airflow). Jin and Zhang (2001) and Zhang et al. (2013) used the finite element method to further promote the "CONTACT" model, which can take into account the influence of the geometry boundary condition outside wheel/rail contact point on the rolling contact behavior.

Starting in the 1980s, the finite element method has been applied to the analysis of wheel/rail rolling contact behavior. Damme et al. (2003) introduced stick-slip contact control conditions and utilized the decomposition method of deformation gradient of liquid layout and the arbitrary Lagrange-Euler method to analyze the 3D elastic bodies in rolling contact, in which fine grid adaptive technology was used. This numerical method is only suitable for solving the steady rolling contact problem of 3D elastic bodies. In the analysis, the rolling speed was low (10 km/h), the non-slip condition was considered, and the contact surfaces were assumed to be smooth (Nackenhorst 1993).

Now, high-speed trains operate at speeds up to 350 km/h, and the test speed reaches 500 km/h. Wheel/rail in service repeatedly encounters the acceleration deceleration, and the excitation by wheel/rail contact surface irregularity (wheel wear polygon, rail corrugation, rail welding joint, rail scratch, wheel flat, turnout, and various irregularities by rail grinding). The wheel/rail rolling contact process is a transient elastic-plastic deformation process in rolling contact. Using the rolling contact theory model discussed above still, cannot explain the wheel/rail unsteady rolling phenomenon and the mechanism of some problems. Zhao and Li (2011), by using the commercial software ANSYS/LS-DYNA, developed the model of 3D elastic-plastic wheel/rail in rolling contact to analyze the dynamical behavior of wheel and rail when the wheel is rolling over the rail with the squat on the rail top. In their model, the actual geometrical sizes of the wheel, the rail and the vehicle's unsprung mass, and the parameters of the track characteristics were considered. The wheel/rail rolling speed in the analysis can be simulated to 40–140 km/h. In this special issue, Zhao et al. (2014) further promotes the development of the calculation model of the 3D elastic-plastic wheel-rail in rolling contact. This model can be used to accurately and reliably analyze the high frequency vibration of the wheel/rail system, and unsteady elastic-plastic rolling contact behavior in the variety of irregular excitation cases (rail welding joint, geometrical and material defects, rail corrugation, wheel/rail contact surface scratches and wear, triangle pit, etc.). Situations of the wheel rolling over the rail at high speed with unsteady rolling-slip, wheel/rail contact spot shape, stick-slip area distribution, wheel/rail creep force, and elastic-plastic stress and strain in the wheel/rail can be also calculated. In the calculation, the simulated rolling speed reaches 500 km/h (Zhao et al. 2014). However, the model calculation speed is very slow, and cannot meet the requirements of high-speed and efficient numerical simulation. This model does not consider the effects of the environmental temperature, the contact surface asperity, the "three media" between wheel and rail, and the material inclusions in the wheel/rail (Wu et al. 2014).

Study on the wheel/rail relationship also includes the wheel/rail matching design theory and technology (Cui et al. 2011), the adhesion theory and adhesion control technology (Chen et al. 2005), the rolling contact fatigue (Wen et al. 2005), the derailment mechanism and control (Xiao et al. 2012), as well as the wheel/rail noise problem (Han et al. 2014a).

At present, wheel/rail studies face many unresolved issues. The best matching design of wheel/rail materials tries to achieve the goal that the wear and fatigue crack growth rates of the wheel/rail materials in service achieve synchronous progress, and then expect that the performance of the selected wheel/rail materials can satisfy the best and longest service cycle through the wheel/rail natural wear eliminating the rolling contact fatigue crack on the surface of the wheel/rail contact, thus reducing wheel/rail maintenance cost. At a running speed under 350 km/h, the wheel/rail noise contribution is still over 50% of the total noise of the whole train/track system. For running safety, there are currently no effective measures to solve the problem of wheel/rail noise from their source. Wheel wear rate increases quickly due to the high operation speed, the high-stiffness track, the wide wheel/rail impact frequency, and large vibration amplitude. Furthermore, the high wheel wear rate causes the conicity and roundness of the wheel to change quickly, and the train cannot keep a good ride quality for a long period. Therefore, the wheels frequently need to be re-profiled and the operating costs rise.

## 4 Vibration and Noise

As speeds increase, the vibration and noise problem has become increasingly prominent. As can be seen from a direct personal sensory comfort index, vibration and noise have become the key factors influencing high-speed train business competition. In general, research on noise and its control has three elements, i.e., source, propagation path, and sound receiver. Controlling the former two is the active measure. The noise sources of the high-speed train can be divided into wheel/rail noise, train aerodynamic noise, pantograph/catenary noise, and auxiliary equipment noise. Their propagation paths include air sound transmission and structure sound transmission (Eade and Hardy 1977).

In practice, the vibration and noise of a high-speed train are related to the source excitation energy and the complex propagation path. For example, some typical vibration and noise problems are: (1) wheel/rail excitation at the passing frequency of the sleepers (e.g., 110 Hz at 250 km/h) can pass to the carriage through the bogie structure, and inspire the local structure modes of the vehicle and the special acoustic modes of interior cavity. In such a situation sound-solid coupling forms in the carriage, and the noise level of the passenger compartment end near the window is 8 dB(A) higher than that of the carriage corridor at the same cross section of the carriage (Zhang et al. 2014a); (2) Due to the carriage special structure design and layout of the built-in dining car, the dining car uses a large number of rigid connection structures and does not benefit from the damping measures at the connections, which leads to the increase of the structural system vibration energy under the external excitation (wheel/rail and airflow). At the same time, the carriage interior decoration uses materials with a low absorption coefficient, which generates the large reverberation noise inside the carriage. Therefore, the dining car interior noise is significantly higher than other carriages, particularly when excited by the wheel wear polygon (Zhang et al. 2013b); (3) The sound insulation and sealing of the carriage structure is not sufficient. At the passenger compartment end above the bogie, due to the toilet layout, there are many pipes installed on the bodywork chassis. The aluminum profile used in the bodywork chassis has many holes where leaking sound directly generates, which leads to the area noise of 3 dB(A) higher than the other end of the carriage where an electrical cabinet is installed; (4) High-frequency sound transmission is through the structures under the carriage. The existing wheel higher-order wear polygon excites the high-frequency vibration response of the wheel/rail system and the resonance of the auxiliary equipment at their natural frequency close to the passing frequency of the wear polygon. The resonance easily passes to the carriage through the suspension systems and the rod pieces under the carriage. Hence, the measured results of the interior vibration and noise show a great peak at the passing frequency of the wheel wear polygon; (5) The car-body surface aerodynamic noise sources mainly include the train head area, the pantograph area and the area between carriages and the bogie area. In these areas, due to the strong aerodynamic effect, the corresponding pneumatic noise and vibration response of the carriage structure are quite high; (6) Some of the latest site test results also show that the air conditioning duct and carriage grid structures can sometimes increase active equipment vibration and then the noise level inside the carriage increases dramatically.

Färm (2000) conducted the testing analysis through equipment transposition and found that the excitation by different stiffness track and different sleeper passing frequency made an 11 dB(A) difference in the total noise level and that SJ50 rail is more likely to provoke vehicle vibration, compared to UIC60 rail. By changing the sleeper spacing, track stiffness and wheelbase vehicle vibration can effectively be eliminated at the sleeper passing frequency. Eade and Hardy (1977), by changing the primary and secondary suspension parameters, concluded that, compared with the second suspension, installing damper at the primary suspension can better improve the interior vibration and noise of the carriage. Through the test and analysis of aerodynamic noise characteristics at the carriage connection and bogie of the French TGV high-speed train, it was concluded that the former is mainly the narrow frequency domain noise in low frequency (500 Hz), and the latter is mainly the high-frequency noise in 500–1000 Hz range (Poisson et al. 2002). The high-speed trains of Japanese Shinkansen adopted streamlined head cars and smooth surfaces in vitro, such that smooth connections between the carriages and other uneven places (side window and side sliding door, etc.), replaced the diamond pantograph for the single-arm pantograph and used pantograph air guide sleeve, and in the area of bogie with bottom cover and apron. These measures can obtain an obvious noise reduction (Akihiko 2003). Further research on the sound source identification technology and application showed that the main noise sources of high-speed trains are, the bogie area, the pantograph/catenary area, the carriage connection area, and train head. The interior noise of 200–800 Hz is dominant (Mellet et al. 2006; Poisson et al. 2008). Research on the mechanism of wheel/rail noise and noise reduction technology is also actively developing (Jones and Thompson 2000; Thompson and Gautier 2006).

In this special issue, He et al. (2014) discusses the experiment and analysis of external noise produced by a Chinese high-speed train traveling at different speeds. The experimental results and their corresponding analysis are very useful for the control and reduction of exterior noise produced by high-speed trains (He et al. 2014).

Zhang et al. (2014b) investigates the effect of wheel wear polygon of a high-speed train on the interior noise of the carriage. The results have a positive role in control measure research of the outside and interior noise of high-speed trains. However, train operation safety is the basic and first priority. Therefore, the noise reduction measures taken in wheel/rail systems are currently very limited at the sound source.

Other research on the vibration and noise includes the evaluation methods of interior noise (Hardy 2000; Zhang et al. 2013a), the sound quality evaluation of the interior carriage (Letourneaux and Guerrand 2000; Parizet et al. 2002), etc. There are many unsolved key issues, including in time domain vibration noise mechanism research, the high-frequency sound transmission mechanism of the bogie structure, lightweight carriage and reliable acoustic design, sound-solid coupling, vehicle sound quality mechanism, and active noise control methods.

## 5 Further Work

This paper has described some of the urgent problems needing a solution and the research progress of them. These problems are still open and need of extensive investigations:

1. For current and future operation, all-round tracking tests and statistical analysis should be carried out, to obtain the relationship between the high-speed wheel/rail service state and the vehicle dynamic performance, and the main factors that influencing the wear and rolling contact fatigue of wheel/rail on lines with different operational speed levels in different geographical areas.
2. Train-track coupling system dynamics theory should fully consider, in the high speed running state, the influence of structure rigid-flexible coupling and fluid-structure interaction on train-track behavior.
3. The design method for wheel/rail geometric profile matching needs to be improved. Research on new wheel/rail modification technology should be conducted.
4. Comprehensive material testing for wheel/rail use should be carried out, with long-term research on wheel/rail new materials and their preparation.
5. Wheel/rail contact surface processing technology needs to be improved with corresponding field test research carried out at the same time.
6. With consideration of the vehicle-track coupling system environment, the 3D elastic-plastic rolling contact model of wheel-rail should be improved. There should be a focus on computational efficiency, the wheel/rail material defect effect, the contact surface state, the wheel/rail environmental impact, and the wheel/rail spin effect.
7. Broad-based and in-depth theory, technology and measures of vibration attenuation and noise reduction for high-speed train and track coupling systems should be developed.

# Appendix

## Highlight

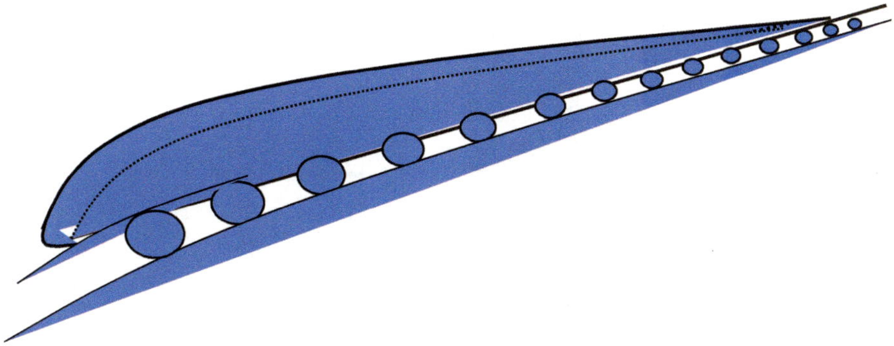

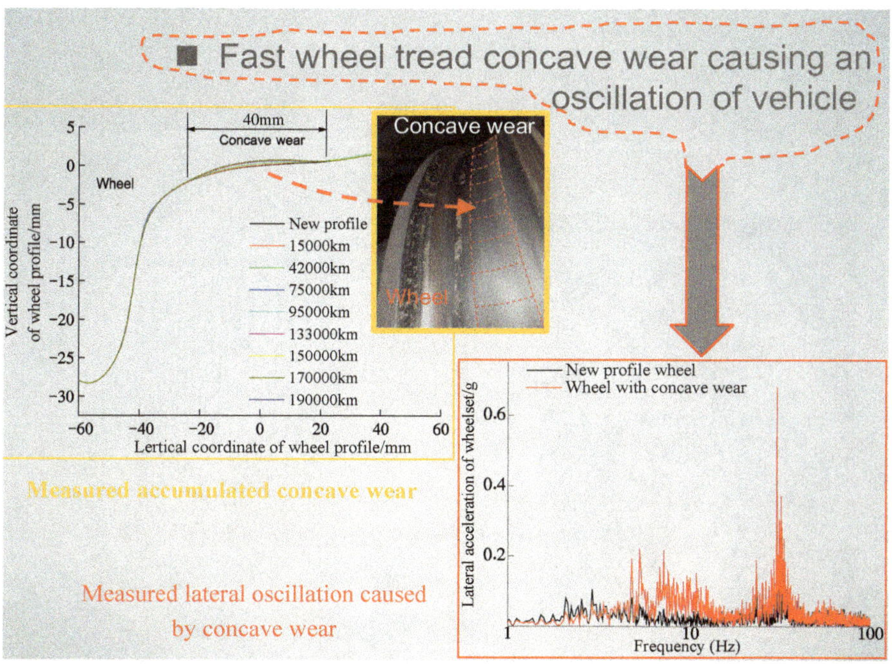

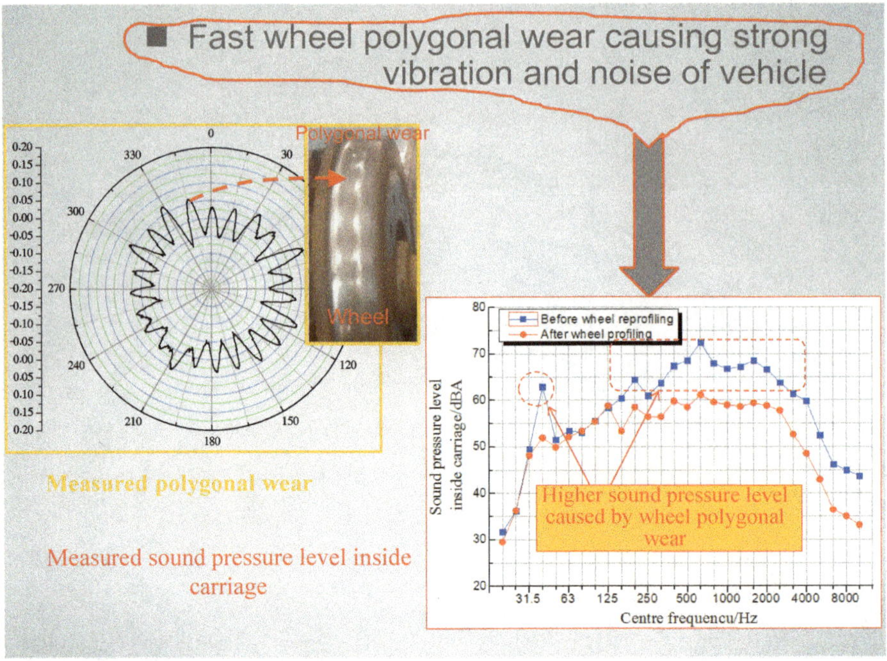
Fast wheel polygonal wear causing strong vibration and noise of vehicle
Polygonal wear
Wheel
Measured polygonal wear
Measured sound pressure level inside carriage
Before wheel reprofiling
After wheel profiling
Sound pressure level inside carriage/dBA
Higher sound pressure level caused by wheel polygonal wear
Centre frequencu/Hz

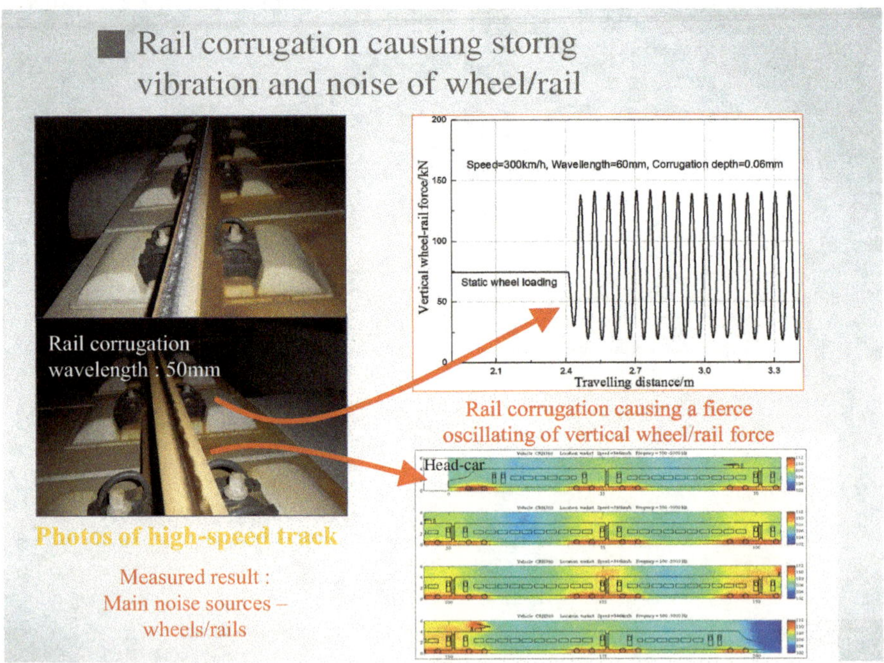
Rail corrugation causting storng vibration and noise of wheel/rail
Rail corrugation wavelength : 50mm
Photos of high-speed track
Measured result : Main noise sources – wheels/rails
Speed=300km/h, Wavelength=60mm, Corrugation depth=0.06mm
Static wheel loading
Vertical wheel-rail force/kN
Travelling distance/m
Rail corrugation causing a fierce oscillating of vertical wheel/rail force
Head-car

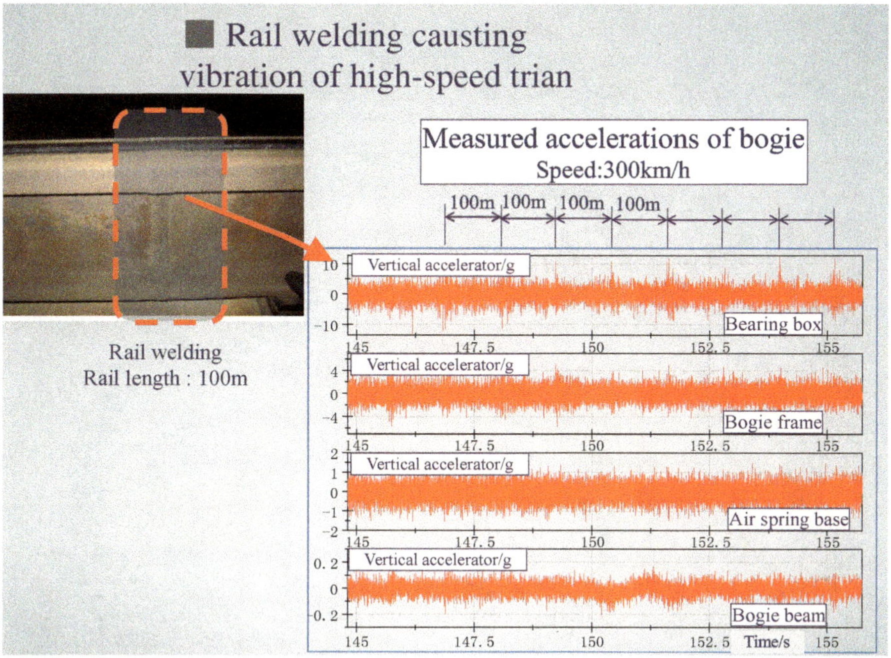
Rail welding causting
vibration of high-speed trian
Measured accelerations of bogie
Speed:300km/h
100m 100m 100m 100m
Vertical accelerator/g
Bearing box
Vertical accelerator/g
Bogie frame
Vertical accelerator/g
Air spring base
Vertical accelerator/g
Bogie beam
Time/s
Rail welding
Rail length : 100m

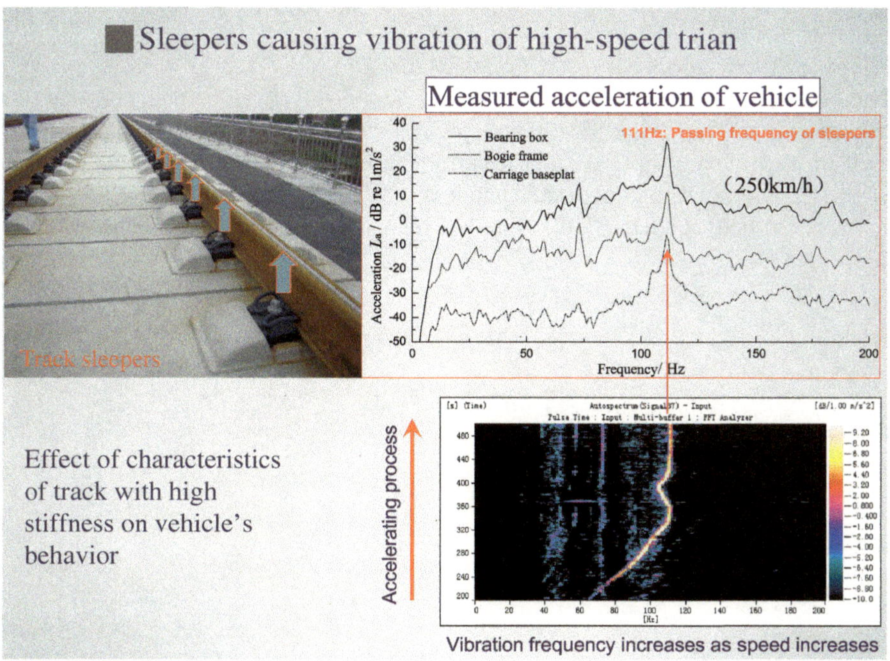
Sleepers causing vibration of high-speed trian
Measured acceleration of vehicle
Bearing box
Bogie frame
Carriage baseplat
111Hz: Passing frequency of sleepers
(250km/h)
Acceleration $L_a$ / dB re 1m/s$^2$
Frequency/ Hz
Track sleepers
Accelerating process
Effect of characteristics of track with high stiffness on vehicle's behavior
Vibration frequency increases as speed increases

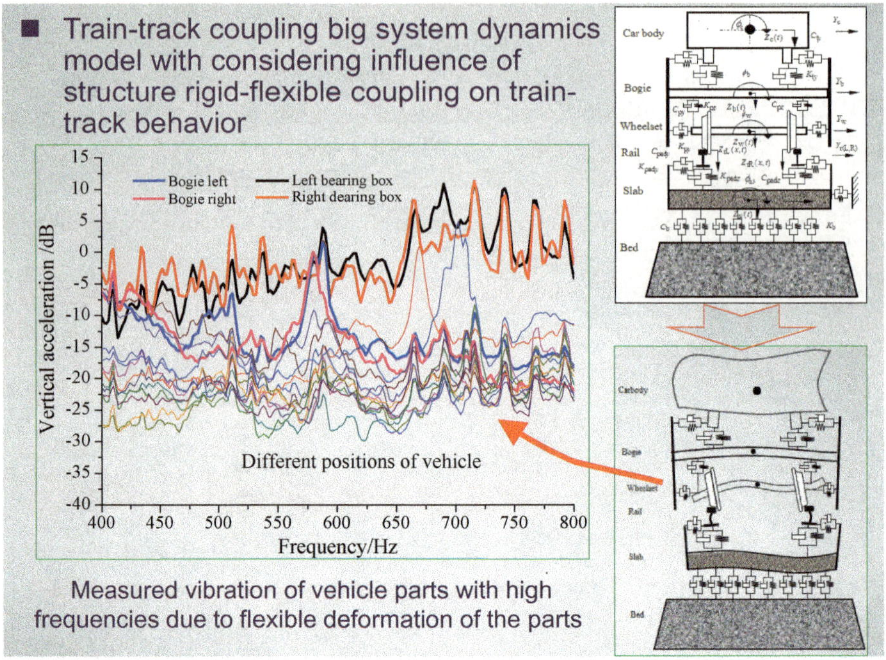


**Purpose of the model**

Train-track coupling big system dynamics model is developed to consider the influence of structure rigid-flexible coupling on train-track behavior. This model is used to help making optimization design of train and track structure matching to inhibit the lateral oscillation of the vehicle due to wheel tread concave wear. Also, it is used to help understanding the mechanism of wheel polygonal wear of high-order.

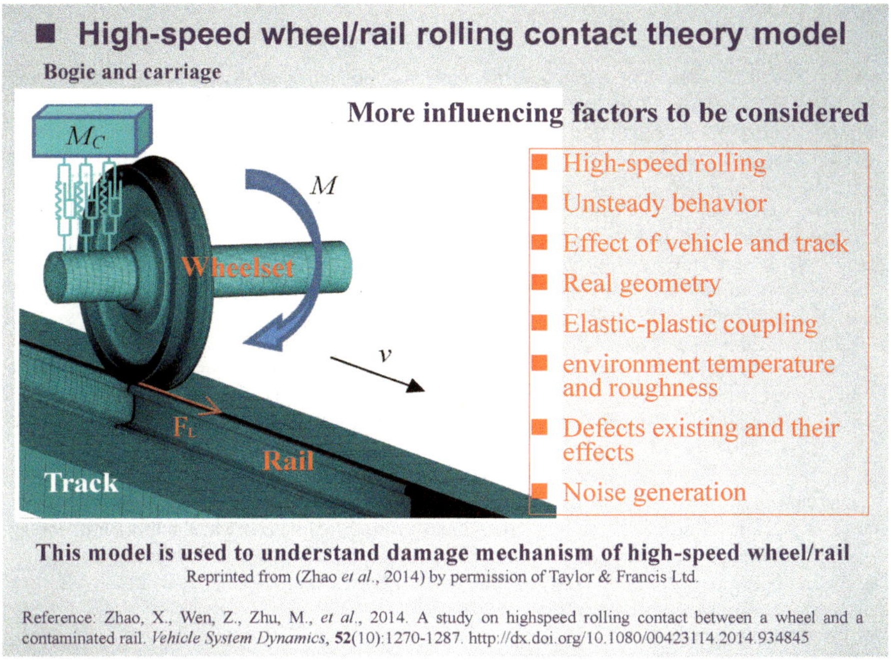


## Conclusions

- For current and future operation of high-speed train, all-round tracking tests and statistical analysis should be carried out, in order to grasp the main factors influencing the wear and rolling contact fatigue of wheel-rail, and train track behavior.
- Train-track coupling big system dynamics theory is developed to consider, in high speed running state, the influence of structure rigid-flexible coupling, and fluid-structure interaction on train-track behavior.
- The design method for wheel/rail geometric profile matching needs to be improved. Research on new wheel/rail modification technology should be conducted. Comprehensive material testing for wheel/rail use should be carried out. Wheel/rail contact surface processing technology needs to be improved with corresponding field test research carried out at the same time.
- 3D elastic-plastic rolling contact model of wheel-rail should be improved. There should be focus on computational efficiency, wheel/rail material defect effect, contact surface, state and wheel/rail environmental impact.
- Broad-based and in-depth theory, technology and measures of vibration attenuation and noise reduction for high-speed train and track coupling systems should be developed.

## References


Akihiko, T. (2003). The noise reduction measures for Shinkansen 700 series train-sets. *Foreign Rolling Stock, 40*(6), 17–19.

Carter, F. (1926). On the action of a locomotive driving wheel. *Proceedings of the Royal Society of London Series A, 112,* 151–157.

Chen, H., Ishida, M., & Nakahara, T. (2005). Analysis of adhesion under wet conditions for three-dimensional contact considering surface roughness. *Wear, 258*(7–8), 1209–1216. doi:10.1016/j.wear.2004.03.031.

Chen, S. Y., & Sun, H. L. (2011). Experimental analysis of long-term stability of soaking subgrade for high speed railway. *Journal of Railway Engineering Society, 12,* 40–44. (in Chinese).

Claus, H., & Schiehlen, W. (2002). Symbolic-numeric analysis of flexible multibody systems. *Mechanics of Structures and Machines, 30*(1), 1–30. doi:10.1081/SME-120001476.

Cui, D. B., Huang, Z. W., Jin, X. S., et al. (2012). A new wheel profile design method for high-speed vehicle. (*Proceedings of the 1st International Workshop on High-speed and Intercity Railways*, 1). *Lecture Notes in Electrical Engineering, 147*, 225–241. doi:10.1007/978-3-642-27960-7_20

Cui, D. B., Li, L., Jin, X. S., et al. (2011). Optimal design of wheel profiles based on weighed wheel rail gap. *Wear, 271*(1–2), 218–226. doi:10.1016/j.wear.2010.10.005.

Cui, D. B., Lin, L., Song, C. Y., et al. (2013). Out of round high-speed wheel and its influence on wheel/rail behavior. *Journal of Mechanical Engineering, 49,* 8–16. (in Chinese).

Damme, S., Nackenhorst, U., Wetzel, A., et al. (2003). On the numerical analysis of the wheel-rail system in rolling contact. In *System dynamics and long-term behaviour of railway vehicles, track and subgrade* (pp. 155–174). Springer, Berlin.

Eade, P. W., & Hardy, A. E. J. (1977). Railway vehicle internal noise. *Journal of Sound and Vibration, 51*(3), 403–415.

Färm, J. (2000). Interior structure-borne sound caused by the sleeper-passing frequency. *Journal of Sound and Vibration, 231*(3), 831–837.

Han, J., Wen, Z. F., Wang, R. Q., et al. (2014a). Experimental study on vibration and sound radiation reduction of the web-mounted noise shielding and vibration damping wheel. *Noise Control Engineering Journal, 62*(3), 110–122. doi:10.3397/1/376211

Han, G. X., Zhang, J., Xiao, X. B., et al. (2014b). Study on high-speed train abnormal interior vibration and noise related to wheel roughness. *Journal of Mechanical Engineering*, *50*(22), 113–121 (in Chinese).

Hardy, A. E. J. (2000). Measurement and assessment of noise within passenger trains. *Journal of Sound and Vibration, 231*(3), 819–829. doi:10.1006/jsvi.1999.2565.

He, B., Xiao, X. B., Zhou, Q., et al. (2014). Investigation into external noise of a high-speed train at different speeds. *Journal of Zhejiang University-SCIENCE A (Applied Physics & Engineering), 15*(12), 1019–1033. doi:10.1631/jzus.A1400307.

Jin, X. S., & Shen, Z. Y. (2001). Development of rolling contact mechanics and its application in research on wheel/rail performances. *Advances in Mechanics, 31*(1), 33–46.

Jin, X. S., & Shen, Z. Y. (2008). *Wheel/rail creep theory and its experiments*. Chengdu, China: Southwest Jiaotong University Press. (in Chinese).

Jin, X. S., Xiao, X. B., Ling, L., et al. (2013). Study on safety boundary for high-speed train running in severe environments. *International Journal of Rail Transportation, 1*(1–2), 87–108. doi:10.1080/23248378.2013.790138.

Jin, X. S., & Zhang, J. Y. (2001). A complementary principle of elastic bodies of arbitrary geometry in rolling contact. *Computers and Structures, 79*(29–30), 2635–2644. doi:10.1016/S0045-7949(01)00087-6.

Jones, C. J. C., & Thompson, D. J. (2000). Rolling noise generated by railway wheels with visco-elastic layers. *Journal of Sound and Vibration, 231*(3), 779–790. doi:10.1006/jsvi.1999.2562.

Kalker, J. J. (1967). *On the rolling contact of two elastic bodies in the presence of dry friction.* PhD Thesis, Delft University, The Netherlands.

Kalker, J. J. (1982). A fast algorithm for the simplified theory of rolling contact. *Vehicle System Dynamics, 11*(1), 1–13. doi:10.1080/00423118208968684.

Kalker, J. J. (1990). *Three-dimensional elastic bodies in rolling contact.* Dordrecht, The Netherlands: Kluwer Academic Publishers.

Knothe, K., & Grassie, S. L. (1993). Modeling of railway track and vehicle/track interaction at high frequencies. *Vehicle System Dynamics, 22*(3–4), 209–262.

Lee, Y. B., Rho, J., Kwak, M. H., et al. (2007). Aerodynamic characteristics of high speed train pantograph with the optimized pan head shape. In *Proceedings of the 7th IASME/WSEAS international conference on fluid mechanics and aerodynamics*, pp. 84–88.

Letourneaux, F., & Guerrand, S. (2000). Assessment of the acoustical comfort in high-speed trains at the SNCF: Integration of subjective parameters. *Journal of Sound and Vibration, 231*(3), 839–846. doi:10.1006/jsvi.1999.2567.

Li, T., Zhang, J. Y., & Zhang, W. H. (2013). A numerical approach to the interaction between airflow and a high-speed train subjected to crosswind. *Journal of Zhejiang University-Science A (Applied Physics & Engineering), 14*(7), 482–493. doi:10.1631/jzus.A1300035.

Ling, L. (2012). *Study on the dynamic behaviour of high-speed train/track coupling system composed of multiple vehicles.* MS thesis, Southwest Jiaotong University (in Chinese).

Ling, L., Xiao, X. B., Xiong, J. Y., et al. (2014). A 3D model for coupling dynamics analysis of high-speed train/track system. *Journal of Zhejiang University-Science A (Applied Physics & Engineering), 15*(12), 964–983. doi:10.1631/jzus.A1400192.

Mellet, C., Letourneau, F., Poissonb, F., et al. (2006). High speed train noise emission: latest investigation of the aerodynamic/rolling noise contribution. *Journal of Sound and Vibration, 293*(3–5), 535–546. doi:10.1016/j.jsv.2005.08.069.

Moon, J. S., Kim, S. W., & Kwon, H. B. (2014). A study on the aerodynamic drag reduction of high-speed train using bogie side fairing. *Journal of Computational Fluids Engineering, 19*(1), 41–46. doi:10.6112/kscfe.2014.19.1.041.

Nackenhorst, U. (1993). *On the finite element analysis of steady state rolling contact. Contact mechanics–Computational techniques* (p. 53). Southampton, Boston: Computational Mechanics Publication,

Oscarsson, J., & Dahlberg, T. (1998). Dynamic train/track/ballast interaction—Computer model and full-scale experiments. *Vehicle System Dynamics, 29*(S1), 73–84. doi:10.1080/00423119808969553.

Parizet, E., Hamzaoui, N., & Jacquemoud, J. (2002). Noise assessment in a high-speed train. *Applied Acoustics, 63*(10), 1109–1124. doi:10.1016/S0003-682X(02)00017-8.

Poisson, F., Gautier, P. E., & Letourneaux, F. (2008). Noise sources for high speed trains: A review of results in the TGV case. *Noise and Vibration Mitigation for Rail Transportation Systems, 99,* 71–77. doi:10.1007/978-3-540-74893-9_10.

Poisson, F., Letourneaux, F., Loizeau, T., et al. (2002). Inside noise of high speed train coaches. Forum Acusticum Sevilla, high speed trains, NOI-03-005-IP.

Ripke, B., & Knothe, K. (1995). Simulation of high frequency vehicle-track interactions. *Vehicle System Dynamics, 24*(S1), 72–85. doi:10.1080/00423119508969616.

Shen, Z. Y., Hedric, K. J., & Elkins, J. (1983). A comparison of alternative creep force models for rail vehicle dynamic analysis. *Vehicle System Dynamics, 12*(1–3), 79–83. doi:10.1080/00423118308968725.

Song, H., Park, J. H., & Song, S. (2008). Numerical analysis on flow characteristics around a cavity with flaps. *Transactions of the Korean Society of Mechanical Engineers B, 32*(9):645–651. doi:10.3795/KSME-B (in Korean).

Thompson, D. J., & Gautier, P. E. (2006). Review of research into wheel/rail rolling noise reduction. *Proceedings of the Institution of Mechanical Engineers, Part F: Journal of Rail and Rapid Transit, 220*(4), 385–408. doi:10.1243/0954409JRRT79.

Tian, J. Y., Yang, Y., Zhang, S., et al. (2013). Gearbox fault diagnosis technology of high-speed trains. *Mechanical Engineer, 6,* 77–78. (in Chinese).

Vermeulen, J. K., & Johnson, K. L. (1964). Contact of non-spherical bodies transmitting tangential forces. *Journal of Applied Mechanics, 31*(2), 338–340. doi:10.1115/1.3629610.

Wen, Z. F., Jin, X. S., & Jiang, Y. Y. (2005). Elastic-plastic finite element analysis of non-steady state wheel-rail rolling contact. *Tribology, ASME, 127*(4), 713–721. doi:10.1115/1.2033898.

Wu, B., Wen, Z. F., Wang, H. Y., et al. (2014). Numerical analysis on wheel/rail adhesion under mixed contamination of oil and water with surface roughness. *Wear, 314*(1–2), 140–147. doi:10.1016/j.wear.2013.11.041.

Xiao, X. B. (2013). *Study on high-speed train derailment mechanism in severe environment*. PhD thesis, Southwest Jiaotong University, Chengdu, China (in Chinese).

Xiao, X. B., Jin, X. S., Deng, Y. Q., et al. (2008). Effect of curved track support failure on vehicle derailment. *Vehicle System Dynamics, 46*(11), 1029–1059. doi:10.1080/00423110701689602.

Xiao, X. B., Ling, L., & Jin, X. S. (2012). A study of the derailment mechanism of a high-speed train due to an earthquake. *Vehicle System Dynamics, 50*(3), 449–470. doi:10.1080/00423114.2011.597508.

Yang G. W., Wei Y. J., & Zhao G. L., et al. (2015). Research progress on the mechanics of high speed rails. *Advances in Mechanics, 45*, 217–460 (in Chinese).

Yu, M. G., Zhang, J. Y., & Zhang, W. H. (2011). Wind-induced security of high-speed trains on the ground. *Journal of Southwest Jiaotong University, 46*(6), 989–995. (in Chinese).

Yu, M. G., Zhang, J. Y., & Zhang, W. H. (2013). Multi-objective optimization design method of the high-speed train head. *Journal of Zhejiang University-SCIENCE A (Applied Physics & Engineering), 14*(9), 631–641. doi:10.1631/jzus.A1300109

Zhai, W. M., Cai, C. B., & Guo, S. Z. (1996). Coupling model of vertical and lateral vehicle/track interactions. *Vehicle System Dynamics, 26*(1), 61–79. doi:10.1080/00423119608969302.

Zhai, W. M., Jin, X. S., & Zhao, Y. X. (2010). Some typical mechanics problems in high-speed railway engineering. *Advances in Mechanics, 40*(4), 358–374. (in Chinese).

Zhai, W. M., Wang, K. Y., & Cai, C. B. (2009). Fundamentals of vehicle-track coupled dynamics. *Vehicle System Dynamics, 47*(11), 1349–1376. doi:10.1080/00423110802621561.

Zhang, J., Han, G. X., Xiao, X. B., et al. (2014b). Influence of wheel polygonal wear on interior noise of high-speed trains. *Journal of Zhejiang University-Science A (Applied Physics & Engineering), 15*(12), 1002–1018. doi:10.1631/jzus.A1400233.

Zhang, S. R., Li, X., Wen, Z. F., et al. (2013a). Theory and numerical method of elastic bodies in rolling contact with curve contact area. *Engineering Mechanics, 30*(2), 30–37. (in Chinese).

Zhang, J., Xiao, X. B., Han, G. X., et al. (2013b). Study on abnormal interior noise of high-speed trains. In *Proceedings of the 11th International workshop on railway noise*, Uddevalla, Sweden, pp. 759–766.

Zhang, J., Xiao, X. B., Han, J., et al. (2014a). Characteristics of abnormal noise at the ends of the coach and acoustic modal analysis of high-speed train. *Chinese Journal of Mechanical Engineering, 50*(12), 97–103 (in Chinese)

Zhang, J., Xiao, X. B., & Zhang, Y. M. (2013a). Noise evaluation in the tourist cabin of high-speed train by using aircraft noise criterion. *Journal of Mechanical Engineering, 49*(16), 33–38. doi:10.3901/JME.2013.16.033 (in Chinese).

Zhang, S. G., Zhang, W. H., & Jin, X. S. (2007). Dynamics of high speed wheel/rail system and its modeling. *Chinese Science Bulletin, 52,* 1566–1575.

Zhao, X., & Li, Z. (2011). The solution of frictional wheel-rail rolling contact with a 3-D transient finite element model: Validation and error analysis. *Wear, 271,* 444–452.

Zhao, X., Wen, Z. F., Wang, H. Y., et al. (2014). Modeling of high-speed wheel-rail rolling contact on a corrugated rail and corrugation development. *Journal of ZhejiangUniversity-Science A (Applied Physics & Engineering), 15*(12), 946–963. doi:10.1631/jzus.A1400191.

Zhong, S. Q., Xiao, X. B., Wen, Z. F., et al. (2013). The effect of first-order bending resonance of wheelset at high speed on wheel-rail contact behavior. *Advances of Mechanical Engineering, 2013,* 1–19.

Zhong, S. Q., Xiong, J. Y., Xiao, X. B., et al. (2014). Effect of the first two wheelset bending modes on wheel-rail contact behavior. *Journal of Zhejiang University-Science A (Applied Physics & Engineering), 15*(12), 984–1001. doi:10.1631/jzus.A1400199.

Zhou, J., Goodall, R., Ren, L., et al. (2009). Influences of car body vertical flexibility on ride quality of passenger railway vehicles. *Journal of Rail and Rapid Transit, 223*(5), 461–471. doi:10.1243/09544097JRRT272.

## Author Biography

**Xue-song Jin** Ph.D., is a professor at Southwest Jiaotong University, Chengdu, China. He is a leading scholar of wheel-rail interaction in China and is a State Council Expert for Special Allowance. He is the author of three academic books and over 200 articles. He is an editorial board member of several journals, and has been a member of the Committee of the International Conference on Contact Mechanics and Wear of Rail/Wheel Systems for more than 10 years. He has been a visiting scholar at the University of Missouri-Rolla for 2.5 years. Now his research focuses on wheel-rail interaction, rolling contact mechanics, vehicle system dynamics, vibration and noise.

# Background of Recent Developments of Passenger Railways in China, the UK and Other European Countries

**Roderick A. Smith and Jing Zhou**

**Contents**



## 1 Early Railway Development in China and the UK

The first rail line in China was built in 1876, during the last days of the Qing Empire, to connect foreigners' settlements in Shanghai with the river docks at Wusong. The trading company Jardine, Matheson facilitated the import of British equipment, but the line was short-lived and was removed less than a year later by an irate local governor. After this false start, rail development was slow. By the start of the twentieth century, China had just 370 miles of track, compared with

R.A. Smith · J. Zhou (✉)
Department of Mechanical Engineering, Imperial College London, London SW7 2AZ, UK
e-mail: zhoujing_zju@126.com

R.A. Smith
e-mail: roderick.smith@imperial.ac.uk

 
Y. Fang and Y.H. Zhang (eds.), *China's High-Speed Rail Technology*, Advances in High-speed Rail Technology, https://doi.org/10.1007/978-981-10-5610-9_3

21,000 miles for Great Britain and more than 182,000 miles in America. However, every modern Chinese government has seen railroads both as a symbol of China's economic independence and as a powerful force for national cohesion; and the railway has grown both physically and in national importance during the post-World War 2 era. Approximately 30,000 km in 1945 has grown to more than 100,000 km today, nearly half of which is electrified. China now has 6% of the world's railways and carries 20% of the world's rail traffic.

The first major passenger railway in Britain was that between the northern cities of Liverpool and Manchester. George Stephenson and his locomotive Rocket played a major role in initiating this first true intercity passenger railway. Essentially, Rocket was a prototype for all the steam engines to come and is now preserved in the Science Museum in South Kensington. Robert, the son of George, was the engineer of the first long intercity line connecting the capital London and the second city Birmingham. In 1847, George became the first President of the Institution of Mechanical Engineers and was followed by Robert his son as the second President. The first author of this paper was proud to be the 126th President in 2011–2012. Mention could be made of many names associated with the heroic era of engineering during the rapid growth of the railways in the UK: perhaps Isambard Kingdom Brunel deserves special mention, not only for his achievements in railways, but also for the design of huge steamships which connected his Great Western Railway with America.

## 2 Further Development of the Railways

The railways in the UK grew rapidly in extent in the nineteenth century and reached a peak in the early twentieth century. The railways had been built and operated by many different private companies with the result that there was duplication, unnecessary competition and incoherence in the system. The two World Wars and the depression of the 1930s left the railways in poor physical shape when they were nationalised and were run by the state from 1948. The network was sharply contacted in the late 1950s and 1960s in order to make it more viable economically in the face of competition from private car ownership which increased strongly in these decades. In 1994, the railways were privatised once again: with very mixed results. Vertical integration was lost as the railway was fragmented into many parts in a most probably misguided attempt to inject competition. Costs have risen dramatically, but passengers have increased in number and are now back to the 1910 peak (Fig. 1).

It is instructive to compare some basis statistics of the railways in China and the UK: Germany and Japan have been added to the comparison. Table 1 shows the basic data: as is well known, China is a very large country and has a huge population. The result is that although it has the largest network in route length in the world, it only amounts to the length of a cigarette, 70 mm, per person. Japan has more than twice this figure and the UK more than three times. The greatest

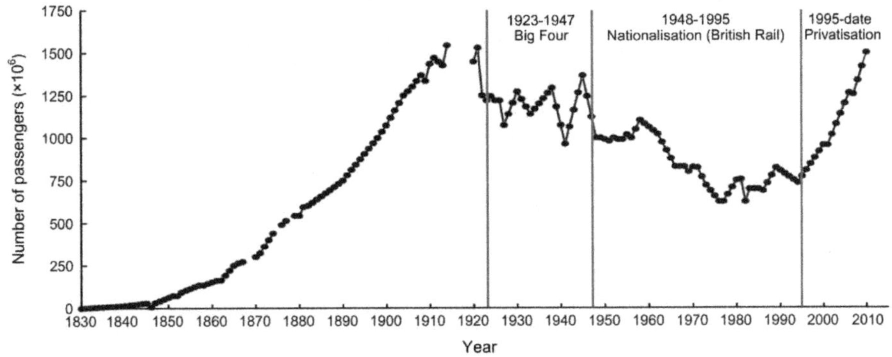


**Fig. 1** Historical development of rail passenger volume in the UK

**Table 1** Some international railway comparisons

| Country | Area ($\times 10^6$ km$^2$) | Population ($\times 10^6$) | Route (km) | Route/person (km/million) | Average journey (km) | Intensity[a] |
|---|---|---|---|---|---|---|
| China | 9.640 | 1351 | $1 \times 10^5$ | 70 | 518 | 26.9 |
| UK | 0.242 | 62.8 | 15,775 | 25 | 40 | 9.3 |
| Germany | 0.349 | 82 | 33,707 | 41 | 41 | 6.3 |
| Japan | 0.378 | 127.5 | 20,852 | 16 | 17 | 51.6 |

[a]Intensity is 1000 passenger km/(route km·d)

difference is in the average length of railway journey: an astonishing 518 km in China (a huge justification for the new high-speed network) compared with only 17 km in Japan, a figure arising from the dense commuter journeys in Tokyo and Osaka in particular. The UK and Germany have more balanced figures of around 40 km. In the case of the UK, the area round London accounts for a high proportion of railways business: very little traffic is generated in Scotland, Wales, the west of England, and north of a line connecting Manchester and Leeds. This geography of population has, of course, implications for future high-speed developments. In contrast, the population of China is more concentrated on the east side of the country and there are many cities with large populations; 160 greater than 1 million including 15 cities in conurbations greater than 5 million. Again these are fertile conditions for high-speed rail.

## 3 Speed-up and High Speed Rail in the UK

Although Britain was the birthplace of the railways, in recent decades it has lagged behind in their development. Nevertheless, it still holds the world speed records for both steam and diesel traction, but its electrification ratio is not as high as that in many other countries. Even at the beginning of the twentieth century, speeds of over

**Fig. 2** The advanced passenger train (*APT*) articulated and tilting; a victim of complexity

160 km/h were reached by steam trains, then the 1930s saw developments in streamlining, which enabled the locomotive Mallard to reach 203 km/h on 3 July 1938, a record which stands to this day. The fifties and sixties were periods of contraction of our railways, as competition with the car intensified. Steam traction was phased out rather rapidly: by 1967, it has ceased and many different types of diesel trains were introduced. Some electrification of main routes, notably the route from London to Manchester, took place in the sixties, which reduced journey time and increased patronage. A bold attempt was made to design and operate a tilting train to cope with the curves on the sinuous routes to Scotland. The so-called advanced passenger train (APT) incorporated many features novel to the railway industry (Fig. 2), through the efforts of aeronautical and other non-railway engineers injected into British Rail's research laboratories in Derby. The complexities were many: the powered tilt system (up to 9°) had a lagging response, the novel hydrokinetic brakes had to deal with stopping distances within the existing block signalling system, the construction was in lightweight aluminium and the carriages were articulated. In 1975, a speed of 245.1 km/h was achieved on unmodified conventional track, but the project was dogged by difficulties. A sound, but perhaps in hindsight, too advanced concept, the train was finally abandoned in 1987. However, the expertise generated by the project fed into the design of what has arguably been the UK's most successful train ever, the InterCity 125 High Speed Train (HST) (Fig. 3). This train, based on diesel power cars at each end hauling various combinations of conventional coaches, was introduced into service in 1976. The modern appeal of its clean aerodynamic lines was bolstered by a publicity campaign. This is the age of the train, rapidly established its popularity in the eyes of the travelling public. In November 1987, it achieved what is still the world speed record for diesel traction when it touched 238 km/h. Even today, this train forms the backbone of several major intercity routes in the UK and is regarded by many, the authors included, as still the best train on the UK network.

A further spin-off from the advanced passenger train project was the InterCity 225, an electric version of the HST, introduced in 1990. Although designed to travel at 225 km/h, the speed in service has been limited to 200 km/h because of the lack of in-cab signalling. Nevertheless, in a test run it achieved a speed of 261.7 km/h. The train is still used on the so-called East Coast Main Line to the north from King's Cross in London to Doncaster, York, Newcastle and Edinburgh.

**Fig. 3** The 125 high-speed train (*HST*): introduced in 1976 and still going strong today. It is the holder of the world's diesel traction speed record of 238 km/h

## 4 Channel Tunnel and Its Link to London

Twenty years ago, on 6 May 1994, the 50.5-km-long Channel Tunnel opened, linking at last Britain to France and onwards to Western Europe by rail. This project had an incredibly long history: a scheme was first mooted in 1802, followed by several others, which in 1881 resulted in physical exploratory works starting on both sides of the Channel, but which were soon abandoned because of doubts about national security. Finally, modern defence technology helped to overcome isolationist views and real work on the project started in 1988. The tunnelling preceded smoothly, the chalk undersea geology between Folkestone and Calais was favourable to tunnelling, but the costs escalated and far exceeded the initial estimates. The scheme of the tunnel is that twin bores connected by a service tunnel and pressure release ducts. The tunnel carries, of course, passenger trains, but also car and freight shuttle wagons onto which vehicles are driven for the passage. Speeds in the tunnel are limited to 160 km/h. When the tunnel was opened, Paris was connected to Calais by new dedicated high-speed line, making speeds of 300 km/h possible. On the British side, trains travelled on convention track at a much more sedate speed (apologists said this was to allow passengers to appreciate the wonderful views of Kent, the so-called Garden of England), to a temporary terminal at London Waterloo station.

Some ten years after the opening of the tunnel, a dedicated high-speed link, the 111-km Channel Tunnel Rail Link (CTRL) was opened: Briton's first dedicated High Speed line allowing 300 km/h running from the tunnel to a newly refurbished St. Pancras station (Fig. 4). The so-called High Speed 1 (HS1) took its name because of its UK pioneering role, subsequent developments take up the name HS2, and so on. London to Paris is now 2.25 h by rail, and air traffic has shrunk to very low volumes, rail having an 85% market share.

Two valuable lessons can be learned from the Channel Tunnel experience. The first is that the hybrid Eurostar train designed to deal with running on dedicated line in France and conventional line in the UK was not one thing or the other: it had to cope with three different current collections systems, including the antiquated low-voltage DC third rail system south and east of London in the UK, and it had three different signalling system and had to operate within the restricted loading gauge of the UK network. All these factors added both to complexity and expense, and to reduced reliability. Secondly, issues relating to safety have arisen. Major fires occurred in the tunnel in 1996 and 2008, both originating from Heavy Goods Vehicles (HGV) on shuttle wagons. There have been several breakdowns resulting in severe disruptions to services. The service tunnel has proved its worth, but the vulnerability of long tunnel systems and the lack of alternative paths should always be uppermost in operators' minds. Finally on HS1, a domestic service is run, by Hitachi Javelin trains which run on both the dedicated high-speed line and on/off to conventional track (Fig. 5). Several of the issues of dual compatibility have arisen, adding to the complexity of the trains.

**Fig. 4** A Eurostar train on Britain's first dedicated HS1 line linking London and the Channel Tunnel; note the smaller footprint compared with the parallel motorway

**Fig. 5** Domestic services provided by the Hitachi Javelin trains: seen here operating on conventional track with a third rail 750 V DC current supply

## 5 Development of High Speed in Europe

The significant openings of dedicated high-speed line symbolised the development of high speed in Europe (Fig. 6). Seventeen years after the opening of the Shinkansen in Japan, France inaugurated the first true high-speed service in Europe when the dedicated line from Paris to Lyon opened in 1981. In many technical aspects, the Train à Grande Vitesse (TGV) differed from the Shinkansen. The line was built with more severe gradients, but cut straight lines across Frances rather sparsely populated countryside. Power cars at front and rear led to rather higher axle loads, 17 t, and public social spaces such as bar areas led to lower passenger densities and, in general, a much higher structural mass per seat. Double-decker or duplex carriages have been introduced to improve capacity on some popular routes. The carbon footprint, however, was very low because of the high nuclear generating capacity in France (interestingly, the original intention was for gas turbine power to be used, but the idea was dropped after the 1973 oil crisis and conventional electric power from overhead supply lines substituted). Over the succeeding years, further lines have been added, basically on a radial pattern from Paris, and more destinations were added to the TGV network through on/off running onto convention tracks. The current dedicated network is just over 2000 km. Although the frequencies of departures of the TGV fall way below the intensive service offered by the Shinkansen, the punctuality leaves something to be desired, an issue compounded by the mixed running. It is also worthy of note that a new generation modified and shortened version of the TGV achieved the world speed record of 574.8 km/h on 3 April 2007: speeds in service are, of course, considerably lower, 320 km/h is the typical maximum. The France efforts earned them considerable recognition in Europe with the result that other countries were persuaded to join the high-speed revolution.

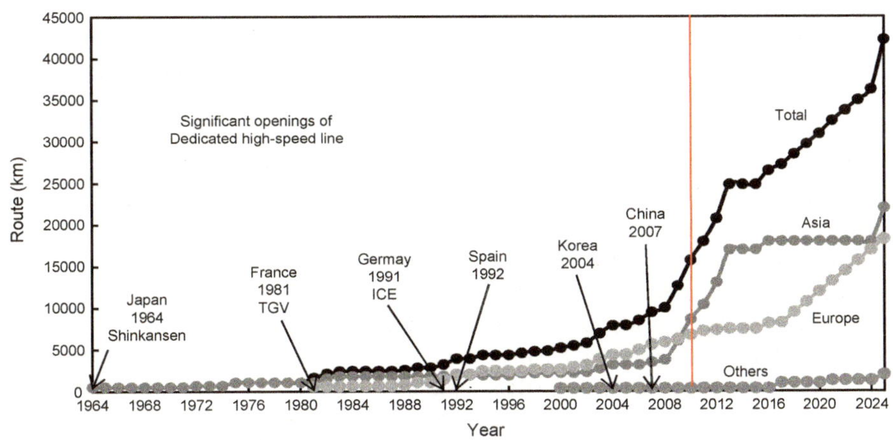


**Fig. 6** Development of dedicated high-speed lines in the world

Germany was the next to enter the High Speed arena with its intercity express (ICE) trains in 1991. But the network is quite different from those of the Shinkansen and the TGV. Germany, the more so after reunification, is a large polycentric country lacking a single city focus like Tokyo or Paris. Much of the running of TGV trains is on conventional track, although upgraded to allow 160 or 230 km/h running. Very little is run at 300 km/h on dedicated lines. Nevertheless, the ICE has captured the public's imagination and is rightly known as the flagship of the German railways. Several services run into adjacent countries: France, Austria, Switzerland, The Netherlands, Belgium and Denmark.

Spain was a later but important entrant. The Madrid Seville service opened on dedicated track in 1992. The Alta Velocidad Española (AVE) translates to Spanish High Speed, but the initials are also a pun on the word *ave*, meaning *bird*. Uniquely, on the line from Madrid to Seville, the service guarantees arrival within 5 min of the advertised time and offers a full refund if the train is late, although only a very small number 0.16% are so delayed. Spain now has over 3000 km of high-speed lines: the second in the world only to China. One might also make particular mention of Turkey where in recent years over 1400 km of track has been cleared for 200 km/h operations.

## 6 Safety of High Speed Trains

Japan has an enviable and remarkable safety record. Not a single fatality due to train accidents since its inauguration and more than 10 billion passengers carried. The Japanese interpretation of this is that, "the system is designed to prevent even the risk of a collision. The key elements of crash avoidance are the use of exclusively dedicated tracks which completely exclude freight and commuter rail being on the same tracks, no at grade crossings of any sort, and an automatic train control (ATC) system, which automatically detects train position and controls the operation of the system". The further claim is that "these elements together maintain safety, advanced energy performance, high speed, large volume, high frequency, and other advantages of operation of the system". Perhaps modesty prevents the claim that this safety record depends too on human factors, particularly the education and continual training of staff and their acceptance as an essential ingredient of the whole system. The major lesson to be drawn here is that high-speed trains are indeed a system comprising infrastructure, vehicles, control and people. That infrastructure, vehicles and operations are best run as a single unit is an obvious conclusion.

Other systems in the world have not maintained an unblemished record, but it is noted that no fatal accidents have occurred on separated high-speed track. The tragedy at Eschede, which killed 101 people on 3 June 1998, involved an ICE train on convention track (Fig. 7). The first author was consulted by Deutsche Bahn in the aftermath: the immediate cause was the fatigue failure of the rim of a resilient wheel. The design had failed to measure the dynamic response of such an unusual

**Fig. 7** Clear signs of the development of a fatigue crack in the fractured wheel rim, which caused the Eschede disaster of June 1998

construction, and laboratory tests had failed to identify the critical failure region of the inside of the rim. Furthermore, no account was taken of the removal of material from the tread of the rim in order to maintain profile and its effect on reducing the bending stiffness of the rim. The major accident at Santiago de Compostela in Spain occurred on 24 July 2013, when an Alvia high-speed train derailed at high speed on a curve about 4 km outside the station. Of the 222 people aboard, around 140 were injured and 79 died. Investigation showed that the train was travelling at twice the allowed maximum for the curve and was operating outside the ATC system of the dedicated track on a length of conventional track as it approached the station. Similarly, on 23 July 2011, two high-speed trains travelling on the Yongtaiwen railway line collided on a viaduct in the suburbs of Wenzhou, Zhejiang province, China, killing 40 passengers (Fang 2011). High speed was not a factor in the accident, since neither train was moving faster than 99 km/h. It has been acknowledged that the accident was caused by a failure of the signalling and control system, but the exact details remain unclear. In the aftermath, maximum speeds were reduced somewhat on the Chinese high-speed system, a sensible measure allowing time for the marriage of human factors into the system and allowing for a sensible bedding in period of the new infrastructure. We note once more that none of these accidents have occurred on dedicated, ATC-controlled high-speed lines.

Much is made of the comparison of the safety records of different modes of transport. Perhaps too much attention is paid to this kind of data. We choose a transport mode for its appropriateness and availability. For example, a short journey to the local shops is undertaken on foot, car or by bicycle and not by airplane, even though flying has in general an enviable safety record and cycling does not! In comparison with cars, trains are much safer per passenger km travelled. A curious statistic from the UK is that in recent years, more people have been killed on the roads per year than the total number of passenger killed on railways since their inception. The price of safety is eternal vigilance, attention from the humans can be and has in many cases been replaced by automated systems, but the standards of maintenance of equipment, vehicles and infrastructure must be extremely high: this is a substantial part of the ongoing costs of running a high-speed railway.

## 7 Railway Research in the UK

When the railways of the UK were integrated under British Rail, the national research laboratories were located in Derby and became a world renowned centre for railway research. Particular successes might be mentioned: the dynamics of bogies, signalling systems, and operations control research.

The privatisation and fragmentation of the railways from 1995 lead to a much more stringent financial view of research based on short termism. The closure of the national laboratories resulted in some parts being sold to private concerns. A few of these survive as consultant partners to the industry, but much of the technical in house expertise and stored knowledge was lost. More recently, a more strategic view has prevailed and an attempt has been made to reintroduce some longer term thinking, mainly based around university research groups. More than 40 universities are joined together in a grouping called Railway Research UK (RRUK), carrying out EPSRC (The Engineering and Physical Sciences Research Council)-funded projects; however, RRUK ended in this form in about 2010. RRUK survives in what is now known as RRUKA (The Rail Research UK Association), which is a grouping of partner universities and industry with some limited funding provided by RSSB (The Rail Safety & Standards Board). Major contributions to these efforts are made at the universities of Birmingham, Huddersfield, Newcastle, Nottingham, Sheffield and Southampton. Again full details can be found on their respective websites. Whilst this research is both useful and commendable, it lacks the synergistic effects of a single centre, and the ability to perform large-scale experimental work from a long-term perspective has been lost.

The authors' own research efforts have been associated with the Future Rail Research Centre at Imperial College London (FRRC), which has been generously supported by Network Rail, the Royal Academy of Engineering and by Hitachi. Some of the major topics, which have been studied over the last decade or so, are outlined and referenced below.

Several Chinese students have worked in the Centre. Not surprisingly, the development of the railway system in China has been of considerable interest (Xue et al. 2002a, b), but these papers predated the rapid development of the high-speed railway in China, which richly deserves a further study. Connections with Japan and its railways have been strong. Several students from FRCC have spent periods working at universities in Japan and at the Railway Technical Research Centre (RTRI) in Tokyo. The first author has spent periods as a Visiting Professor at Tokyo University, supported by JR (Japan Railway) Central Railway Company, Kyushu University and at the Tokyo Institute of Technology. A paper reviewing the birth of the Japanese Shinkansen (Smith 2003a) proved to be a catalyst for the UK railway when a Minister of Transport was impressed by the story it told. Bringing this up to date, a new paper has reviewed the history of the Shinkansen (Smith 2014) to mark its 50th anniversary in 2014.

The first author studied fatigue and structural integrity for many years at the beginning of his research career. There are many applications in railways: in fact, fatigue came to be studied in the early days of the railways because of broken axles

(Smith and Hillmansen 2004). This theme has been taken up and revisited applying modern developments in fracture mechanics (Hillmansen and Smith 2004). In the early days of the privatisation of the railways in the UK, issues involving rolling contact fatigue of rails caused a significant accident, which led to considerable efforts to rediscover issues originally well understood by British Rail Research. The first author was involved in investigating technical aspects of the accident in question, and some of the advances in understanding were discussed in a review (Zerbst et al. 2009).

Recent years have seen increasing attention being paid to global warming: therefore, the energy and emission performance of the railways had come under scrutiny (Inderwildi and King 2013). The widely held view is that railways are one of the most energy efficient modes of transport (Smith 2003a, b), but there are several caveats to this simple view (not the least of which is the load factor of passengers). Issues such as driving style (Read et al. 2011) and alternative energy sources (Read et al. 2009) have been studied.

In principal, the faster we go, the more energy we use. Thus, the topic of energy use of high-speed trains raises many questions (Zhou and Smith 2013). Many aspects have been studied using the FRRC train energy calculator and many sensitivity studies have been computed (Zhou and Smith 2013; Gao et al. 2016).

Switches and crossing are very costly in terms of maintenance, and the dynamics of the passage of a train through a crossing causes many interesting dynamic loads, which lead to several types of deterioration mechanisms. A major study has been performed for Network Rail, starting with a statistical analysis of failures on the working railway (Cornish and Smith 2011), experimental measurements of loads and stresses in crossings in service (Cornish et al. 2012) and theoretical considerations of the complex series of contacts between wheel and rail during the traverse of a crossing (Coleman et al. 2012).

More than 20 years ago, the first author was Chairman of a major project undertaken by British Rail Research to develop crashworthiness standards for railway vehicles. The basic mechanism was the absorption of energy by controlled plastic deformation to the ends of rail carriages and a limit on the deceleration experienced by passengers. The project involved verification of computed results by the crushing of full-scale vehicle ends and a full-scale highly instrumented train collision. The results were incorporated in the UK and European standards for rail vehicle design. It would probably be impossible for the UK rail industry to carry out such a project today. However, detailed research is ongoing although of a computational rather than experimental nature (Xue et al. 2005, 2007).

Several studies of a more general nature have been undertaken, including a study of ways of introducing new technology to the railway industry (Lovell et al. 2011) and ways of increasing innovation into an essentially conservative industry (Lovell and Smith 2008).

Finally, although railways are considerably safer than many other modes of transport, safety is both costly and requires eternal vigilance. It is therefore a worth topic of study (Santos-Reyes et al. 2005).

## 8 Development of High Speed in China

In Asia, Korea opened its Seoul–Busan line 40 years after the opening of the Shinkansen. However, the most remarkable developments in Asia have occurred more recently in China. The development of high-speed rail surged in 2010 (Fig. 6), and the main contribution has to be attributed to China. The pace of China's development has been absolutely astonishing. The medium-to-long-term railway network development target is to cover China's major economic areas with eight high-speed rail corridors, four running north–south and four going east–west. China now has by far the world's longest network of dedicated high-speed lines, much of which has been built in the last seven years and more networks are promised in the next two years, bringing the total length to about 18,000 km. Currently, over 1.33 million passengers travel on the Chinese network every day.

Before the introduction of high-speed rail, China's overall railway technology and equipment were similar to that of developed countries in 1970s due to historical reasons. China initially conducted independent attempts to develop high-speed rail technology domestically, and some notable results include the China Star. However, if to use only their own resources and expertise, China might need a decade or even longer to catch up with developed nations in high-speed rail technology. China's early high-speed trains were initially imported or built under technology transfer agreements with train-makers in other countries, including Siemens, Bombardier and Kawasaki Heavy Industries. In 2004, the Chinese State Council and the Ministry of Railways decided to introduce advanced technology from other countries, in order to bring the joint design and production, and finally to build Chinese brand.

Chinese engineers then redesigned train systems and built indigenous trains based on self-developed key techniques. To unite the academic and industrial research interests and resources, the Chinese Ministry of Science and Technology and the Ministry of Railways set up a "China's High Speed Train Innovation Joint Action Plan" in 2008. On 30 June 2011, the World's longest high-speed line linking Beijing and Shanghai was completed, with a world record speed of 486.1 km/h (Fang 2011).

The development of high-speed rail has risen research interests from many aspects in China. Till now, contributions have been made from infrastructure (Duan et al. 2011; Yang and Shi 2014), traction system (Chang 2014), operation strategy (Lu et al. 2011), aerodynamics (Li et al. 2011; Yu et al. 2013; Cui et al. 2014), etc. On a recent trip from Hangzhou to Beijing, the first author was astonished by the excellent ride quality of Chinese trains operating at speeds over 300 km/h, and by the passenger service and comfort matching anything offered in other regions of the world. One can only stand back in astonishment and applaud China's remarkable achievement.

However, from the author's personal perspective, some issues may still need a bit further consideration: the design life of slab track is usually around 60 years; however, no slab track system in the world currently lives long enough to examine its end-of-life stage. Whether the system is able to withstand the long period depends hugely on the quality of concrete. Periodical renewal of components is required to maintain the riding quality. After a period of service, the strength of the infrastructure decreases and crack may initiate in some components; thus, the rail infrastructure is more vulnerable to sudden external impacts, like natural disasters. It is likely for the infrastructure to fail before the predicted end of its service life. For safety reasons, non-destructive testing method could be needed to examine the cracks, which might be embedded in the core of a slab.

## 9 Future High Speed Rail in the UK

Plans are now being made to extend the high-speed rail system of the UK: a project known as High Speed Two (HS2). The principal drivers are that the existing conventional network is overcrowded and has seen a huge increase in patronage in recent years (an increase in passenger km of 100%, doubling from 750 million in 1995 to 1500 million now, leading to levels of traffic as great as any since about 1910 but on a network of approximately half the size), the population of England is growing rapidly, what was 52 million in 2008 will become 60 million in 2033 and 70 million by 2050, and there is a pressing need to develop the industrial towns of the north by better interconnections between themselves and with London. Although the proposed network will eventually cover most of the country, the initial plan is for a Y-shaped network, from London to Birmingham then onwards to Manchester and Leeds. There is considerable opposition from people who live near the proposed line: as there was for HS1 to the Channel Tunnel, but the consensus then was that when the line was built, the results were by no means as discomforting as the opponents originally feared. There is also opposition of the grounds of costs: Could more be achieved by improving the existing network, or spending the money on other "good" things: schools, hospitals, social services, etc.? Currently, there is a consensus amongst the major political parties that the project should go ahead and appropriate Bill is being steered through Parliament: a process which may take two or more years. The timetable may appear leisurely to many people: the plan is to start construction in 2016/2017 and to open the first sections by 2026. This is a constraint of a democratic parliamentary system but simultaneously is a frustration to the achievement of a long overdue addition to the transport infrastructure of the country. It is unlikely that any procurement of rolling stock will take place until around 2020: By then, many innovations and improvements will have been made, but we are assured that all offers both domestic and international will receive fair and open consideration.

## 10 Concluding Remarks

As the 50th anniversary of the opening of the world's first dedicated High Speed railway approaches, we can reflect on how the Japanese Shinkansen has become a catalyst for the development and rebirth of railways in many other countries in the world. In Europe, France and Germany led the way, but the most astonishing aspect has been the more recent development of an extensive High Speed system in China, surpassing by far on scale all previous developments and still growing rapidly.

It has become a priority to increase the efficiency and convenience of transport without adding to carbon generation. The High Speed train system offers considerable advantages in this area, all the more so if renewable sources of electricity can be employed. Although not discussed in this paper, the development of new High Speed transport hub stations presents an opportunity to move away from existing congested city centres and so improve the quality of life in urban areas, which have become choked with the ubiquitous automobile.

## Appendix

### Highlight

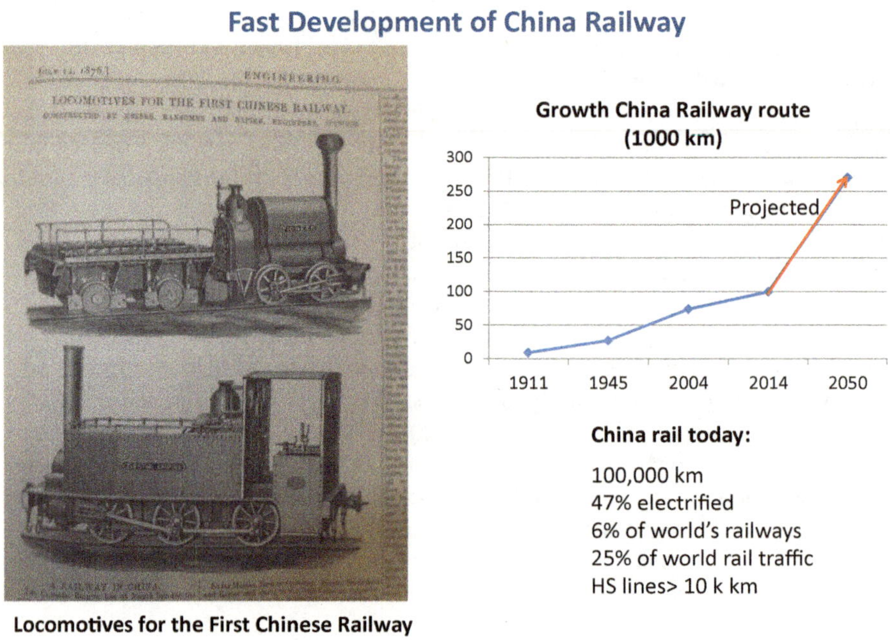

**Historical development of rail passenger volume in the UK**

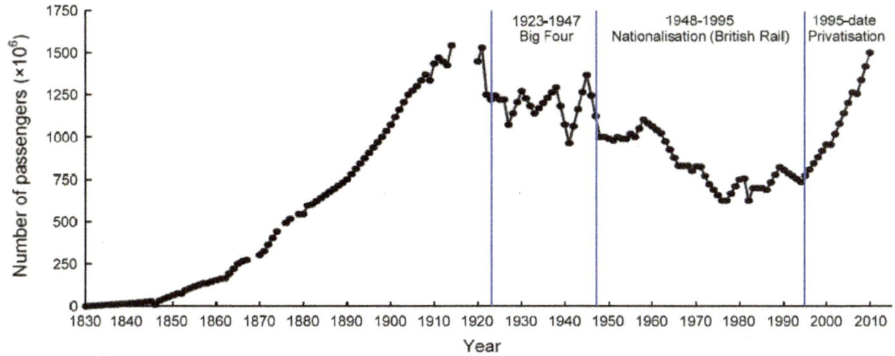


The cuts in number of railway passengers came during a period of rapid expansion of car ownership in the 1960s. This graph of ridership shows the strong return of huge numbers of passengers to the railway in the last decade or so. It is not clear what the reason for this is, but congestion and uncertain journey times on the crowded road network certainly have played a part.

**Some international railway comparisons**

| Country | Area ($\times 10^6$ km$^2$) | Population ($\times 10^6$) | Route (km) | Route/person (km/million) | Average journey (km) | Intensity* |
|---|---|---|---|---|---|---|
| China | 9.640 | 1351 | $1\times10^5$ | 70 | 518 | 26.9 |
| UK | 0.242 | 62.8 | 15 775 | 25 | 40 | 9.3 |
| Germany | 0.349 | 82 | 33 707 | 41 | 41 | 6.3 |
| Japan | 0.378 | 127.5 | 20 852 | 16 | 17 | 51.6 |

Notice that the huge route length in China is shared by many people, so the length of route per capita is only about ¼ that of the UK and 1/6 that of Germany. The average journey length in China is huge 518 km reflecting the importance of the intercity network: notice the low figure of 17 for Japan reflecting the huge commuter network: largely absent in China but developing strongly with new metro and subway builds. The intensity, defined in the figure, is a measure of passenger utilisation of route: Japan is the most intense, China is a large multiple better than the UK and other European countries.

**Fatality risk of passenger using different mode of transport**

(EU-27 in 2008-2010)

| Mode | Fatalities per billion passenger kilometres |
|---|---|
| **Airline passenger** | **0.1** |
| **Railway passenger** | **0.16** |
| **Car occupant** | **4.45** |
| **Bus/Coach occupant** | **0.43** |
| **Powered two-wheelers** | **52.59** |
| **Vessels passenger** | **N/A** |

*Source* ERA.

It is well established that rail is a relatively safe mode of transport; these European figures illustrate the point. Notice that the airlines have an extremely impressive safety record.

**Complex technical system**

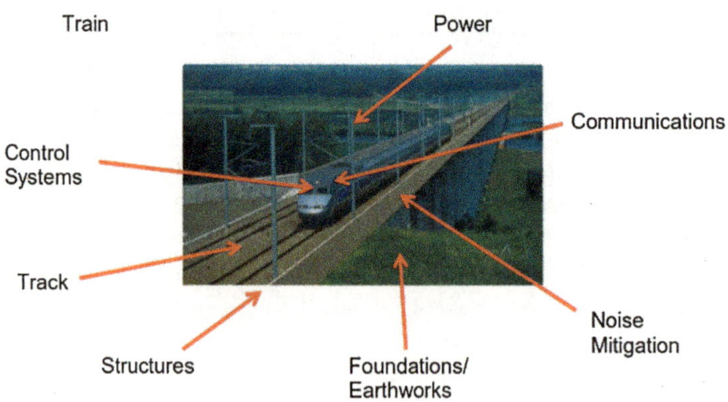


+ people, both staff and passengers

## Brief Bibliography

The followings are particularly useful books on the British, Japanese and German railways respectively:

Herrmann, H. (1996). *ICE high-tech on rails* (3rd ed.). Darmstadt, Germany: Hestra-Verlag.

Hood, C. P. (2006). *Shinkansen: From bullet train to symbol of modern Japan*. London, UK: Routledge.

Simmons, J., & Biddle, G. (1997). *The Oxford companion to British Railway history*. Oxford, UK: Oxford University Press.

A comprehensive account of the origins, development, building and operation of the Chinese high speed rail system is awaited. Considerable material can be accessed on the internet, with the usual health warnings for accuracy.

The annual report of various railway undertakings are available online and contain much useful information. The Japan Railway and Transport Review (JRTR) is an English-language transport magazine published quarterly since March 1994 by East Japan Railway Culture Foundation. It is available online at http://www.jrtr.net/ and is an invaluable source.

## References

Chang, Y. (2014). Optimal harmonic filters design of the Taiwan high speed rail traction system of distributer generation system with specially connected transformers. *International Journal of Electrical Power & Energy Systems, 62,* 80–89. doi:10.1016/j.ijepes.2014.02.014.

Coleman, I., Kassa, E., & Smith, R. (2012). Wheel-rail contact modelling within switches and crossings. *International Journal of Railway Technology, 1*(2), 45–66. doi:10.4203/ijrt.1.2.3.

Cornish, A., & Smith, R. (2011). Investigation of failure statistics for switches and crossings in the UK. In *Proceedings railway engineering conference*. London: University of Westminster.

Cornish, A., Kassa, E., & Smith, R. (2012). Field experimental studies of railway switches and crossings. In *International Conference on Railway Technology: Research, Development and Maintenance*.

Cui, T., Zhang, W., & Sun, B. (2014). Investigation of train safety domain in cross wind in respect of attitude change. *Journal of Wind Engineering and Industrial Aerodynamics, 130,* 75–87. doi:10.1016/j.jweia.2014.04.006.

Duan, Y. F., Zhang, R., Zhao, Y., et al. (2011). Smart elasto-magneto-electric (EME) sensors for stress monitoring of steel structures in railway infrastructures. *Journal of Zhejiang University-Science A (Applied Physics & Engineering), 12*(12), 895–901. doi:10.1631/jzus.A11GT007

Fang, Y. (2011). On China's high-speed railway technology. *Journal of Zhejiang University-Science A (Applied Physics & Engineering), 12*(12), 883–885. doi:10.1631/jzus.A11GT000.

Gao, Y., Zhou, J., Fang, Y., et al. (2016). Analysis of the energy use of various high speed trains and comparison with other modes of transport. In *Ecological vehicles and renewable energies (EVER), 2016 Eleventh International Conference on*. Monte Carlo, Monaco. doi:10.1109/EVER.2016.7476385.

Hillmansen, S., & Smith, R. A. (2004). The management of fatigue crack growth in railway axles. *Proceedings of the Institution of Mechanical Engineers, Part F: Journal of Rail and Rapid Transit, 218*(4), 327–336. doi:10.1243/0954409043125879.

Li, X., Deng, J., Chen, D., et al. (2011). Unsteady simulation for a high-speed train entering a tunnel. *Journal of Zhejiang University-Science A (Applied Physics & Engineering), 12*(12), 957–963 doi:10.1631/jzus.A11GT008.

Lovell, K., & Smith, R. A. (2008). All shook up: An exploratory study of innovation routes for UK railway rolling stock. In G. K. K. Duysters (Ed.). Wuhan, China: Wuhan University Technology Press.

Lovell, K., Bouch, C., & Smith, R. A. (2011). Introducing new technology to the railway industry: System-wide incentives and impacts. *Proceedings of the Institution of Mechanical Engineers, Part F: Journal of Rail and Rapid Transit, 225*(2), 192–201. doi:10.1177/09544097JRRT383.

Lu, Q., Wang, B., Huang, X., et al. (2011). Simulation software for CRH2 and CRH3 traction driver systems based on SIMULINK and VC. *Journal of Zhejiang University-Science A (Applied Physics & Engineering), 12*(12), 945–949. doi:10.1631/jzus.A11GT006.

Read, M. G., Smith, R. A., & Pullen, K. R. (2009). Are flywheels right for rail? *International Journal of Railway, 2*(4), 139–146.

Read, M. G., Griffiths, C., & Smith, R. A. (2011). The effect of driving strategy on hybrid regional diesel trains. *Proceedings of the Institution of Mechanical Engineers, Part F: Journal of Rail and Rapid Transit, 225*(2), 236–244. doi:10.1177/09544097JRRT374.

Santos-Reyes, J., Beard, A., & Smith, R. A. (2005). A systemic analysis of railway accidents. *Proceedings of the Institution of Mechanical Engineers, Part F: Journal of Rail and Rapid Transit, 219*(2), 47–65. doi:10.1243/095440905X8745.

Smith, R. A. (2003a). The Japanese Shinkansen: Catalyst for the renaissance of rail. *The Journal of Transport History, 24*(2), 222–237. doi:10.7227/TJTH.24.2.6.

Smith, R. A. (2003b). Railways: How they may contribute to a sustainable future. *Proceedings of the Institution of Mechanical Engineers, Part F: Journal of Rail and Rapid Transit, 217*(4), 243–248. doi:10.1243/095440903322712847.

Smith, R.A. (2013). Energy for Railways. In O. Inderwildi & D. King (Eds.), *Energy, transport, and the environment* (pp. 561–576). Elsevier.

Smith, R. A. (2014). The Shinkansen—World leading high-speed railway system. *Japan Railway and Transport Review, 64,* 38–49.

Smith, R. A., & Hillmansen, S. (2004). A brief historical overview of the fatigue of railway axles. *Proceedings of the Institution of Mechanical Engineers, Part F: Journal of Rail and Rapid Transit, 218*(4), 267–277. doi:10.1243/0954409043125932.

Xue, X., Schmid, F., & Smith, R. A. (2002a). An introduction to China's rail transport part 1: History, present and future of China's railways. *Proceedings of the Institution of Mechanical Engineers, Part F: Journal of Rail and Rapid Transit, 216*(3), 153–163. doi:10.1243/095440902760213585.

Xue, X., Schmid, F., & Smith, R. A. (2002b). An introduction to China's rail transport part 2: Urban rail transit systems, highway transport and the reform of China's railways. *Proceedings of the Institution of Mechanical Engineers, Part F: Journal of Rail and Rapid Transit, 216*(3), 165–174. doi:10.1243/095440902760213594.

Xue, X., Smith, R. A., & Schmid, F. (2005). Analysis of crush behaviours of a rail cab car and structural modifications for improved crashworthiness. *International Journal of Crashworthiness, 10*(2), 125–136. doi:10.1533/ijcr.2005.0332.

Xue, X., Schmid, F., & Smith, R. A. (2007). Analysis of the structural characteristics of an intermediate rail vehicle and their effect on vehicle crash performance. *Proceedings of the Institution of Mechanical Engineers, Part F: Journal of Rail and Rapid Transit, 221*(3), 339–352. doi:10.1243/09544097JRRT77.

Yang, X., & Shi, G. (2014). The effect of slab track on wheel/rail rolling noise in high speed railway. *Intelligent Automation and Soft Computing, 20*(4), 575–585. doi:10.1080/10798587.2014.934597.

Yu, M., Zhang, J., & Zhang, W. (2013). Multi-objective optimization design method of the high-speed train head. *Journal of Zhejiang University-Science A (Applied Physics & Engineering), 14*(9), 631–641. doi:10.1631/jzus.A1300109

Zerbst, U., Lunden, R., Edel, K., et al. (2009). Introduction to the damage tolerance behaviour of railway rails—A review. *Engineering Fracture Mechanics, 76*(17), 2563–2601. doi:10.1016/j.engfracmech.2009.09.003.

Zhou, J., & Smith, R. A. (2013). Energy use of high speed trains: Case studies on the effect of route geometry. In *12th International conference on Railway engineering*, London, UK.

## Author Biographies

**Roderick A. Smith, Ph.D.** , is Professor in the Department of Mechanical Engineering, Imperial College London. He has had an academic career at the Universities of Oxford, Cambridge, Sheffield and Imperial College. He has been Head of the Departments of Mechanical Engineering at both Sheffield and Imperial College London, and a Fellow of Queens' College Cambridge and St John's College Oxford. He holds visiting Chairs in several countries. He is a Fellow of Royal Academy of Engineering, Fellow of the Institution of Mechanical Engineers and Fellow of the Institute of Materials. He was the Chief Scientific Advisor to the Department of Transport, UK, 2013–2014 and the President of the Institution of Mechanical Engineers, 2011–2012.

**Jing Zhou, Ph.D.** , graduated from Xi'an Jiaotong University with a Bachelor's degree in 2011 and then received a Ph.D. from Imperial College London in 2014. Then, she continued post-doctoral research in Zhejiang University, and her research mainly focuses on multi-field coupled analysis and wireless power transfer.

# Comparison of the Technologies of the Japanese Shinkansen and Chinese High-Speed Railways

**Satoru Sone**

## Contents



---

S. Sone (✉)
Kogakuin University, Tokyo 113-8677, Japan
e-mail: sonesatoru@gmail.com





Y. Fang and Y.H. Zhang (eds.), *China's High-Speed Rail Technology*,
Advances in High-speed Rail Technology, https://doi.org/10.1007/978-981-10-5610-9_4

# 1 Brief History of Japanese and Chinese High-Speed Railways

The world's first high-speed railway (HSR), the Japanese Tokaido Shinkansen, was inaugurated in 1964. This obviously means naturally that Japan has the longest experience of HSR but also by far the greatest experience of problems on HSR because Japan's traffic volume was, and still is, much denser than any other European countries, that followed Japan in the field of HSR (Smith 2014; Takai 2015).

On the other hand, China started internal discussion of an increase in train speed as late as the 1990s and the first real attempt took place in 1997 at a maximum speed of 140 km/h, while Japan was beginning 300 km/h operation on the Sanyo Shinkansen (Ehara 2015). Up to 2003, research and development (R&D) of this field in China was based on China's own knowledge and skill but was revealed to be too slow to catch up with the national fast-growing economy, so the government decided to change its own method of realising HSR based on its own internal development to the introduction of technologies from the developed countries. Thus, based on mainly Japanese and German technologies in addition to China's strategic plan of introduction and abundant resources, the first substantial increase in train speed took place in 2007, as the sixth increase in train speed (countries with major technical and manufacturing centres for the origins of CRH1, CRH2, CRH5 and CRH3 are Germany (not Canada), Japan, Italy (not France) and Germany, respectively). At this time-point, many railway people around the world paid attention suddenly with much surprise, because the route length of HS tracks exceeded 6000 single-track km, which was longer than those of any other countries, including Japan and France.

2015, only eight years after 2007, see Table 1 of route length of more than 16,000 km in China, which exceeds the total length of all other HSRs put together with the second to fifth being Spain (3400 km), Japan (2800 km), France (2100 km) and Germany (1800 km), respectively.

# 2 Comparison of HSRs in Japan and China by Basic Transportation Figures

To compare the figures and to understand the differences, some major data are shown in Table 2.

**Table 1** Basic transportation figures of high-speed railways in Japan and in China[a]

| Fiscal (year) | Japan | | | China[b] | | |
|---|---|---|---|---|---|---|
| | Route (km) | Passengers carried (×$10^4$) | Transported passenger (km) (×$10^8$) | Route (km) | Passengers carried (×$10^4$) | Transported passenger (km) (×$10^8$) |
| 1965 | 553 | 3097 | 107 | 0 | | |
| 1975 | 1177 | 15,722 | 278 | 0 | | |
| 1985 | 2012 | 17,983 | 554 | 0 | | |
| 1995 | 2037 | 27,590 | 708 | 0 | | |
| 2005 | 2387 | 30,140 | 779 | 0 | | |
| 2008 | 2387 | 31,024 | 817 | 1077 | 734 | 16 |
| 2010 | 2620 | 32,444 | 769 | 5133 | 13,323 | 463 |
| 2012 | 2620 | 32,200 | 860 | 9356 | 38,815 | 1446 |
| 2014 | 2848 | 33,969 | 910[c] | 16,456 | 70,378 | 2815 |

[a]The figure is based on commercial route length, which is sometimes up to 15% longer than the actual length both in China and in Japan for various reasons, and Japan's figure of 2848 commercial-km corresponds to an actual length of 2765 km
[b]Data are provided by Prof. Zhong-ping YANG from Beijing Jiaotong University, China
[c]Not yet officially disclosed and estimated by the author

**Table 2** Differences in scale of China versus Japan to understand HSRs

| Expressed by Chinese figure/Japanese figure of each item | | | | | | |
|---|---|---|---|---|---|---|
| Land area | Population | Route (km) | Passengers carried | Transported passenger (km) | Number of vehicles | Maximum number of trains per day[a] |
| 25 | 11 | 5.8 | 2.1 | 3.1 | 3.3 (15,536/4755) | 1.0 (average 330) |

[a]Comparison with the Tokaido Shinkansen (Tokyo–Osaka) and Beijing-Shanghai HSR

# 3 Comparison of Technological Achievements of HSRs in Japan and in China

## *3.1 Outline*

### 3.1.1 Original Technologies of Japan and of China

It can be thought that all or almost all technologies of HSR are of Japanese origin because Japan was the pioneer in the world for HSRs, but actually this was not the case. When Japan decided to construct a Shinkansen between Tokyo and Osaka, the conventional Tokaido line, which was and still is a double-tracked electrified railway with comparatively short quadruple track sections near Tokyo and near Osaka, was approaching its limit of transportation capacity and the quick addition of two more tracks was an absolutely necessary requirement. The majority of the then Japanese National Railways (JNR) managers intended to add a double-track route alongside the conventional narrow-gauge tracks. To add two more tracks on the same route was thought extremely difficult in the already densely populated area in a short period of five targeted years, so an alternative idea to construct a separate route with necessary interchanges was the most viable. But there were a few people who had strongly wanted to improve the JNR by changing the narrow gauge to standard gauge since the end of nineteenth century, less than 30 years after the first railway opened in Japan. The then president of JNR, Shinji SOGO (十河 信二) was one of them, but all other influential JNR people were not. Shinji SOGO and Hideo SHIMA (島 秀雄) who had been retired after severe accident on the JNR and recalled by SOGO as Chief Engineer together proposed strongly to add a standard gauge new double-track line instead of a quadrupled conventional line. SOGO and SHIMA were able to narrowly persuade the rest of JNR headquarters, and the Shinkansen was therefore successfully initiated. In this situation, the Shinkansen had to realise enough traffic capacity for both expanding passenger and freight traffic together with the conventional line with enough availability and by the targeted opening date. Thus, all construction and trains together with the operating systems design were very conservative without any important failure allowed. Conservative design means not too innovative, rather heavy cars, less speedy movement, and a tendency to have an excess margin in vehicle safety and a small potential to increase further the train speed in future track design.

### 3.1.2 Original Design of Japan's Cars: Development of Lightweight Bogies with Stable Operation at High Speed (Sone 2014)

The Shinkansen's car was based on a very innovative super express (SE) trainset of the Odakyu Electric Railway (OER), which established a world speed record on a 1067 mm gauge railway of 145 km/h in 1957 and a tare weight of 147 t with 108 m length. The Shinkansen's Series 0 cars were comparatively heavy with an average 56 t tare weight and each was 25 m long. To avoid a further increase in mass, a wheel base of 2.5 m and a wheel diameter of 0.91 m were chosen and proven to have enough stability margin to run up to 210 km/h. JNR intended to reduce each car by 4 t, to realise a maximum axle load of 15 t, but failed mainly because of the mass of the traction equipment with enough margin. This meant an axle load of 16 t was narrowly met including the fully seated passengers.

### 3.1.3 Original Design of Japan's Cars: Distributed Traction System with All Axles Motored

Based on the then Japanese Private Railways (JPR) standard design of high-performance electric multiple units (EMUs), the traction system of paired motor cars with eight traction motors controlled by a traction controller was a natural and unique solution and never a result of the optimised design of a number of cars, motored and non-motored, i.e., trailer cars. After Japan's success with the HSR, Britain and France followed in 1976 and in 1981, respectively, but with concentrated traction systems of different types. Until the 1990s, when AC motor traction[1] was available, concentrated traction systems were thought by many railway engineers other than Japanese to be superior to distributed traction with less maintenance of traction motors and more maintenance of non-motored axles' braking system. But in 1992, Japan's Shinkansen succeeded in introducing the world's first high-speed EMU with AC regenerative braking, as the Shinkansen's Series 300 (Fig. 1). In R&D works for this innovative trainset (the author was one of the team), a very important discovery was revealed; a non-motored axle with a necessary brake has more mass (total and unsprung mass) with more heat and more maintenance works required than a motored axle mainly because of the theoretical results of the braking system requiring kinetic energy to heat energy conversion. Thus, in Japan the next Series 500 was changed from 10M6T formation to 16M (0T), mainly to reduce the unsprung and total mass. This important knowledge was presented in 1994 at the World Congress on Railway Research (WCRR) in Paris by the author (Sone 1994).

**Fig. 1** Series 300 of Central Japan Railways; the world first high-speed trainset adopting regenerative braking

[1]Excluding widely used AC commutator motors used on 16.7 Hz electrified lines, which are similar to DC motors.

**Fig. 2** Series N700 (14M2T) of Central Japan Railways; this can run at 270 km/h at 2500 m radius sharp curve by body inclination control. Figs. 2, 4 and 5 were provided by Mr. Chun-xiao LUO from Editorial Office of *Railway Knowledge*, China

### 3.1.4 Chinese Design Concept

When China decided in 2003 to introduce HS trainsets from developed countries, it excluded concentrated traction systems such as the French TGV (Train à Grande Vitesse), German ICE-1 (InterCity Express) and Italian ETR500; this sound and clever decision was supposed to be based on the above-mentioned Japanese result as well as China's own experience of difficulties in the R&D of China Star (中華之星), with concentrated traction, against Changbaishan (長白山), with distributed traction trainsets. All of the introduced CRH1, CRH2, CRH5 and CRH3 have distributed traction systems but there are two different types: CRH2s with driving trailers and the rest of CRH1, CRH5 and CRH3s all of which are of European origin, with driving motors (Zhang 2009a). This difference is, from Japanese experience, very important from the following two viewpoints: from the electromagnetic compatibility (EMC) point of view, a driving motor whose front bogie has traction motors brings much more trouble to signalling circuits, and from the adhesion point of view as well. Wheels of the front bogie tend to slip and slide because they are still wet in rainy condition. On the contrary, a formation with driving trailers and intermediate motors does not require sophisticated adhesion control and if there is no brake applied at the front bogie, the wheels can be used as a reliable speed and position sensor because they do not slide at all (Fig. 2).

At present, Chinese-designed HS trainsets, CRH380s, are of the above-mentioned two types: CRH380A and AL consist of driving trailers with all intermediate cars motored while the rest of CRH380B, C and D have driving motors and intermediate motors and trailers with the same numbers of motors and trailers: 4M4T or 8M8T (Yang 2015). If Chinese railways want to standardise the traction systems, the author strongly proposes driving trailers with as many numbers of motors as are economically viable. Even if a sophisticated and high-performance re-adhesion control is used, small slip and slide are inevitable under adverse conditions, meaning that the acquired speed and position have many errors. If CRH380AL's 14M2T formation is thought to provide too much margin and is too expensive compared with CRH380A's 6M2T formation using the same traction motors, the 12M4T formation may be the best solution for construction costs, but still the running cost may be better for the 14M2T because of much better braking performance.

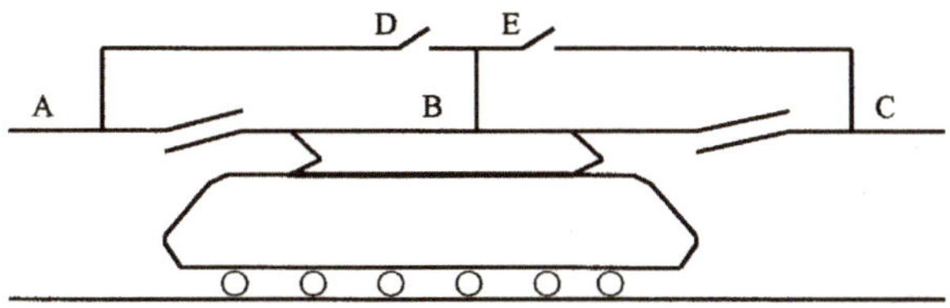


**Fig. 3** Quasi-continuous power feeding system. Before entering to *B*, switch *D* is kept on, after the whole train with two interconnected pantographs comes into *B*, switch *D* turns off and about 300 ms after that switch *E* turns on. Interruption of 300 ms was necessary in the technology of 1960s by various reasons when slow action contact switch is only available. The train can run through in either mode of acceleration, coasting or regenerative braking

## *3.2 Comparison of Achievements of the HSR in Japan and in China Relating to Subsystems Used*

### 3.2.1 Power Feeding System

Among Japanese original technologies, one of which that has not yet been adopted in other countries is the quasi-continuous power feeding system using two consecutive catenary sections and wayside changeover switch (Ishikawa et al. 2015). The system's action is as follows (Fig. 3): suppose the train moves from A to C via B, the switch D is kept closed until the whole train enters the section B, when D is off and a very short time, typically 0.3 s, after opening D, switch E is closed. This system together with simple but superior current collection using two interconnected pantographs achieves nearly perfect current collection because of no interrupted acceleration and deceleration using a regenerative brake and stable electrical connection even when one pantograph loses mechanical contact (Study Group of High-speed Railways 2003). At the moment, the European system prohibits the use of a parallel connected two pantograph system to prevent short circuiting from two power sources in different phases. The changeover switch was improved from the original air blast type and then changed to a vacuum switch and now gradually to a semiconductor type.

China studied using a quasi-continuous power feeding system but has abandoned it for the moment. The reason was supposed to be easy acceptance of the European type trainset, and changeover switches are still being improved from vacuum breaker type to semiconductor ones.

### 3.2.2 Electromagnetic Compatibility

In Japan, there are many special designs for avoiding EMC problems based on experienced difficulties. Contrary to the above-mentioned standard practice, Series N700 of 8 car formation in the Kyushu and the Sanyo Shinkansen are not 6M2T with driving trailers but 8M0T with driving motors to cope with restarting conditions after stopping on a steep gradient section of 35‰ with half the units cut-off (Fukunaga 2015). To cope with EMC problems due to driving motors, traction motors and bogies are of slightly different designs for the front bogie, rear bogie of the front car and of intermediate cars both from an emission and immunity point of view. As far as adhesion is concerned, acceleration and deceleration forces of the front two bogies are squeezed while the rest, including the rearmost cars, produce full force.

### 3.2.3 Phase Balancing Measures After Introduction of Regenerative Trains

The author should touch upon a practice and its changes in phase balancing measures relating to a single-phase high-power load from a three-phase power grid. At the start of the Tokaido Shinkansen in 1964, three-phase AC was converted to rectangular two-phase AC using Scott-connected transformers. One phase was fed to down tracks and the other to up tracks. Thus, on average on most occasions a better balance was established because average power was almost balanced between down and up trains. The situation was changed in two ways: in a change from a booster transformer (BT) feeding system to an auto-transformer (AT) system and the introduction of regenerative trains. A BT was used to pick up return current flowing in the running rails to a negative feeder located about the same height as the catenary wires so that induced electrical noise to nearby communication wires was cancelled. This was good for avoiding EMC but for current collection, the BT section was a weak point where the terminal voltage of BT should be broken by leaving pantograph producing electrical arc every time a train passed across the BT section. To cope with this problem, the BT feeding system was changed to an AT feeding system and according to this, the power balance to be expected was changed from between up trains and down trains to between up and down trains running on one side of a substation and those on the other side.

The other important change in the situation was the introduction of regenerative trains in 1992. The principle of three-to-two-phase conversion by a traction transformer is that when the power sum of one of the two phases is equal to that of the other phase, the three-phase side is balanced. Before the introduction of regenerative braking, the worst case is maximum power at one phase and zero at the other phase. In this worst case, three-to-two-phase conversion has no effect; the same as taking the total power from one of the three phases alone, but after the introduction of a regenerative brake the worst case is positive power on one side and negative power on the other side; in this case, three-to-two-phase conversion has a negative effect; the situation is much worse than taking both positive and negative power from one of the three phases.

To cope with this adverse effect together with a voltage stabiliser both on the grid side and the train side, additional equipment called a railway static power conditioner (RPC) was introduced where it was thought necessary. An RPC can transfer active power from one side to the other and absorb necessary reactive power on each side independently (Ishikawa et al. 2015).

From a purely technical point, an RPC can be seen as sometimes necessary but from a totally economical design of the whole system, it is doubtful whether or not it is justifiable.

### 3.2.4 Result of Lightweight Design of Cars

Know-how in constructing lightweight cars is by far the most advanced in Japan mainly because of the following three factors: (1) weak roadbed and infrastructures, (2) many privately owned train operators and car manufacturers, and (3) not too much buffer strength required as is necessary for heavy freight trains. After World War II passenger traffic demand expanded very rapidly between big cities and their suburbs, and the majority of this demand was carried on the suburban lines of the JPRs. Thus, lightweight and high-performance EMUs were developed and manufactured from the early 1950s by JPRs, which are the origin of the Shinkansen basic vehicle, with the other origin being AC electrification developed by the JNR. Some of the European car manufacturers think Japan's trainsets of small mass mean there is not enough strength in a crash and some others think there is not enough rigidity against vibration. The former criticism has been proven wrong on various occasions: one of the typical cases is the Wenzhou collision on 2011-07-23. In this case, a lightweight aluminium-bodied CRH2 hit a comparatively heavy stainless steel-bodied CRH1 and damage was much greater to the CRH1. The latter criticism is half true and half incorrect; it is a matter of the design's balance between riding comfort and energy saving. The mass per axle of less than 12 t of the E2 of the Tohoku Shinkansen, the origin of the CRH2, is the result of Japanese balance (Yang 2012) and the roughly 15 t of the CRH380A is the result of Chinese balance (Yang 2015).

### 3.2.5 Cab Signal ATC

Although there had been some cab signal systems in smaller urban railways, automatic train control (ATC) using a cab signal adopted by a major railway was a world first for the Shinkansen (Kotsu Kyoryoku Kai Foundation 2015) and since its inauguration more than half a century has passed but still it is not perfect nor is there a final version of such a system. Japan's ATC is not intended for driverless operation, but the braking operation is almost automatic with manual operation left for final stopping at a designated position at a train speed of below 30 km/h. The French TGV has a different philosophy; an automatic train protection (ATP) system should help the driver's operation by showing a forthcoming situation and only if the driver fails to react safely does the ATP system intervene. German philosophy is a little different to that of France but does not differ that much.

China has introduced several signalling systems from Europe and Japan and at the moment several systems are coexisting on the same track according to the several different trainsets. Coexistence of several signalling systems on the same track is not desirable from the standpoints of drivers' confusion, more maintenance work, and possible interference between the systems.

To solve this problem in the future, there will be much potential for solutions such as sorting the systems into categories by area, line groups, speed range or climate conditions, etc. The process and results of the Chinese solution will be paid strong attention to by the developed countries, none of which has had such a need as yet.

### 3.2.6 Poor Tracks and Infrastructure in Japan

The Shinkansen's weak points are often seen in the infrastructure; too little spacing between adjacent lines of 4.2 m (in the case of the Tokaido Shinkansen) or 4.3 m (Sanyo Shinkansen and all thereafter), a small cross section in double track tunnels of about 64 $m^2$, a non-reversible signalling system, unique in the world, and a speed restriction of 70 km/h at turnouts into and out of stopping stations. The poor infrastructure was justifiable for the Tokaido Shinkansen because it was the pioneer, who cannot learn from others, and because of the fact that Japan was not economically strong enough at the start of the construction in 1959. After minimal improvement of the Sanyo Shinkansen, further improvements in line with world standards have not been realised.

In contrast to this, China can choose from many products around the world and a traditionally strong infrastructure for heavy load has been in due course further improved for the HSR. Some Japanese infrastructure specialists say that poor turnouts from the viewpoint of restricted speed should be justified to maintain necessary reliability by prohibiting two or more shifting motors, which is thought the main reason of turnout troubles.

# 4 Other Developed Countries' HSR Technologies (Akiyama 2014)

## 4.1 *Britain*

British Rail was the first railway to realise 200 km/h operation successfully in 1976. Although the success of the High-Speed Train (HST) had greatly surprised Japanese traction engineers, who were confident of the need to have many driving axles to cope with the adhesion limit, the influences to Japan and many other countries were neglected or not strong because the HST was for non-electrified lines. The HST technology was exported only to the XPT in Australia; all other important HSRs have been electrified.

## 4.2 France and Germany

After experimental introduction of locomotive hauled trains with 200 km/h operation, with just a few trains on the same line, substantial HSR started in France in 1981 and in Germany in 1991 using a fixed formation concentrated traction system with locomotives at both ends, as the French TGV and the German ICE-1, respectively. The French TGV was built with much technology learned from the Japanese Shinkansen leading to a much improved HS system in French eyes. This success, together with a very high-speed record established of 380 km/h before regular services started, caused much attention in Japan, especially relating to current collection, which was one of the major sources of technical troubles in Japan. French railway specialists learned about poor Japanese current collection and moved to realise the use of a gas turbine, which can avoid current collection itself, but the world oil crisis brought the French back to electric power again. The TGV used only one pantograph over the rear locomotive to avoid troublesome problems of current collection at high speed, using over-the-roof high tension cables to feed current to the front locomotive. Japan in turn learnt from this but by using the same method with improvement, after careful tests of the numbers and positions of pantographs, electrically connected as the French ones. It was decided that the best position was not at the front or back of the train, where there is a whirling airflow, but at a rectified airflow position, and the best number of raised pantographs was two, almost always electrically connected even if one of the two jumped from the catenary. The same collection system cannot be adopted in Europe because of feeding sections and European rules prohibit the use of two or more interconnected pantographs.

## 4.3 French Versus Japanese Technologies

French technologies were very different to Japanese ones. This was natural because the French had learnt from the experiences in Japan so there were many ways in which the French were superior to the Japanese such as the following:

Current collection: Many equally spaced individual pantographs versus rear pantograph feeding its own locomotive and feeding the front traction unit with over-the-roof cable.

ATP: As described above, much more freedom was given to the driver. For example, under the French nominal maximum speed limit of 260 km/h in 1981 actual speed could be raised as high as to 285 km/h.

Train length: An allowed maximum of 300 m, twelve 25-m-long car fixed formation in 1964 or 400 m with a sixteen 25-m-long car formation after 1970 in Japan or a half the maximum length formation of 200 m with a dual formation train as required (i.e., two 200-m-long trainsets coupled together).

Adaptation to different demands in the train scheduling: The Japanese fixed pattern scheduling for the heaviest demand with surplus trains cancelled at lighter demand day and/or time versus the French basic train scheduling for the minimum demand period with increased formation, i.e., dual trainsets coupled, and/or additional trains running before the advertised train. The decision of the operation scheduling was made about one month prior to actual operation in order to sell reserved seat tickets in Japan versus up to about two days before travel according to the data of the sold tickets in France.

Among the different methods adopted by the French TGV, which were thought to be improvements by French railways, the following examples were thought not to be improvements by the Japanese.

Concentrated traction system: Less maintenance work required for a smaller number of higher power traction motors versus more maintenance work required for many unmotored axles. In Japanese practical operations, increased maintenance work due to brake friction should substantially be avoided as is the additional mass and the possibility of fire.

Articulated configuration: No noisy bogies under passengers' feet and a smaller total mass versus a smaller number but bigger axle load. In Japan, where track is weaker than in France, the priority is less maximum axle load.

Air suspension: The French explained that compared with commuter trains in which ratio of fully loaded versus tare weights is big, the TGV does not require variable parameter suspension because the ratio is small.

## *4.4 French Versus Japanese Technologies in Recent Years*

Thirty-five years have passed since the TGV was first introduced in 1981. Most of the above-mentioned differences have been judged by railway engineers worldwide.

Current collection was once one of the severest problems of the Shinkansen; the TGV escaped from occasional troubles of current collection but later Japanese engineers could make further improvements, i.e., two parallel pantograph current collection system as described above.

ATP: The basic principles are still different from each other, but the Japanese ATC has been improved so that riding comfort and punctual operation are realisable by both well trained and experienced French drivers and by all Japanese drivers.

Trainset length: The Tokaido Shinkansen is very special Shinkansen where the maximum transportation capacity is still not enough to meet the maximum traffic demand. For this reason, JR Central, the operator of the Tokaido Shinkansen, does not wish to construct other than 16 car formations nor improve the amenities with less passenger capacity. But all other operators, JR East, the operator of Tohoku, Joetsu and Hokuriku Shinkansen, JR West, the operator of the Sanyo Shinkansen and part of the Hokuriku Shinkansen, and JR Kyushu, the operator of the Kyushu Shinkansen, have different formations adaptable to different demands. The practice of half the maximum length formation with dual trainset coupled together as required is not widely adopted nowadays in Japan because of possible difficulty in evacuation in case of fire; in Japan escape from fire is thought necessary in both directions, forward and back. In order to realise this, passengers must be able to walk through coupled trainsets; but to realise gangways between highly streamlined front ends is not practical on the Shinkansen. The Chinese practice of building both an 8 car formation and a 16 car formation as seen with the CRH380A and CRH380AL, respectively, seems to be learnt from original TGV's practice. JR West had once the same practice but in recent years two 8 car formations coupled together are no longer used.

Distributed traction: When DC traction motors were used, there was keen discussion between distributed versus concentrated traction systems, but once AC traction motors, which are much lighter and require almost no maintenance work, were used, everybody, except some of the French, was confident that distributed traction is much superior.

Air suspension: The TGV also introduced air suspension after realising the practical, not theoretical, superiority of air suspension for riding comfort.

Articulation: The super express (SE) trainset of OER, made in 1957, which provided many technical ideas and parts and practice to the Shinkansen's Series 0 trainset, was also of articulated configuration. JNR and JR East compared and sometimes built test cars with partially articulated trainsets and the results were not substantially superior nor inferior.

## 5 Technologies of Chinese Origin

Introduction of distributed traction only: Apparently, it is the same as in Japan but the reason was completely different. Japan introduced the system as the only practical solution for realising high-speed operation, which was denied by the British HST in 1976, the French TGV in 1981 and the German ICE-1 in 1991. Japan could not persuade European railway engineers of the superiority of distributed traction systems until the 1990s when traction motors changed from DC motors, with maintenance required commutators, to AC motors with much less mass and no maintenance required. The Chinese decision, made in 2003, was thought wise enough because at that time not a few people were confident that the TGV was the world's best. This decision was supposed to be made based on experiences prior to 2003, when developments were made without license, and by discerning leaders who could evaluate many papers including the author's (Sone 1994).

Coexistence of several different systems: this is because of the introduction of both Japanese and European practices. This has two effects, both favourable and unfavourable: the former which is favourable is to be able to judge from Chinese experience and the latter, unfavourable, is that the system is complicated and expensive, and sometimes difficult to maintain safety because of interference between the systems. Among many subsystems, power feeding/current collection and signalling system are important. The Japanese practice of power feeding/current collection is superior, with practically no interruption of power, which enables continuous powering and regenerative braking beyond different phase areas and also taking auxiliary power directly from a main transformer, not through a bulky, costly and heavy battery-assisted AC-DC-AC converter, which is necessary with the 16.7 Hz catenary but not for 50 or 60 Hz catenary. Systems should be in multiple both on board and/or at the wayside if using trains with different signalling systems on one line or using the same trainset to run through to different signalling sections. In addition to costly construction, different systems sometimes interfere each other; preliminary elimination of the interference is almost impossible because the designer of system A does not know in detail about system B, and vice versa.

In near future, China must conquer this problem either by unification or separation as far as possible: either way has merit and demerit; nationwide unification tends to stop further improvement and separation tends to restrict necessary through operation or joint operation.

Chinese origins in the narrower sense: The fluid dynamic design of really HS trainset is thought to be one of the typical examples of this category (Fig. 4). The world's first HSR developed in Japan used tunnels of too small a diameter and too small distance between adjacent tracks. To cope with these poor conditions, Japanese HS trainsets are very long with strange front shape (Oguri 2015) (Fig. 5).

To contribute worldwide, it is necessary for Chinese engineers to show an optimum, or nearly optimum, set of construction parameters as a function of the required maximum speed. Although the basic size of an HS trainset with 3.4 m wide and with 1.3 m floor height is not the solution to optimisation, it is common in the Japanese Shinkansen because of historical links between Japan and China. In other Asian countries, traffic demand is not as high as in China or in Japan. These countries may require the optimum size as well as the number of cars in the formation.

**Fig. 4** CRH380A of Chinese Railways, based on Series 500 (*front shape*), E2 (traction system), and ICE-3 (*front cabin design*) totally improved by Chinese Railway designers

**Fig. 5** Series E5 of East Japan Railways; Japan's highest speed operation at 320 km/h on 4000 m radius curve assisted by body inclination control realised together with E6, seen at the rear of the photo

# 6 Further Discussion on Technologies of Japan and China

## *6.1 Optimum Motored to Non-motored Cars (MT) Ratio and Brake System Design*

Japan has established the optimum design of motored cars in a fixed formation; if the formation is long enough such as 10–16 cars, all intermediate cars should be motored with the end cars non-motored. This is sometimes thought to provide surplus power and be too expensive but it is not so as described above. This means that even if surplus power is used only in braking, this is better than creating friction or any other types of kinetic energy to heat conversion brakes. It is very easy to use that installed surplus power with a reduced current so that there is not too much burden on the substation with a much reduced probability of wheel slip on acceleration.

For a short formation such as 6 or 8 cars, on a steep gradient section in some restricted situations such as a one traction unit cut-off, all axles should be motored with reduced traction effort in order not to create wheel slip, which can cause considerable errors in train speed and position in the front car. Also, special care should be taken not to cause the EMC problem as described in Sect. 3.2.2.

Contrary to Japanese practice, the German ICE-3, which was changed from ICE-1 and ICE-2 concentrated trainsets in order to meet the French axle load limit of 17 t on the HS line, the LGV, has the formation of a driving motor and intermediate motors and trailers based on European locomotive technology. To develop CRH380B and CRH380BL, Chinese HS engineers met some EMC problems due to emissions from the front motored car.

## 6.2 Safety and Reliable Design and Operation

From the HS trainset design point of view, the most important factor is a safe, reliable and high-performance braking system (Kumagai 2015). Although the final braking performance must be guaranteed by mechanical braking, taking multiple failures of electric systems into consideration, service braking as well as emergency braking should be dependent on electric braking in regenerative mode that produces no substantial heat. If some of the electrical subsystems are not reliable, such as interruption of supply for a considerably long time and/or wide area near the boundary of feeding areas, electric braking should be assisted by rheostatic braking as well, which produces substantial heat.

From the train operation's point of view, Japan's practice has been to build enough margin in the following areas of running time, adhesion, rise in temperature, train headway, supply voltage fluctuation, etc. What can be seen as too much surplus power and safety, such as in the 14M2T formation rather than the 12M4T trainsets, compared with Europeans the Japanese engineers are confident they can use a more reliable and safe operation without almost any negative effects by taking surplus current from substations; the squeezing of catenary current is very easy to achieve. Thus, when 1 unit (2M) out of the total 7 units is cut-off, the remaining 6 units can operate the train in normal mode, rather than squeezed mode.

## 6.3 Design of Automatic Train Protection (ATP)

The design philosophy of the Shinkansen and the TGV has been different from each other with respect to the crew's role. In earlier years, the TGV had better riding comfort due to deceleration, energy consumption and in some cases average train speed. The Shinkansen's ATP, called ATC, has by several steps progressed to be today's digital ATC with continuous, not stepwise, braking dependable on the trainset's performance. But the author and some other railway engineers do not think it is the final version of such a protection system. The reason is as follows: As a protection device, applicable braking performance should be very low because even in the worst conditions, such as deep snow on the running rails, the train must be protected from collision, and in normal conditions the running time with manual driving is apparently shorter than when ATC is in operation. Actually on a snowy day in February 2014, a Toyoko Line's train using high-performance continually controlled ATC, using the same philosophy with a different apparatus in detail from the Shinkansen's ATC, collided with the preceding train.

## 7 Peculiarity of Japanese and Chinese Railways

Among many countries which have HSRs, Japan alone has almost no medium-speed railways: between the Shinkansen, whose network of 2,910 km with 240 km/h or higher top speeds, and conventional lines of many operators totalling 24,900 km whose maximum speeds do not exceed 130 km/h, only 10 km of Keisei Electric Railway's section, 10 km near Narita International Airport, can be operated at maximum speed of 160 km/h.

One of the most important social roles of Japan's railways is the suburban commuter traffic. Almost all of the commuter traffic inside the Tokyo metropolitan area, in about a 50 km radius and with a total population of around 26 million, is made by rail traffic sometimes combined with transport such as bus, bicycle or private car between home and the train station, by kiss and ride, not park and ride mode. Tokyo's high density of population and efficient economic operation are kept moving thanks to this high-density rail traffic.

On the contrary, China has many medium-speed railways but no efficient suburban railways. This is very strange if seen from the outside because both countries have many big cities, suitable for connection by railways either by HSR or by medium-speed railways and some big cities like Beijing, Shanghai and Chongqing have a big population in suburban areas that could be connected by efficient commuter railways. Some routes of railways in Chinese urban areas are provided by underground railways but actually in suburban areas most are elevated lines, which are independent from national railway network and not at all efficient from the standpoint of positionings of stations, spacing between stops and also for train scheduling patterns. Among the world's big cities, the S-Bahn networks, urban and suburban railways, in Germany, Austria and German speaking Swiss areas, and the RER, a similar in French speaking area, are better than Chinese urban networks but inferior to the Japanese urban railways operated jointly by JPRs and JRs as well as local government railways.

## 8 Important Relationship to Be Kept Between Japanese and Chinese HSRs

As analysis was made in the previous sections showing that under similar social conditions and economic activities and based on modern railway technologies, the present status of HSRs, medium-speed lines and suburban lines is too different from each other.

In the HSR area, if we combine Japanese experience and the related know-how with highly reliable and safe components and Chinese efficient construction and manufacturing ability and highly progressive development, the results will contribute much to HSRs across the world (Zhang 2009b).

No existing medium-speed lines in Japan can hardly be justified socially. The so-called Yamagata- and Akita-Shinkansen whose trainsets can run very fast on the Shinkansen tracks are only allowed to travel at up to 130 km/h on conventional lines, converted from narrow gauge to standard gauge, because of safety at level crossing and for some other technical reasons. Practical know-how established in China will contribute to Japan upgrading the required conventional lines into medium-speed railways.

Chinese suburban railways operated by local governments are different from the Chinese national network operator's lines in many ways. In Japan, after 1960 through operations in the Greater Tokyo area started between private railways and both local government railways including underground lines and national railways (the then JNR) by mutual discussion and the making of a through operations agreement. This was a very big success, and the private railways' know-how about train operations has been gradually prevailing in Japan's conurbations. Important points relating this know-how include constructing many stations that are accessible on foot, and how to combine fast, limited stop and slow trains, stopping at every station, and how to organise cross-platform transfer at major stations. If this is adopted in China, the travelling time between home and the office can be dramatically shortened, taking access times into consideration.

## 9 Conclusions

Both countries, Japan and China, have their good points and weak points. In such a situation, to combine the good points, one country with those of the other will create a big profit for the railway industry and passengers in both countries in the following area: HSR, medium-speed lines and suburban lines. If the best combination in these important areas is realised, the de facto standard in the two countries can contribute to the rest of the world's passenger rail services.

## Appendix

### Highlight
### Aim of Shinkansen

- Aim of Shinkansen was NOT the world first high-speed railway nor the world fastest railway BUT to increase capacity of Tokaido main line

- Chosen from the three alternatives: (1) quadrupling of the existing narrow-gauge double track, (2) addition of (partly) separated double track with necessary interchanges and (3) construction of standard gauge double-track line with better performance
- Inside Japanese National Railways (JNR) majority supported (1) or (2)
- President SOGO (十河) and Chief Engineer SHIMA (島) were among minority
- To persuade majority, freight traffic was to deal with Shinkansen as well but never realised.

**Success of Shinkansen**

- Success of Shinkansen is mainly due to systematic approach under targeted date of inauguration using proven technologies
- JNR's Railway Technical Research Institute (RTRI) had already disclosed its confidence to realise connection Tokyo and Osaka within three hours at its 50 years commemorative presentation in 1957
- Technical roots of the success were (1) bullet train project in 1940, which was only partly started and discontinued by the war, (2) lightweight emu car realised jointly by a private railway and RTRI and (3) AC electrification using industrial frequency developed by JNR
- All important technologies were thought proven.

**Chinese first approach**

- Chinese approach is quite different from the case in Japan
- No concrete targets of date, speed or other requirements
- Voluntary developments learned from developed countries but not license agreements were revealed to be too slow and lack of availability to meet the requirement of rapid developing society at the year 2003
- Change of ways of realising high-speed railways taken by clever authority in 2003 to be mentioned later.

**European approach**

- European approach started just after Japanese success was anticipated
- Questions of Japanese traction systems, knowledge of adhesion, current collection, etc., were motive forces of development of different trainsets from Japanese
- British HST and French TGV's success surprised engineers in Japan very much but Japanese practice was not changed substantially.

**Shinkansen's Series 300**

- Shinkansen's Series 300 was really innovative; ac motor traction with regenerative braking, lightweight, energy and maintenance saving and environmental friendly
- The then world best high-speed trainset, TGV, changed suddenly to be obsolete, but French engineers did not realise the fact soon

- Development of AGV, the successor of TGV, was too slow and they lost the chance to export trainset to China, except for Italian-designed CRH 5, which was not for the need of Chinese high-speed railway.

**Shinkansen's Series 300**

**Chinese change of ways of realising high-speed railways**

- Chinese change of ways of realising high-speed railways in 2003 was very clever to catch up with developed countries and to become world best high-speed railway holder
- CRH 1 and CRH 5 were base of developing medium-speed trainsets
- CRH 2 and CRH 3 were base of developing high-speed trainsets of Chinese origin, CRH380 series
- Loading gauge of Japanese and Chinese HSRs is common for both countries, but Chinese track design are much favourable for operation at high speed.

**Necessary and important relationship of Japan and China**

- Coordination of advantages of both sides and early escape from disadvantages of each side
- Required coordination of Japan and China in other fields of high-speed railways
- Medium-speed railways; Japan has almost nothing although latent demand for it is very high
- Urban railway systems; Japan has the world best practice, but Chinese high demand has been still subconscious.

## References

Akiyama, Y. (2014). High speed rail worldwide-past and future. Special issue 50 years of high-speed railways. *JRTR, 64*, 36–47.

Ehara, G. (2015). History of the Sanyo Shinkansen and the opening of the Hokuriku Shinkansen. *Japanese Railway Engineering, 55*(2), 18–27.
Fukunaga, H. (2015). Technology and outlook for the Kyushu Shinkansen which has been in service for 10 years. *Japanese Railway Engineering, 55*(2), 39–43.
Ishikawa, S., Ito, M., Inoue, Y., et al. (2015). The 50 year history of the Tokaido Shinkansen. *Japanese Railway Engineering, 55*(2), 9–17.
Kotsu Kyoryoku Kai Foundation. (2015). *50 Years of the Shinkansen*. Japan: Transportation News Co. (in Japanese).
Kumagai, N. (2015). Creation of value using Shinkansen technology—Strategies for improved safety and speed. *Japanese Railway Engineering, 55*(2), 5–8.
Oguri, A. (2015). Commercial operation at 320 km/h of the East Japan Railway Company's Shinkansen and transition of technology. *Japanese Railway Engineering, 55*(2), 28–38.
Smith, R. A. (2014). The Shinkansen—World leading high-speed railway system. *JRTR, 64,* 6–17.
Sone, S. (1994). Distributed versus concentrated traction systems—Which can better meet the needs of customers and operators? *World Congress on Railway Research, Paris, 1,* 239–244.
Sone, S. (2014). Technological history of Shinkansen for 50 years—Progress and future of high-speed railways. Kodansha Blue Backs Series B1863 (in Japanese).
Study Group of High-speed Railways (Ed.) (2003). *Shinkansen—All about High-speed Railway Technologies*. Sankaido, October (in Japanese).
Takai, H. (2015). Historical tracing of research and development on the Shinkansen. *Japanese Railway Engineering, 55*(2), 49–52.
Yang, Z. P. (2012). *All about Shinkansen—Japanese high-speed rail technology* (2nd ed.). China: China Railway Publishing House. (in Chinese).
Yang, Z. P. (2015). *Technical outline of high-speed railways*. Beijing, China: Tsinghua University Press. (in Chinese).
Zhang, S. G. (2009a). *Study on design methods of high-speed trains*. China Railway Publishing House, China (in Chinese).
Zhang, S. G. (2009b). *Study on improving Beijing to Shanghai high-speed railway system*. China Railway Publishing House (in Chinese).

## Author Biography

**Satoru Sone** Ph.D., in Electrical Engineering, is Professor of Open College, Kogakuin University, arranging lecture series in railway technologies for adult education since 2000. In 1984–2000, he was a professor in various fields of electrical, electronic and communication engineering, Univ. of Tokyo, and in 2000–2007, he built up railway engineering R&D units in Kogakuin Univ. He also served on various bodies in railway technologies including the Wessex Institute for COMPRAIL, the Japan Railway Engineers' Association, the Ministry of Transport and the West Japan Railway Company.

# Part II
# Aerodynamics of High-Speed Rail

# Unsteady Simulation for a High-Speed Train Entering a Tunnel

**Xin-hua Li, Jian Deng, Da-wei Chen, Fang-fang Xie and Yao Zheng**

**Contents**



## 1 Introduction

When a train enters or leaves a tunnel, finite amplitude pressure waves are generated and propagate in the tunnel. At the tunnel exit, the waves are partially reflected. The reflected waves of opposite pressure return into the tunnel and are partly emitted outside the tunnel as a microwave. The air pressure change in and around the tunnel has a great impact on the safe operation of trains, staff comfort, and environment around the tunnel (Joseph 2001; Tian 2007). A 1D numerical simulation method was first used to study the pressure variations on the train and in the tunnel (Woods and Pope 1979; Mei et al. 1995; William-Louis and Tournier 2005). In the twentieth century, a series of experiments concerning the pressure changes, microwave, and aerodynamic drag was carried out on the Japanese

X. Li · J. Deng (✉) · F. Xie · Y. Zheng
School of Aeronautics and Astronautics, Zhejiang University,
Hangzhou 310027, China
e-mail: zjudengjian@zju.edu.cn

D. Chen
National Engineering Laboratory for System Integration
of High-Speed Train (South), CSR Qingdao Sifang Co., Ltd, Qingdao 266111, China

 
Y. Fang and Y.H. Zhang (eds.), *China's High-Speed Rail Technology*,
Advances in High-speed Rail Technology, https://doi.org/10.1007/978-981-10-5610-9_5

Shinkansen. In recent years, Japanese researchers have simulated the changes of the pressure and microwave using 3D models and model experiments (Ogawa and Fujii 1997; Tanaka et al. 2003; Sato and Sassa 2005; Iida et al. 2006). In Europe, model experiments were performed on the compression wave generated by a train entering a tunnel with different configurations (flared portals, vented portals, vented hoods, etc.) (Ricco et al. 2007; Anthoine 2009). In China, 3D numerical simulation and model experiments have been used, and the influences of running speed, blocking ratio, train head shape, and railroad condition on the aerodynamic effects have been analyzed (Luo et al. 2004; Zhao et al. 2006; Zhao and Li 2009; Li et al. 2010; Liu et al. 2010a, b). To date, studies have mainly focused on a speed of 250 km/h and few have examined the aerodynamic effects above 300 km/h. Moreover, 3D effects are rarely taken into account.

In this paper, we solve the 3D Reynolds average Navier–Stokes equations of a viscous compressible fluid, and the dynamic grid technique is employed for moving bodies. The unsteady aerodynamic effects of a high-speed train entering a tunnel are studied to obtain the changes of the pressure and the aerodynamic force. In order to find the relationship between the train speed, and the pressure changes and microwave, we calculate the models at different speeds of 200, 250, 300, 350, 400, 450, and 500 km/h.

## 2 Fundamental Flow Equations

When a train passes through a tunnel, the flow field around the train is compressible, viscous, unsteady turbulent flow. For more details of the turbulent flow equations, please refer to (Ferzieger and Peric 1996). The Reynolds average Navier–Stokes equations are presented as below.

Continuity equation:

$$\frac{\partial \rho}{\partial t} + \frac{\partial(\rho u_j)}{\partial x_j} = 0, \quad j = 1, 2, 3, \tag{1}$$

where $t$ is the time, $\rho$ is the density, $x_j$ is the cartesian coordinate, and $u_j$ is the absolute fluid velocity component in direction $x_j$.

Momentum equation:

$$\frac{\partial(\rho u_i)}{\partial t} + \frac{\partial(\rho u_i u_j)}{\partial x_j} = \rho f_i - \frac{\partial p}{\partial x_i} + \frac{\partial \tau_{ij}}{\partial x_j}, \tag{2}$$

where $p$ is the static pressure, $u_i$ is the velocity component in direction $x_i$, $f_i$ is the mass force component in direction $x_i$, and $\tau_{ij}$ is the stress tensor with the specific expression as

$$\tau_{ij} = \mu\left[\left(\frac{\partial u_j}{\partial x_i} + \frac{\partial u_i}{\partial x_j}\right) - \frac{2}{3}\frac{\partial u_k}{\partial x_k}\delta_{ij}\right], \tag{3}$$

where $\mu$ is the viscosity coefficient of the air and $\delta_{ij}$ is a unit tensor.

Turbulent kinetic energy ($k$) equation:

$$\begin{aligned}\frac{\partial(\rho k)}{\partial t} + \frac{\partial(\rho k u_i)}{\partial x_i} = &\frac{\partial\left[\left(\mu + \frac{\mu_t}{\sigma_k}\right)\frac{\partial k}{\partial x_j}\right]}{\partial x_j} \\ &+ G_k - \rho\varepsilon + G_b - Y_M + S_k.\end{aligned} \tag{4}$$

Turbulent dissipation rate ($\varepsilon$) equation:

$$\begin{aligned}\frac{\partial(\rho\varepsilon)}{\partial t} + \frac{\partial(\rho\varepsilon u_i)}{\partial x_i} = &\frac{\partial\left[\left(\mu + \frac{\mu_t}{\sigma_\varepsilon}\right)\frac{\partial \varepsilon}{\partial x_j}\right]}{\partial x_j} \\ &+ C_{\varepsilon 1}\frac{\varepsilon}{k}(G_k + C_{\varepsilon 3}G_b) - C_{\varepsilon 2}\rho\frac{\varepsilon^2}{k} + S_\varepsilon.\end{aligned} \tag{5}$$

## 3 Computational Domain and Mesh Generation

A simplified geometric model of electric multiple units (EMU) is used in the numerical simulation (Fig. 1). This model removes the bogies, pantograph, and other details of the train. The front shape of the model and the real train match well. The total length of the train model is 76.5 m.

Figure 2 shows the grids used in the numerical simulation of the train entering a tunnel. The grids are divided into two parts: one part is the moving mesh, which contains the semicylindrical region around the train; the other part is the fixed part, which includes the area around the tunnel wall. The sliding interface technology is used to solve the relative motion between the train and the tunnel. The moving mesh moves at the speed of the train. On the direction of the motion, the front boundary deletes grids, while the opposite one adds grids layer-by-layer to make the grids move. The grids have about 1.2 million grid hexahedral elements.

In order to simulate the impact of the turbulent boundary layer, the standard turbulent wall function is adopted. Therefore, the first layer near the wall should meet the requirement of $30 < y+ < 300$ ($y+$ is the nondimensional parameter of the

**Fig. 1** Simplified geometric model of electric multiple units (EMU)

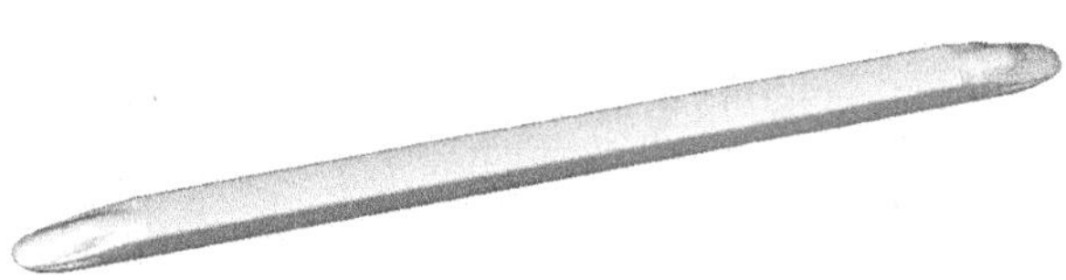

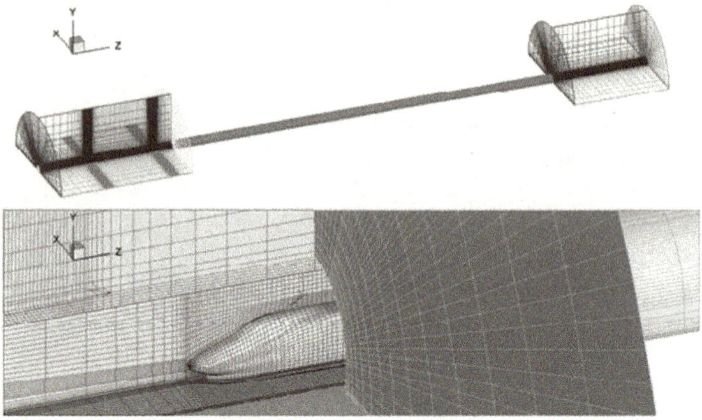

Fig. 2 Grids of a train entering a tunnel

wall) to include it in the logarithmic area of the turbulent boundary layer. According to the velocity and size of the train, the height of the first layer near the wall should be 0.005 m.

## 4 Results

### 4.1 Description of the Wave Propagation Process

Figure 3 shows 25 monitor points that are set on the train. The maximum pressure differences of the monitor points 1–11 and 15–25 are also shown in this figure. The differences of the pressure amplitude between different monitor points are small, and the maximum value of the differences is 147 Pa. Due to the train passing through a double-track tunnel, the distribution of the pressure differences is asymmetrical, and the pressure differences on the side that is closer to the tunnel wall are higher than that of the other side. About the longitudinal distribution of the train, the two sides of the train take on an entirely different distribution. For the side closer to the tunnel wall, the pressure difference amplitude on the middle of the train is smaller than that on the two ends. However, on the other side of the train, the pressure difference amplitude on the middle of the train is higher than that on the two ends.

Figure 4 shows the pressure change in the tunnel wall and on the train surface when the EMU passes through a tunnel at the speed of 200 km/h. Figure 4a shows the Mach wave propagation diagram and Fig. 4b shows the pressure history of the measured point 6 on the train surface. This point is presented by the dash-and-dot line in Fig. 4a. Now, we focus on the pressure changes on the train surface: point *A* indicates that the initial expansion wave of the train tail reached the point 6; point

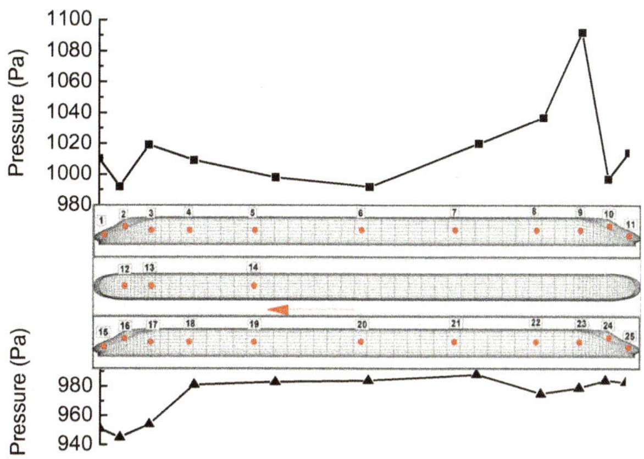


**Fig. 3** Twenty-five monitor points set on the train surface and the pressure amplitude with the location of the monitor points

*B* denotes that the first reflection expansion wave of the train head reached the point 6; point *C* presents that the first reflection compression wave of the train tail reached the point 6; point *D* shows that the second reflection compression wave of the train head reached the point 6; point *E* indicates that the second reflection expansion wave of the train tail reached the point 6; and, points *F–J* can be handled in the same manner. Point *L* indicates that the compression wave of the train head leaving the tunnel reached the point 6. When a compression wave passes through the measuring point, the pressure of the point increases; and when an expansion wave passes through the point, the pressure of the point decreases. We can infer that when the time reaches points *A*, *B*, *E*, *F*, *I*, and *J* in Fig. 4b, the pressure at these points decreases, and when the time reaches points *C*, *D*, *G*, *H*, *K*, and *L* in Fig. 4b, the pressure at these points increases.

Figure 4c shows the pressure history of the measured point at 250 m from the tunnel entrance, and it is represented by the $x = 250$ m line in Fig. 4a. Point 1 indicates that the initial compression wave of the train head goes through the measured point, the pressure of the measured point increases; point 2 indicates that the initial expansion wave of the train tail goes through the measured point, the pressure of the measured point decreases; point 3 shows that the train head goes through the measured point, the pressure of the measured point decreases; point 4 indicates that the first reflection expansion wave of the train head goes through the measured point, the pressure of the measured point decreases; point 5 reveals that the train tail goes through the measured point, the pressure of the measured point increases; point 6 indicates that the second reflection compression wave of the train head and the first reflection compression wave of the train tail go through the measured point, while the pressure of the measured point increases; and, point 7 denotes that the compression wave of the train head leaving the tunnel goes through the measured point, the pressure of the measured point decreases.

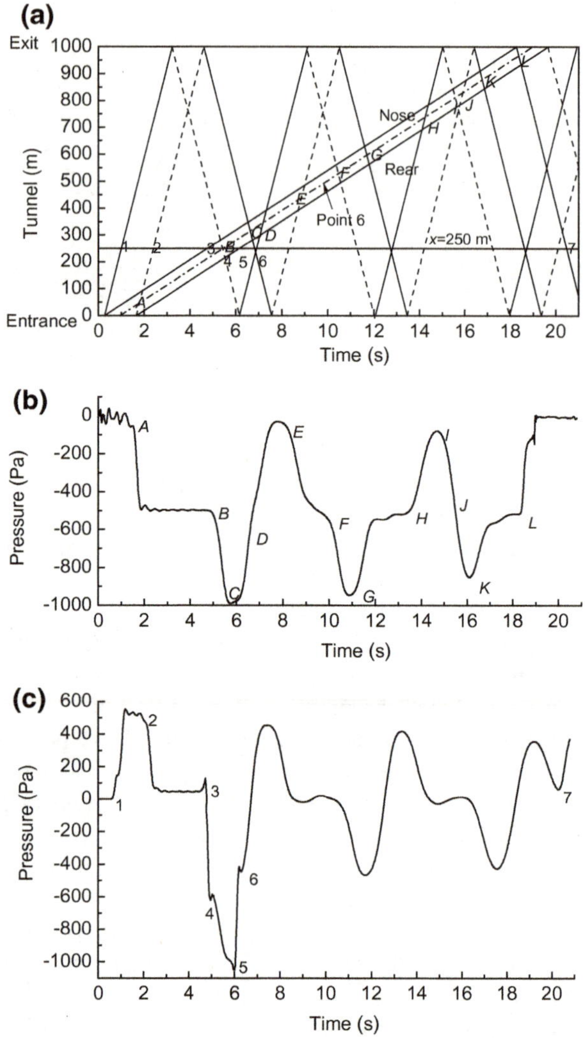


**Fig. 4** Wave diagram and pressure transients due to EMU passing through a tunnel at the speed of 200 km/h. **a** Wave diagram; **b** Pressure changes at point 6 on the train; **c** Pressure changes at 250 m in the tunnel

## 4.2 *Pressure Difference Amplitudes on the Train Surface and the Tunnel Wall*

Table 1 shows the results of the pressure difference amplitudes on the train surface and the tunnel wall of the train passing through the tunnel at different speeds. The fitting formulas between the pressure difference amplitude ($y$) and the train speed ($x$) for these two cases are, respectively, $y = 0.017x^{2.08}$ and $y = 0.013x^{2.15}$, which indicate that, when the train operates at a higher speed, greater than 200 km/h, the amplitudes of the pressure change on the tunnel wall and the surface of the train are approximately proportional to the square of the train speed.

**Table 1** Pressure difference amplitudes on the train surface and tunnel wall

| Train speed (km/h) | Pressure difference amplitude (Pa) | |
|---|---|---|
| | Train surface | Tunnel wall |
| 200 | 1027 | 1169.8 |
| 250 | 1613 | 1874.5 |
| 300 | 2335 | 2840.7 |
| 350 | 3199 | 3716.8 |
| 400 | 4202 | 4937.6 |
| 450 | 5380 | 6377.5 |
| 500 | 6751 | 8264.0 |
| Fitting formula | $y = 0.017x^{2.08}$ | $y = 0.013x^{2.15}$ |

Figure 5 compares the numerical simulation results with the on-site test results of pressure change magnitude on the train surface with the train speed. The numerical simulation results approximately conform to the on-site test results. However, the numerical result is still a little higher than the on-site test results.

## 4.3 *Microwave*

The microwave is radiated from the portal toward the outside. Therefore, it is connected with the pressure amplitude of the tunnel exit. Figure 6a shows the pressure changes at 970 m in the tunnel, which is 30 m away from the tunnel exit. The pressure at this point is taken to stand for the pressure of the tunnel exit. In Fig. 6a, the first wave crest is produced by the initial compression wave of the train head and the first reflection expansion wave of the train head, and the first wave trough is produced by the initial expansion wave of the train tail and the first reflection compression wave of the train tail. The first peak of the microwave is from the first wave crest of the pressure, and the first trough of the microwave is from the first wave trough of the pressure.

The amplitude of the first pressure wave of the tunnel exit and the amplitude of the first microwave are compiled as shown in Fig. 7. The fitting lines of the microwave, and changes with the pressure of the tunnel exit, are also presented in

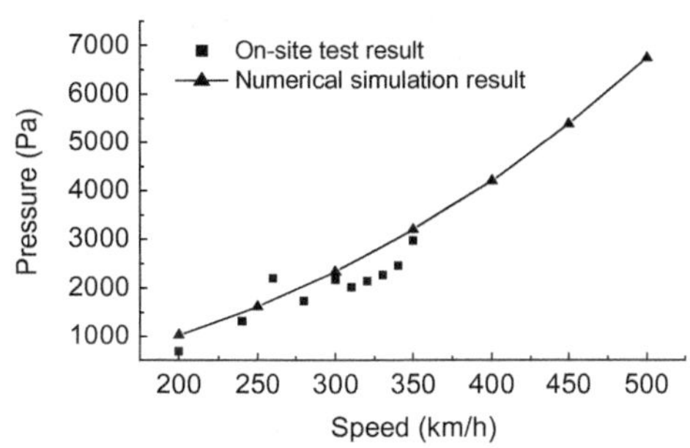


**Fig. 5** Pressure difference amplitude on the train surface

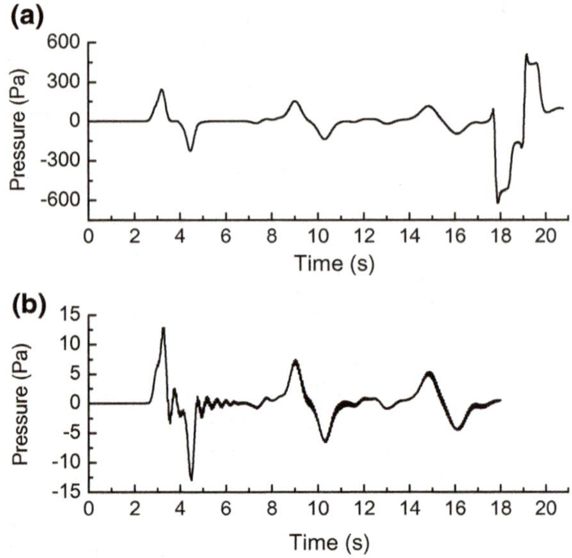


**Fig. 6** Pressure changes at 970 m in the tunnel (a) and the microwave history of the tunnel (b)

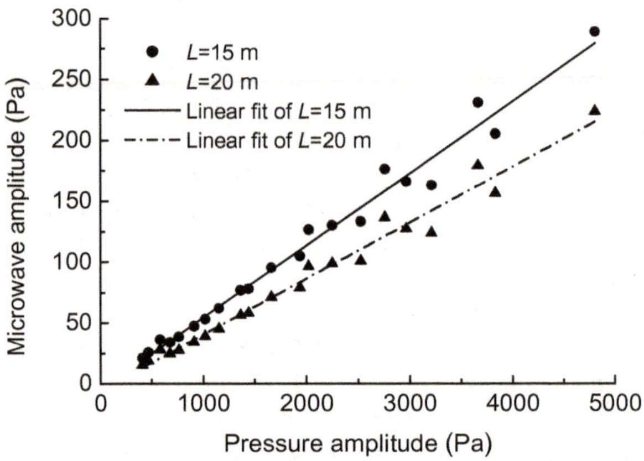


**Fig. 7** Microwave change amplitudes with the initial magnitude of tunnel exit pressure

Fig. 7. The fitting lines between the microwave amplitude and the amplitude of the pressure are $y = -4.6 + 0.059x$ ($L = 15$ m, where $L$ is the distance between the test point and the tunnel exit) and $y = -5.6 + 0.046x$ ($L = 20$ m), respectively. The microwave is approximately proportional to the pressure of the tunnel exit.

The relationships between the microwave and train speed are shown in Fig. 8. The figure shows that the microwave is approximately proportional to the square of train speed. The fitting formulae between the microwave amplitude and the train speed are $y = 1.5 \times 10^{-4}x^{2.28}$ ($L = 15$ m) and $y = 7.9 \times 10^{-4}x^{2.34}$ ($L = 20$ m), respectively.

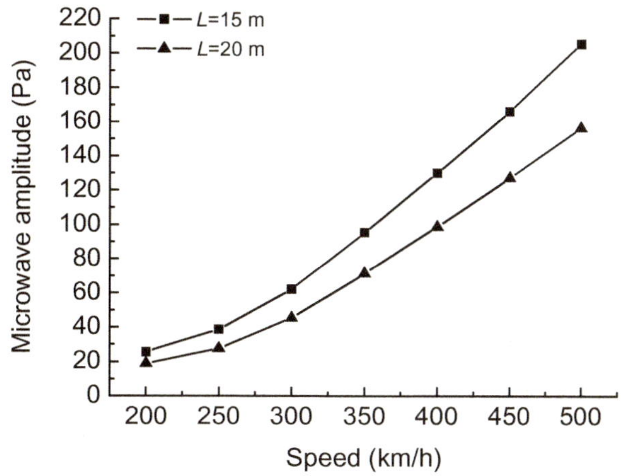


**Fig. 8** Microwave change amplitude with the train speed

## *4.4 Aerodynamic Forces of the Train*

Due to the pressure wave propagation and airflow in the tunnel, the aerodynamic forces are different between the train passing through the double-track tunnel and running on the open line. Figure 9 shows the results of aerodynamic forces change of the train passing through the tunnel. It can be concluded that the lift force of the train exhibits no significant difference between the cases running on the open line and in the tunnel. The lateral force of the train increases when the train enters the tunnel, and then remains constant after the train entering. When the train leaves the tunnel, the lateral force increases rapidly, and then decreases rapidly to a steady state. The aerodynamic drag increases rapidly when the train enters the tunnel, then fluctuates about a certain value. Compared with the changes of the lateral force and

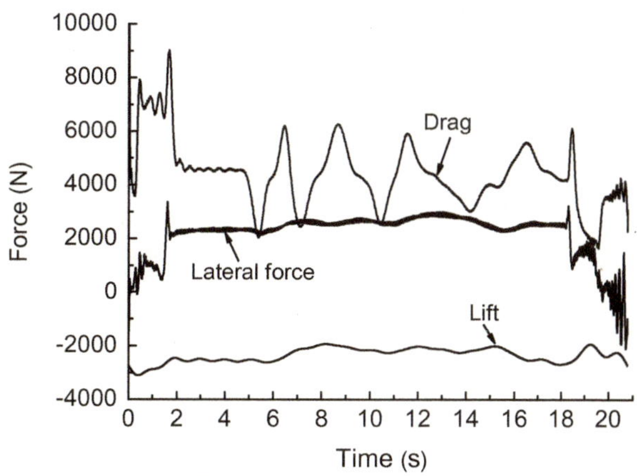


**Fig. 9** Aerodynamic forces of the train passing through the tunnel

the lift force of the train, the changes of the drag of the train act a complex manner. A detailed analysis on the change of the drag for the train passing through the tunnel is as follows.

In Fig. 10, point *A* stands for the train head beginning to enter the tunnel. The pressure of the train head increases, and the drag of the train increases rapidly; point *B* stands for the train tail entering the tunnel. When the whole train enters the tunnel, the pressure of the train tail increases, and the drag of the train decreases; at point *C*, the initial expansion wave of the tail reaches the train head, the pressure of the train head decreases, and the drag of the train decreases; point *D* stands for the first reflection expansion wave of the train head meeting the train head, when the pressure of the train head decreases, and the drag of the train decreases; point *E* stands for the first reflection expansion wave of the train head reaching the train

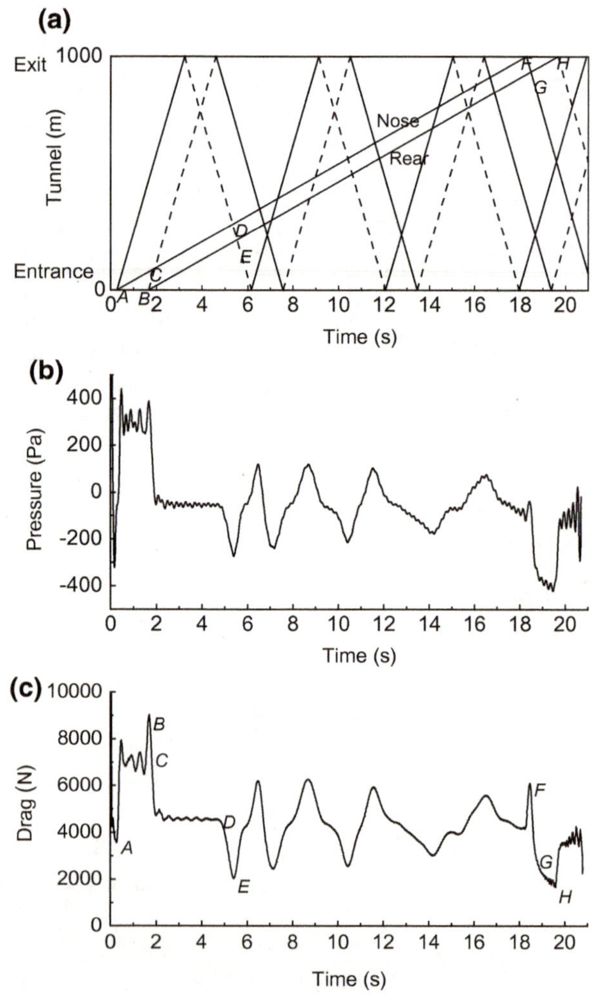


**Fig. 10** Wave diagram and drag changes due to EMU passing through a tunnel at the speed of 200 km/h. **a** Wave diagram; **b** Pressure difference between the monitor points 1 and 11; **c** Drag history of the train

**Fig. 11** Drag change with the train speed

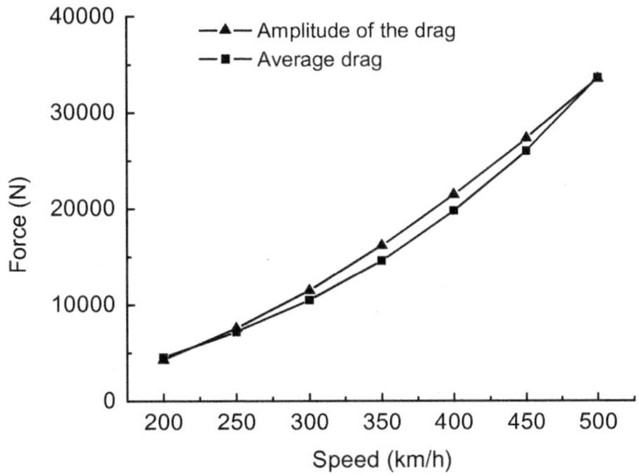


tail, when the pressure of the train tail decreases, and the drag of the train increases; point *F* stands for the train head running out the tunnel, when the pressure of the train head decreases, and the drag of the train decreases; point *G* stands for the compression wave of the train leaving the tunnel meeting the train tail, when the pressure of the train tail increases, and the drag of the train decreases; and, point *H* stands for the whole train leaving the tunnel, when the pressure of the train tail decreases, the drag of the train increases. The expansion wave reaches the train head, the pressure of the train head decreases, and the drag of the train decreases. The compression wave reaches the train head, the pressure of the train head increases, and the drag of the train increases. The expansion wave reaches the train tail, the pressure of the train tail decreases, and the drag of the train increases. The compression wave reaches the train tail, the pressure of the train tail increases, and the drag of the train decreases.

Most researchers analyze the drag of the train varying with the speed of the train. The maximum value of the drag is used, and the amplitude of the drag force is neglected. In this study, we combine the amplitude of the drag and the average value of the drag to analysis the drag variation with train speed. The drag of the train can be represented by Drag = DA $\pm$ (DB/2), where DA is the average value of the drag, and DB is the amplitude of the drag. Figure 11 shows the average value of the drag and the amplitude of the drag change with the speed of the train. The fitting formulas between the drag and train speed are $y = 0.025x^{2.3}$ and $y = 0.065x^{2.1}$, respectively, which show that the average value of the drag and the amplitude of the drag are all approximately proportional to the square of train speed.

## 5 Conclusions

Via numerical simulation of the train passing through a tunnel, the following conclusions are made:

1. The amplitude of the pressure change on the tunnel wall and in the train surface is approximately proportional to the square of the train speed.
2. The microwave is approximately proportional to the square of train speed.
3. The drag of the train is represented by Drag = DA ± (DB/2), where DA is the average value of the drag, and DB is the amplitude of the drag. Both the average value of the drag and the amplitude of the drag are approximately proportional to the square of train speed.

**Acknowledgements** The authors are grateful for the assistance in all the numerical computations by the Center for Engineering and Scientific Computation (CESC), Zhejiang University, China.

## References

Anthoine, J. (2009). Alleviation of pressure rise from a high-speed train entering a tunnel. *AIAA Journal, 47*(9), 2132–2142. doi:10.2514/1.41109.

Ferzieger, J. L., & Peric, M. (1996). *Computational methods for fluid dynamics* (pp. 265–328). Heidelberg: Springer.

Iida, M., Kikuchi, K., & Fukuda, T. (2006). Analysis and experiment of compression wave generated by train entering tunnel entrance hood. *JSME International Journal Series B, 49*(3), 761–770. doi:10.1299/jsmeb.49.761.

Joseph, A. S. (2001). Aerodynamics of high-speed trains. *Annual Review of Fluid Mechanics, 33* (3), 371–414.

Li, R. X., Zhao, J., Liu, J., & Zhang, W. H. (2010). Influence of air pressure pulse on side windows of high-speed trains passing each other. *Chinese Journal of Mechanical Engineering*, *46*(4), 87–98 (in Chinese). doi:10.3901/JME.2010.04.087.

Liu, T. H., Tian, H. Q., & Liang, X. F. (2010a). Aerodynamic effects caused by trains entering tunnels. *Journal of Transportation Engineering, 136*(9), 846–853. doi:10.1061/(ASCE)TE. 1943-5436.0000146.

Liu, T. H., Tian, H. Q., & Liang, X. F. (2010b). Design and optimization of tunnel hoods. *Tunnelling and Underground Space Technology, 25*(3), 212–219. doi:10.1016/j.tust.2009.12.001.

Luo, J. J., Gao, B., Wang, Y. X., & Zhao, W. C. (2004). Numerical simulation of unsteady three dimensional flow induced by high-speed train entering tunnel with shaft. *Journal of Southwest Jiaotong University, 39*(4), 442–446. (in Chinese).

Mei, Y. G., Zhao, H. H., & Liu, Y. Q. (1995). Numerical analysis about pressure wave of high-speed train. *Journal of Southwest Jiaotong University, 30*(6), 667–672. (in Chinese).

Ogawa, T., & Fujii, K. (1997). Numerical investigation of three-dimensional compressible flows induced by a train moving into a tunnel. *Computers & Fluids, 26*(6), 565–585. doi:10.1016/ S0045-7930(97)00008-X.

Ricco, P., Baron, A., & Molteni, P. (2007). Nature of pressure waves induced by a high-speed train travelling through a tunnel. *Journal of Wind Engineering and Industrial Aerodynamics, 95*(8), 781–808. doi:10.1016/j.jweia.2007.01.008.

Sato, T., & Sassa, T. (2005). Prediction of the compression pressure wave generated by a high-speed train entering a tunnel. *International Journal of Computational Fluid Dynamics, 19* (1), 53–59. doi:10.1080/10618566412331286328.

Tanaka, Y., Iida, M., & Kikuchi, K. (2003). Method to simulate generation of compression wave inside a tunnel at train entry with a simple geometry model. *Transactions of the Japan Society of Mechanical Engineers Series B*, *69*(683), 1607–1614 (in Japanese). doi:10.1299/kikaib.69.1607.
Tian, H. Q. (2007). Train aerodynamics (pp. 268–303). Beijing, China: China Railway Publishing House (in Chinese).
William-Louis, M., & Tournier, C. (2005). A wave signature based method for the prediction of pressure transients in railway tunnels. *Journal of Wind Engineering and Industrial Aerodynamics, 93*(6), 521–531. doi:10.1016/j.jweia.2005.05.007.
Woods, W. A., & Pope, C. W. (1979). On the range of validity of simplified one dimensional theories for calculating unsteady flows in railway tunnels. In The 3rd International Symposium on the Aerodynamics and Ventilation of Vehicle Tunnels, Sheffield, UK (pp. 115–150).
Zhao, J., & Li, R. X. (2009). Numerical analysis of aerodynamics of high-speed trains running into tunnels. *Journal of Southwest Jiaotong University, 44*(1), 96–100. (in Chinese).
Zhao, W. C., Gao, B., & Wang, Y. X. (2006). Numerical investigation of micro pressure wave radiated from the exit of a railway tunnel. *Journal of Basic Science and Engineering, 14*(3), 444–453. (in Chinese).

## Author Biographies

**Jian Deng** Ph.D., is an associate professor at the Department of Mechanics, School of Aeronautics and Astronautics, Zhejiang University, China. From Dec. 2012 to Dec. 2014, he was a visiting scholar at the Department of Applied Mathematics and Theoretical Physics at the University of Cambridge, and was elected to a By-Fellowship at Churchill College, the University of Cambridge. His research interests include computational fluid dynamics, flow instability, and biomimetic hydrodynamics.

**Yao Zheng** Ph.D., is a Cheung Kong Chair Professor with Zhejiang University (2001–). He had been a Senior Research Scientist for NASA Glenn Research Center, Cleveland, Ohio (1998–2002), a Senior Software Scientist with CD-adapco (Computational Dynamics-Analysis & Design Application Company), New York (1997–1998), and had studied and worked at the University of Wales Swansea, UK, for 8 years. His research has spanned computational mechanics, numerical simulation, combustion and propulsion, and flight vehicle design. He has authored or co-authored four books, two translation books, and over 240 papers.

# Aerodynamic Modeling and Stability Analysis of a High-Speed Train Under Strong Rain and Crosswind Conditions

**Xue-ming Shao, Jun Wan, Da-wei Chen and Hong-bing Xiong**

**Contents**



## 1 Introduction

When high-speed trains running under strong rain and crosswind conditions, especially at exposed locations such as bridges or embankments, the aerodynamic forces and moments may increase significantly and result in the train instability. It was well known that the strong crosswind may increase the aerodynamic drag force, side force and yawing moment. If the rain and the crosswind coexist, the aerodynamic performance of train will deteriorate more severely, which may cause

X. Shao · J. Wan · H. Xiong (✉)
Department of Mechanics, Zhejiang University, Hangzhou 310027, China
e-mail: hbxiong@zju.edu.cn

D. Chen
National Engineering Laboratory for System Integration
of High-Speed Train (South), CSR Qingdao Sifang Co., Ltd, Qingdao 266111, China

 
Y. Fang and Y.H. Zhang (eds.), *China's High-Speed Rail Technology*,
Advances in High-speed Rail Technology, https://doi.org/10.1007/978-981-10-5610-9_6

train delays, shutdowns, derailments and even overturning. In 2007, trains derailed with 11 cars under high wind conditions, causing a serious accident on the Xinjiang Railway South Line and a train overturned in Tianjin (Fig. 1a). In Japan, there have been about 30 wind-induced accidents to date (Fig. 1b).

Numerous studies have been done on the aerodynamic performance of trains under crosswind conditions. Ma et al. (2009) investigated the aerodynamic characteristics of a train on a straight line at 350 km/h with a crosswind. More systematic researches on numerical simulation of a train traveling along a straight line and curves have been done (Liang and Shen 2007; Yang et al. 2010). In addition to wind tunnel experiments, numerical simulation analyzes the effects of crosswinds in more detail with results consistent with those from experiments (Christina et al. 2004; Javier et al. 2009; Sanquer et al. 2004; Masson et al. 2009). However, less attention has been paid to the effects of combined strong rain and crosswind on the aerodynamic characteristics and safety of high-speed trains. This paper deals with the influence of strong rain and crosswind conditions on high-speed train aerodynamics, based on numerical simulations. A quasi-static stability analysis based on the moment balance is also used to determine the limit safety speed of a train under different rain and wind levels, which provides some guidance for the train operation safety.

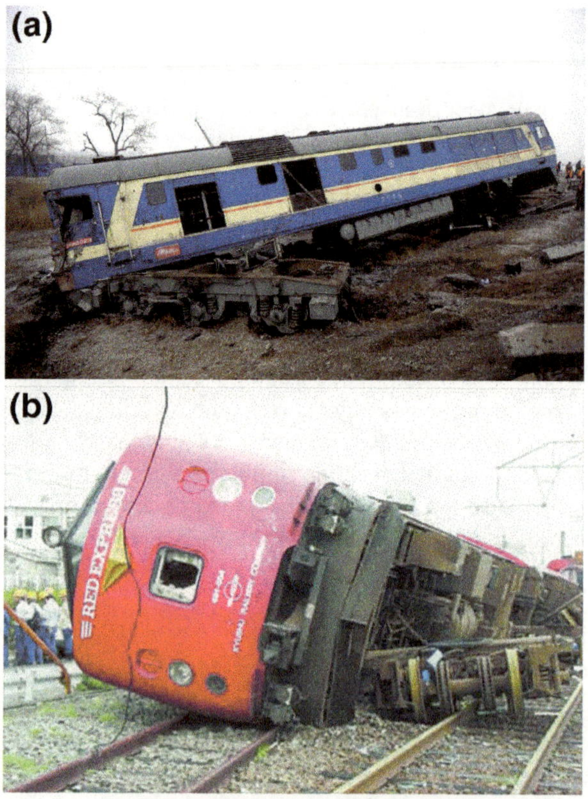


**Fig. 1** Train overturning in China (**a**) and Japan (**b**)

## 2 Numerical Simulation

### 2.1 Computational Model

Computational fluid dynamics (CFD) software FLUENT is used for numerical simulation in this study. For multiphase flow problems, there are mainly two types of multiphase flow models: one is the discrete particle model (DPM) proposed by Crowe and Smoot (1979); the other one is Eulerian–Eulerian model proposed by Gidaspow (1994). The fluid phase in DPM is solved using the Eulerian method, while the granular phase is tracked by the Lagrangian method. Though providing more detail, DPM is not suitable for large-scale engineering simulation with numerous dispersed particles, due to the limitation of finite memory capacity and CPU efficiency. The Eulerian–Eulerian approach is more efficient and usually more complex. Each phase is treated as a continuous medium that may interpenetrate with other phases, and is described by a set of equations with regard to momentum, continuity, and energy. The Eulerian–Eulerian approach has been successfully applied to the simulation of gas-particle multiphase flow with a large number of particles in large equipment. For example, Liu et al. (2006) studied the liquid-solid slurry transport within a pipeline, and Cao et al. (2005) simulated the bubble growth, integration, and the flow characteristics in a fluidized bed. Due to these advantages, the Eulerian–Eulerian model is used to simulate the example in this study.

The flow around the train is viscous, turbulent, and gas-droplet two-phase flow. Thus, the Eulerian–Eulerian multiphase model coupling with the $k$-$\varepsilon$ turbulence equation is used for the gas-droplet two-phase flow field around the train. Details of the Eulerian–Eulerian multiphase model and its validation could be found in another paper of this issue (Xiong et al. 2011) conducted in our group.

### 2.2 Computational Domain

A simplified model of CRH2 (China Railway High Speed 2) is studied, including three coaches of the head, middle, and tail. To fully develop flow around the train and to ensure the accuracy of results, a large semicylindrical numerical wind tunnel was established as the computational domain (Fig. 2). The distances of the semi-cylindrical computational domain in the vertical, horizontal and vertical directions are 600, 400, and 200 m, respectively. The length of vehicle is 76 m. The distance between nose of vehicle and the inlet boundary is 100 m. Direction of the flow field is at the positive $x$-axis.

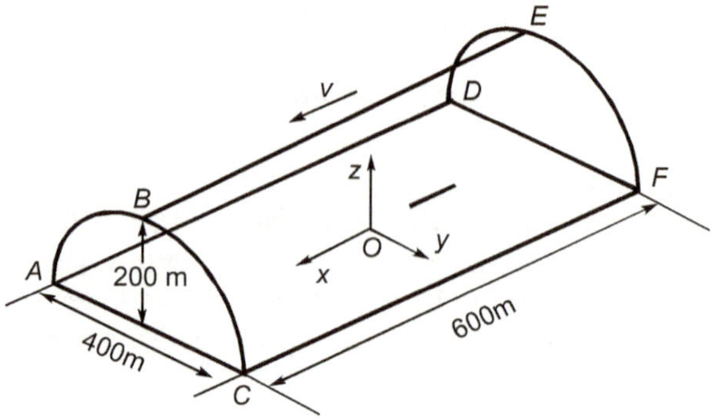


**Fig. 2** Computational domain

## 2.3 Computational Mesh

The computational domain is meshed with the hexahedral structured grid, with refine mesh on the front and rear of the train and surrounding areas as shown in Fig. 3. To validate the train model via numerical simulation, grid dependency has been conducted with three kinds of grid generation: 330,000, 600,000, 980,000 grids, as shown in Table 1, where the change rate Δ is defined as

$$\Delta = \frac{F_i - F}{F}, \tag{1}$$

where $F_i$ is the aerodynamic force (or moment) for different mesh models, and $F$ is the aerodynamic force with 98,000 grids. The results of 980,000 grids are very close to that of 60,000 grids. Therefore, we can conclude that 980,000 grids are acceptable for the numerical simulation of the train.

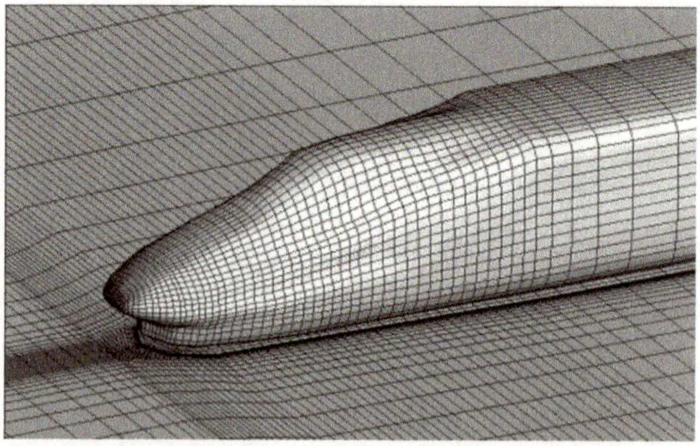

**Fig. 3** Computational mesh

**Table 1** Aerodynamic comparison in different grid models

| Mesh model | Drag force (N) | Δ | Side force (N) | Δ | Lift force (N) | Δ |
|---|---|---|---|---|---|---|
| 330,000 | 19,945.18 | 7.6% | 56.96 | 60.3% | 7686.27 | −1.2% |
| 600,000 | 19,018.51 | 2.6% | 36.27 | 2.1% | 7747.46 | −0.4% |
| 980,000 | 18,528.52 | 0 | 35.53 | 0 | 7782.27 | 0% |
| Mesh model | Rolling moment (N·m) | Δ | Pitching moment (N·m) | Δ | Yawing moment (N·m) | Δ |
| 330,000 | 4552.40 | −1.8% | −565,823 | 5.9% | 13,900.92 | 12.9% |
| 600,000 | 4616.65 | −0.5% | −541,518 | 1.4% | 12,571.86 | 2.1% |
| 980,000 | 4637.63 | 0 | −534,160 | 0 | 12,317.19 | 0 |

## 2.4 Boundary Conditions

Planes DEF and ABCFED in the computational domain (Fig. 2) are given as the boundaries of velocity inlet. Plane ABC is a pressure outlet boundary with a static pressure of 0. Train surfaces are stationary, with nonslip boundary condition.

Plane ACFD adopts a moving boundary with speed equal to flow velocity. The crosswind direction is on the positive *y*-axis, perpendicular to the train. The continuous phase and the granular phase sets are as follows: the continuous phase uses inlet boundary velocity of 360 km/h; the granular phase uses inlet boundary with *x*-axis velocity equal to the gas inlet velocity and negative *z*-axis velocity of raindrop of 5 m/s. Turbulent kinetic energy $k$ and turbulent dissipation rate $\varepsilon$ are determined by

$$k = 0.004u_{\mathrm{m}}^2 \quad (2)$$

$$\varepsilon = 0.09\frac{k^{1.5}}{0.03R'} \quad (3)$$

where $u_{\mathrm{m}}$ is the mean flow velocity, and $R$ is the turbulence length scale. The phase-coupled-simple algorithm is used for solving the coupling between pressure and velocity effects (Moukalled et al. 2003).

# 3 Problem Description

In this study, the train running speed is set as 360 km/h. The crosswind speed ranges from 0 to 40 m/s, and the direction of the wind is perpendicular to the running direction of the train. Under the crosswind and heavy rainfall computing conditions, rainfall rate (rainfall intensity per hour) is 60 mm/h. A raindrop is regarded as spherical with a constant falling velocity of 5 m/s and diameter of 0.002 m.

## 4 Results and Discussion

### *4.1 Pressure Distribution on the Train Surface*

The drag force on the train consists of pressure drag and friction drag. The pressure drag increases dramatically when running at high speed. Due to the impact of rainfall, the pressure distribution around the train may be more complex when the train runs in the rain conditions. By analyzing the pressure distribution around the train, it is helpful to understand the mechanism of the aerodynamic characteristics on the train.

The pressure distribution of the train under crosswind of 30 m/s and rainfall rate of 60 mm/h is shown in Fig. 4. For the head vehicle (Fig. 4a), the maximum pressure occurs at the front of the nose and the windward side of the head vehicle. The negative pressure appears at the leeward side of the vehicle window, which will cause a side force along the wind direction. If the side force is large enough, it may lead to a train derailment. Pressure on the tail vehicle is opposite to the head vehicle as shown in Fig. 4b. The largest negative pressure occurs on the windward side of the rear window transition zone, and the positive pressure occurs on the leeward side of the window transition zone and tail vehicle nose. This makes side force of tail vehicle reverse the direction of head train. As the side forces of head vehicle and tail vehicle are in the opposite direction, the train will have a yawing moment, and this may bring a risk of train derailment.

### *4.2 Aerodynamic Force of Train Under Strong Rain and Crosswind Conditions*

Side force under no rain and rain conditions are shown in Fig. 5a for different crosswind speeds. It could be seen that the side force increases with wind speed. Due to the effect of rain, the side force under the rain condition is larger than that under no rain condition. Side force is mainly caused by the pressure difference on the two sides of the train. The effect of rain on side force is significant and increases the risk of derailment.

The lift force under no rain and rain conditions are shown in Fig. 5b. It can be seen that the lift force increases with wind speed. For the rain condition, the lift force is larger than that under no rain condition, increasing the risk of derailment.

Drag force under no rain and rain conditions are shown in Fig. 5c. Drag force increases first, reaching a maximum value when the wind speed reaches about 15 m/s, and then decreases with wind speed. This is due to the drag force of the first vehicle decreasing with wind speed, and even changing force direction as negative drag, though the drag force of the tail vehicle always increases with wind speed.

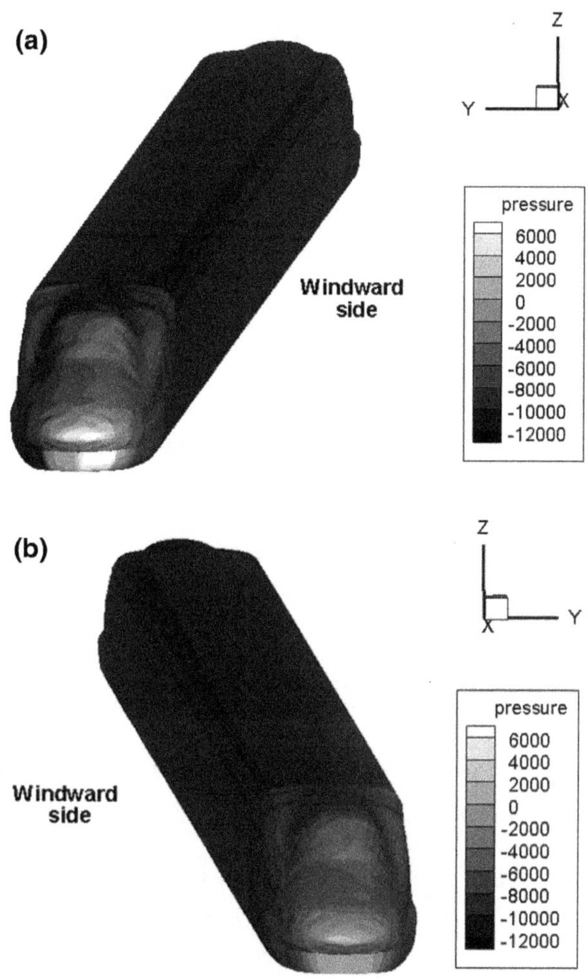


**Fig. 4** Pressure on the train. **a** Front view; **b** Rear view (Unit: Pa)

These three force components have different effects on the train stability and safety. Generally, the side force increases the wheel-track load on the leeward side and the wheel-rail contact force. Large side forces worsen the wear of the wheel and rail, and may cause train derailment, or even overturning. The negative lift force increases axle load and exacerbates the wear of track and wheel; the positive lift force floats the train, and large positive lift force may cause train derailment. Drag force increases rapidly with an increase of train speed, requiring more energy consumption and increasing the air noise.

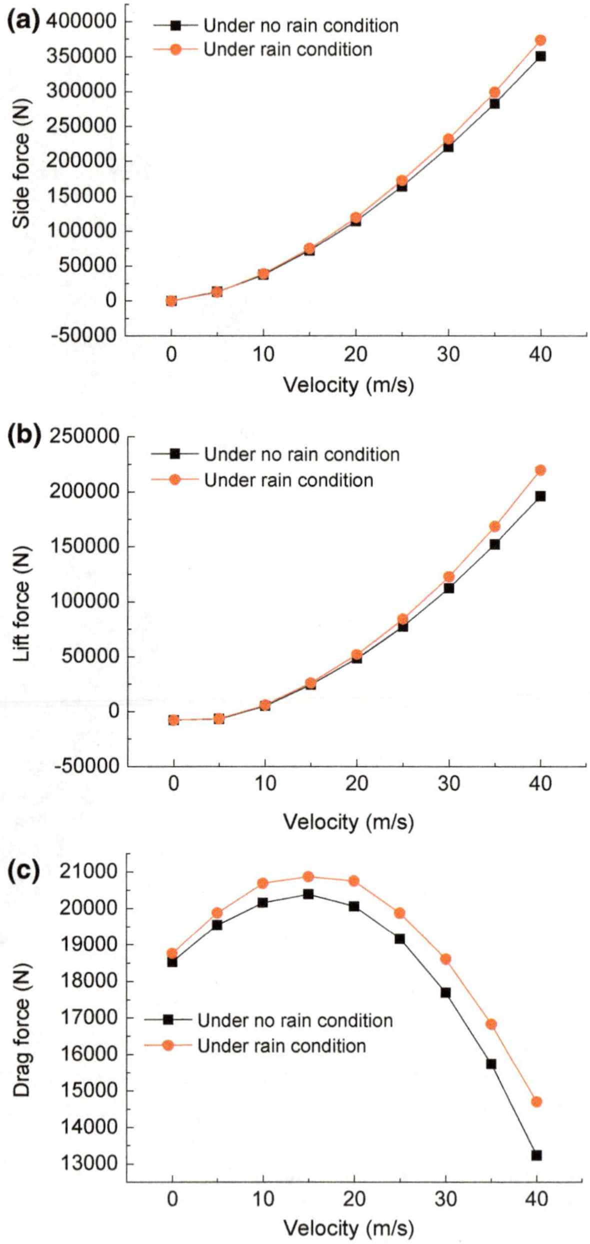


**Fig. 5** Side (**a**), lift (**b**) and drag force (**c**) of the train

### 4.3 Aerodynamic Moment of Train Under Strong Rain and Crosswind Conditions

Figure 6a illustrates the train rolling moments under no rain and rain conditions. Rolling moment increases with wind speed and has a larger absolute value if considering the influences of rain. This means that the train derailments are more likely to happen under heavy rainfall conditions. Pitching moments under the heavy rainfall and no rainfall conditions are plotted in Fig. 6b. Results show that the pitching moments, in the direction of nose-down, increase slightly with wind speed under 10 m/s, and then decrease when the wind speed continually increases. The pitching moments reverse their direction as nose-up at the wind speed of about 30 m/s, and then monotonically increase with wind speed. Figure 6c illustrates the train yawing moments under no rain and rain conditions. The yawing moment has no detectable change in the conditions of rain and no rain. In both cases, the yawing moments increase with the wind speed, mainly resulting from the large side force in the windward direction.

In the three components of train moment, the rolling moment is mostly important for the train safety, which increases the risk of overturning. The less important one is the yawing moment, which makes the train swing around the axle and increases the tendency to derail. If the train is subject to intermittent wind gusts, the large yawing moment may cause train vibration and worsen passenger comfort. The minor parameter for train safety is the pitching moment, which makes the train move up and down and affects passenger comfort mainly.

### 4.4 Stability Analysis of the Train Under Rain and Crosswind Conditions

Stability analysis of the train overturning is used to evaluate the train speed limit when running on the straight or curved rails considering the effects of strong rain and crosswind. When running on curved rails, the outer rail has an excessive height compared to the inner rail and an unbalanced centrifugal force will be generated. There are three kinds of turnover: (1) Running on a straight rail, the train turn over in the windward direction; (2) Running on a curved rail, the train turn over toward the outside rail with wind blowing from the inside rail; (3) Running on a curved rail, the train turn over toward the inside rail with wind blowing from the outside rail.

The running stability of the train concludes the shape of the train, size of the train, the mass of the train, the height of the center of gravity, running speed, and so on. The relationship of limit running speed and crosswind speed can be derived from the dynamic torque balance principle. The method and formulae have been studied (Gao and Tian 2004; Tian 2007). This method is used in this study to calculate the limit running speed of train under rain conditions.

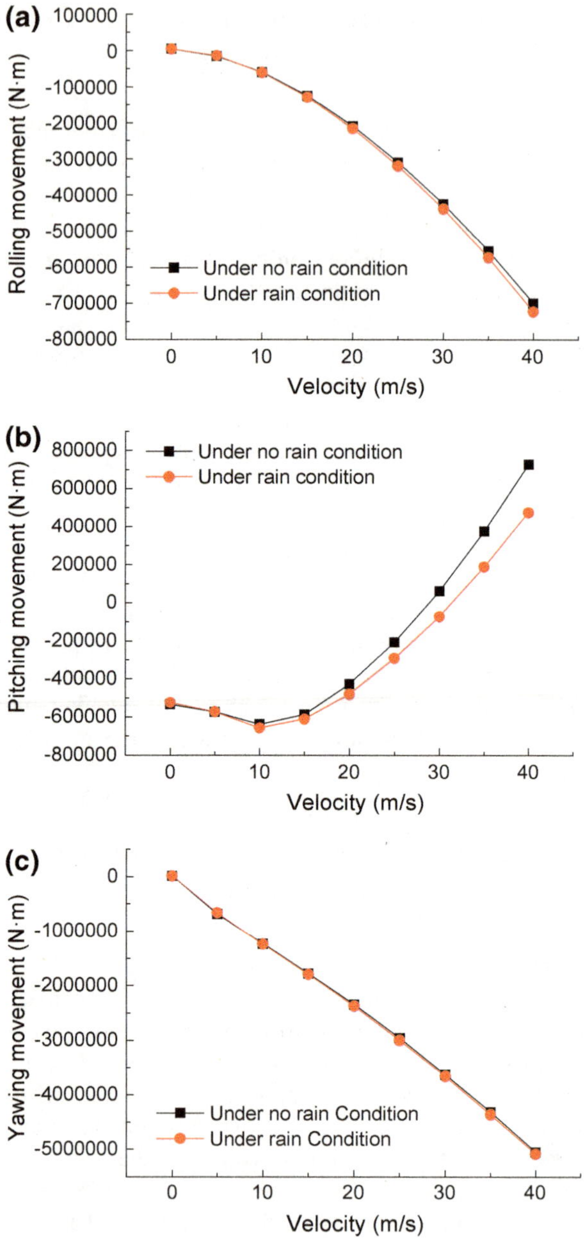


**Fig. 6** Rolling (**a**), pitching (**b**) and yawing moment (**c**) of the train

According to the cases listed in Table 2, we calculated the lift force coefficient $c_l$, side force coefficient $c_s$, and rolling moment coefficient $c_m$ at different yaw angles. These data are then processed to fit the relationship between yaw angle ($\alpha$) and $c_l$, $c_s$, and $c_m$ as follows:

**Table 2** Computing conditions at different yaw angles

| Working condition | Yaw angle (°)[a] | Rainfall rate (mm/h) |
|---|---|---|
| 1 | 0 | 70 |
| 2 | 30 | 70 |
| 3 | 45 | 70 |
| 4 | 60 | 70 |
| 5 | 75 | 70 |

[a]Yaw angle is defined as the angle between the train velocity vector and the resultant velocity vector

Under no rain condition:

$$\begin{aligned} c_l &= -1.2901 \times 10^{-6}\alpha^3 + 1.5166 \times 10^{-4}\alpha^2 + 2.9972 \times 10^{-3}\alpha, \\ c_s &= -1.2310 \times 10^{-6}\alpha^3 + 1.1188 \times 10^{-4}\alpha^2 + 8.2266 \times 10^{-3}\alpha, \\ c_m &= -2.6019 \times 10^{-7}\alpha^3 - 5.7588 \times 10^{-6}\alpha^2 + 7.2329 \times 10^{-3}\alpha. \end{aligned}$$

Under rain condition:

$$\begin{aligned} c_l &= -1.2183 \times 10^{-6}\alpha^3 + 1.3945 \times 10^{-4}\alpha^2 + 3.8124 \times 10^{-3}\alpha, \\ c_s &= -9.4652 \times 10^{-7}\alpha^3 + 7.4161 \times 10^{-5}\alpha^2 + 9.8655 \times 10^{-3}\alpha, \\ c_m &= -9.4652 \times 10^{-7}\alpha^3 + 7.4161 \times 10^{-5}\alpha^2 + 9.8655 \times 10^{-3}\alpha. \end{aligned}$$

Finally, taking these formulae into the moment balance analysis, the limit speeds of the train running in the straight and curve rails were calculated. Results are listed in Table 3 for no rain condition and Table 4 for rain condition.

We could see from Table 4 that under heavy rain and crosswind, the yaw angle increases as the wind speed increases, and the critical speed of the train decreases. When the train is running on a curve, the critical running speed of the train caused by wind blows from the outside is smaller than that caused by wind blows from the inside, which means it is more easily to overturn on the inner curve track. The critical running speed of a train running on a straight line is larger than that on a curve line when the

**Table 3** Relationship between train running safety speed and side wind speed under no rain condition

| α (°) | Overturning on line | | Overturning on outer curve line | | Overturning on inner curve | |
|---|---|---|---|---|---|---|
| | Wind speed (m/s) | Train speed (m/s) | Wind speed (m/s) | Train speed (m/s) | Wind speed (m/s) | Train speed (m/s) |
| 15 | 20.600 | 76.922 | 26.346 | 98.376 | 17.355 | 64.805 |
| 30 | 27.883 | 48.325 | 30.987 | 53.703 | 25.451 | 44.110 |
| 45 | 32.454 | 32.480 | 34.728 | 34.756 | 30.484 | 30.508 |
| 60 | 35.326 | 20.420 | 37.193 | 21.500 | 33.635 | 19.443 |
| 75 | 36.934 | 9.923 | 38.599 | 10.370 | 35.393 | 9.509 |
| 90 | 37.601 | 0.030 | 39.227 | 0.031 | 36.092 | 0.029 |

**Table 4** Relationship between train running safety speed and side wind speed under heavy rain

| α (°) | Overturning on line | | Overturning on outer curve line | | Overturning on inner curve | |
|---|---|---|---|---|---|---|
| | Wind speed (m/s) | Train speed (m/s) | Wind speed (m/s) | Train speed (m/s) | Wind speed (m/s) | Train speed (m/s) |
| 15 | 17.617 | 65.783 | 20.897 | 78.031 | 15.425 | 57.596 |
| 30 | 23.484 | 40.701 | 25.338 | 43.913 | 21.913 | 37.977 |
| 45 | 27.006 | 27.027 | 28.381 | 28.404 | 25.753 | 25.774 |
| 60 | 29.101 | 16.822 | 30.239 | 17.480 | 28.032 | 16.204 |
| 75 | 30.149 | 8.100 | 31.161 | 8.372 | 29.183 | 7.840 |
| 90 | 30.411 | 0.024 | 31.383 | 0.025 | 29.479 | 0.023 |

wind blows from the outside rail to the inside rail, but smaller than that on a curve line when the wind blows from the inside rail to the outside rail. Comparing to the condition of crosswind without rain, the train limit speeds decrease about 10–20%, which means that the train is more easily to overturn under strong rain condition.

## 5 Conclusions

Research studies have been focused on the train aerodynamic behaviors under rain and crosswind conditions, using the Eulerian–Eulerian two-phase model. Compared to no rain condition, the drag force, side force, and lift force increase under rain conditions. For high crosswind velocities, the effects of rain on aerodynamic behavior are more apparent. The rolling moment in the rain is greater than that without the rain. The pitching moment changes obviously under rain condition. The yawing moment has no detectable change in the conditions of rain and no rain. The main factors affecting the train safety considerations are the side force, lift force, and rolling moment. A quasi-static stability analysis using the moment balance is used to determine the limit safety speed of train under different rain and wind levels. Results show that the train is more easily to overturn on the inner curve track. The train limit speeds under rain decrease about 10–20% than that under no rain, which means that the train is more easily to overturn under strong rain condition.

## References

Cao, Y., Li, X., & Yan, J. (2005). Numerical simulation of dense gas-solid flow in pipe-type fluidized bed. *Journal of Zhejiang University (Engineering Science), 39*(6), 1–6. (in Chinese).

Christina, R., Thomas, R., & Wu, D. (2004). *Computational modeling of cross-wind stability of high speed trains* (pp. 1–20). Jyväskylä: European Congress on Computational Methods in Applied Sciences and Engineering.

Crowe, C. T., & Smoot, L. D. (1979). Multicomponent Conservation Equation. In *Pulverized-Coal Combustion and Gasification*. New York: Plenum Press.

Gao, G. J., & Tian, H. Q. (2004). Effect of strong cross-wind on the stability of trains running on the Lanzhou-Xinjiang railway line. *Journal of the China Railway Society, 26*(4), 36–41. (in Chinese).

Gidaspow, D. (1994). *Multiphase Flow and Fluidization*. Boston, America: Academic Press.

Javier, G., Jorge, M., Antonio, C., & Aitor, B. (2009). Comparison of experimental and numerical results for a reference CAF train exposed to cross winds. In *Euromech Colloquium 509 Vehicle Aerodynamics* (pp. 82–92). Berlin, Germany.

Liang, X. F., & Shen, X. Y. (2007). Side aerodynamic performances of maglev train when two trains meet with wind blowing. *Journal of Central South University, 38*(4), 1–6. (in Chinese).

Liu, Y. B., Chen, J. Z., & Yang, Y. R. (2006). Numerical simulation of liquid-solid two-phase flow in slurry pipeline transportation. *Journal of Zhejiang University (Engineering Science), 40*(5), 1–6. (in Chinese).

Ma, W. H., Luo, S. H., & Song, R. R. (2009). Influence of cross-wind on dynamic performance of high-speed EMU on straight track. *Journal of Chongqing Institute of Technology, 23*(3), 1–5. (in Chinese).

Masson, E., Allain, E., & Paradot, N. (2009). CFD analysis of the underfloor aerodynamics of a complete TGV high speed train set at full scale. In *Euromech Colloquium 509 Vehicle Aerodynamics* (pp. 188–202). Berlin, Germany.

Moukalled, F., Darwish, M., & Sekar, B. (2003). A pressure-based algorithm for multi-phase flow at all speeds. *Journal of Computational Physics, 190*(2), 550–571. doi:10.1016/ S0021-9991 (03)00297-3.

Sanquer, S., Barré, C., de Virel, M. D., & Cléon, L. M. (2004). Effect of cross winds on high-speed trains: development of a new experimental methodology. *Journal of Wind Engineering and Industrial Aerodynamics, 92*(7–8), 535–545. doi:10.1016/j.jweia.2004.03.004.

Tian, H. Q. (2007). *Train Aerodynamics*. Beijing, China (in Chinese): China Railway Publishing House.

Xiong, H. B., Yu, W. G., Chen, D. W., & Shao, X. M. (2011). Numerical study on the aerodynamic performance and safe running of high-speed trains in sandstorms. *Journal of Zhejiang University-SCIENCE A (Applied Physics and Engineering), 12*(12), 971–978. doi:10. 1631/jzus.A11GT005.

Yang, Z. G., Ma, J., & Chen, Y. (2010). The unsteady aerodynamic characteristics of a high-speed train in different operating conditions under cross wind. *Journal of the China Railway Society, 32*(2), 1–6. (in Chinese).

## Author Biographies

**Xue-ming Shao** is a professor in the Department of Mechanics, Zhejiang University, Hangzhou, China. He has held this position since 2006. He is an executive member of the editorial board of the *Journal of Hydrodynamics* and member of the tenth council of the Chinese Society of Theoretical and Applied Mechanics as well as of the editorial board of the *Journal of Zhejiang University (Engineering Science)*. Prof. Shao is also the leader of the hydrodynamics group of CSTAM.

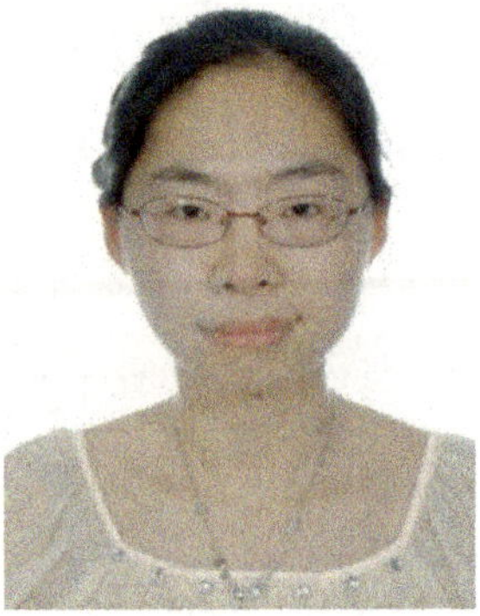

**Hong-bing Xiong** is an associate professor and vice director at the Institute of Fluid Engineering at Zhejiang University, China. She received her Ph.D. from the State University of New York at Stony Brook and her B.S. from University of Science and Technology of China. She specializes in numerical fluid and thermal analysis.

# Numerical Study on the Aerodynamic Performance and Safe Running of High-Speed Trains in Sandstorms

**Hong-bing Xiong, Wen-guang Yu, Da-wei Chen and Xue-ming Shao**

## Contents



## 1 Introduction

China Railways has made a great leap forward during the last decade, and now being the largest exporter and licensor of the high-speed railways in the world. The most significant progress has been made with regard to the rapid growth of railway construction and the innovation of the high-speed train. With train speed increasing, some extreme weather condition, for example, a crosswind or sandstorm, may severely limit the speed and affect the safe running of a railway train, especially in

H. Xiong · W. Yu · X. Shao (✉)
Department of Mechanics, Zhejiang University, Hangzhou 310027, China
e-mail: mecsxm@zju.edu.cn

D. Chen
National Engineering Laboratory for System Integration
of High-Speed Train (South), CSR Qingdao Sifang Co., Ltd, Qingdao 266111, China

 
Y. Fang and Y.H. Zhang (eds.), *China's High-Speed Rail Technology*,
Advances in High-speed Rail Technology, https://doi.org/10.1007/978-981-10-5610-9_7

north China. Due to the dry climate and desertification in that region, severe winds and sandstorms occur all year round (Qiu 2011; Zhang and Zhang 2011). In strong sandstorm weather, wind speed, and sand concentration are high. The running of trains during such weather conditions may cause derailment, overturning, or other serious train accidents. For example, many rollover accidents have occurred on the Lanzhou–Xinjiang Railway Line (Fig. 1). The north China has a dense railway network, so train safety in sandstorms is an important issue to study.

Many Chinese scholars have studied the effects of strong crosswinds on trains in recent years. Several wind tunnels have been constructed to carry out various experiments including studies of the aerodynamic performance of trains, tests of trains passing by each other and tests of stability in crosswinds (Tian and Liang 1998; Tian and Lu 1999; Gao et al. 2007). Besides these experiments, numerical simulation provides another efficient tool to study these issues (Tian and He 2001; Tian and Gao 2003; Chen et al. 2009). Similar research has also been carried out in some other countries with well-developed railways, such as Japan, Germany, France, and Sweden. These studies included train safety under crosswinds (Sanquer et al. 2004; Baker 2010), train slipstream effect, and drag reduction and aerodynamic performance (Watkinsa et al. 1992). However, for trains under sandstorm

**Fig. 1** Overturning accidents on Lanzhou–Xinjiang Railway Line on Feb. 28, 2007 (reported by China Daily)

conditions, less attention has been paid and their effects on the aerodynamic characteristics and train safety are unknown.

A sandstorm is a typical two-phase flow problem with gas air and solid sand particles. Currently, this kind of flow is modeled mainly using two approaches: the Eulerian–Lagrangian (EL) and Eulerian–Eulerian (EE). In EL (Kuo et al. 2002; Lun 1984), the fluid is treated as a continuum phase by solving the Navier–Stokes equation, while the dispersed-phase is solved by tracking a large number of individual particles. Though providing more details of the particle phase, the EL approach is not suitable for large-scale engineering simulation with numerous dispersed particles, due to the limitations of finite memory capacity and CPU efficiency. The EE approach is more efficient and usually more complex (Ding 1990; Gidaspow 1994). In EE, each phase is treated as a continuous medium that may interpenetrate other phases, and is described by a set of equations with regard to momentum, continuity, and energy. EE has been successfully applied to the simulation of pneumatic transport, fluidized beds, and other gas-solid multiphase flows with large numbers of particles in large equipment (van Wachem et al. 1998; Zha 2000). EE has also been used to simulate the influence of sandstorms on high-rise buildings (Wang and Wu 2009). The train simulation in this study was complicated by the large computational domain and a large number of particles. Therefore, the EE approach was chosen instead of EL.

In this paper, the EE two-phase model is used to simulate the sandstorm flow around a train. First, the numerical model is validated using published data. Then, the train aerodynamic performance under different sandstorm levels and no sand conditions are investigated. Finally, an equation of train stability is derived using the theory of moment balance from the view of dynamics, and a recommended speed limit for the train is calculated for different sandstorm levels.

# 2 Mathematical Model and Numerical Simulation

## 2.1 Eulerian–Eulerian Multiphase Model

The continuity equation for the $k$th phase is

$$\frac{\partial}{\partial t}(f_k\rho_k) + \nabla(f_k\rho_k U_k) = 0, \tag{1}$$

where the subscript $k$ represents different phases: when $k$ = g, it represents gas, when $k$ = s, it represents the solid particle phase; $f_k$ represents the volume fraction of the $k$ phase. In current EE models, $f_k + f_s = 1$. $\rho_k$ represents the density of the $k$ phase, $\rho_g = 1.225$ kg/m$^3$, $\rho_s = 2500$ kg/m$^3$; and $U_k$ represents the velocity of the $k$ phase. The momentum equation of the gas phase is

$$\frac{\partial}{\partial t}(f_g \rho_g U_g) + \nabla(f_g \rho_g U_g U_g) \\ = -f_g \nabla p + \nabla \tau_g + f_g \rho_g g + K_{sg}(U_s - U_g), \tag{2}$$

where $p$ represents the pressure of the gas phase and solid phase. $\tau_g$ is the viscous shear tensor of gas. $g$ is acceleration due to gravity, $g = 9.8$ m/s$^2$, and $K_{sg} = K_{gs}$ represents the gas-solid momentum exchange term.

The momentum equation of solid phase is

$$\frac{\partial}{\partial t}(f_s \rho_s U_s) + \nabla(f_s \rho_s U_s U_s) \\ = -f_s \nabla p - \nabla p_s + \nabla \tau_s + f_s \rho_s g + K_{gs}(U_g - U_s), \tag{3}$$

where $p_s$ is the pressure of solid: $p_s = \alpha_s \rho_s \theta_s + 2\rho_s(1 + e_s)\alpha_s^2 g_0 \theta_s$, $e_s = 0.7$ is the coefficient of restitution, and $g_0$ is the radical distribution function, which represents the probability of collision between the sand. $\theta_s$ is the temperature of the sand and varies in direct ratio to the random kinetic energy of sand. The same model has been used in our group to study the train aerodynamic performance and safe operation under strong rainstorm condition (Shao et al. 2011).

## 2.2 Computational Simulation

The train in this study is a simplified model of the CRH2 (China Railway High-Speed 2), including three coaches: the head, middle, and tail (Zhang 2008). To fully develop the flow around the train and to ensure the accuracy of the results, a large semicylindrical numerical wind tunnel was established as the computational domain. The tunnel was 400 m wide in the horizontal direction, 600 m long in the axial direction and 200 m high in the vertical direction (Fig. 2a). The train had a length of 76.2 m, and was located on the centerline of the tunnel with an axial distance of 200 m from the train centroid to the tunnel inlet. In the preprocessing of mesh generation, a tetrahedral/hexahedral hybrid grid was adopted (Fig. 2b), and the number of grids was about 1.26 million.

The velocity inlet and pressure outlet boundary conditions were adopted for both gas and solid phases. Owing to the small volume of the sand fraction, constant velocities of fluid and solid were specified equally with given phase fractions. For the gas phase, no slip wall boundary condition was adopted for the train walls. For the solid phase, a partial slip boundary condition was adopted with a specified specularity coefficient of 0.1.

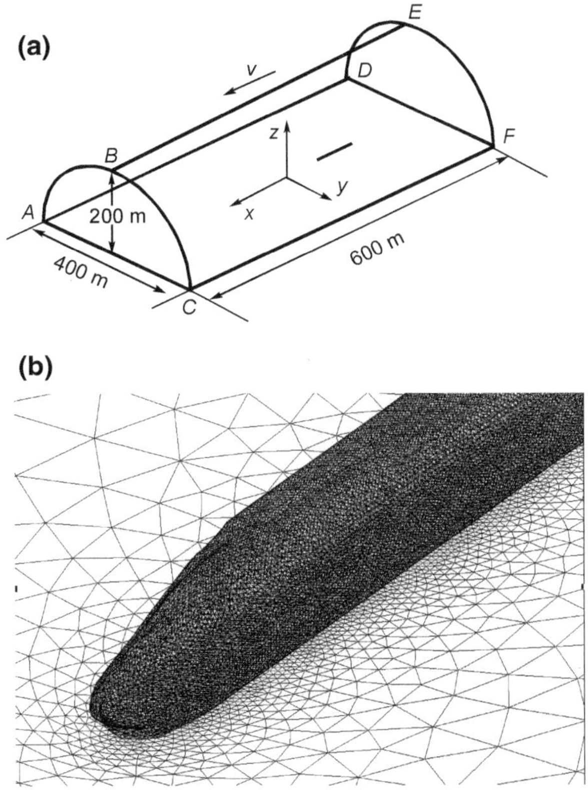


**Fig. 2** Computional domain (**a**) and grid (**b**)

## 2.3 *Validation of the Numerical Model*

### 2.3.1 Verification of Eulerian Two-Phase Model

To verify the accuracy of the Eulerian two-phase model in the external flow, we first used a Eulerian two-phase mode to simulate the aerodynamic characteristics of the airfoil NACA64-210 in a rain environment, and compared the numerical results with published experimental data (Bezos et al. 1992). The drag force ($C_d$) at a different angle of attack ($\alpha_a$) was illustrated in Figs. 3 and 4 for experiments and simulations, respectively. Our simulation results agreed reasonably well with the experimental data.

### 2.3.2 Verification of the Train Model

To validate the train model of numerical simulation, grid dependency was conducted with three sizes of grid generation: 1.95, 1.26, and 0.54 million, respectively. Comparing the train aerodynamic performance, we found that the difference

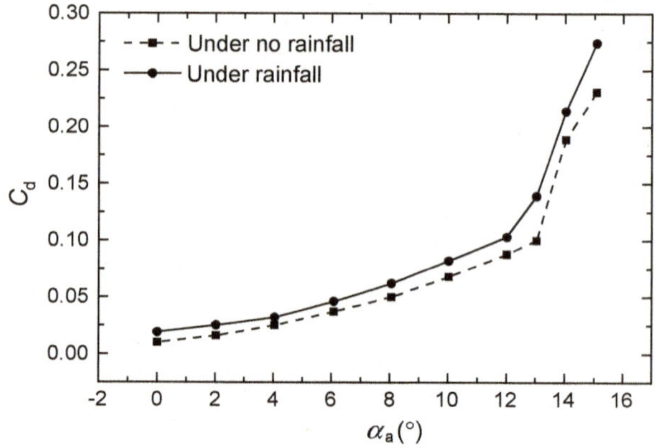


**Fig. 3** Experimental results for the drag coefficient of an airfoil under no rain and rain environments

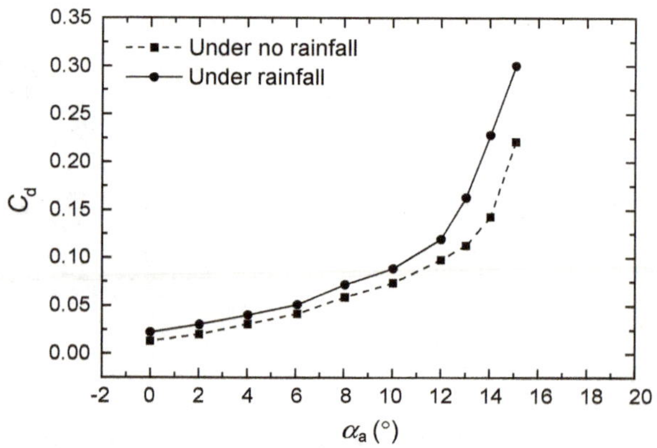


**Fig. 4** Numerical results for the drag coefficient of an airfoil under no rain and rain environments

between the results from 1.26 million grids and from 1.95 million grids was only about 2%, while the result from 0.54 million grids was very different from those two cases. Therefore, considering the accuracy and efficiency of calculation, 1.26 million grids were adopted.

In this study, we reduced the geometrical complexity and used a simplified train model without the bogies, pantograph, and windshield, for computational efficiency. In our previous study, the aerodynamic characteristics of the complete train model were calculated and gave a drag coefficient of $C_x = 0.385$, compared with the experimental data result of $C_x = 0.393$ (Zhang and Xiong 2011). The good correspondence of these results demonstrates that our numerical model for the CRH2 train is reliable.

# 3 Results and Discussion

Sandstorm weather is divided into six levels according to its intensity. The typical values of sand concentration and wind velocity for different sandstorm levels are shown in Table 1.

The aim of this study was to simulate the aerodynamic performance of the train in different wind speeds and different sandstorm levels. Detailed cases were as follows:

Under sand conditions, the speed of the train is $V_t$ = 300 km/h, and the sandstorm levels are floating dust, blowing dust, weak sandstorm, medium sandstorm, strong sandstorm, and particularly strong sandstorm (Table 1).

Under no sand conditions, the speed of the train is $V_t$ = 300 km/h, and the wind speeds ($V$) are 5, 10, 15, 20, 25, and 30 m/s, respectively, corresponding to the sandstorm levels.

Among the aerodynamic forces, the drag force $F_x$, lift force $F_y$, side force $F_z$, and overturning moment $M_x$ plays vital roles in train aerodynamics and safe running in sandstorm conditions. Therefore, we will first focus on the influence of sand on these forces.

## 3.1 *Influence of Sand on the Drag Force*

The drag force of the train under sand conditions and no sand conditions are shown in Fig. 5. $F_{x0}$ and $F_{x1}$ represent the drag under no sand conditions and sand conditions, respectively. From the curve $F_{x0}$, we find that the drag increases when there is a crosswind. Note that the drag decreases at a relatively large wind speed, though drag force increases with the wind speed. The reason is that the drag is based on the sum of train and wind velocity. The trend curve of $F_{x1}$ is similar to that of $F_{x0}$, but the values of $F_{x1}$ are much bigger than those of $F_{x0}$, due to the effects of the sand. So we can conclude that the sand increases the drag force of the train, and the higher the level of sandstorm, the higher the increase in drag force.

**Table 1** Wind speeds and sand concentrations for different sandstorm levels

| Sandstorm | TSP[a] (mg/m$^3$) | Volume fraction of sand | Wind speed (m/s) |
|---|---|---|---|
| Floating dust | 0.4 | $1.6 \times 10^{-10}$ | 5 |
| Blowing dust | 1.2 | $4.8 \times 10^{-10}$ | 10 |
| Weak sandstorm | 6 | $2.4 \times 10^{-9}$ | 15 |
| Medium sandstorm | 30 | $1.2 \times 10^{-8}$ | 20 |
| Strong sandstorm | 90 | $3.6 \times 10^{-8}$ | 25 |
| Particularly strong sandstorm | 270 | $1.08 \times 10^{-7}$ | 30 |

[a]TSP: total suspended particulate

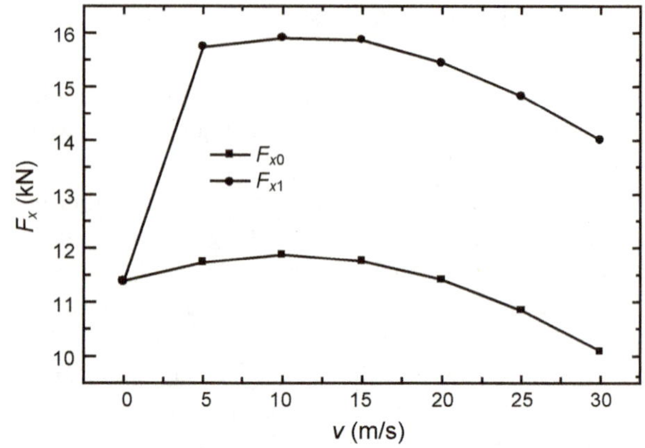


**Fig. 5** Drag force of the train under no sand and sand environments

## 3.2 Influence of Sand on the Lift Force

The lift force of a train is an important factor that affects its safety and comfort. A large lift force will bring instability and discomfort to the passengers, and may even cause the train to overturn. If the lift force is positive with an upward direction, the adhesive force between the wheels and the railway lines reduces so that the possibility of derailment increases. A negative lift force with a small value is favorable as it will increase the contact force between the wheels and the railway lines.

The lift forces of the train under sand conditions ($F_{z1}$) and no sand conditions ($F_{z0}$) are shown in Fig. 6. The trend of these two curves is the same: both $F_{z1}$ and $F_{z0}$ grow with the wind speed corresponding to sandstorm level. But the lift force under the sand is bigger than that under no sand, and the difference grows with the sandstorm level. The force under particularly strong sandstorm conditions is about

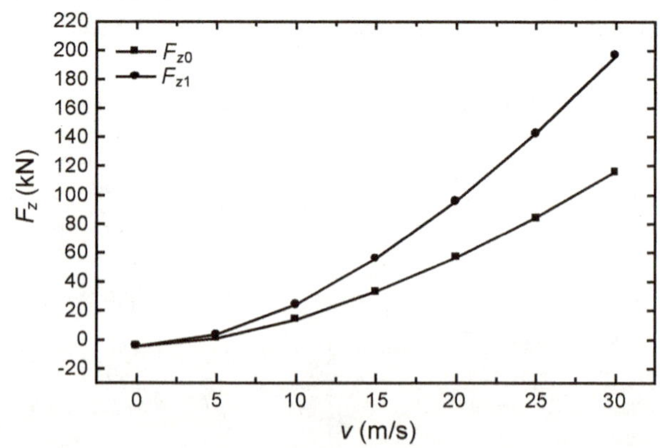


**Fig. 6** Lift forces of the train under no sand and sand environments

50% larger than that under no sand conditions. So the sand increases the value of the lift force significantly: the stronger is the sandstorm, the higher the lift force, and the lower the train's stability.

### 3.3 Influence of Sand on the Side Force

Owing to the asymmetric structure of flow around the train, the pressure distribution on the surface of the train is asymmetric, which leads to a differential pressure side force. Viscous shear stress acts on the surface that forms the side friction force. The sum of the differential pressure side force and the side friction force is the total side force $F_y$ (Fig. 7).

As predicted, the side force increases with the speed of the crosswind. The force under sand conditions ($F_{y1}$) is larger than that under no sand ($F_{y0}$). The difference between them grows with the level of sandstorm. The side force under particularly strong sandstorm conditions increases by 33.5%. The sand increases the side force and may reduce the train's stability.

### 3.4 Influence of Sand on the Overturning Moment

The overturning moment, $M_x$, is the moment generated by the aerodynamics of the train around the vertical axis ($x$-axis) and tending to overturn the train (Fig. 8). The overturning moment significantly affects the stability of the train. $M_x$ increases dramatically with crosswind, and increases with wind speed. The value of overturning moment under sand conditions ($M_{x1}$) is larger than that under no sand conditions ($M_{x0}$). The difference between $M_{x0}$ and $M_{x1}$ increases with the growing level of the sandstorm. So the sand reduces the stability of trains in a crosswind.

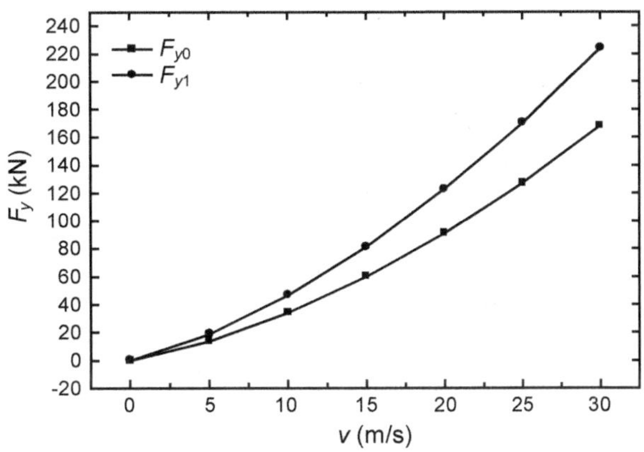


Fig. 7 Side forces of the train under no sand and sand environments

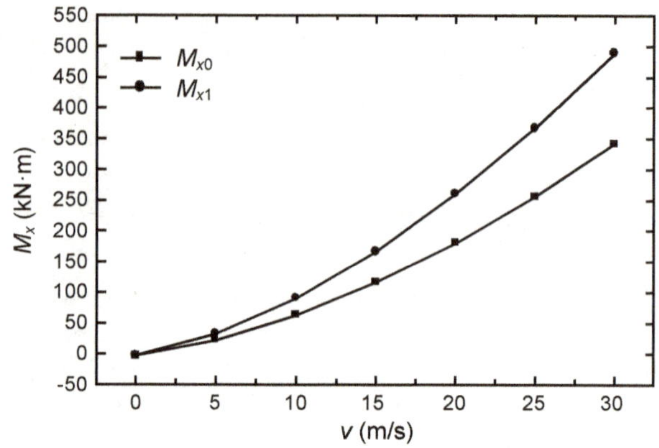


**Fig. 8** Overturning moments of the train under no sand and sand environments

## 3.5 *Influence of Sand on the Train's Safety and Recommended Speed Limit Under Crosswind*

The stability of a train against overturning or derailment is usually used to evaluate safety performance when a train runs on straight or curved rails in a crosswind. When running on curved rails, the outer rail is higher than the inner rail and an unbalanced centrifugal force will be generated. The centrifugal force combined with the side force generated by a crosswind, and the inertia force of transversal vibration increase the possibility that the train will turn over. There are three ways in which a train may turn over: (1) Running on straight rails, the train will turn over in the direction of the crosswind; (2) Running on curved rails, the train will turn over in the direction from the inside rail to the outside rail; (3) Running on curved rails, the train will turn over in the direction from the outside rail to the inside rail.

The running stability of a train depends on several factors including its shape, size, and mass, the height of its center of gravity, and its running speed. The relationship between the running speed limit and the crosswind speed can be derived from the dynamic torque balance principle. The method and formulas have been studied (Gao and Tian 2004; Tian 2007). This method is used in this study to calculate the speed limit of a train under different speeds of crosswind and sand conditions. The influence of sand on the speed limit was then analyzed by comparing the results with those obtained under no sand conditions. The procedure to derive the relationship between the speed limit and wind speed is as follows.

First, the aerodynamic performance, under sand, and no sand conditions, at different yaw angles $\alpha$ were simulated in the CFD software FLUENT. The yaw angle, defined as the angle between the train velocity vector and the resultant velocity vector, varied from 0° to 180°. In this study, we assumed that the direction of the crosswind was perpendicular to the velocity of the train. The three aerodynamic coefficients $C_y$, $C_z$, and $C_{mx}$ were calculated. These coefficients are defined as follows:

$$C_i = \frac{F_i}{\frac{1}{2}\rho V_{\mathrm{TW}}^2 S}, \quad C_{\mathrm{m}i} = \frac{M_i}{\frac{1}{2}\rho V_{\mathrm{TW}}^2 SL}, \tag{4}$$

where $i = x, y, z$ for the force and moment components $F_i$ and $M_i$ in different direction, $\rho = 1.225\ \mathrm{kg/m^3}$, $S = 268.34\ \mathrm{m^2}$, $L = 76.2$ m, and $V_{\mathrm{TW}}$ represents the resultant velocity of train and wind speed. The curves of $C_y - \alpha$, $C_z - \alpha$, and $C_{\mathrm{m}x} - \alpha$ under two weather conditions were plotted, respectively (Fig. 9), and the fitting results of the relationships between the aerodynamic coefficients and yaw angle $\alpha$ were:

Under no sand conditions:

$$\begin{aligned} C_{\mathrm{m}x} &= -2.72777 \times 10^{-9}\alpha^3 - 4.9575 \times 10^{-5}\alpha^2 + 9.005 \times 10^{-3}\alpha, \\ C_y &= -2.8445 \times 10^{-9}\alpha^3 - 8.2909 \times 10^{-5}\alpha^2 + 1.501 \times 10^{-2}\alpha, \\ C_z &= -2.8308 \times 10^{-9}\alpha^3 - 9.5106 \times 10^{-5}\alpha^2 + 1.721 \times 10^{-2}\alpha. \end{aligned}$$

Under sand conditions:

$$\begin{aligned} C_{\mathrm{m}x} &= -2.7789 \times 10^{-9}\alpha^3 - 6.4072 \times 10^{-5}\alpha^2 + 1.1615 \times 10^{-2}\alpha, \\ C_y &= -2.918 \times 10^{-9}\alpha^3 - 1.07220 \times 10^{-4}\alpha^2 + 1.9386 \times 10^{-2}\alpha, \\ C_z &= -3.1229 \times 10^{-9}\alpha^3 - 1.229 \times 10^{-4}\alpha^2 + 2.2213 \times 10^{-2}\alpha. \end{aligned}$$

Finally, taking the results above to the moment balance formulas, the speed limits of the train running on straight or curved rails under different crosswind speeds and sand conditions were calculated (Tables 2 and 3).

Form Tables 2 and 3, we find that, at the same yaw angle, the speed limit of train is the highest on curved rails when the train is to turn over from the outside to the inside rails; the limit speed is the lowest when the train is to turn over from the inside to the outside rails; the limit speed on straight rails is intermediate. At the

**Fig. 9** Aerodynamic coefficients under no sand and sand conditions

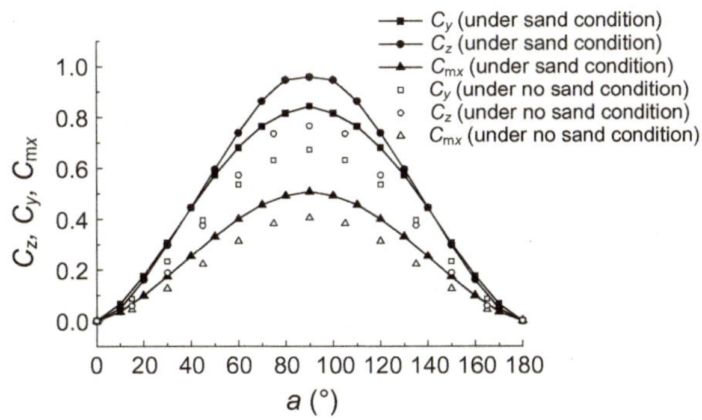

Table 2 Relationship between the train speed limit and crosswind speed under no sand conditions

| α (°) | Straight line | | Turning form outside to inside | | Turning form inside to outside | |
|---|---|---|---|---|---|---|
| | $V_w$ (m/s) | $V_t$ (m/s) | $V_w$ (m/s) | $V_t$ (m/s) | $V_w$ (m/s) | $V_t$ (m/s) |
| 15 | 17.799 | 66.461 | 20.799 | 77.666 | 15.718 | 58.691 |
| 30 | 25.494 | 44.184 | 27.490 | 47.643 | 23.797 | 41.243 |
| 45 | 31.022 | 31.047 | 32.689 | 32.716 | 29.513 | 29.536 |
| 60 | 34.892 | 20.170 | 36.406 | 21.045 | 33.485 | 19.356 |
| 75 | 37.205 | 9.996 | 38.644 | 10.382 | 35.853 | 9.632 |
| 90 | 37.979 | 0.030 | 39.389 | 0.0314 | 36.640 | 0.029 |

$V_t$: limit speed of train; $V_w$: limit speed of wind

Table 3 Relationship between the train speed limit and crosswind speed under sand conditions

| α (°) | Straight line | | Turning form outside to inside | | Turning form inside to outside | |
|---|---|---|---|---|---|---|
| | $V_w$ (m/s) | $V_t$ (m/s) | $V_w$ (m/s) | $V_t$ (m/s) | $V_w$ (m/s) | $V_t$ (m/s) |
| 15 | 15.669 | 58.507 | 17.561 | 65.573 | 14.215 | 53.081 |
| 30 | 22.444 | 38.898 | 23.721 | 41.112 | 21.297 | 36.910 |
| 45 | 27.312 | 27.334 | 28.375 | 28.398 | 26.312 | 26.333 |
| 60 | 30.722 | 17.759 | 31.678 | 18.312 | 29.801 | 17.227 |
| 75 | 32.764 | 8.802 | 33.667 | 9.045 | 31.886 | 8.567 |
| 90 | 33.442 | 0.027 | 34.330 | 0.027 | 32.577 | 0.0259 |

$V_t$: limit speed of train; $V_w$: limit speed of wind

same yaw angle, the limit speed of the train under sand conditions is smaller than that under no sand conditions, decreasing about 10–15%. In other words, a sandstorm reduces the safe speed limit of trains at a given crosswind speed. The safe running performance is reduced in sandstorm weather.

## 4 Conclusions

In this paper, an EE multiphase model was used to study the performance of a train running under different levels of sandstorm weather, and the results were compared with that from no sand conditions. The influence of sand on aerodynamic performance was analyzed parametrically. Results indicate that the drag, lift, side forces, and overturning moment increase in sandstorm weather, and the sand effect increases with the level of sandstorm. Finally, according to the quasi-static analysis method of moment balance, the safe running speed limit of the train was calculated under sandstorm conditions and no sand conditions. The results indicated that a sandstorm reduces the speed limit of the train.

## References


Baker, C. J. (2010). The simulation of unsteady aerodynamic cross wind forces on trains. *Journal of Wind Engineering and Industrial Aerodynamics, 98*(2), 88–99.

Bezos, G. M., Dunham, R. E., Gentry, G. L., & Melson, W. E. (1992). Wind tunnel aerodynamic characteristics of a transport-type airfoil in a simulated heavy rain. *Environment Technical Report, NASA Technical Paper, 3184,* 66–87.

Chen, R. L., Zeng, Q. Y., Zhong, X. G., Xiang, J., Guo, X. G., & Zhao, G. (2009). Numerical study on the restriction speed of train passing curved rail in cross wind. *Journal of Science in China Series, 52*(7), 2037–2047. doi:10.1007/s11431-009-0202-5.

Ding, J. M., & Gidaspow, D. (1990). A bubbling fluidization model using kinetic theory of granular flow. *AIChE Journal, 36*(4), 523–538.

Gao, G. J., & Tian, H. Q. (2004). Effect of strong crosswind on the stability of trains running on the Lanzhou-Xinjiang railway line. *Journal of the China Railway Society, 26*(4), 36–41.

Gao, G. J., Tian, H. Q., & Miao, X. J. (2007). Research on the stability of box-car on Qinghai-Tibet Railway Line under strong cross wind. *Science Paper Online, 2*(9), 684–987.

Gidaspow, D. (1994). *Multiphase Flow and Fluidization.* Boston: Academic Press.

Kuo, H. P., Knight, P. C., & Tsuji, Y. (2002). The influence of DEM simulation parameters on the particle behaviour in a V-mixer. *Chemical Engineering Science, 57,* 3621–3638.

Lun, C. K. K., Savage, S. B., Jeffrey, D. J., & Chepurniy, N. (1984). Kinetic theories for granular flow: inelastic particles in Couette flow and slightly inelastic particles in a general flowfield. *Journal of Fluid Mechanics, 140,* 223–256.

Qiu, X. F., Zeng, Y., & Miu, Q. L. (2011). Temporal-spatial distribution as well as tracks and source areas of sand-dust storms in China. *Actageographica Sinica, 56*(3), 318–319.

Sanquer, S. S., Barréa, C., de Virel, M. D., & Cléon, L. M. (2004). Effect of cross winds on high-speed trains: development of a new experimental methodology. *Journal of Wind Engineering and Industrial Aerodynamics, 92*(7–8), 535–545.

Shao, X. M., Wan, J., Chen, D. W., & Xiong, H. B. (2011). Aerodynamic modeling and stability analysis of a high-speed train under strong rain and crosswind conditions. *Journal of Zhejiang University SCIENCE-A (Applied Physics and Engineering), 12*(12), 964–970. doi:10.1631/jzus.A11GT001.

Tian, H. Q. (2007). *Train Aerodynamics*. Beijing, China: China Railway Publishing House.

Tian, H. Q., & Liang, X. F. (1998). Test research on crossing air pressure pulse of quasi-high-speed train. *Journal of the China Railway Society, 4*(7), 1–8.

Tian, H. Q., & Lu, Z. Z. (1999). Air tunnel test and research on the control car in 200 km/h electrical passenger train unit. *Journal of the China Railway Society, 12*(7), 15–18.

Tian, H. Q., & He, D. X. (2001). 3-D numerical calculation of the air pressure pulse from two trains passing by each other. *Journal of the China Railway Society, 23*(3), 18–22.

Tian, H. Q., & Gao, G. J. (2003). The analysis and evaluation on the aerodynamic behavior of 270 km/h high-speed train. *Journal of the China Railway Society, 24*(2), 14–18.

van Wachem, B. G. M., Schouten, J. C., Krishna, R., & van den Bleek, C. M. (1998). Eulerian simulations of bubbling behaviour in gas-solid fluidised beds. *Computers and Chemical Engineering*, *22*(Supp.), 299–307.

Wang, S. T., & Wu, S. Z. (2009). Numerical simulation of drifting sand flow field on high-rise buildings. MS thesis, Lanzhou University, China (in Chinese).

Watkinsa, S., Saundersa, J. W., & Kumara, H. (1992). Aerodynamic drag reduction of goods trains. *Journal of Wind Engineering and Industrial Aerodynamics, 40*(2), 147–178.

Zha, X. D., Fan, J. R., Sun, P., & Cen, K. F. (2000). Numerical simulation on dense gas-particle riser flow. *Journal of Zhejiang University SCIENCE, 1*(1), 29–38.

Zhang, S. G. (2008). *The CRH-2 high-speed train*. Beijing, China: China Railway Publishing House.
Zhang, M., & Xiong, H. B. (2011). Effects of different components on the aerodynamics of high-speed train. *Journal of Manufacturing Automation, 33*(4), 202–205.
Zhang, X. L., & Zhang, Y. F. (2011). Causes of sand-dust storm in northern China in recent years and its control. *Journal of Catastrophology, 16*(3), 70–76.

## Author Biographies

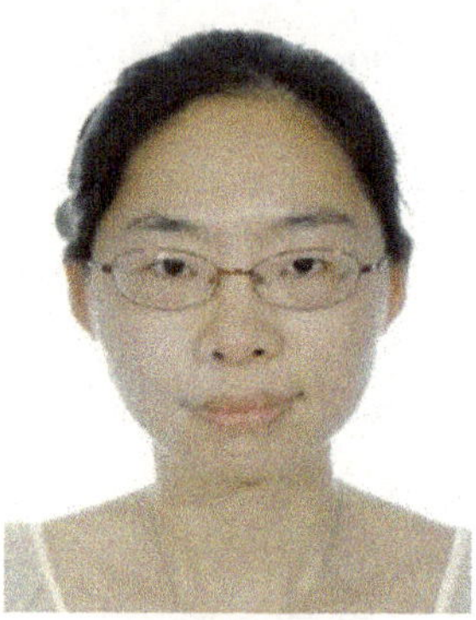

**Hong-bing Xiong** is associate professor and vice director at the Institute of Fluid Engineering at Zhejiang University, China. She received her Ph.D. from the State University of New York at Stony Brook and her B.S. from University of Science and Technology of China. She specializes in numerical fluid and thermal analysis.

**Xue-ming Shao** is a professor in the Department of Mechanics, Zhejiang University, Hangzhou, China. He has held this position since 2006. He is an executive member of the editorial board of the *Journal of Hydrodynamics* and member of the tenth council of the Chinese Society of Theoretical and Applied Mechanics as well as of the editorial board of the *Journal of Zhejiang University (Engineering Science)*. Prof. Shao is also the leader of the hydrodynamics group of CSTAM.

# Influence of Aerodynamic Braking on the Pressure Wave of a Crossing High-Speed Train

Meng-ling Wu, Yang-yong Zhu, Chun Tian and Wei-wei Fei

**Contents**



## 1 Introduction

Aerodynamic braking, as one of the non-adhesion braking methods, uses the pressure difference between two sides of the brake wing to generate resistance force by opening the braking wings on the roof. The force can be a part of high-speed train braking force, which is proportional to the square of the velocity. As a result, this non-adhesion braking method functions very well under high speeds (Tian 2006). At the same time, when aerodynamic braking works, the braking wings can change the flow field around the train. Some studies have shown that aerodynamic impulsion caused by crossing high-speed trains can result in serious impact on the comfort and safety (Wang 2004; Qiu et al. 2005; Lu 2006; Fei et al. 2009; Qi 2010), including oversize deformation, loud noise, and windows being broken (Raghunathan et al. 2002; Tian 2007). Therefore, this paper focuses on the characteristics of the pressure pulse of crossing high-speed trains.

M. Wu (✉) · Y. Zhu · C. Tian · W. Fei
Research Institute of Railway & Urban Mass Transit, Tongji University, 200092 Shanghai, China
e-mail: wuml_sh@163.com

 
Y. Fang and Y.H. Zhang (eds.), *China's High-Speed Rail Technology*, Advances in High-speed Rail Technology, https://doi.org/10.1007/978-981-10-5610-9_8

Three typical cases of crossing events are studied and the air flow field and crossing air pressure waves for the three cases are simulated by the sliding mesh method in the computational fluid dynamics software FLUENT. By comparing the pressure waves of high-speed trains with or without aerodynamic braking, the influence of the braking system is determined.

## 2 Model of a High-Speed Train and Grids Partition

### 2.1 *Geometric Model of a High-Speed Train with Aerodynamic Braking*

The calculation model of the train is a simplification model of the China Railway High-speed (CRH) train, which is composed of the head, the tail, and six middle vehicles. Meanwhile, the model removes the protrusions of the train, so that the surface is smooth, and it also removes the complex components at the bottom of the train, so that there forms a slit between the train and the ground. The basic parameters of the train are shown in Table 1.

The brake wings are installed on the train and operated as a rotary type. The brake wings are drawn back in the vehicle body when the train runs normally. The geometric appearance of the brake wings is shown in Fig. 1. Through numerical optimization calculation, we install seven couples of brake wings in the train, which is composed of eight vehicles. The brake wings distribute symmetrically among the longitudinal middle section of the train and the positions of the brake wings are shown in Fig. 2 (Fei et al. 2009).

### 2.2 *Calculation Area and the Mesh Partition*

#### 2.2.1 Selected Section for the Calculation Area

We use the following calculation area: the length is 800 m, the width is 52.5 m, the height is 100 m, the track space is 5 m (Qi 2010), and the slit between the bottom of the train and the ground is 0.2 m. The calculation begins when the distance between the two head vehicles is 50 m and then ends when the two tail vehicles are 50 m away. The calculation area is shown in Fig. 3.

**Table 1** Geometric parameters of the train

| Parameter | Value (mm) |
|---|---|
| Length of the head (tail) vehicle | 25,530 |
| Length of the middle vehicle | 24,175 |
| Width of the vehicle | 3255 |
| Height of the vehicle | 3680 |

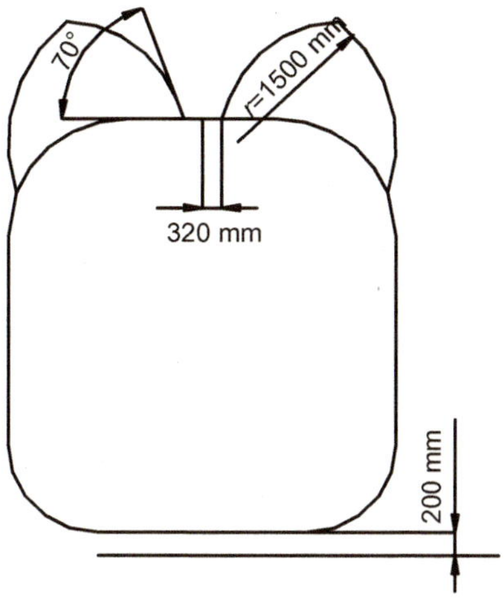


**Fig. 1** Shape of the brake wing

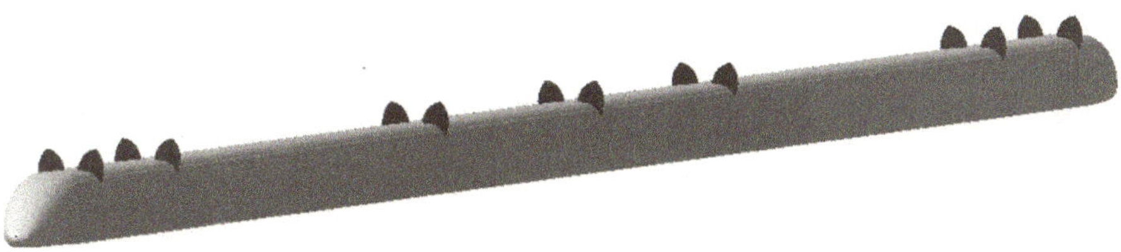

**Fig. 2** Positions of the brake wings on the train. Reprinted from Fei et al. (2009), with permission from the Central South University Press

**Fig. 3** Calculation area of the initial meeting time

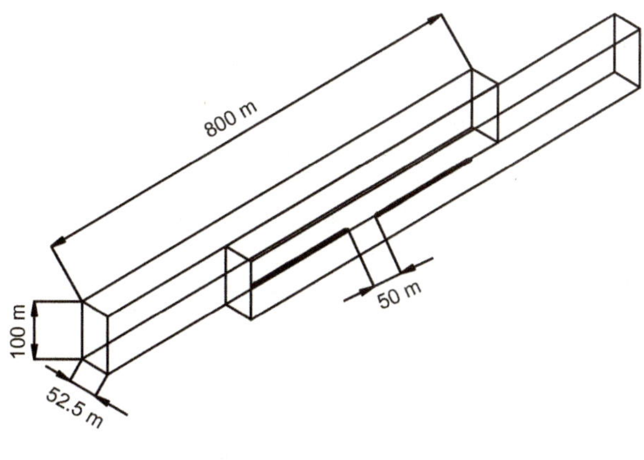

**Fig. 4** Area meshes of the train and brake wings

### 2.2.2 Mesh Partition of a Single Train

This study uses the ICEM CFD software to partition the structured mesh and adopts O-grid structure to partition the vehicle and the brake wing. The meshes are shown in Fig. 4. The mesh number of the single train with the brake wing is approximately 6.5 million and without the brake wing is generally 5.5 million.

### 2.2.3 Sliding Mesh

The sliding mesh technology is adopted to exchange the mesh information between two trains, and the sliding mesh model allows the adjacent meshes to slide. We set the mesh area around the two trains to the sliding mesh type and then match the sliding velocity to the vehicle's running speed before calculation. In order to simulate the real condition, we set a short calculation time step. However, the train is so long that the meeting time during the crossing process will be extended. Changing the unsteady time step is required to solve the problem; that is, we set the time step relatively long before the meeting time and relatively short during the meeting time. The long time step value is 0.018 s and the short one is 0.0045 s, which will not only shorten the calculation time, but also simulate the key points of the crossing process more closely.

**Fig. 5** Border area

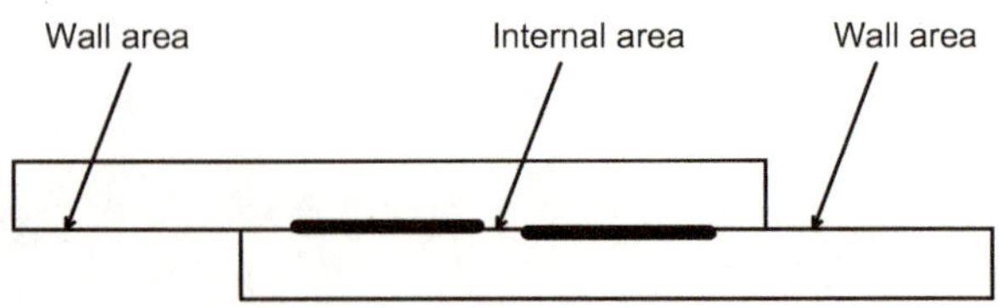

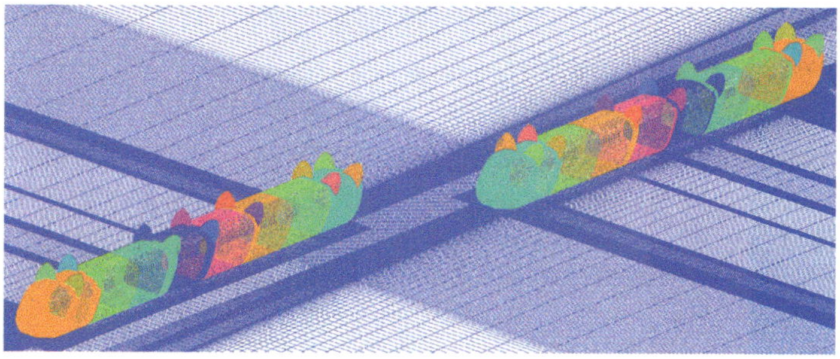

**Fig. 6** Model of the meeting trains with brake wings

There exists a border surface between the two train sliding meshes, and the border surface brings an internal area and two wall areas (Fig. 5). The overlap boundary surface area is the corresponding internal area. The number of pieces of border surface area varies due to the relative motivation of every boundary surface during the crossing process. The flow of the border surface will be calculated according to the face, which is born from the two boundary surfaces. The numerical simulation meshes that we adopt are shown in Fig. 6, and the mesh number is approximately 13 million.

## 3 Numerical Simulation of the Aerodynamic Status of Crossing High-Speed Trains

According to the simulation model, the speed of the virtual high-speed train is 400 km/h. The study focuses on the characteristics of the flow field of crossing high-speed trains.

Condition 1 The crossing of two trains without aerodynamic braking.
Condition 2 The crossing of two trains, one with and one without aerodynamic braking.
Condition 3 The crossing of two trains with aerodynamic braking.

The crossing situation of two trains is shown in Fig. 7. The initial distance is set to be 50 m, and the trains are in the crossing period after 20 steps. On the 60th step, one train's head is passing the other's center. On the 148th step, the crossing period ends. Considering the influence of the wake flow of the trains, they keep on running until the distance between their tails is up to 50 m. The whole calculation process lasts 160 steps.

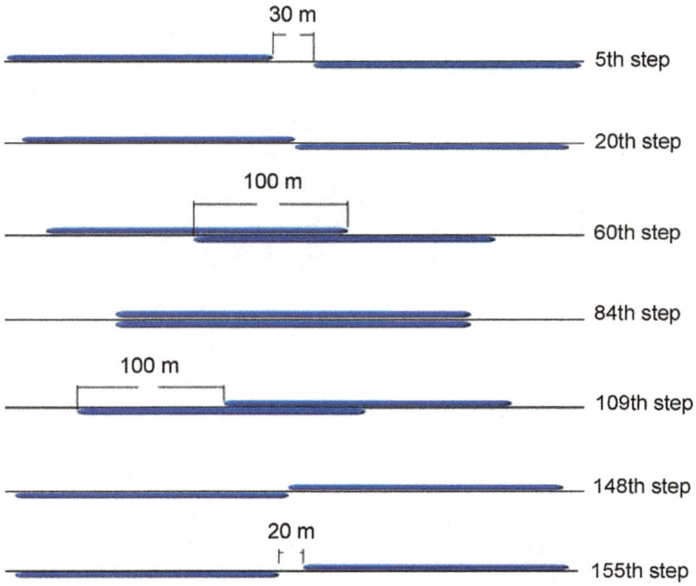


**Fig. 7** Key moments during the crossing of high-speed trains

## *3.1 Characteristic of the Air Flow Field*

The following is the qualitative analysis of the flow field characteristics of a high-speed train with aerodynamic braking.

Figure 8 shows different pressure contours on different steps of the crossing of high-speed trains. Comparing Fig. 8a, b, it is clear that the positive pressure areas on the windward surface of the brake wing are enlarged, while the areas of negative pressure on the other side are also increasing. As shown in Fig. 8c, the pressure wave on one train's head can produce a positive pressure area on the other train's body. Similarly, the pressure waves around one train's brake wing can also cause similar positive pressure areas, which are depicted in Fig. 8e.

## *3.2 Influence of Aerodynamic Braking on the Crossing Train*

The observation point is set on the surface of the train body near the 4th brake wing and is 2.5 m above the ground. The data of the observation point in conditions 1–3 with or without aerodynamic braking are compared.

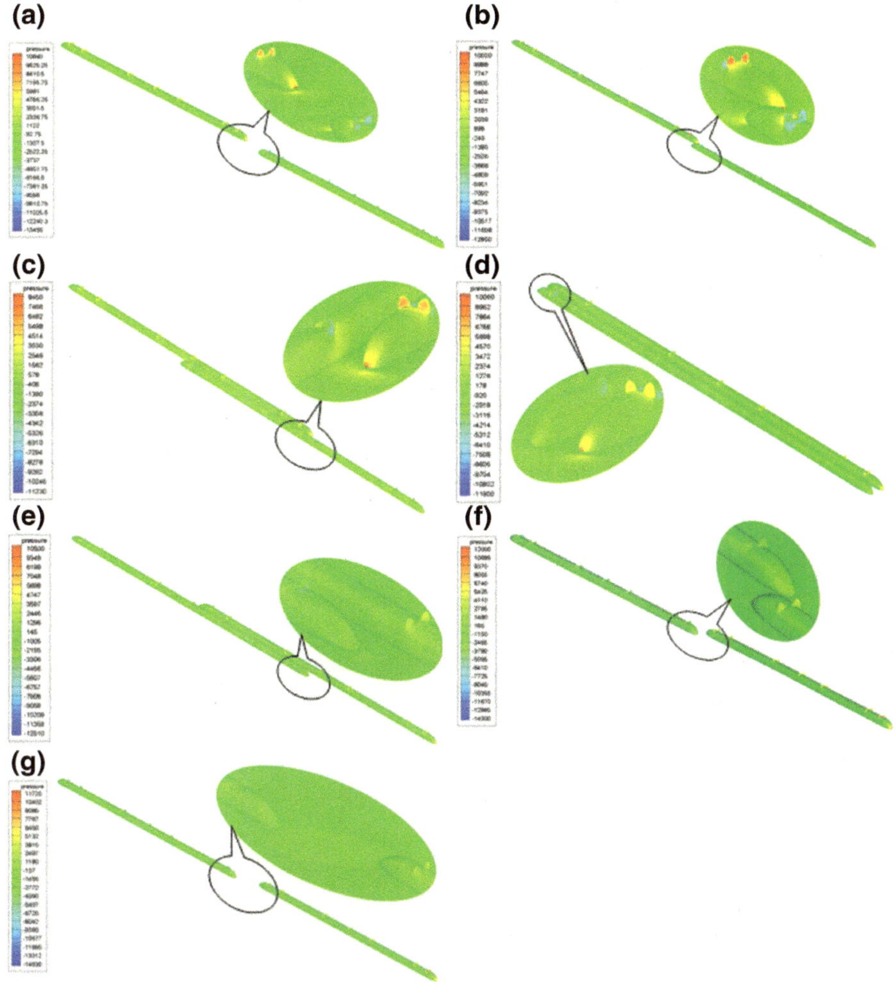


**Fig. 8** Pressure contours of crossing trains with aerodynamic braking (Pa). **a** 5th step; **b** 20th step; **c** 60th step; **d** 84th step; **e** 109th step; **f** 148th step; **g** 155th step

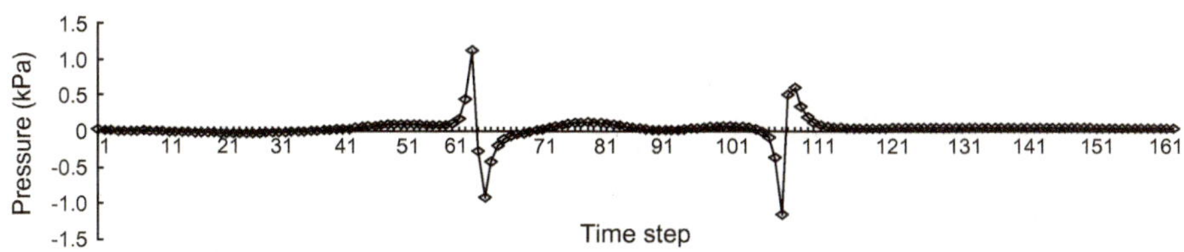


**Fig. 9** Pressure waves of the observation point of the two trains without aerodynamic braking

In condition 1, as shown in Fig. 9, there are one positive and one negative pressure fluctuations around the observation points, which are in accordance with previous studies (Raghunathan et al. 2002).

The wave chart of the observation points in condition 2 is illustrated in Fig. 10. The pressure wave of the observation points on the high-speed train with aerodynamic braking is similar to its counterpart in condition 1. However, the pressure wave on the other train indicates additional positive pressure fluctuation. Studying the moments of the occurrence of various pressure fluctuations, we can see that:

The 1st wave crest is on the 59th step when the head of one train is approaching the observation point.
The 2nd wave crest is on the 61st step when the 1st brake wing is near the observation point.
The 3rd wave crest is on the 64th step when the 2nd brake wing is near the observation point.
The 4th wave crest is on the 76th step when the 3rd brake wing is near the observation point.
The 5th wave crest is on the 82nd step when the 4th brake wing is near the observation point.
The 6th wave crest is on the 89th step when the 5th brake wing is near the observation point.
The 7th wave crest is on the 102nd step when the 6th brake wing is near the observation point.
The 8th wave crest is on the 105th step when the 7th brake wing is near the observation point.
The 9th wave crest is on the 109th step when the tail of the train is approaching the observation point.

We can see that distances between the pressure pulses are equal to the longitudinal distances of the brake wings. Thus, we can conclude that brake wings lead to the pressure pulses. In addition, Fig. 10 also shows that the amplitude of the pressure pulses caused by brake wings is far less than those at the head and the tail.

The wave chart of the observation points in condition 3 is illustrated in Fig. 11. As we can see from the figure, the pressure wave of the observation point on the high-speed train with aerodynamic braking is similar to its counterpart in

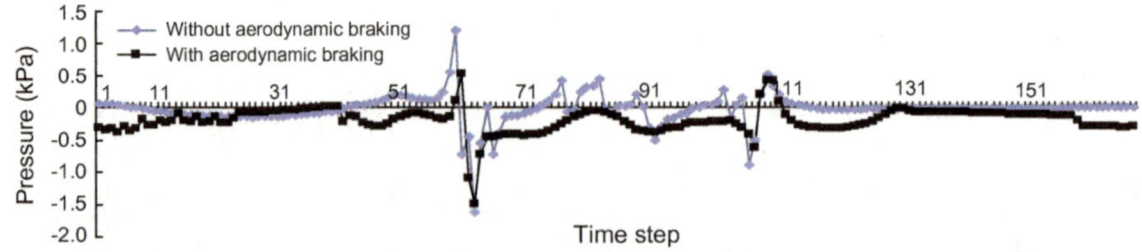


**Fig. 10** Pressure waves of the observation point of the two trains with and without aerodynamic braking

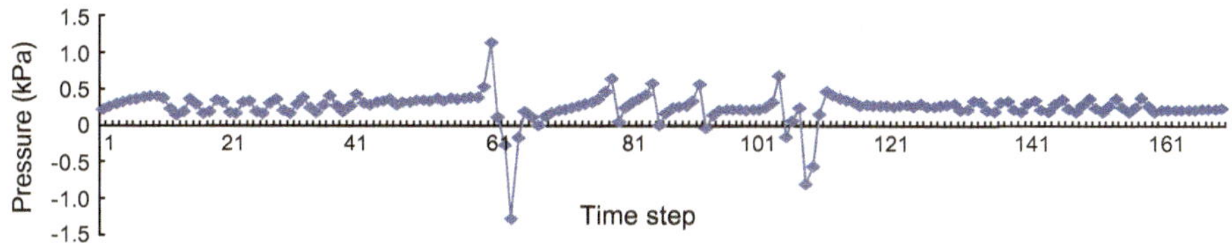


**Fig. 11** Pressure waves of the observation point of the two trains with aerodynamic braking

condition 2. The pressure pulses of the observation points appear not only at the head and tail of the train, but also in the middle. During the crossing period, the highest crossing air pressure is 2 kPa. The glass of the train windows can endure 8.28 kPa (Qi 2010), as glass is pasted to the windows, and it will not impact the safety when the track space is 5 m or more.

As shown in Figs. 9, 10 and 11, if the crossing train is not equipped with aerodynamic braking, pressure pulses will not appear, except at the head and tail of the train. Thus, the train without aerodynamic braking will not impact the crossing train.

## 4 Conclusions

Three typical cases in the crossing events are studied, and the air flow field and crossing air pressure wave under the three cases are simulated by the sliding mesh method in the computational fluid dynamics software FLUENT. By comparing the pressure waves of high-speed trains with or without aerodynamic braking, the influence is determined. The result shows that:

1. When two trains with aerodynamic braking pass by each other, the highest crossing air pressure is 2 kPa.
2. When a train passes by another one equipped with aerodynamic braking, the air pressure pulse around the train head will cause a positive low pressure area. Analogously, the air pressure pulse around the brake wing will cause a pressure-oscillation area on the surface of the train body.
3. When two trains equipped with aerodynamic braking pass by each other, the pressure pulses of observation points appear not only at head and tail of the train, but also in the middle. The middle pulses result from the brake wings. And the amplitude of the pressure pulses caused by brake wings is far less than those at the head and the tail.
4. If the crossing train is not equipped with aerodynamic braking, pressure pulses will not appear except at the head and tail of the train. Thus, a train without aerodynamic braking will not impact the crossing train.

## References

Fei, W. W., Tian, C., Wu, M. L. (2009). Research on brake type of high-speed train based on air dynamics. In The National Industrial Aerodynamic Academic Conference, Changsha, China, pp. 238–244 (in Chinese).

Lu, G. D. (2006). The aerodynamics points of high speed trains. *Rolling Stock*, *44*(6):1–3, 44–45 (in Chinese).

Qi, Z. D. (2010). Aerodynamics research on high-speed train passing each other. MS Thesis, Southwest Jiaotong University, Chengdu, China (in Chinese).

Qiu, Y. Z., Xu, Y. G., & Wang, Y. L. (2005). An aerodynamic model of bottom structures of high-speed trains. In Fluent Chinese User Conference, pp. 151–155 (in Chinese).

Raghunathan, R. S., Kim, H. D., & Setoguchi, T. (2002). Aerodynamics of high speed railway train. *Progress in Aerospace Sciences, 38*(6–7), 469–514. doi:10.1016/S0376-0421(02)00029-5.

Tian, H. Q. (2006). Study evolvement of train aerodynamics in China. *Journal of Traffic and Transportation Engineering, 6*(1), 1–9. (in Chinese).

Tian, H. Q. (2007). *Train Aerodynamics*. Beijing, China (in Chinese): China Railway Press.

Wang, F. J. (2004). *Computational Fluid Dynamics (CFD) software analysis principle and application* (pp 67–70). Beijing, China: Tsinghua University Press. (in Chinese).

## Author Biography

**Meng-ling Wu** born in 1959 in Hangzhou, Zhejiang Province, China, is a professor, doctoral supervisor, and the director of the Braking Technology Institute, Research Institute of Rail Transit, Tongji University, China. His current research focuses on rail vehicle braking and security technology. He has published more than 40 papers and written a number of books. Prof. Wu has been involved in much key state-funded research and received the 1st class award of the National Technology of China Railway Society.

# A Numerical Approach to the Interaction Between Airflow and a High-Speed Train Subjected to Crosswind

**Tian Li, Ji-ye Zhang and Wei-hua Zhang**

**Contents**



## 1 Introduction

The crosswind stability of railway vehicles has been studied for several decades, motivated by overturning accidents (Cheli et al. 2010; Li et al. 2011a). Most studies on crosswind stability have focused on aerodynamic issues.

Aerodynamics of railway vehicles subjected to crosswind can be investigated by experiments in wind tunnels and numerical simulations. Wind tunnel experiments have played an important role in determining the aerodynamic characteristics of trains for years. Experimental results for aerodynamic forces were reported by Orellano and Schober (2006), and comparisons were made with regard to different

T. Li (✉) · J. Zhang · W. Zhang
State Key Laboratory of Traction Power, Southwest Jiaotong University,
Chengdu 610031, China
e-mail: litian2008@home.swjtu.edu.cn



Y. Fang and Y.H. Zhang (eds.), *China's High-Speed Rail Technology*,
Advances in High-speed Rail Technology, https://doi.org/10.1007/978-981-10-5610-9_9

levels of geometric complexity addressing the issues of bogies and spoilers. There was only a small effect for the coefficients, which contributed most to overturning, i.e., the roll moment and the side force. Cheli et al. (2010) presented a numerical experimental procedure for the aerodynamic optimization of AnsaldoBreda EMUV250. Wind tunnel experiments (Suzuki et al. 2003; Bocciolone et al. 2008) were performed to evaluate the aerodynamic characteristics of railway vehicles on typical infrastructures, such as bridges and embankments. To date, the only full-scale experiments were performed on a coastal site at Eskmeals in Cumbria (northwest England) (Baker et al. 2004). Thorough reviews were given by Baker (1991).

In addition, several numerical simulations of aerodynamic characteristics of railway vehicles subjected to crosswind were performed. The unsteady aerodynamic forces of a high-speed train were analyzed in a variety of unsteady crosswinds (Xu and Ding 2006; Baker 2010; Thomas et al. 2010; Baker et al. 2011; Shao et al. 2011). A 3D source/vortex panel method (Chiu 1995) was developed to predict the aerodynamic loads on an idealized railway train model in crosswind at large yaw angles. Numerical results of the airflow passing a simplified train under different yawing conditions were summarized (Khier et al. 2000). Diedrichs (2003) and Diedrichs et al. (2007) discussed the aerodynamic characteristics of the airflow passing a train and found that the 6-m-high embankments reduced the permissible crosswind velocity by approximately 20%. Moreover, numerical simulations were carried out using the commercial code FLUENT or STAR-CD (Diedrichs et al. 2007; Cheli et al. 2010).

The dynamic response of a railway vehicle subjected to crosswind was calculated using a vehicle model. Baker (1991) and Ding et al. (2008) used the simple vehicle models to evaluate the running safety against crosswind. In fact, the vehicle and track subsystems were coupled through wheel/rail interaction (Zhai et al. 1996). A detailed vehicle-track coupling model was established for the first time (Zhai et al. 1996); the model has been widely applied.

The crosswind stability in the aforementioned studies was evaluated using an off-line simulation method. Firstly, aerodynamic forces of a train subjected to crosswind were calculated by computational fluid dynamics (CFD), and then, the dynamic response of the train subjected to crosswind was simulated using a vehicle or vehicle-track model. The influence on the aerodynamic forces of displacements was neglected using the off-line simulation. In fact, aerodynamic forces and dynamic performances of railway vehicles were coupled. On the one hand, the aerodynamic forces change the displacements of the train. On the other hand, the displacements affect the aerodynamic forces. To date, there has been no research to simulate the interaction between airflow and a high-speed train.

In this paper, a numerical approach to the interaction between airflow and a high-speed train was presented. The vehicle-track coupling model was adopted to calculate the dynamic response of the train subjected to crosswind. Dynamic performances of a high-speed train subjected to crosswind were discussed in detail.

## 2 Governing Equations

### 2.1 Equations of Fluid Dynamics

The flow field around a train subjected to crosswind can be considered as a 3D incompressible viscous turbulent flow. To describe the flow field around the train, a $k$-$\varepsilon$ turbulence model is adopted. The equations of the standard $k$-$\varepsilon$ two-equation model are written as

$$\begin{aligned} &\frac{\partial}{\partial t}\int_V \rho\varphi \mathrm{d}V + \int_A \boldsymbol{n}(\rho(\boldsymbol{u} - \bar{\boldsymbol{u}})\boldsymbol{\varphi})\mathrm{d}A \\ &= \int_A \boldsymbol{n}(\boldsymbol{\Gamma}\mathrm{grad}\boldsymbol{\varphi})\mathrm{d}A + \int_V \boldsymbol{S}\mathrm{d}V, \end{aligned} \tag{1}$$

where $V$ is an arbitrary control volume, $A$ is the surface of the volume $V$, $t$ is the time, $\rho$ is the air density, $\boldsymbol{u}$ is the velocity vector of the flow, $\bar{\boldsymbol{u}}$ is the velocity vector of the surface $A$, $\boldsymbol{\varphi}$ is the vector of fluxes, $\boldsymbol{n}$ is the normal direction vector of the surface $A$, $\boldsymbol{S}$ is the source term, and $\boldsymbol{\Gamma}$ is the generalized diffusion coefficient. The vector $\boldsymbol{\varphi}$ includes the flow velocity $\boldsymbol{u}_i$, the turbulent kinematics energy $k$, and turbulent dissipation $\varepsilon$.

### 2.2 Equations of Vehicle-Track Coupling Dynamics

Vehicle-track coupling dynamics mainly consist of vehicle dynamics, track dynamicsx, and wheel–rail interaction.

A four-axle railway vehicle with two suspension systems, which is a common railway vehicle used in China, was chosen in this study. The vehicle shown in Fig. 1a consists of a carbody, two bogies, four wheelsets, and two suspension systems. The carbody, bogies, and wheelsets are regarded as rigid components, and their elastic deformations are neglected. Every rigid component is assigned five degrees of freedom: the vertical displacement, the lateral displacement, the roll displacement, the yaw displacement, and the pitch displacement with respect to its mass center.

The track subsystem shown in Fig. 1b is modeled as an infinite Euler beam, which is supported on a discrete-elastic foundation consisting of three layers. The sleeper is assigned three degrees of freedom: the vertical displacement, the lateral displacement, and the roll displacement. To account for the shearing continuity of the particles between the adjacent ballasts, linear springs and dampers are introduced to model the shear coupling effects. Moreover, the ballasts are assigned only one degree of freedom: the vertical displacement.

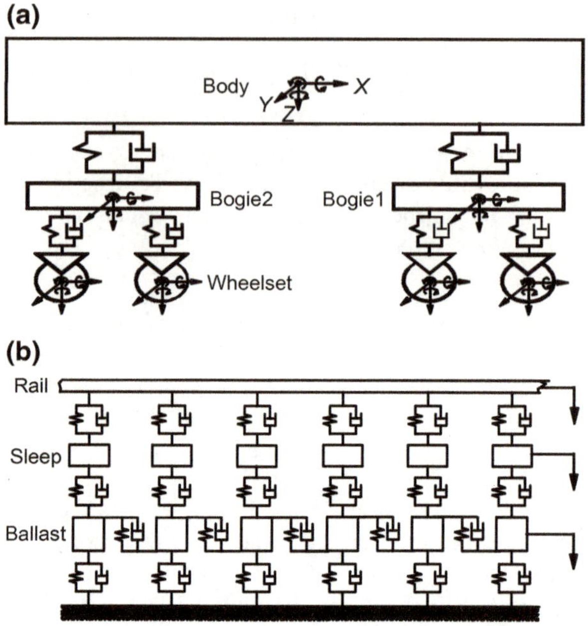


**Fig. 1** Vehicle-track coupling system. **a** Vehicle system; **b** Track system

Wheel–rail interaction involves two basic issues: the geometric relationship and the contact forces. The Hertz contact theory is used to solve the vertical normal forces. The creep forces are calculated using the Shen–Hedrick–Elkins theory. More detailed descriptions of the normal and creep forces can be found in (Zhai et al. 1996). In this study, rail irregularities are measured from high-speed railways in China.

The equation of the vehicle-track dynamics (Zhai et al. 1996) is written as

$$\boldsymbol{M}\ddot{\boldsymbol{X}} + \boldsymbol{C}\dot{\boldsymbol{X}} + \boldsymbol{K}\boldsymbol{X} = \boldsymbol{F}, \tag{2}$$

where $\boldsymbol{M}$, $\boldsymbol{C}$, and $\boldsymbol{K}$ are the mass, damping, and stiffness matrices of the vehicle-track system, respectively. $\boldsymbol{X}$, $\dot{\boldsymbol{X}}$, and $\ddot{\boldsymbol{X}}$ are the generalized displacement, velocity, and acceleration vectors of the system, respectively. $\boldsymbol{F}$ is the generalized force vector including wheel rail contact forces and aerodynamic forces.

# 3 Numerical Approach to the Interaction

## 3.1 Vehicle-Track Dynamics Solution Technique

The equations of vehicle-track dynamics were solved using the Zhai method (Zhai et al. 1996). The integral format of the Zhai method is written as

$$\begin{cases} \boldsymbol{X}_{n+1} = \boldsymbol{X}_n + \dot{\boldsymbol{X}}_n \Delta t + (1/2 + \mu)\ddot{\boldsymbol{X}}_n \Delta t^2 - \mu \ddot{\boldsymbol{X}}_{n-1} \Delta t^2, \\ \dot{\boldsymbol{X}}_{n+1} = \dot{\boldsymbol{X}}_n + (1 + \lambda)\ddot{\boldsymbol{X}}_n \Delta t - \lambda \ddot{\boldsymbol{X}}_{n-1} \Delta t, \end{cases} \tag{3}$$

where $\lambda$ and $\mu$ are integral parameters, $\Delta t$ is the time step and the subscript $n$ denotes the iteration number of the time step.

Equation (3) at the time $t = (n + 1)\Delta t$ is given by

$$\boldsymbol{M}\ddot{\boldsymbol{X}}_{n+1} + \boldsymbol{C}\dot{\boldsymbol{X}}_{n+1} + \boldsymbol{K}\boldsymbol{X}_{n+1} = \boldsymbol{F}_{n+1}. \tag{4}$$

Substituting Eq. (3) into Eq. (4), $\ddot{\boldsymbol{X}}_{n+1}$ can be obtained.

The code for the vehicle-track coupling dynamics was written in the programming language Fortran and verified to be trustworthy (Li et al. 2011a).

## 3.2 Dynamic Mesh Technique

Re-mesh and spring analogy methods (Li et al. 2011b) were adopted to renew the mesh of CFD. If the spring analogy method fails to renew the mesh, the re-mesh method is adopted. The spring analogy method is simple but highly efficient. The stiffness of a given edge $i$–$j$ is defined as $K_{ij} = 1/r_{ij}$, where $K_{ij}$ is the stiffness and $r_{ij}$ is the distance between the $i$th and $j$th nodes.

The displacements of nodes are obtained:

$$\sum_{j}^{N_i} K_{ij} \Delta \boldsymbol{r}_j = 0, \tag{5}$$

where $N_i$ is the total number of nodes connected to the $i$th node and $\Delta \boldsymbol{r}_j$ is the displacements of the $j$th node.

The new position of the $i$th node is determined by

$$\bar{\boldsymbol{r}}_i = \boldsymbol{r}_i + \Delta \boldsymbol{r}_i. \tag{6}$$

## 3.3 Solution Strategies

Co-simulation means that aerodynamic forces and dynamic performances of a high-speed train are calculated alternatively; namely, the interaction between airflow and a high-speed train is considered.

Figure 2 shows a schematic diagram of the numerical approach to the interaction between airflow and a high-speed train. The aerodynamic forces $\boldsymbol{F}$ include the side force, the lift force, the roll moment, the pitch moment, and the yaw moment. The displacements of the train $\boldsymbol{D}$ include the lateral displacement $y$, the vertical

Fig. 2 Numerical procedure

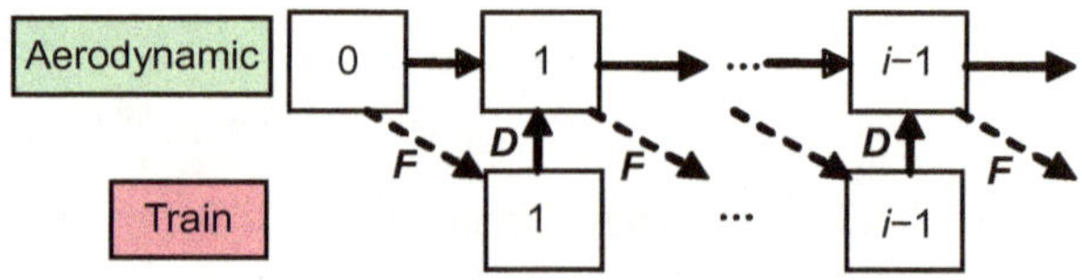


displacement $z$, the roll displacement $\theta$, the yaw displacement $\psi$, and the pitch displacement $\beta$.

$\Delta t_f$ and $\Delta t_v$ represent the time step sizes in the computational fluid dynamics and vehicle-track dynamics, respectively. Generally, the orders of magnitude of $\Delta t_f$ and $\Delta t_v$ are $1.0 \times 10^{-3}$ s and $1.0 \times 10^{-5}$ s, respectively. As shown in Fig. 2, the procedure for the numerical approach to the interaction between airflow and a high-speed train can be summarized as follows:

1. Calculate the aerodynamic forces of the train in an initial static status until the forces reach relatively steady values, such as the fluctuation is less than 3%.
2. Transfer the message of aerodynamic forces to the user-defined function (UDF), which is an interface of the commercial code FLUENT, and then invoke the code for the vehicle-track coupling dynamics.
3. Load the above aerodynamic forces onto the vehicle-track model and calculate the dynamic response of the vehicle. The number of time step iterations for the vehicle-track dynamics is $\Delta t_f/\Delta t_v$.
4. Transfer the message of displacements to the commercial code FLUENT.
5. Renew the mesh of CFD using the dynamic mesh technique described in Sect. 3.2.
6. Keep solving Eq. (1) and calculate the aerodynamic forces of the train in the above displacements.
7. Repeat steps 2–6 until the number of time step iterations reaches the expected value.

## 4 Computational Model and Domain

A schematic diagram of the computational domain of a high-speed train subjected to crosswind is shown in Fig. 3. The ballast, sleeper, and rails are neglected in the computational fluid dynamics. Unstructured tetrahedral meshes were adopted.

The running speed of the train $u_t$ and crosswind velocity $u_w$ were specified at the inlet boundary, i.e., $u_{inlet} = -u_t$; $v_{inlet} = u_w$. In addition, traction-free and symmetry conditions were specified at the outlet and top boundaries, respectively. The slip condition was specified at the wall boundary, i.e., $u_{wall} = -u_t$. Although natural wind was turbulent and a boundary layer existed near the ground, the effect of the boundary layer was neglected and a uniform inflow was adopted. The calculation was performed for the high-speed train at $u_t$ = 350 km/h and subjected to crosswind

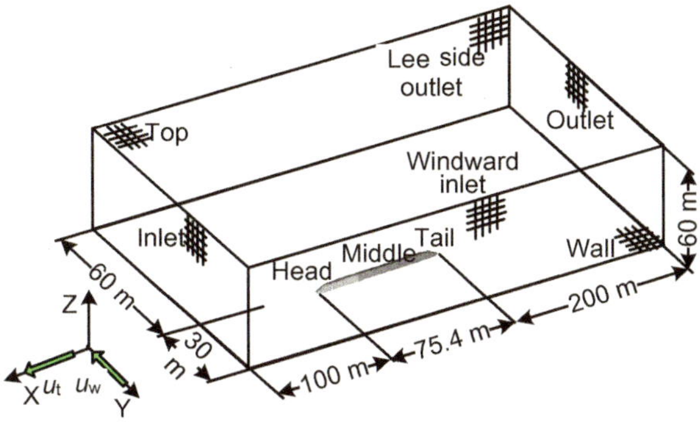


**Fig. 3** Schematic diagram of computational domain

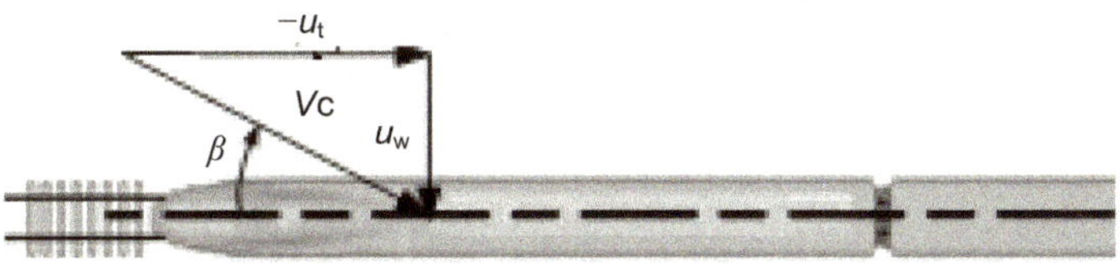


**Fig. 4** Schematic diagram of the combined wind velocity. $\beta$: yaw angle; $V_c$: combined velocity

$u_w$ = 13.8 m/s. The combined wind velocity was 98.2 m/s and the yaw angle was 8.08°, as shown in Fig. 4.

Figure 5 shows a schematic diagram of aerodynamic forces and moments. Aerodynamic forces include the side force $F_y$, the lift force $F_z$, the roll moment $M_x$, the pitch moment $M_y$, and the yaw moment $M_z$. The reference points of the moment of head, middle, and tail coaches are (−13.2, 0.0, 1.72), (−38.2, 0.0, 1.72), and (−63.2, 0.0, 1.72), respectively.

In order to describe the flow field around a train and the wall stress on the train surface, a $k$-ε turbulence model with a wall function treatment was adopted. The Reynolds-averaged Navier–Stokes (RANS) equations were used to simulate the flow field around the train and solved by the finite volume method (FVM). The velocity-pressure coupling was achieved by using the SIMPLE (semi-implicit method for pressure-linked equations) scheme. The techniques for solving the fluid dynamics in this study were similar to those in (Diedrichs et al. 2007). Numerical simulations were carried out using the commercial code FLUENT.

Table 1 shows the effect of aerodynamic forces on the maximum mesh sizes of the train surface. It is shown that aerodynamic forces are almost stable when the maximum mesh sizes of head, middle, and tail coaches are 80, 100, and 80 mm, respectively. Therefore, these sizes were chosen.

Table 2 shows the effect of aerodynamic forces on the computational domain including the length $L$, width $W$, and height $H$. For example, the length 75 + 150 means that the distance between the inlet surface and nose of the head coach is 75 m, and the distance between the outlet surface and the nose of the tail coach is

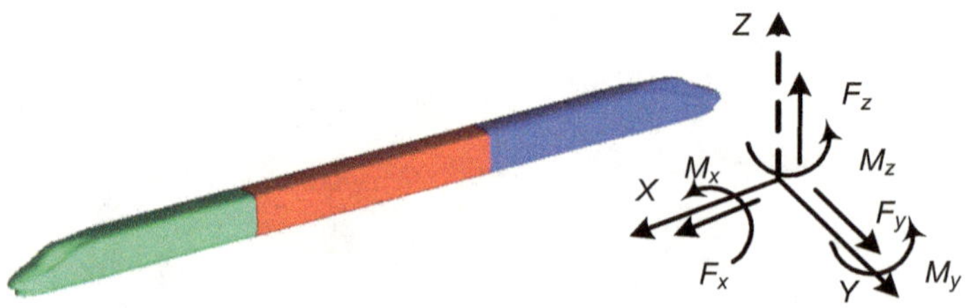


**Fig. 5** Schematic diagram of aerodynamic forces

**Table 1** Effect of aerodynamic forces on mesh sizes

| Mesh sizes (mm) | | | $F_y$ (kN) | | $F_z$ (kN) | |
|---|---|---|---|---|---|---|
| Head | Mid | Tail | Head | Tail | Head | Tail |
| 120 | 150 | 120 | −49.22 | −16.87 | 4.46 | 11.97 |
| 100 | 120 | 100 | −48.72 | −16.06 | 5.42 | 11.30 |
| 80 | 100 | 80 | −47.22 | −15.25 | 6.38 | 10.83 |
| 60 | 80 | 60 | −46.71 | −14.99 | 6.46 | 10.63 |

**Table 2** Effect of aerodynamic forces on domain

| Domain (m) | | | $F_y$ (kN) | | $F_z$ (kN) | |
|---|---|---|---|---|---|---|
| $L$ | $W$ | $H$ | Head | Tail | Head | Tail |
| 75 + 150 | 30 + 40 | 30 | −48.47 | −16.34 | 7.63 | 11.97 |
| 75 + 200 | 30 + 50 | 40 | −47.35 | −15.42 | 6.52 | 11.02 |
| 100 + 200 | 30 + 60 | 60 | −47.22 | −15.25 | 6.38 | 10.83 |

150 m. The aerodynamic forces are almost stable when the length, width, and length of the domain are 375.4, 90, and 60 m, respectively. Therefore, that domain was chosen.

Table 3 shows the effect of aerodynamic forces on turbulent models including standard $k$-$\varepsilon$, re-normalization group (RNG) $k$-$\varepsilon$, realizable $k$-$\varepsilon$, standard $k$-$\omega$, and shear stress transports (SST) $k$-$\omega$ models. It is shown that the differences between standard $k$-$\varepsilon$ model and the other models are small except the standard $k$-$\omega$ model. Therefore, the standard $k$-$\varepsilon$ model was chosen.

The time step sizes in CFD and vehicle-track dynamics are $2.0 \times 10^{-3}$ s and $5.0 \times 10^{-5}$ s, respectively. Calculations were performed using 4 CPUs of X5650 on a DELL 7500 mainframe. Approximately 30 time step iterations per hour were achieved for the current meshes.

The vector of aerodynamic load $\hat{\boldsymbol{F}}_{\mathrm{a}}$ loaded to the vehicle-track model is described as

$$\hat{\boldsymbol{F}}_{\mathrm{a}}(t) = \begin{cases} \boldsymbol{F}_{\mathrm{a}}(t), & t \geq t_0, \\ \frac{t}{t_0}\boldsymbol{F}_{\mathrm{a}}(t), & t < t_0, \end{cases} \tag{7}$$

where $\boldsymbol{F}_{\mathrm{a}}$ is the vector of aerodynamic forces, $t_0$ is a constant, and $t_0 = 4.5$ s.

**Table 3** Effect of aerodynamic forces on turbulence models

| Method | $F_y$ (kN) | | | $F_z$ (kN) | |
|---|---|---|---|---|---|
| | Head | Mid | Tail | Head | Tail |
| Standard $k$-$\varepsilon$ | −47.22 | −8.62 | −15.25 | 6.38 | 10.83 |
| RNG $k$-$\varepsilon$ | −47.02 | −8.65 | −15.14 | 6.32 | 10.91 |
| Realizable $k$-$\varepsilon$ | −47.16 | −8.56 | −15.27 | 6.08 | 10.92 |
| Standard $k$-$\omega$ | −48.20 | −9.65 | −15.77 | 5.79 | 10.54 |
| SST $k$-$\omega$ | −47.22 | −8.61 | −15.37 | 6.09 | 10.92 |

# 5 Numerical Simulation

The calculation was performed for a high-speed train at $u_t$ = 350 km/h and subjected to crosswind $u_w$ = 13.8 m/s.

## *5.1 Aerodynamics and Displacements*

In this section, we describe the aerodynamic characteristics and displacements of the train subjected to crosswind.

### 5.1.1 Head Coach

Figure 6a–j shows a comparison of the responses of the head coach calculated by off-line simulation and co-simulation methods. When the interaction between airflow and a high-speed train is considered, the changes are as follows:

1. The magnitude of the side force increases by approximately 5 kN, which is over 10% greater than that calculated by the off-line simulation method. The magnitude of the lateral displacement increases by approximately 15 mm.
2. The magnitude of the lift force and the vertical displacement increases by approximately 4 kN and 6 mm, respectively. The trends are consistent because the coordinate systems of the vehicle-track coupling dynamics and the aerodynamics are different in the $y$ direction.
3. The magnitude of the roll moment decreases by approximately 3.5 kN m. The variation trend of the roll moment depends on the reference point of the moment. However, the magnitude of the roll displacement increases by approximately 0.3°, which is over 25% greater than that calculated by the off-line simulation method.
4. The magnitude of the pitch moment and the pitch displacement decreases.
5. The magnitude of the yaw moment and the yaw displacement decreases.

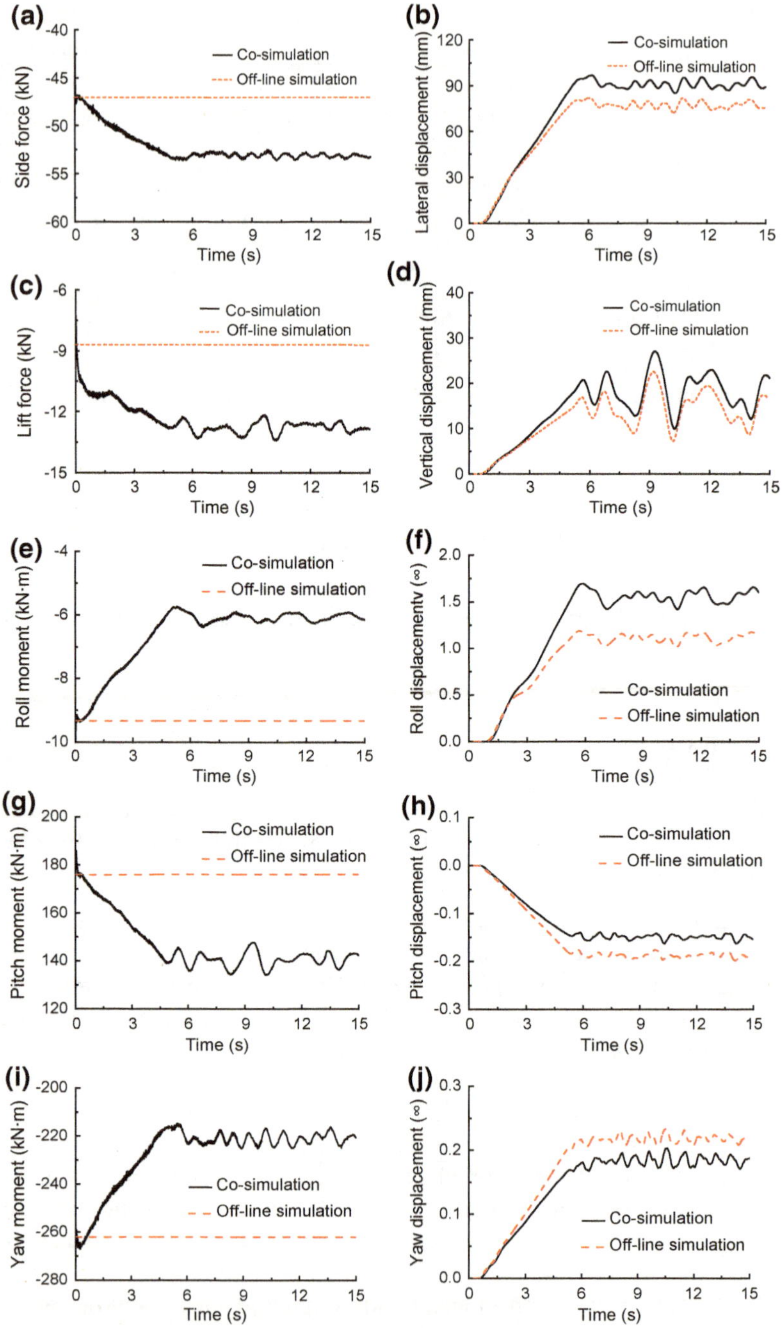


**Fig. 6** Responses of the head coach calculated by off-line simulation and co-simulation methods. **a** Side force; **b** Lateral displacement; **c** Lift force; **d** Vertical displacement; **e** Roll moment; **f** Roll displacement; **g** Pitch moment; **h** Pitch displacement; **i** Yaw moment; **j** Yaw displacement

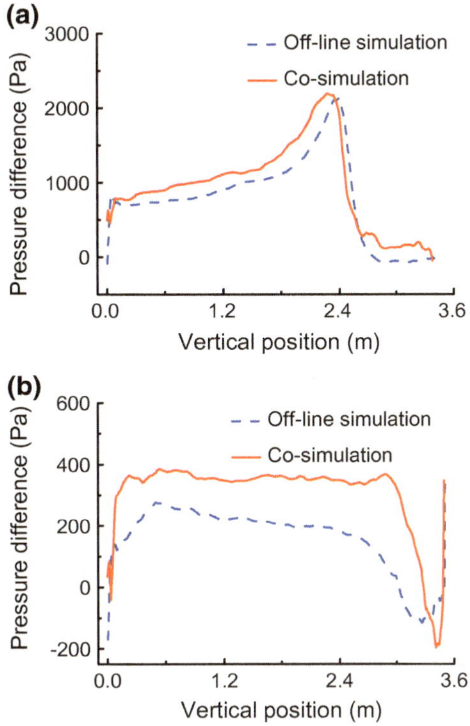


**Fig. 7** Pressure differences along the vertical dimension of the head coach: $d = 5$ m (**a**) and $d = 18$ m (**b**)

Baker et al. (2004) conducted that the magnitude of side and lift forces would increase with the consideration of the rolling motion of the carbody. The trends in the numerical results in this study are consistent with those of Baker et al. (2004)'s experiments.

The magnitude of the side force and the roll moment is closely related to pressure differences between windward and leeward sides of the train. Figure 7a, b shows the pressure differences along the vertical dimension of the head coach, where $d$ indicates the longitudinal distance between the cross section and nose. There are two cross sections: one at $d = 5$ m located in the streamlined nose and the other one at $d = 18$ m located in the non-streamlined region. Pressure differences at almost all vertical positions are higher in the co-simulation as the pressure on the windward side of the head coach increases. Therefore, the magnitude of the side force of the head coach increases by 10%. Consequently, the magnitude of the roll moment of the head coach increases.

The magnitude of the lift force is closely related to pressure differences between the top and bottom of the train. Figure 8a, b shows pressure differences along the lateral dimension of the head coach. The pressure on the bottom of the head coach increases because of the rolling motion of the train body. The integral of pressure differences is greater in the co-simulation. As a result, the magnitude of the lift force increases.

### 5.1.2 Middle Coach

Figure 9a–j shows a comparison of the responses of the middle coach calculated by off-line simulation and co-simulation methods. When the interaction between airflow and a high-speed train is considered, the changes are as follows:

1. The magnitude of the side force and the lateral displacement increases by approximately 2 kN and 5 mm, respectively.
2. The magnitude of the lift force and the vertical displacement changes slightly.
3. The magnitude of the roll moment and the roll displacement increases by approximately 0.5 kN m and 0.15°, respectively.
4. The magnitude of both the yaw and pitch moments increases.

The pressure on two different cross sections of the middle coach is in agreement because of the steady flow field around the middle coach. The pressure on the windward side of the middle coach almost increases and the pressure on the leeward side almost decreases. It is seen that the magnitude of the side force on the middle coach increases. According to Fig. 10a, b, pressure differences along the lateral dimension increase and the magnitude of the lift force on the middle coach decreases to some extent.

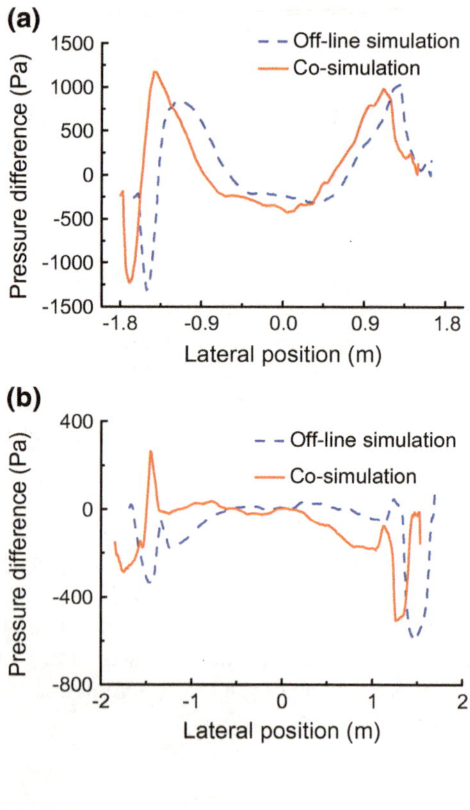


**Fig. 8** Pressure differences along the lateral dimension of the head coach: $d$ = 5 m (**a**) and $d$ = 18 m (**b**)

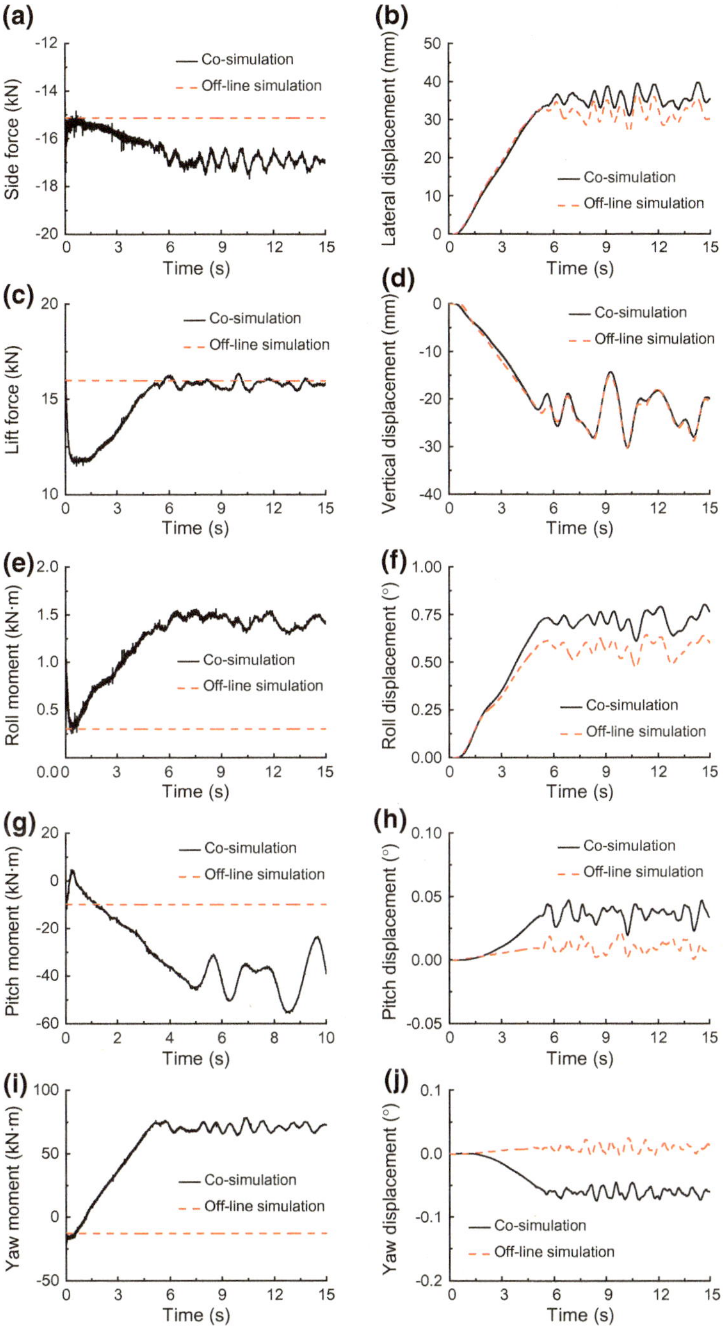


**Fig. 9** Responses of the middle coach calculated by off-line simulation and co-simulation methods. **a** Side force; **b** Lateral displacement; **c** Lift force; **d** Vertical displacement; **e** Roll moment; **f** Roll displacement; **g** Pitch moment; **h** Pitch displacement; **i** Yaw moment; **j** Yaw displacement

### 5.1.3 Tail Coach

Figure 11a–j shows a comparison of the responses of the tail coach calculated by off-line simulation and co-simulation methods. When the interaction between airflow and a high-speed train is considered, the changes are as follows:

1. The magnitude of the side force and the lateral displacement increases by approximately 3 kN and 10 mm, respectively. Besides, the direction of the side force is opposite to those of the head and middle coaches.
2. The magnitude of the lift force increases and the magnitude of the vertical displacement increases to some extent.
3. The magnitude of the roll moment decreases; however, the magnitude of the roll displacement increases by approximately 0.3°.
4. The magnitude of the yaw and pitch moments decreases. Correspondingly, the magnitude of the yaw and pitch displacements decreases.

In the streamlined region, pressure differences between windward and leeward sides are much less than zero. Therefore, the direction of the side force on the tail coach is windward. Moreover, the pressure on the windward side of the tail coach increases. Figure 12a, b shows pressure differences along the vertical dimension of the tail coach. There are two cross sections: one at $d = 65$ m located in the

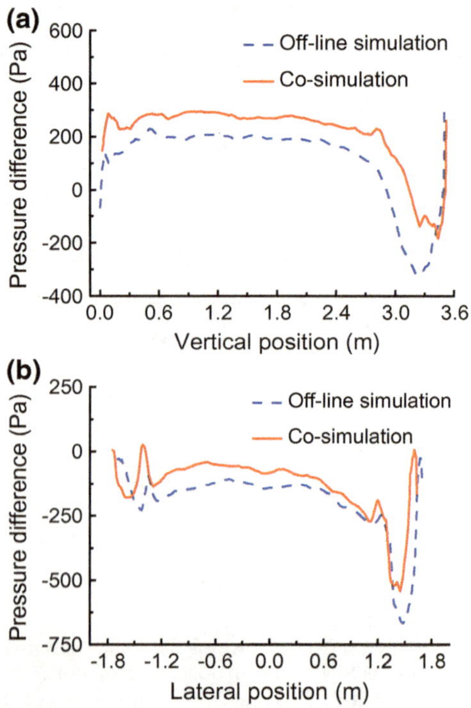


**Fig. 10** Pressure differences along vertical and lateral dimensions of the middle coach: vertical (**a**); lateral (**b**)

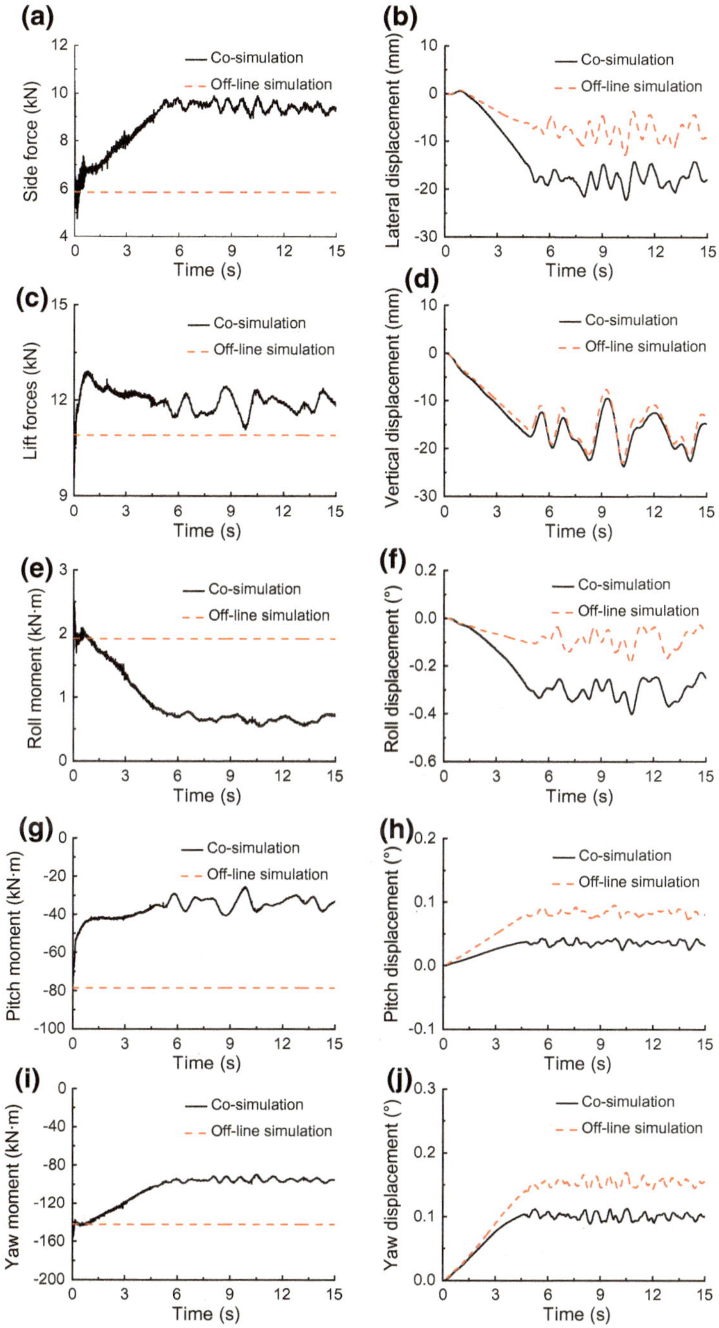


**Fig. 11** Responses of the tail coach calculated by off-line simulation and co-simulation methods. **a** Side force; **b** Lateral displacement; **c** Lift force; **d** Vertical displacement; **e** Roll moment; **f** Roll displacement; **g** Pitch moment; **h** Pitch displacement; **i** Yaw moment; **j** Yaw displacement

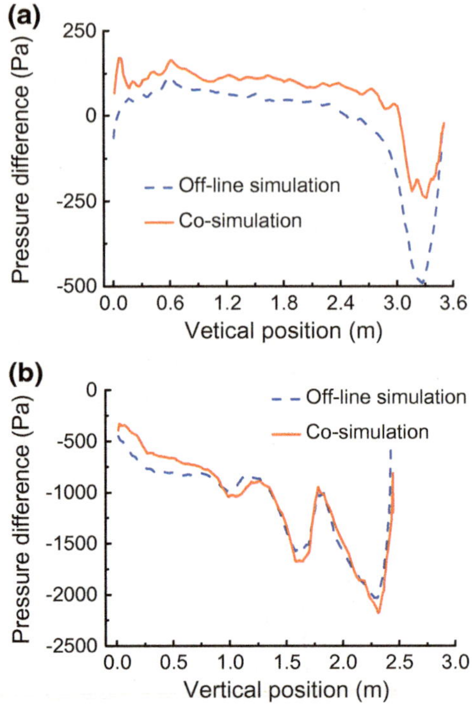


**Fig. 12** Pressure differences along the vertical dimension of the tail coach: $d = 65$ m (**a**) and $d = 72.5$ m (**b**)

non-streamlined region and the other at $d = 72.5$ m located in the streamlined nose. In the non-streamlined region, pressure differences at almost all vertical positions are greater in the co-simulation. Figure 13a, b shows pressure differences along the lateral dimension of the tail coach at different cross sections. In the streamlined region, the pressure differences are smaller in the co-simulation. As a result, the magnitude of the lift force increases.

## 5.2 *Dynamic Performances of Vehicle Track*

In this section, we describe the running safety of the train subjected to crosswind. Table 4 shows comparisons of the displacement calculated by different simulation methods. Displacements of the train include the lateral displacement $y$, the vertical displacement $z$, the roll displacement $\theta$, the yaw displacement $\psi$, and the pitch displacement $\beta$. The values given in the table are the maximum or minimum values of the indexes. The lateral displacements of the head and middle coaches are toward the leeward side, which are opposite to that of the tail coach. Each component of the performance indexes of the head coach is the largest among the three coaches. Besides, the magnitude of displacements increases when the interaction is considered.

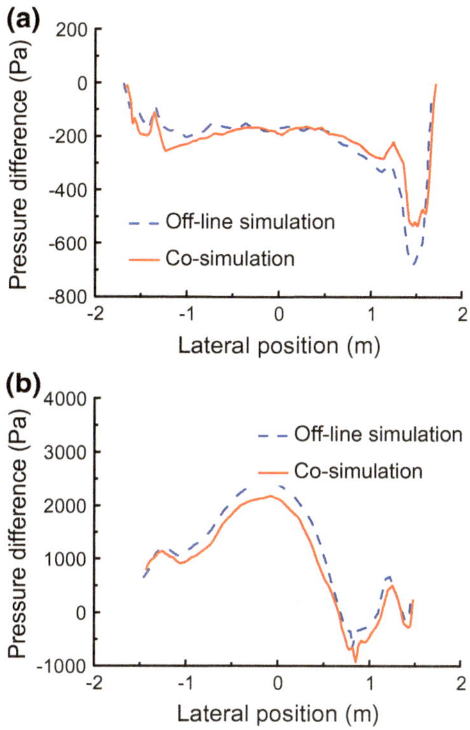


**Fig. 13** Pressure differences along the lateral dimension of the tail coach: $d$ = 65 m (**a**) and $d$ = 72.5 m (**b**)

**Table 4** Comparisons of displacements calculated by different simulation methods

| Train type | Method | $y$ (mm) | $z$ (mm) | $\theta$ (°) | $\beta$ (°) | $\psi$ (°) |
|---|---|---|---|---|---|---|
| Head | Off-line simulation | 82.03 | 22.52 | 1.17 | −0.16 | 0.23 |
| | Co-simulation | 95.33 | 27.14 | 1.65 | −0.19 | 0.20 |
| Middle | Off-line simulation | 36.87 | −30.17 | 0.65 | 0.04 | 0.02 |
| | Co-simulation | 39.53 | −30.27 | 0.79 | 0.02 | −0.07 |
| Tail | Off-line simulation | −12.88 | −7.55 | −0.19 | 0.09 | 0.17 |
| | Co-simulation | −22.18 | −9.48 | −0.40 | 0.04 | 0.11 |

The most important indexes of vehicle-track dynamics are safety indexes, including the wheel/rail vertical force, the lateral wheelset force, the derailment coefficient, and the wheel unloading rate. Table 5 shows comparisons of safety indexes calculated by different simulation methods. Among the three coaches, the head coach shows the largest increments in the safety indexes. The magnitude of the wheel unloading rate increases. The running safety of the train becomes worse with the consideration of the rolling motion of the carbody. Besides, the magnitude of the wheel set unloading rate and the derailment coefficient of the head coach increase by approximately 0.14 and 0.02, respectively.

**Table 5** Comparisons of safety indexes calculated by different simulation methods

| Train type | Method | Wheel/rail vertical force (kN) | Lateral wheelset force (kN) | Derailment | Wheel unloading |
|---|---|---|---|---|---|
| Head | Off-line simulation | 92.80 | 28.69 | 0.32 | 0.65 |
| | Co-simulation | 96.40 | 32.84 | 0.34 | 0.79 |
| Middle | Off-line simulation | 77.51 | 13.32 | 0.19 | 0.44 |
| | Co-simulation | 85.32 | 15.20 | 0.20 | 0.45 |
| Tail | Off-line simulation | 73.95 | 12.30 | 0.17 | 0.39 |
| | Co-simulation | 81.92 | 13.89 | 0.19 | 0.40 |

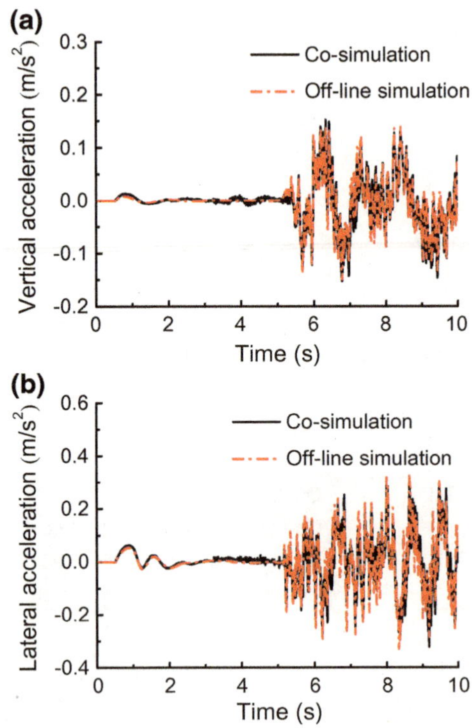


**Fig. 14** Vertical (**a**) and lateral (**b**) accelerations of the head coach

Figure 14 shows a comparison of the accelerations of the head coach calculated by different simulation methods. There is a slight difference in accelerations of the head coach.

## 6 Conclusions

A numerical approach to the interaction between airflow and a high-speed train is presented in this paper.

1. The interaction between airflow and a high-speed train significantly affects displacements and aerodynamic forces of the train subjected to crosswind.
2. The running safety of the train subjected to crosswind deteriorates when the interaction between airflow and a high-speed train is considered; however, there is a slight difference in accelerations of the head coach.
3. Among the three coaches, each component of the dynamic performance indexes of the head coach is the largest, and the head coach shows the largest increments in the safety indexes.
4. It is necessary to consider the interaction between airflow and a high-speed train subjected to crosswind.

## References

Baker, C. J. (1991). Ground vehicles in high cross winds part III: The interaction of aerodynamic forces and the vehicle system. *Journal of Fluids and Structures, 5*(2), 221–241. doi:10.1016/0889-9746(91)90478-8.

Baker, C. J. (2010). The simulation of unsteady aerodynamic crosswind forces on trains. *Journal of Wind Engineering and Industrial Aerodynamics, 98*(2), 88–99. doi:10.1016/j.jweia.2009.09.006.

Baker, C. J., Jones, J., Lopez-Calleja, F., & Munday, J. (2004). Measurements of the cross wind forces on trains. *Journal of Wind Engineering and Industrial Aerodynamics, 92*(7–8), 547–563. doi:10.1016/j.jweia.2004.03.002.

Baker, C. J., Hemida, H., Iwnicki, S., Xie, G., & Ongaro, D. (2011). Integration of crosswind forces into train dynamic modelling. *Proceedings of the Institution of Mechanical Engineers, Part F: Journal of Rail and Rapid Transit, 225*(2), 154–164. doi:10.1177/2041301710392476.

Bocciolone, M., Cheli, F., Corradi, R., Muggiasca, S., & Tomasini, G. (2008). Crosswind action on rail vehicles: Wind tunnel experimental analyses. *Journal of Wind Engineering and Industrial Aerodynamics, 96*(5), 584–610. doi:10.1016/j.jweia.2008.02.030.

Cheli, F., Ripamonti, F., Rocchi, D., & Tomasini, G. (2010). Aerodynamic behaviour investigation of the new EMUV250 train to cross wind. *Journal of Wind Engineering and Industrial Aerodynamics, 98*(4–5), 189–201. doi:10.1016/j.jweia.2009.10.015.

Chiu, T. W. (1995). Prediction of the aerodynamic loads on a railway train in a cross-wind at large yaw angles using an integrated two- and three-dimensional source/vortex panel method. *Journal of Wind Engineering and Industrial Aerodynamics, 57*(1), 19–39. doi:10.1016/0167-6105(94)00099-Y.

Diedrichs, B. (2003). On computational fluid dynamics modeling of crosswind effects for high-speed rolling stock. *Proceedings of the Institution of Mechanical Engineers, Part F: Journal of Rail and Rapid Transit, 217*(3), 203–226. doi:10.1243/095440903769012902.

Diedrichs, B., Sima, M., Orellano, A., & Tengstrand, H. (2007). Crosswind stability of a high-speed train on a high embankment. *Proceedings of the Institution of Mechanical Engineers, Part F: Journal of Rail and Rapid Transit, 221*(2), 205–225. doi:10.1243/0954409JRRT126.

Ding, Y., Sterling, M., & Baker, C. J. (2008). An alternative approach to modeling train stability in high cross winds. *Proceedings of the Institution of Mechanical Engineers, Part F: Journal of Rail and Rapid Transit, 222*(1), 85–97. doi:10.1243/09544097JRRT138.

Khier, W., Breuer, M., & Durst, F. (2000). Flow structure around trains under side wind conditions: A numerical study. *Computers & Fluids, 29*(2), 179–195. doi:10.1016/S0045-7930(99)00008-0.

Li, T., Zhang, J. Y., & Zhang, W. H. (2011a). Performance of vehicle-track coupling dynamics under crosswinds. *Journal of Traffic and Transportation Engineering, 9*, 55–60. (in Chinese).

Li, T., Zhang, J. Y., & Zhang, W. H. (2011b). Nonlinear characteristics of vortex-induced vibration at low Reynolds number. *Communications in Nonlinear Science and Numerical Simulation, 16*(7), 2753–2771. doi:10.1016/j.cnsns.2010.10.014.

Orellano, A., & Schober, M. (2006). Aerodynamic performance of a typical high-speed train. In *Proceedings of the 4th WSEAS International Conference on Fluid Mechanics and Aerodynamics* (pp. 18–25). Elounda, Greece.

Shao, X. M., Wan, J., Chen, D. W., & Xiong, H. B. (2011). Aerodynamic modeling and stability analysis of a high-speed train under strong rain and crosswind conditions. *Journal of Zhejiang University-Science A (Applied Physics & Engineering), 12*(12), 964–970. doi:10.1631/jzus.A11GT001.

Suzuki, M., Tanemoto, K., & Maeda, T. (2003). Aerodynamic characteristics of train/vehicles under cross winds. *Journal of Wind Engineering and Industrial Aerodynamics, 91*(1–2), 209–218. doi:10.1016/S0167-6105(02)00346-X.

Thomas, D., Diedrichs, B., Berg, M., & Stichel, S. (2010). Dynamics of a high-speed rail vehicle negotiating curves at unsteady crosswind. *Proceedings of the Institution of Mechanical Engineers, Part F: Journal of Rail and Rapid Transit, 224*(6), 567–579. doi:10.1243/09544097JRRT335.

Xu, Y. L., & Ding, Q. S. (2006). Interaction of railway vehicles with track in cross-winds. *Journal of Fluids and Structures, 22*(3), 295–314. doi:10.1016/j.jfluidstructs.2005.11.003.

Zhai, W. M., Cai, C. B., & Guo, S. Z. (1996). Coupling model of vertical and lateral vehicle/track interactions. *Vehicle System Dynamics, 26*(1), 61–79. doi:10.1080/00423119608969302.

## Author Biographies

**Tian Li** holds Ph.D. and is a Research Associate in the State Key Laboratory of Traction Power, Southwest Jiaotong University, Chengdu. He received his Doctor of Philosophy in vehicle operation engineering from Southwest Jiaotong University in 2012. He has been active in computational fluid dynamics, fluid–structure interaction, and high-speed train aerodynamics.

**Ji-ye Zhang** Ph.D. is currently a Professor at State Key Laboratory of Traction Power, Southwest Jiaotong University, Chengdu, China. He is the principal author/co-author of over 140 journal and conference papers, as well as a book. His research interests include dynamics and control of high-speed train, stability and control of complex system, and hybrid power vehicular and transportation systems.

**Wei-hua Zhang** holds the post of Executive Deputy Director of the National Laboratory for Rail Transit (currently in the preparatory stage) and is a member of the National Teaching Advisory Board for Higher Education Institutions for Transportation Engineering and Engineering Disciplines. He is honored as Distinguished Professor of the "Cheung Kong Scholars" Program of China and is winner of the "National Science Funds for Distinguished Young Scholars." He has been awarded the first prize twice (both ranking the second) and the second prize twice (both ranking the first) for the National Scientific and Technological Progress Award. His paper "A Study on Dynamic Behaviour of Pantographs by Using Hybrid Simulation Method" was acclaimed as the Best Paper Award of the Year of 2005 by the Institution of Mechanical Engineers (IMechE), making him the first Chinese winner of the Thomas Hawksley Gold Medal.

# Multi-objective Optimization Design Method of the High-Speed Train Head

**Meng-ge Yu, Ji-ye Zhang and Wei-hua Zhang**

## Contents



## 1 Introduction

With a number of technical advantages of its fast speed, heavy transport capacity, low energy consumption, and slight pollution, the high-speed railway has become a common trend of the development of world railway transport. The high-speed train, which is the core of modern high-speed railway, has overcome a series of technical difficulties and is developing rapidly. With the increase of the train speed, the dynamic environment of the train turns out to be aerodynamic domination. The aerodynamic problem is becoming the key technology of the high-speed train (Schetz 2001; Raghunathan et al. 2002; Shao et al. 2011; Li et al. 2013). The aerodynamic drag is proportional to the square of the train speed. The proportion of


M. Yu (✉) · J. Zhang · W. Zhang
State Key Laboratory of Traction Power, Southwest Jiaotong University, Chengdu 610031, China
e-mail: yumengge0627@163.com





Y. Fang and Y.H. Zhang (eds.), *China's High-Speed Rail Technology*,
Advances in High-speed Rail Technology, https://doi.org/10.1007/978-981-10-5610-9_10

aerodynamic drag in the total resistance is small when the train speed is low. However, the aerodynamic drag could take a much greater proportion of the total resistance at a higher train speed, e.g., when the train speed reaches 250–300 km/h, the aerodynamic drag could take 75% of the total resistance (Brockie and Baker 1990). Thus, the aerodynamic drag has become one of the key factors to restrain the further increase of the train speed and energy conservation. As a result, the reduction of the aerodynamic drag is of great importance to the design of the high-speed train head. However, the reduction of the aerodynamic drag may increase other aerodynamic forces (moments), possibly deteriorating the operational safety of the high-speed train. For example, the upward lift would reduce the wheel–rail contact, which will easily lead to the train derailment due to the excessive upward lift. The effect of aerodynamic forces (moments) on the operational safety of the high-speed train can be described through the operational safety indicators (such as the load reduction factor). Thus, to reduce the aerodynamic drag and meanwhile to improve the operational safety of the high-speed train has become one of the key issues in the optimization design of the high-speed train head.

Currently, the main design methods of the high-speed train head are wind tunnel tests and numerical simulation. The general design idea of the high-speed train head design is as follows: the first step is to map out various head types, the next step is to compare and pick out the best head type through wind tunnel tests or numerical simulation, and the last step is to improve the design according to the operational conditions. Maeda et al. (1989) gave some suggestion for the purpose of aerodynamic drag reduction based on the aerodynamic drag comparison of 0 series, 100 series and 200 series on Shinkansen, Japan. Kikuchi et al. (2001) studied nine kinds of train heads (the combination of three types of nose section configuration and three different nose lengths) using the 3D boundary element method, and found out that the nose section configuration resembling a wedge could effectively reduce the air pressure pulse due to train passage. Hemida and Krajnović (2010) analyzed the effect of nose length on the flow field and aerodynamic force of the high-speed train. The calculation results showed that the flow structure and aerodynamic force of the high-speed train with a long nose were much different from those with a short nose. The short nose represented more transient and 3D characteristics. Essentially, the methods adopted mentioned above belong to the optimum seeking method which is heavily dependent on engineering experience, and only the relationships between a single optimization design variable and optimization objectives are obtained. As a result, the final selected head may not be the optimal one.

To get the global optimal head shape, the direct optimization method should be adopted. The direct optimization design means using mathematical methods to seek for the minimum or maximum (such as the minimum of the aerodynamic drag or the minimum of the load reduction factor) of some design goals while at the same time satisfying certain constraint conditions. Therefore, the optimization design problem of the high-speed train head can be transformed into a multi-objective constrained optimization problem. Optimization design variables are extracted from the parametric modeling of the high-speed train, which can be automatically updated through the multi-objective optimization algorithm. Optimization objectives can be obtained by the calculation of aerodynamics and vehicle system dynamics of the high-speed train. Currently, very few studies on multi-objective optimization design of train head can be found. Kwon et al. (2001) studied the influence of the nose shape on the intensity of the pressure gradient of the compression wave at the tunnel entrance, where the response surface method was used as a basis for the optimization of nose shape of high-speed trains. The analytical results showed that the front 20% part of the train nose played the most important role in the minimization of the maximum pressure gradient. Lee and Kim (2008) developed a proper approximate metamodel to deal with the nose shape design of the high-speed train so as to minimize the maximum micro-pressure wave and suggested an optimal nose shape that was an improvement over the current design in terms of micro-pressure wave. Sun et al. (2010) combined genetic algorithms and arbitrary shape deformation techniques to optimize the head shape of the China Railways High-speed3 (CRH3). Ku et al. (2010) used the Broyden-Fletcher-Goldfarb-Shanno (BFGS) algorithm and response surface model to minimize the micro-pressure wave. The cross-sectional area distribution of high-speed trains with different nose lengths was selected as an optimization design variable to conduct the single-objective optimization design. Ikeda et al. (2006) and Suzuki et al. (2008) used B-spline curve to set up a parametric model of cross-sectional panhead and optimized the shape of the cross-sectional contour of the panhead. Yao et al. (2012) adopted a new parametric approach called local shape function based on the free form surface deformation and a new optimization method based on the response surface method of genetic algorithm-general regression neural network (GA-GRNN). After optimization, the aerodynamic drag for a three carriage train was reduced by 8.7%.

In the present paper, a multi-objective optimization design process of the high-speed train head is proposed to carry out the automatic optimization design of the head shape, with the optimization objectives of aerodynamic drag and load reduction factor. This optimization design process mainly involves the following aspects: (1) 3D parametric model design; (2) the aerodynamic mesh generation and the aerodynamic calculation of the high-speed train; (3) the calculation of vehicle system dynamics; and (4) the multi-objective optimization algorithm. In the optimization process, the 3D parametric model of the high-speed train is established using CATIA, with which the train head can be generated and deformed automatically. The aerodynamic mesh is divided automatically by ICEM. FLUENT and SIMPACK are used for the automatic numerical calculation of aerodynamics and vehicle system dynamics of the high-speed train, respectively. The improved non-dominated sorting genetic algorithm II (NSGA-II) is used for the automatic optimization design of the high-speed train head.

## 2 Basic Concepts and Optimization Process

### *2.1 Basic Concepts of Multi-objective Optimization*

To get a clear understanding of the multi-objective optimization, a brief introduction of some basic concepts of multi-objective optimization is provided (Aguilar Madeira et al. 2005).

Multiple objectives are made to reach the optimization at the same time, which is known as the multi-objective optimization problem, and the mathematical expressions are

$$
\begin{array}{lll}
 & \min f_m(\boldsymbol{x}), & m = 1, 2, \ldots, M, \\
\text{s.t.} & g_j(\boldsymbol{x}) \le 0, & j = 1, 2, \ldots, J, \\
 & h_k(\boldsymbol{x}) = 0, & k = 1, 2, \ldots, K, \\
 & x_i^{\mathrm{L}} \le x_i \le x_i^{\mathrm{U}}, & i = 1, 2, \ldots, N,
\end{array}
\tag{1}
$$

where $M$ is the total number of the objective functions, $N$ is the total number of design variables, $J$ is the total number of inequality constraints, $K$ is the total number of equality constraints, $x_i$ is the design variable, $x_i^{\mathrm{L}}$ is the lower bound of $x_i$, $x_i^{\mathrm{U}}$ is the upper bound of $x_i$, $f_m(\boldsymbol{x})$ is the $m$th objective function, $g_j(\boldsymbol{x})$ is the $j$th inequality constraint, and $h_k(\boldsymbol{x})$ is the $k$th equality constraint.

In most cases, the objectives are contradictory to each other, and it is not possible for several objectives to achieve the optimal solution at the same time. Otherwise, it does not belong to the category of the multi-objective optimization. The ultimate goal of solving the multi-objective problem is to coordinate the compromises and trade-off between various objectives so that each of the objectives reaches the optimization as far as possible.

The French economist V. Pareto was the first person to study the multi-objective optimization problem within the field of economics, and proposed the concept of the Pareto-optimal set.

Suppose $\boldsymbol{x} \in \boldsymbol{X}$ ($\boldsymbol{X}$ is the feasible region for the design variables), if and only if there is no $\boldsymbol{x}' \in \boldsymbol{X}$ so that $f_m(\boldsymbol{x}') \le f_m(\boldsymbol{x})$, $m = 1, 2, \ldots, M$, and at least one strict inequality holds, then $\boldsymbol{x}$ is a Pareto-optimal solution of the multi-objective optimization.

A collection of all Pareto-optimal solutions is called Pareto-optimal set. The Pareto-optimal set in the objective function space is called Pareto-optimal front.

To solve the multi-objective optimization problem is to find the Pareto-optimal set. Then a compromise solution needs to be made by the decision-makers in accordance with the relevant information and requirements.

## 2.2 *Multi-objective Optimization Process*

The multi-objective optimization design of the high-speed train head is the core technology of the high-speed train design. For some original head shape, users are often required to improve the aerodynamic performance (drag coefficient, lift coefficient, etc.), to reduce the energy consumption and improve the operational safety of the high-speed train. The automatic optimization design of the high-speed train head is of great significance. The multi-objective optimization design of the high-speed train head mainly involves the following aspects: the 3D parametric model design of the high-speed train, the aerodynamic calculation of the high-speed train (including mesh generation), the vehicle system dynamic calculation of the high-speed train, multi-objective optimization algorithms, system integration framework, and so on. The multi-objective optimization design process of the high-speed train head is shown in Fig. 1. The commercial 3D geometric modeling software or parametric design program can be used to set up the 3D parametric model. The commercial computational fluid dynamics software or self-programming can be used for the aerodynamic calculation of the high-speed train. The commercial multi-body system dynamics software or self-programming can be used for the vehicle system dynamic calculation of the high-speed train. Genetic algorithms or neural networks can be used for the optimization design of the high-speed train head. The commercial integration framework or batch program can be used for system integration.

# 3 3D Parametric Model of the Train

The 3D parametric model of the high-speed train is established by CATIA. To achieve the automatic deformation of the head shape, the following three tasks need to be done successively:

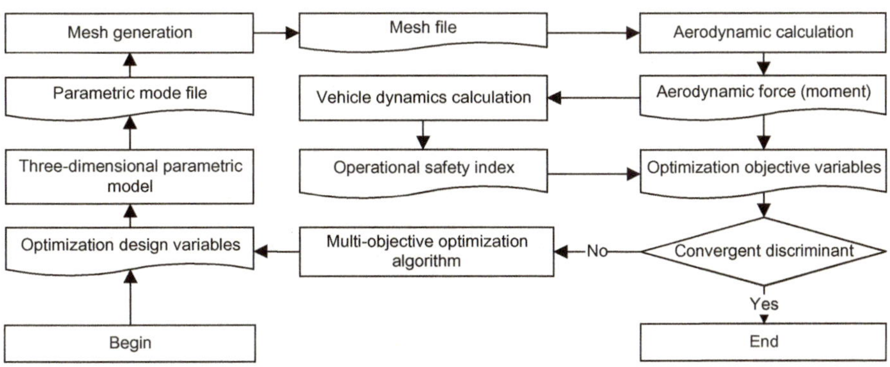


**Fig. 1** Overall design flow for optimization

1. Establish the entity model of the left half of a train head;
2. Parameterize the left half of the train head using the script file of CATIA;
3. Modify the parameter values in the script file of CATIA using a MATLAB program, and perform the deformation of the high-speed train head by running the script file.

## 3.1 Entity Model of the Left Half of the Train Head

As the head shape of the high-speed train has a good symmetry, only the left half (or right half) portion of the train head needs to be modeled. The head shape of the high-speed train is quite complex, which cannot be described by simple analytic surfaces, but can be described by continuous splicing of some sub-surfaces. In this study, a number of B-spline surfaces are used to approximate the outer surface of the left half portion of the train head. B-spline surfaces are constituted by a series of B-spline curves, which are generated by a series of control points on the surface of the train head.

According to the head shape of a high-speed train, 162 control points are set up on the surface of the train head, which are used to build 12 B-spline curves. Then, the 12 B-spline curves can be used to build seven B-spline surfaces. After that, the left half of the train head is established, as shown in Fig. 2. To facilitate the later analysis, the B-spline curves are numbered C1–C12, respectively.

**Fig. 2** Left-half model of the train head

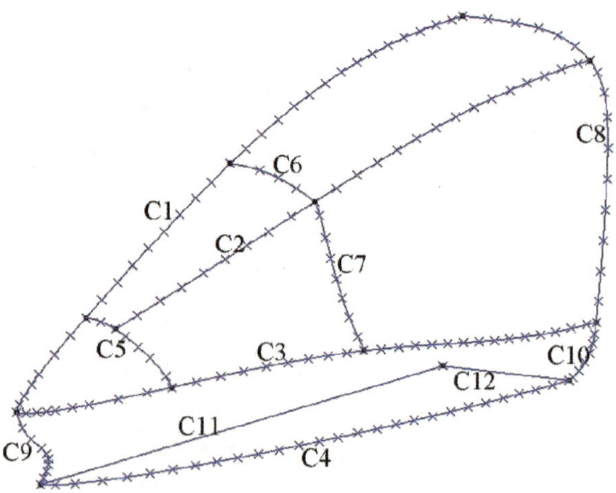

## 3.2 Parametric Model of the High-Speed Train

Based on the entity model of the left half of the train head, a parametric model of the left half of the train head is established by the script file of CATIA. The coordinates of 162 control points of the left half portion can be recorded automatically to the script file by CATIA, and then the deformation of the head shape can be achieved by modifying the coordinates of 162 control points.

Based on the parametric model of the train head's left half, the parametric model of a high-speed train with three carriages can be built through translation, symmetry, and so on, which can be recorded to the script file of CATIA.

## 3.3 Optimization Design Variables

To optimize the head shape of the high-speed train, five optimization design variables are selected, which correspond to the longitudinal symmetry line C1, the maximum horizontal contour line C3, the bottom horizontal contour line C4, the central auxiliary control line C7, and the nose height, respectively. With the increase of the streamlined length, the aerodynamic performance of the train will significantly be improved. Therefore, on the basis of a fixed streamlined length, the external shape of the train head is optimized to improve the aerodynamic performance and vehicle dynamic performance of the high-speed train.

The deformation of C1 is carried out by changing the vertical coordinates of the control points of C1. The vertical coordinate of the midpoint of C1 is varying with $dz_1$, while the vertical coordinates of both ends of C1 remain unchanged, i.e., the variation is 0. As to the points between the midpoint and the two end points, the variation of the vertical coordinates is in accordance with the linear law. Figure 3 shows the deformation of C1, where the original form represents the initial form of C1, the upward movement means that $dz_1$ is positive, and the downward movement means that $dz_1$ is negative.

The deformation of C3 is carried out by changing the horizontal coordinates of the control points of C3. The horizontal coordinate of the midpoint of C3 is varying by $dy_3$, while the horizontal coordinates of both two ends of C3 remain unchanged, i.e., the variation is 0. As to the points between the midpoint and the two end points, the variation of the horizontal coordinates is in accordance with the linear law. Figure 4 shows the deformation of C3, where the original form represents the initial form of the C3, the inward movement (i.e., close to the longitudinal symmetry plane) means that $dy_3$ is positive, and the outward movement (i.e., away from the longitudinal symmetry plane) means that $dy_3$ is negative.

The deformation of C4 is similar to that of C3, and the optimization design variable is $dy_4$, which will not be described in detail here.

The deformation of C7 is carried out to adjust concavity and convexity of the curve; therefore, the two ends of C7 need to be fixed and the deformation is

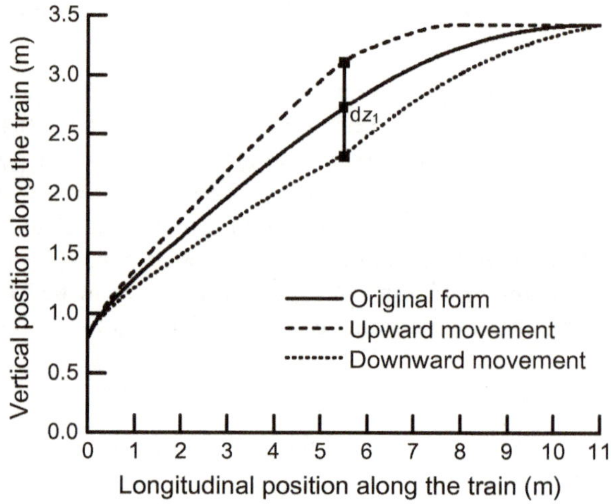


**Fig. 3** Deformation of the longitudinal symmetry line

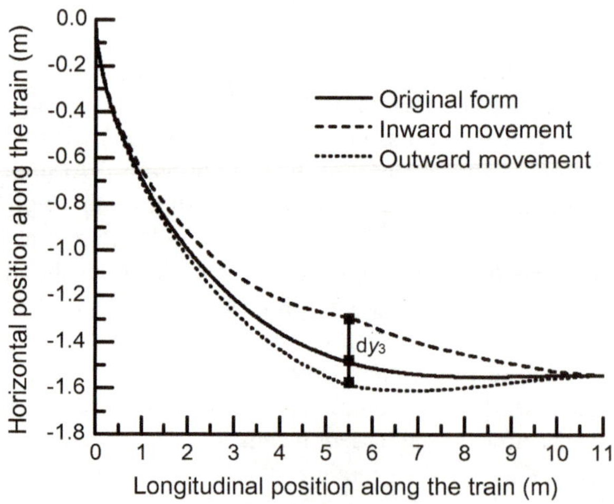


**Fig. 4** Deformation of the horizontal maximum control line

becoming greater from the two end points to the midpoint. The following equation is adopted for the deformation:

$$y_{7,\text{new}}(i) = y_{7,\text{old}}(i) \times \left(1 + \frac{\mathrm{d}y_7(i-1)(n_7-i)}{(i-1)(i-1)+(n_7-i)(n_7-i)}\right), \tag{2}$$

where $n_7$ is the number of the control points of C7, $y_{7,\text{old}}(i)$ is the value of the horizontal coordinate before the deformation, $y_{7,\text{new}}(i)$ is the value of the horizontal coordinate after the deformation, and $i$ is the number of the control point of C7.

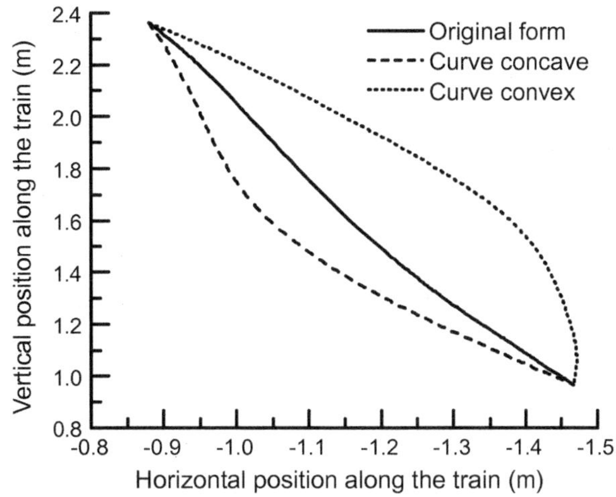


**Fig. 5** Deformation of the central auxiliary control line C7

Figure 5 shows the deformation of C7. The curve is convex when $\mathrm{d}y_7 > 0$ and concave when $\mathrm{d}y_7 < 0$.

As to the variation of the nose height, the vertical coordinates of the control points of C9 need to be multiplied by a coefficient nscale. When nscale is greater than 1, the nose height becomes larger, and when nscale is less than 1, the nose height becomes smaller.

Note that when these curves mentioned above perform deformation, the relevant curves also need to be changed to ensure that the surface of the train head is continuous and smooth.

When optimization design variables are determined, the coordinates of the 162 control points in the script file of CATIA are modified by the MATLAB program, and then a new head shape of the high-speed train can be produced by running the script file.

# 4 Aerodynamic Model

As a high-speed train running on the open track, the operating speed is generally not more than 400 km/h. The impact of the air density on the flow can be ignored without taking into account the trains passing each other or going through a tunnel. Therefore, the incompressible steady flow is adopted to simulate the flow field around the train, and the standard *k*-ε turbulence model is adopted, then the control equation of which can be expressed as follows (Versteeg and Malalasekera 2007):

$$\mathrm{div}(\rho \boldsymbol{u} \boldsymbol{\varphi}) = \mathrm{div}(\Gamma \mathrm{grad} \boldsymbol{\varphi}) + \boldsymbol{S}, \tag{3}$$

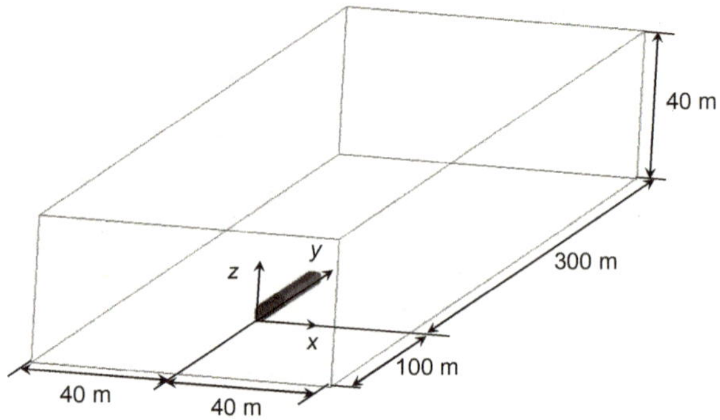


**Fig. 6** Flow-field computational region

where $\rho$ is air density, $\boldsymbol{u}$ is velocity vector, $\boldsymbol{\varphi}$ is the flow flux, $\boldsymbol{S}$ is the source item, and $\boldsymbol{\Gamma}$ is the diffusion coefficient.

The train model for computing the flow field around the train is the 3D parametric model established in Sect. 3. The flow-field computational domain is shown in Fig. 6. The left of the computational domain is set as the velocity inlet boundary, the right as the pressure outlet boundary, the two sides and the top as the symmetric boundary, and the train surface as the stationary wall boundary condition. The ground is set as the slip wall boundary condition, and the slip velocity as the train speed in order to simulate the ground effect. The triangle mesh is generated on the train surface, and the tetrahedral mesh is used for spatial meshes.

To perform the automatic optimization design of the high-speed train head, the mesh generation and the aerodynamics calculation of the high-speed train should be executed automatically. The script files of ICEM and FLUENT are used for the automatic mesh generation and aerodynamics calculation, respectively. The script files can be performed by the batch command.

## 5 Vehicle System Dynamics Model

The vehicle system dynamics mainly includes vehicle dynamics and wheel–rail contact. It is assumed that the carbody, bogie, and wheelsets are rigid, and their elastic deformation can be neglected. The equation of the vehicle system dynamics is

$$\boldsymbol{M}\ddot{\boldsymbol{X}} + \boldsymbol{C}\dot{\boldsymbol{X}} + \boldsymbol{K}\boldsymbol{X} = \boldsymbol{F}, \tag{4}$$

where $\boldsymbol{M}$, $\boldsymbol{C}$, and $\boldsymbol{K}$ are the mass, damping, and stiffness matrices of the train system, respectively. $\boldsymbol{X}$ is the generalized displacement vector of the system, $\dot{\boldsymbol{X}}$ is the generalized velocity vector of the system, $\ddot{\boldsymbol{X}}$ is the generalized acceleration vector of the system, and $\boldsymbol{F}$ is the generalized load vector of the system, including the track excitation and aerodynamic loads.

A multi-body system dynamics model of the high-speed train is constructed by SIMPACK. The multi-body model of a single car is composed of one carbody, two bogies, four wheelsets, and eight tumblers. The rigid carbody, bogie, and wheelsets have six degrees of freedom each, while the tumbler has one. The dynamics model of a single car has 50 degrees of freedom. The train dynamics model with three carriages of "trailer car-motor car-trailer car" is built in this study, which has a total of 150 degrees of freedom. As to the trailer car and motor car, the degrees of freedom, the connection and constraints of the various components, the structure and most of the suspension parameters are exactly the same. The difference only exists in some local parameters such as the body mass, center of gravity height, and body rotational inertia. The wheel rail contact is in general the core of a railway model. LMA tread and T60 rail are used in this model. Track irregularities complicate the evaluation of the wheel unloading. Here, we adopt the measured track spectrum of a high-speed railway in China as the track irregularity.

The aerodynamic loads are dealt with as the external loads on the multi-body system dynamics model to analyze the operational safety of the high-speed train. Due to the translation and equivalent of the force, the pressure distribution can be simplified to a given point to obtain the concentrate forces and moments. The batch command can be used to call the file of SIMPACK named profile.ksh to realize the automatic calculation of vehicle system dynamics and the automatic output of the calculation results.

## 6 Multi-objective Optimization Algorithm

Currently, there are two major methods to solve the multi-objective optimization problems: the normalization approach and the non-normalized approach. The normalization approach transforms multiple objectives into a single objective so that the single-objective optimization methods can be used directly. When taking different weights, different solution sets can be computed to approximate the Pareto-optimal set. The normalization approach which has a poor computational efficiency is quite sensitive to the shape of the Pareto-optimal front. The non-normalization approach deals with the multi-objective optimization problems directly using the Pareto mechanism. The multiple objectives do not need to be converted to a single objective, and the shortcomings of the normalization approach are thus overcome. The non-normalization approach enables the forefront of the solution set to reach the Pareto front as close as possible and tries to evenly cover the Pareto front. There are two major classes of the non-normalization algorithms, which are evolutionary multi-objective optimization (EMO) and direct search method (DSM) algorithms (Custódio et al. 2011; 2012; Zhou et al. 2011). Some commonalities exist in the design of DSM and EMO algorithms, such as searching in the neighborhood of existing solutions in order to find improvement, Pareto non-dominance, diversity maintenance strategies, and so on. However, there are also remarkable differences between DSM and EMO algorithms. DSM algorithms are deterministic, which can present a well-established convergence analysis. EMO algorithms are randomized, and the convergence will be of probabilistic nature, also addressing global optimums (Custódio et al. 2012).

In this study, the algorithm NSGA-II is adopted, which is a widely used EMO method. NSGA-II proposes a fast non-dominated sorting approach with an elitist strategy and replaces the sharing function approach with a crowded-comparison approach, which does not require any user-defined parameter for maintaining diversity among population members. The main loop of NSGA-II is described as follows (Deb et al. 2002):

1. A random parent population $P_0$ of size $N$ is created. The population is sorted based on the non-domination. Each solution is assigned a fitness (or rank) equal to its non-domination level. Thus, minimization of fitness is assumed. Then, the usual binary tournament selection, recombination, and mutation operators are used to create an offspring population $Q_0$ of size $N$. Let $t = 0$;
2. At the $t$th iteration, the combination of the random parent population $P_t$ and the offspring population $Q_t$ is defined as the combined population $R_t$, viz. $R_t = P_t \cup Q_t$, and the size of $R_t$ is 2 $N$. Then the population $R_t$ is sorted according to non-domination to get non-dominated front $F_1$, $F_2$, …;
3. Sort all $F_i$ based on the crowded-comparison operator in descending order, and select the best $N$ solutions to form the new population $P_{t+1}$;
4. The new population $P_{t+1}$ of size $N$ is used for selection, crossover, and mutation to create a new population $Q_{t+1}$ of size $N$;
5. If the termination condition is true, the procedure ends. Otherwise, $t = t+1$, and then turn to step 2.

## 7 Numerical Simulation

There are 12 initial sample points in the design and 25 generations used in the optimization so that 300 designs of the head shape optimization are obtained after the optimization.

The histories of optimization design variables and optimization objectives for all the designs are presented in Fig. 7. Figure 7a shows the history of $dz_1$ for all the designs, and Fig. 7b illustrates the history of aerodynamic drag $F_d$ for all the designs. We have the dot notation to present the Pareto-optimal solutions in the optimization process. As shown in Fig. 7, through repeated iterative calculation, the optimization design variables and optimization objectives tend to converge along with the optimization process, and the Pareto-optimal set and Pareto-optimal front are obtained.

Figure 8 shows the correlation coefficients between optimization objectives and optimization design variables. As shown in Fig. 8a, a positive correlation between the variables $dz_1$, nscale, and the objective $F_d$ within a certain range can be found, which means that a more concave longitudinal symmetry line or a shorter nose height would lead to a lower aerodynamic drag. A negative correlation between the

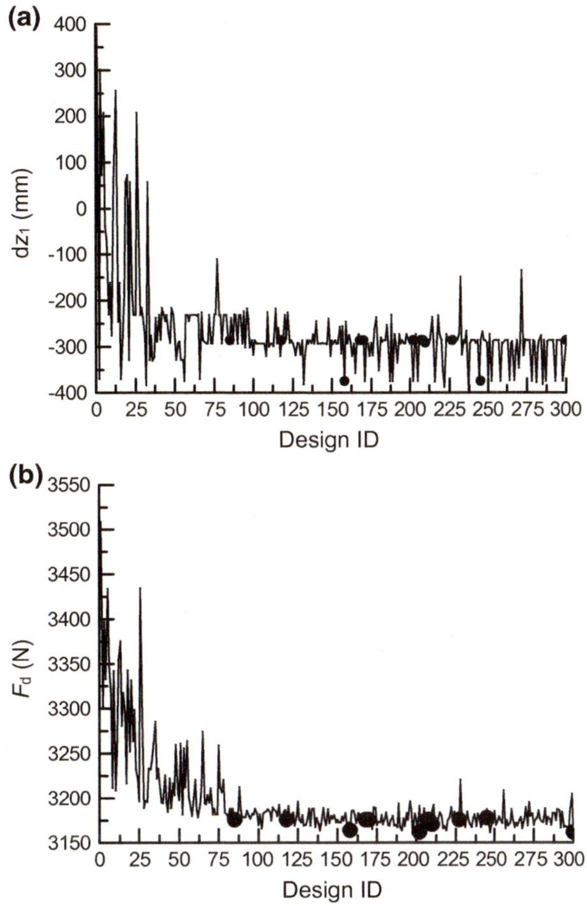


**Fig. 7** Histories of $dz_1$ (**a**) and $F_d$ (**b**)

variables $dy_3$, $dy_4$ and the objective $F_d$ can be found, which means that when the horizontal maximum control line or the bottom horizontal contour line moves to the longitudinal symmetry, the aerodynamic drag would decrease. As shown in Fig. 8b, a positive correlation between each design variable (except the central auxiliary control line) and the load reduction factor within a certain range can be found. There is little impact of the variable $dy_7$ on the aerodynamic drag or the load reduction factor.

The obvious significant factors which affect the aerodynamic drag and load reduction factor are successively $dy_4$, $dy_3$, $dz_1$, and nscale. In addition, from Fig. 8, each design variable has a bigger impact on the aerodynamic drag than that on the load reduction factor.

To further dig the nonlinear relationship between optimization objectives and optimization design variables, according to the analysis above, the variables $dy_4$ and $dy_3$, which are the most influential parameters, are chosen to conduct the response surface analysis with the aerodynamic drag. A 3D response surface of aerodynamic

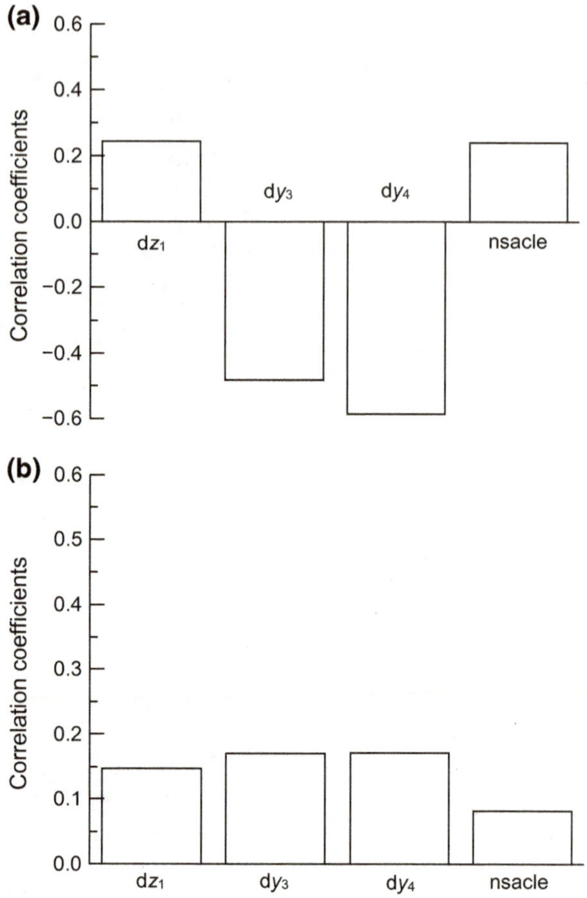


**Fig. 8** Correlation between aerodynamic drag (**a**), load reduction factor (**b**), and optimization design variables

drag $F_d$, the variables $dy_4$ and $dy_3$ is shown in Fig. 9. No pure linear relationship exists among $F_d$, $dy_4$, and $dy_3$, which can never be attained by a usual optimum seeking method (Fig. 9). The aerodynamic drag $F_d$ of the high-speed train shows a decreasing trend with the increase of $dy_4$ or $dy_3$, which is coincident with the results of Fig. 8.

The convergence of the optimization variables in the image space for all the designs is shown in Fig. 10. In Fig. 10, $(P_0 - P)/P_0$ indicates the load reduction factor, the curve connected by the dot notation "●" indicates the Pareto-optimal front of the multi-objective optimization of the high-speed train head, the pentagram "★" represents the aerodynamic drag force and load reduction factor corresponding to the initial head shape, and the square "■" denotes the aerodynamic drag force and load reduction factor corresponding to designs in the optimization process. It can be concluded that, after the multi-objective optimization design of the head shape, the performances of both the aerodynamic drag and load reduction factor have been improved. Compared with the initial head shape, the aerodynamic

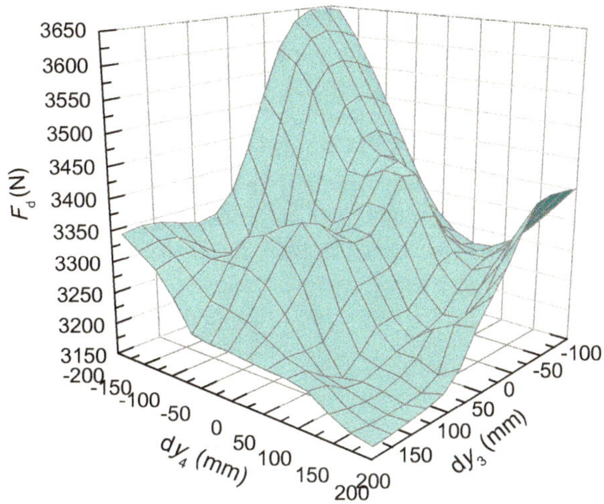


**Fig. 9** Three-dimensional response surface

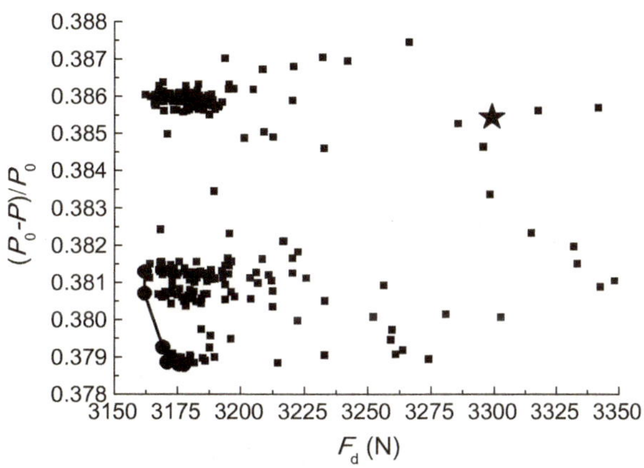


**Fig. 10** Pareto-optimal front of the head shape optimization

drag is reduced by up to 4.15% and the load reduction factor is reduced by up to 1.72% after optimization.

## 8 Conclusion

A parametric model of the high-speed train head is established in the present paper. The aerodynamic performance and vehicle dynamic performance of the high-speed train are calculated through the batch commands and script files. The multi-objective optimization algorithm NSGA-II is used for the automatic multi-objective optimization of the head shape, with the optimization objectives of the aerodynamic drag and load reduction factor. The proposed method can greatly reduce the design cycle of the head shape, and obtain a better head shape with good aerodynamic performance and vehicle dynamic performance.

The computational results show that a positive correlation between the variables $dz_1$, nscale, and the objective $F_d$ within a certain range can be found, and a negative correlation between the variables $dy_3$, $dy_4$, and the objective $F_d$, as well. There is a positive correlation between these four variables and the load reduction factor within a certain range. Through optimization, $dy_4$ and $dy_3$ are found to be the most influential parameters, and no linear relationship exists among the aerodynamic drag force and these two variables. After optimization, the aerodynamic drag is reduced by up to 4.15% and the load reduction factor is reduced by up to 1.72%.

## References


Aguilar Madeira, J. F., Rodrigues, H., & Pina, H. (2005). Multi-objective optimization of structures topology by genetic algorithms. *Advances in Engineering Software, 36*(1), 21–28. doi:10.1016/j.advengsoft.2003.07.001.

Brockie, N. J. W., & Baker, C. J. (1990). The aerodynamic drag of high speed train. *Journal of Wind Engineering and Industrial Aerodynamics, 34*(3), 273–290. doi:10.1016/0167-6105(90)90156-7.

Custódio, A. L., Madeira, J. F. A., Vaz, A. I. F., & Vicente, L. N. (2011). Direct multisearch for multiobjective optimization. *SIAM Journal on Optimization, 21*(3), 1109–1140. doi:10.1137/10079731X.

Custódio, A. L., Emmerich, M., & Maderia, J. F. A. (2012). Recent developments in derivative-free multiobjective optimization. *Computational Technology Reviews, 5,* 1–30. doi:10.4203/ctr.5.1.

Deb, K., Agrawal, S., Pratap, A., & Meyarivan, T. (2002). A fast and elitist multiobjective genetic algorithm: NSGA-II. *IEEE Transactions on Evolutionary Computation, 6*(2), 182–197. doi:10.1109/4235.996017.

Hemida, H., & Krajnović, S. (2010). LES study of the influence of the nose shape and yaw angles on flow structures around trains. *Journal of Wind Engineering and Industrial Aerodynamics, 98*(1), 34–46. doi:10.1016/j.jweia.2009.08.012.

Ikeda, M., Suzuki, M., & Yoshida, K. (2006). Study on optimization of panhead shape possessing low noise and stable aerodynamic characteristics. *Quarterly Report of RTRI, 47*(2), 72–77. doi:10.2219/rtriqr.47.72.

Kikuchi, K., Tanaka, Y., Iida, M., Yamauchi, N., Yoshida, Y., Nakanishi, M., et al. (2001). Countermeasures for reducing pressure variation due to train passage in open sections. *Quarterly Report of RTRI, 42*(2), 77–82. doi:10.2219/rtriqr.42.77.

Ku, Y. C., Rho, J. H., Yun, S. H., Kwak, M. H., Kim, K. H., Kwon, H. B., et al. (2010). Optimal cross-sectional area distribution of a high-speed train nose to minimize the tunnel micro-pressure wave. *Structural and Multidisciplinary Optimization, 42*(6), 965–976. doi:10.1007/s00158-010-0550-6.

Kwon, H. B., Jang, K. H., Kim, Y. S., Yee, K. J., & Lee, D. H. (2001). Nose shape optimization of high-speed train for minimization of tunnel sonic boom. *JSME International Journal Series C: Mechanical Systems Machine Elements and Manufacturing, 44*(3), 890–899. doi:10.1299/jsmec.44.890.

Lee, J., & Kim, J. (2008). Approximate optimization of high-speed train nose shape for reducing micropressure wave. *Structural and Multidisciplinary Optimization, 35*(1), 79–87. doi:10.1007/s00158-007-0111-9.

Li, T., Zhang, J. Y., & Zhang, W. H. (2013). A numerical approach to the interaction between airflow and a high-speed train subjected to crosswind. *Journal of Zhejiang University-Science A (Applied Physics & Engineering), 14*(7), 482–493. doi:10.1631/jzus.A1300035.

Maeda, T., Kinoshita, M., Kajiyama, H., & Tanemoto, K. (1989). Aerodynamic drag of Shinkansen electric cars (series 0, series 200, series 100). *Railway Technical Research Institute, Quarterly Report, 30*(1), 48–56.

Raghunathan, R. S., Kim, H. D., & Setoguchi, T. (2002). Aerodynamics of high-speed railway train. *Progress in Aerospace Sciences, 38*(6–7), 469–514. doi:10.1016/S0376-0421(02)00029-5.

Schetz, J. A. (2001). Aerodynamics of high-speed trains. *Annual Review of Fluid Mechanics, 33* (1), 371–414. doi:10.1146/annurev.fluid.33.1.371.

Shao, X. M., Wan, J., Chen, D. W., & Xiong, H. B. (2011). Aerodynamic modeling and stability analysis of a high-speed train under strong rain and crosswind condition. *Journal of Zhejiang University-Science A (Applied Physics & Engineering), 12*(12), 964–970. doi:10.1631/jzus.A11GT001.

Sun, Z. X., Song, J. J., & An, Y. R. (2010). Optimization of the head shape of the CRH3 high speed train. *Science China Technological Sciences, 53*(12), 3356–3364. doi:10.1007/s11431-010-4163-5.

Suzuki, M., Ikeda, M., & Yoshida, K. (2008). Study on numerical optimization of cross-sectional panhead shape for high-speed train. *Journal of Mechanical Systems for Transportation and Logistics, 1*(1), 100–110. doi:10.1299/jmtl.1.100.

Versteeg, H. K., & Malalasekera, W. (2007). *An introduction to computational fluid dynamics. The finite volume method* (2nd ed.). New Jersey, America: Prentice Hall.

Yao, S. B., Guo, D. L., & Yang, G. W. (2012). Three-dimensional aerodynamic optimization design of high-speed train nose based on GA-GRNN. *Science China Technological Sciences, 55*(11), 3118–3130. doi:10.1007/s11431-012-4934-2.

Zhou, A., Qu, B. Y., Li, H., Zhao, S. Z., Suganthan, P. N., & Zhang, Q. F. (2011). Multiobjective evolutionary algorithms: A survey of the state of the art. *Swarm and Evolutionary Computation, 1*(1), 32–49. doi:10.1016/j.swevo.2011.03.001.

## Author Biographies

**Ji-ye Zhang**, Ph.D., is currently a Professor at State Key Laboratory of Traction Power, Southwest Jiaotong University, Chengdu, China. He is the principal author/co-author of over 140 journal and conference papers, as well as a book. His research interests include dynamics and control of high-speed train, stability and control of complex system, and hybrid power vehicular and transportation systems.

**Wei-hua Zhang** holds the post of Executive Deputy Director of the National Laboratory for Rail Transit (currently in the preparatory stage) and is a member of the National Teaching Advisory Board for Higher Education Institutions for Transportation Engineering and Engineering Disciplines. He is honored as Distinguished Professor of the "Cheung Kong Scholars" Program of China and is winner of the "National Science Funds for Distinguished Young Scholars". He has been awarded the first prize twice (both ranking second) and the second prize twice (both ranking first) for the National Scientific and Technological Progress Award. His paper "A Study on Dynamic Behaviour of Pantographs by Using Hybrid Simulation Method" was acclaimed as the Best Paper Award of the Year of 2005 by the Institution of Mechanical Engineers (IMechE), making him the first Chinese winner of the Thomas Hawksley Gold Medal.

# Study on the Safety of Operating High-Speed Railway Vehicles Subjected to Crosswinds

**Xin-biao Xiao, Liang Ling, Jia-yang Xiong, Li Zhou and Xue-song Jin**

## Contents



## 1 Introduction

With the rapid development of high-speed railways around the world, the operating safety of high-speed trains has become one of the major concerns of current railway research. Fatal railway accidents, which are the catastrophic consequences of unsafe operating conditions, should be prevented (Evans 2011; Silla and Kallberg 2012). Strong crosswinds are among the extreme forces of nature that threaten the safe operation of trains. Many railway vehicles have been blown over by extreme


X. Xiao · L. Ling · J. Xiong · L. Zhou · X. Jin (✉)
State Key Laboratory of Traction Power, Southwest Jiaotong University,
Chengdu 610031, China
e-mail: xsjin@home.swjtu.edu.cn





Y. Fang and Y.H. Zhang (eds.), *China's High-Speed Rail Technology*,
Advances in High-speed Rail Technology, https://doi.org/10.1007/978-981-10-5610-9_11

crosswinds in locations around the world. As shown in Fig. 1, on the 28th of February, 2007, a train from Urumqi to Aksu was blown off its track by strong winds in Turpan, Xinjiang Uygur Autonomous Region of China (Xinhua News Agency 2007). Four people were killed, and more than 30 were injured. To date, more than 30 strong crosswind-induced accidents have been reported in Japan (Fujii et al. 1999; Gawthorpe 1994). Most of these accidents occurred on narrow-gauge tracks (Fujii et al. 1999).

Three characteristics of high-speed trains, i.e., their lightweight construction, high driving velocities, and distributed traction (Fujii et al. 1999), have significant influences on their operational safety when subjected to crosswinds. In recent years, the crosswind safety of railway vehicles has been of great interest to researchers and railway industries. Many railway vehicle safety standards, such as EN 14067-6 (CEN 2010) and TSI/HS-RST-L64-7/3/2008 (OJEU 2008), have been proposed to evaluate the dynamic response of trains to crosswind action and ensure their operational safety. Reviews of recent international work in this field were presented by Carrarini (2006) and Baker et al. (2009).

Crosswind stability analysis of railway vehicles involves two issues. The first is the flow field around a train in operation and the aerodynamic forces acting on the car body. The second is the resultant dynamic response and crosswind stability of the train-track coupling system and its safety assessment. Most of the previous studies on this subject have focused on the first issue. A large number of full-scale wind tunnel tests and computational fluid dynamics (CFD) simulations have been carried out to examine the airflow around high-speed trains in crosswind scenarios (Baker et al. 2004; Diedrichs 2005; Cheli et al. 2010). The second issue, which was investigated in this study, has not received much attention in previous studies. Many efforts have been made to use multi-body dynamic models to study the characteristic wind curves, which represent critical crosswind speeds, at which the selected derailment criteria reach their limits and vehicle overturning occurs (Orellano and Schober 2003; Cheli et al. 2006; Xu and Ding 2006). Typically, quasi-steady approaches are proposed for use in calculating the wheel loading reduction caused by crosswind forces. Such approaches are based on the equilibrium of the steady aerodynamic forces and the restoring forces on the railway vehicle and do not take into account the transient response that occurs when a vehicle is subjected to a crosswind (RSSB 2000; Carrarini 2006).

To investigate the operating safety of high-speed railway vehicles subjected to strong crosswinds, a vehicle–track model that considers the crosswind effect was

**Fig. 1** Train overturned by crosswinds (Xinhua News Agency 2007)

developed and was used in a numerical analysis carried out in a time domain. In this approach, the vehicle is modeled as a nonlinear multi-body system, and the track is modeled as a three-layer system. The rails are modeled as Timoshenko beams supported by discrete sleepers. The coupling of the vehicle and the track is simulated by the track moving with respect to the vehicle operating at a constant speed, which permits consideration of the effects of periodic discrete rail supports on the vehicle–track interaction. The rolling contact of the wheel-rail system reflects the geometric relationship and contact forces between the wheels and rails. The wheel-rail geometric relationship is solved spatially and evaluated on-line using a new wheel-rail contact model (Chen and Zhai 2004). The wheel-rail contact forces include normal and tangential forces. The normal forces of the wheel-rail system are calculated using the Hertzian contact theory, and their tangential forces are calculated using the nonlinear creep theory proposed by Shen et al. (1983). In the analysis conducted in this study, the crosswind was assumed to be steady, and the aerodynamic forces due to the crosswind were modeled as ramp shape external forces exerted on the car body. The crosswind forces considered included the side force, the lift forces, the roll moment, the pitch moment, and the yaw moment. The numerical analysis was conducted to investigate the dynamic response and derailment mechanism of a high-speed vehicle in a strong crosswind scenario. The effects of the crosswind speed, the crosswind attack angle, and the vehicle speed on the operational safety of the vehicle were examined in detail. The operational safety boundaries of a high-speed vehicle subjected to crosswinds were determined from dynamic simulations of vehicle–track coupling and existing safety assessment criteria.

## 2 Dynamic Model of Coupled Vehicle–Track System in Crosswinds

The causes of derailment or overturn of railway vehicles operating in strong crosswinds are not easy to identify, and it is very difficult to recreate accidents in site tests or laboratory experiments. Numerical modeling is an effective means of studying the causes of derailments under extreme conditions, such as in strong crosswinds and earthquakes. Numerical simulation is a very convenient, highly efficient, and low-cost approach to investigating the effects of one or more factors in derailment. An advanced vehicle–track interaction model can be used to characterize derailment of railway vehicles in strong crosswinds. Based on the theories of coupled vehicle–track dynamics (Zhai et al. 1996), a spatial model of a coupled vehicle–track system was developed in this study to simulate vehicle–track interaction for a train operating in crosswind scenarios. The model consists of four subsystems: the vehicle, the track, the wheel-rail contact, and the aerodynamic forces on the vehicle. These subsystems are described in Sects. 2.1–2.5, respectively.

## 2.1 Vehicle Model

The vehicle–track model is shown in Fig. 2a, b. A high-speed railway vehicle used in China, which consists of a car body, a pair of two-axle bogies, and four wheelsets, was modeled in this study. The primary suspension connects the wheelsets and the bogie frames, and the car body is supported on the bogie through the secondary suspension.

The vehicle is modeled as a nonlinear multi-body system. The structural elastic deformations of the vehicle components are ignored. The vehicle model includes seven rigid bodies, and each body has five degrees of freedom: the lateral ($Y$), vertical ($Z$), roll ($\phi$), pitch ($\beta$), and yaw ($\psi$) motions. Thus, the total number of degrees of freedom of the vehicle model is 35. All rotational motions of the vehicle parts are considered to be small, which allows linearization of the motion equation for the vehicle parts. Three-dimensional (3D) spring-damper elements are used to represent the primary and secondary suspensions, and the nonlinear dynamic characteristics of the suspension systems are considered. The vehicle speed is assumed to be constant. Therefore, the longitudinal accelerations of the centers of all the parts are always zero. However, the vehicle model considers the relative longitudinal motion of the suspension systems, due to the yaw motions of the car body, the bogie frames, and the wheelsets.

The following are the differential equations of the car body:

$$M_c\ddot{Y}_c = F_{ybL1} + F_{ybR1} + F_{ybL2} + F_{ybR2} + F_{wy}, \tag{1}$$

$$M_c\ddot{Z}_c = -F_{zbL1} - F_{zbR1} - F_{zbL2} - F_{zbR2} + M_c g + F_{wz}, \tag{2}$$

$$\begin{aligned} I_{cx}\ddot{\phi}_c = & -(F_{ybL1} + F_{ybR1} + F_{ybL2} + F_{ybR2})H_{cB} \\ & + (F_{zbL1} - F_{zbR1} + F_{zbL2} - F_{zbR2})d_s + M_{wx}, \end{aligned} \tag{3}$$

$$\begin{aligned} I_{cy}\ddot{\beta}_c = & (F_{zbL1} + F_{zbR1} - F_{zbL2} - F_{zbR2})l_c \\ & - (F_{xbL1} + F_{xbR1} + F_{xbL2} + F_{xbR2})H_{cB} + M_{wy}, \end{aligned} \tag{4}$$

$$\begin{aligned} I_{cz}\ddot{\psi}_c = & (F_{ybL1} + F_{ybR1} - F_{ybL2} - F_{ybR2})l_c \\ & - (F_{xbL1} - F_{xbR1} + F_{xbL2} - F_{xbR2})d_s + M_{wz}. \end{aligned} \tag{5}$$

The following are the differential equations of the bogie $i$ ($i$ = 1, 2):

$$\begin{aligned} M_b\ddot{Y}_{bi} = & F_{yfL(2i-1)} + F_{yfL(2i)} - F_{ybLi} \\ & - F_{ybRi} + F_{yfR(2i-1)} + F_{yfR(2i)}, \end{aligned} \tag{6}$$

$$\begin{aligned} M_b\ddot{Z}_{bi} = & F_{zbLi} - F_{zfL(2i-1)} - F_{zfL(2i)} \\ & + F_{zbRi} - F_{zfR(2i-1)} - F_{zfR(2i)} + M_b g, \end{aligned} \tag{7}$$

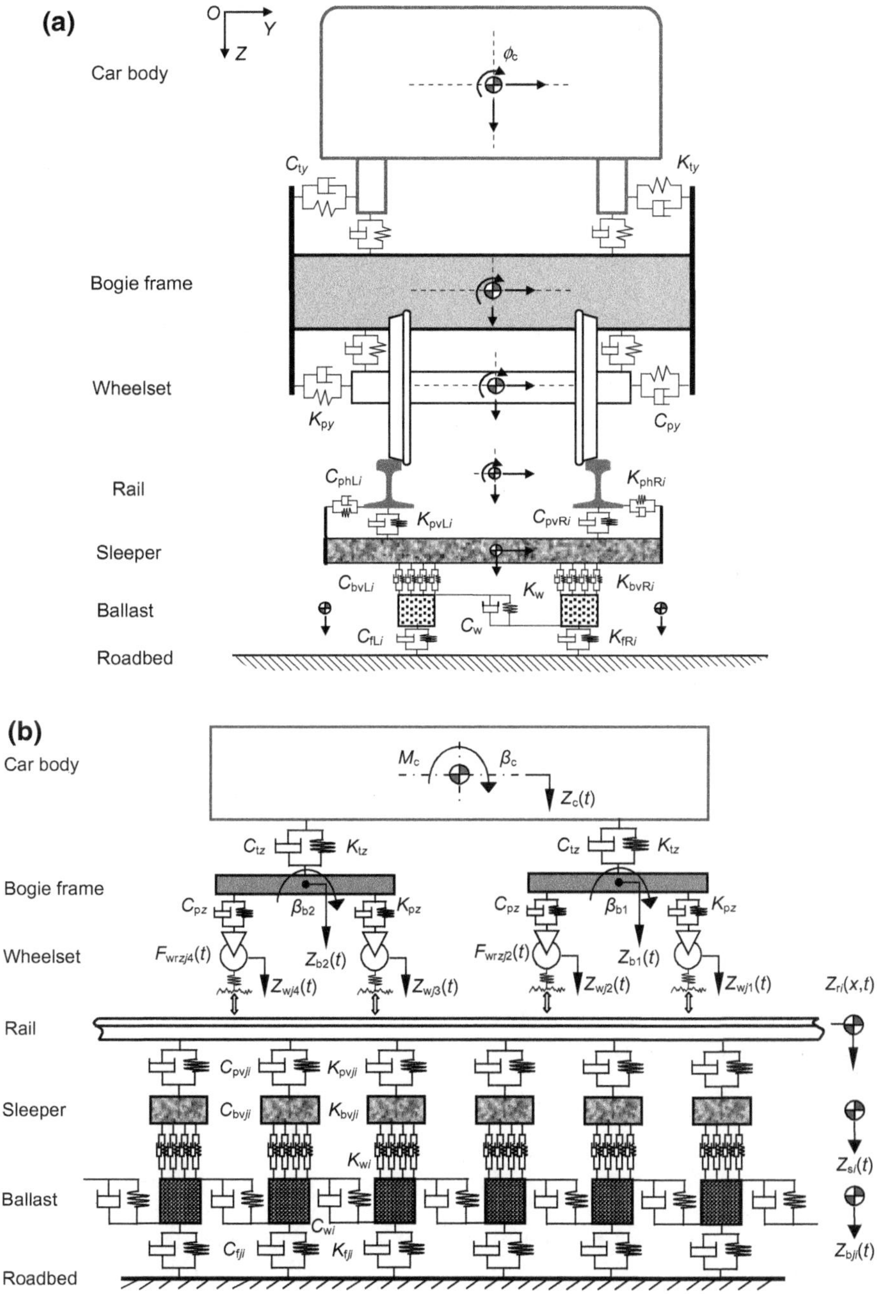


**Fig. 2** Coupled vehicle–track model: **a** elevation and **b** side elevation

$$\begin{aligned} I_{bx}\ddot{\phi}_{bi} = & -[F_{yfL(2i-1)} + F_{yfR(2i-1)} + F_{yfL(2i)} + F_{yfR(2i)}]H_{tw} \\ & + [F_{zfL(2i-1)} + F_{zfL(2i)} - F_{zfR(2i-1)} - F_{zfR(2i)}]d_w \\ & + (F_{zbRi} - F_{zbLi})d_s - (F_{ybLi} + F_{ybRi})H_{Bt}, \end{aligned} \tag{8}$$

$$\begin{aligned} I_{by}\ddot{\beta}_{bi} = & [F_{zfL(2i-1)} + F_{zfR(2i-1)} - F_{zfL(2i)} - F_{zfR(2i)}]l_b \\ & - [F_{xfL(2i-1)} + F_{xfR(2i-1)} + F_{xfL(2i)} + F_{xfR(2i)}]H_{tw} \\ & - (F_{xbLi} + F_{xbRi})H_{Bt}, \end{aligned} \tag{9}$$

$$\begin{aligned} I_{bz}\ddot{\psi}_{bi} = & [F_{yfL(2i-1)} + F_{yfR(2i-1)} - F_{yfL(2i)} - F_{yfR(2i)}]l_b \\ & - [F_{xfL(2i-1)} + F_{xfL(2i)} - F_{xfR(2i-1)} - F_{xfR(2i)}]d_w \\ & + (F_{xbLi} - F_{xbRi})d_s. \end{aligned} \tag{10}$$

The following are the differential equations of the wheelset $i$ ($i$ = 1, 2, 3, 4):

$$M_w\ddot{Y}_{wi} = -F_{yfLi} - F_{yfRi} + F_{wryLi} + F_{wryRi}, \tag{11}$$

$$M_w\ddot{Z}_{wi} = F_{zfLi} + F_{zfRi} - F_{wrzLi} - F_{wrzRi} + M_w g, \tag{12}$$

$$\begin{aligned} I_{wx}\ddot{\phi}_{wi} = & d_L F_{wrzLi} - d_R F_{wrzRi} - r_{Li}F_{wryLi} \\ & - r_{Ri}F_{wryRi} + d_w(F_{zfRi} - F_{zfLi}), \end{aligned} \tag{13}$$

$$\begin{aligned} I_{wy}\ddot{\beta}_{wi} = & r_{Li}F_{wrxLi} + r_{Ri}F_{wrxRi} + r_{Li}\psi_{wi}F_{wryLi} \\ & + r_{Ri}\psi_{wi}F_{wryRi} + M_{wryLi} + M_{wryRi}, \end{aligned} \tag{14}$$

$$\begin{aligned} I_{wz}\ddot{\psi}_{wi} = & (d_L F_{wrxLi} - d_R F_{wrxRi}) + (d_L F_{wryLi} - d_R F_{wryRi})\psi_{wi} \\ & + d_w(F_{xfLi} - F_{xfRi}) + M_{wrzLi} + M_{wrzRi}. \end{aligned} \tag{15}$$

The definitions of the symbols used in Eqs. (1)–(15) are given in Table 1, and the detailed expressions of the mutual forces between the vehicle's components are presented in Xiao et al. (2011).

## 2.2 Track Model

The ballast track model presented by Xiao et al. (2008), a three-layer model consisting of rails, sleepers, and ballasts, as shown in Fig. 2, was used in this study. The gauge of the tangent track was 1435 mm, the rail cant was 1:40, and the sleeper pitch was 600 mm. The rails were modeled as having a mass of 60 kg/m (CN60) to represent a rail type that is widely used on high-speed rail lines in China. The track, except for the rails, was also modeled as a rigid multi-body dynamic system. The

**Table 1** Notations for equations of vehicle system

| Notation | Description |
|---|---|
| $M_c$ | Car body mass |
| $M_{bi}$ | The *i*th bogie mass |
| $M_{wi}$ | The *i*th wheelset mass |
| $I_{bx}$, $I_{by}$, $I_{bz}$ | Bogie body roll, pitch, and yaw moments of inertia, respectively |
| $I_{cx}$, $I_{cy}$, $I_{cz}$ | Car body roll, pitch, and yaw moments of inertia, respectively |
| $I_{wx}$, $I_{wy}$, $I_{wz}$ | Wheelset body roll, pitch, and yaw moments of inertia, respectively |
| $V$ | Forward speed of vehicle |
| $g$ | Gravity acceleration |
| $r_L$, $r_R$ | Left and right rolling radii |
| $H_{cB}$ | Height of the car body center from the secondary suspension location |
| $H_{Bt}$ | Height of the secondary suspension from the bogie center |
| $H_{tw}$ | Height of the bogie center from the wheelset center |
| $l_c$ | Half of the distance between bogie centers |
| $l_b$ | Half distance between the two axles of the bogie |
| $d_s$ | Half distance between the secondary suspension systems of the two sides of the bogie |
| $d_w$ | Half distance between the two primary suspensions of the two sides of the bogie |
| $F_{wy}$, $F_{wz}$ | Side and lift forces applied to the vehicle body |
| $M_{wx}$, $M_{wy}$, $M_{wz}$ | Roll, pitch, and yaw moments applied to the vehicle body |
| $F_{xbji}$, $F_{ybji}$, $F_{zbji}$ ($i$ = 1 or 2, $j$ = L or R) | Forces between the car body and the bogie frame in $x$, $y$, and $z$ directions |
| $F_{xfji}$, $F_{yfji}$, $F_{zfji}$ ($i$ = 1, 2, 3, 4, $j$ = L or R) | Forces between the bogie frame and the wheelset in $x$, $y$, and $z$ directions |
| $F_{wrxji}$, $F_{wryji}$, $F_{wrzji}$ ($i$ = 1, 2, 3, 4, $j$ = L or R) | Forces between the wheels and rails in $x$, $y$, and $z$ directions |
| $M_{wryji}$, $M_{wrzji}$ ($i$ = 1, 2, 3, 4, $j$ = L or R) | Spin moment components between the wheels and rails in $y$ and $z$ directions |

rails were modeled as Timoshenko beams on an elastic point-supporting foundation. The lateral and vertical bending deformations and twisting of the simply supported beams were taken into account.

According to the Timoshenko beam theory, the equations of bending deformations of the rails can be written as follows.

Lateral bending deformation:

$$\begin{cases} \rho_r A_r \frac{\partial^2 y}{\partial t^2} + \kappa_{ry} G_r A_r \left( \frac{\partial \psi_y}{\partial x} - \frac{\partial^2 y}{\partial x^2} \right) \\ = \sum_{i=1}^{N_W} F_{wryi}(t)\delta(x - x_{wi}) - \sum_{j=1}^{N_S} F_{rsyj}(t)\delta(x - x_{sj}), \\ \rho_r I_{rz} \frac{\partial^2 \psi_y}{\partial t^2} - E_r I_{rz} \frac{\partial^2 \psi_y}{\partial x^2} + \kappa_{ry} G_r A_r \left( \psi_y - \frac{\partial y}{\partial x} \right) = 0. \end{cases} \tag{16}$$

Vertical bending deformation:

$$\begin{cases} \rho_r A_r \frac{\partial^2 z}{\partial t^2} + \kappa_{rz} G_r A_r \left( \frac{\partial \psi_z}{\partial x} - \frac{\partial^2 z}{\partial x^2} \right) \\ = \sum_{i=1}^{N_W} F_{wrzi}(t)\delta(x - x_{wi}) - \sum_{j=1}^{N_S} F_{rszj}(t)\delta(x - x_{sj}), \\ \rho_r I_{ry} \frac{\partial^2 \psi_z}{\partial t^2} - E_r I_{ry} \frac{\partial^2 \psi_z}{\partial x^2} + \kappa_{rz} G_r A_r \left( \psi_z - \frac{\partial z}{\partial x} \right) = 0. \end{cases} \tag{17}$$

Torsion:

$$\begin{aligned} \rho_r I_{r0} \frac{\partial^2 \phi}{\partial t^2} - G_r K_r \frac{\partial^2 \phi}{\partial x^2} &= \sum_{i=1}^{N_W} M_{wri}(t)\delta(x - x_{wi}) \\ &- \sum_{j=1}^{N_S} M_{rsj}(t)\ \delta(x - x_{sj}). \end{aligned} \tag{18}$$

In Eqs. (16)–(18), $y$, $z$, and $\phi$ are the lateral, vertical, and torsional deflections, respectively, of the rail; $\psi_y$ and $\psi_z$ are the slopes of the deflection curve of the rail with respect to the $z$ and $y$ axes, respectively; $\rho_r$, $G_r$, and $E_r$ are the density, shear modulus, and Young's modulus of the rail, respectively; $m_r$ and $A_r$ are the mass per unit longitudinal length and the area of the cross section of the rail, respectively; $I_{ry}$ and $I_{rz}$ are the second moments of the area around the $y$ and $z$ axes, respectively; $I_{r0}$ is the polar moment of inertia; and $\kappa_{ry}$, $\kappa_{rz}$, and $K_{rs}$ are the shear coefficients of the lateral and vertical bending deformation and torsion, respectively. The subscript $i$ indicates wheelset $i$; $j$ indicates sleeper $j$; $\delta(x)$ is the Dirac delta function; $x_{wi}$ and $x_{fj}$ are the longitudinal positions of wheel $i$ and sleeper $j$, respectively; $N_W$ and $N_S$ are the total numbers of wheelsets and sleepers on the analyzed rail, respectively; $M_{wri}(t)$ and $M_{rsj}(t)$ are the equivalent moments acting on the rail; $F_{wryi}(t)$ and $F_{wrzi}(t)$ are the wheel-rail forces on wheel $i$ in the lateral and vertical directions, respectively; and $F_{rsyj}(t)$ and $F_{rszj}(t)$ are the lateral and vertical forces, respectively, between the rails and sleepers.

The sleepers were modeled as rigid rectangular beams. The lateral and vertical translational motions and the roll motion of each sleeper were considered. The lateral, vertical, and rolling motion equations of sleeper $i$ can be written as

$$M_s \ddot{Y}_{si} = (F_{yi\mathrm{L}} + F_{yi\mathrm{R}}) - F_{y\mathrm{sb}i}, \tag{19}$$

$$M_s \ddot{Z}_{si} = (F_{zi\mathrm{L}} + F_{zi\mathrm{R}}) - (F_{zi\mathrm{bL}} + F_{\mathrm{zb}i\mathrm{R}}), \tag{20}$$

$$\begin{aligned} I_s \ddot{\phi}_{si} = {} & d_\mathrm{b}(F_{\mathrm{zb}i\mathrm{R}} - F_{\mathrm{zb}i\mathrm{L}}) \\ & + d_\mathrm{r}(F_{zi\mathrm{L}} - F_{zi\mathrm{R}}) - b_\mathrm{s}(F_{yi\mathrm{L}} + F_{yi\mathrm{R}}), \end{aligned} \tag{21}$$

where $M_s$ is the sleeper mass; $I_s$ is the moment of inertia of the sleeper in the rolling direction; $F_{yi\mathrm{L}}$ and $F_{yi\mathrm{R}}$ are the lateral forces between the sleeper $i$ and the left and right rails, respectively; $F_{zi\mathrm{L}}$ and $F_{zi\mathrm{R}}$ are the vertical forces between the sleeper $i$ and the left and right rails, respectively; $F_{y\mathrm{sb}i}$ is the lateral force between the sleeper $i$ and the ballasts; $F_{\mathrm{zb}i\mathrm{L}}$ and $F_{\mathrm{zb}i\mathrm{R}}$ are the vertical forces between the sleeper $i$ and the left and right equivalent ballast bodies, respectively; $d_\mathrm{b}$ is the half distance between the centers of the left and right ballast bodies; $d_\mathrm{r}$ is the half distance between the left and right rails; and $b_\mathrm{s}$ is the half thickness of the sleeper.

The ballast bed is assumed to be composed of equivalent rigid ballast bodies. Only the vertical motion of the ballast body is considered. The motion equations of the left and right ballast bodies $i$ in the vertical direction can be written as

$$M_{\mathrm{bL}} \ddot{Z}_{\mathrm{bL}i} = F_{\mathrm{bzL}i} + F_{\mathrm{zrL}i} + F_{\mathrm{zLR}i} - F_{\mathrm{zgL}i} - F_{\mathrm{zfL}i}, \tag{22}$$

$$M_{\mathrm{bR}} \ddot{Z}_{\mathrm{bR}i} = F_{\mathrm{bzR}i} + F_{\mathrm{zrR}i} - F_{\mathrm{zLR}i} - F_{\mathrm{zgR}i} - F_{\mathrm{zfR}i}, \tag{23}$$

where $F_{\mathrm{zgL}i}$ and $F_{\mathrm{zgR}i}$ are the vertical support forces due to the roadbed, and $F_{\mathrm{zfL}i}$, $F_{\mathrm{zfR}i}$, $F_{\mathrm{zrL}i}$, $F_{\mathrm{zrR}i}$, and $F_{\mathrm{zLR}i}$ are the vertical shear forces between neighboring ballast bodies. This equivalent model can represent the two vertical rigid modes of the ballasts in the vertical-lateral plane of the track. Uniformly viscoelastic elements are used to simulate the roadbed beneath the ballast bed, and the motion of the roadbed is neglected. The rails and the sleepers, the sleepers and the ballast bodies, and the discrete ballast bodies and the roadbed are connected with equivalent springs and dampers.

## 2.3 *Wheel-Rail Contact Model*

Wheel-rail contact generates the necessary conditions for a railway vehicle to run stably on a track. In the analysis of transient dynamics and derailment (or overturning) of high-speed railway vehicles under strong crosswinds, accurate and fast calculation of the wheel-rail contact is important. The rolling contact of the wheel-rail system depends on the geometric relationship and the contact forces between the wheels and rails. A new wheel-rail contact model (Chen and Zhai 2004) was used in this study to characterize the geometry of the wheel-rail rolling contact, and this model is able to consider the separation of wheels and rails.

The wheel-rail contact forces include the normal load and the tangential forces. The normal load is calculated using the following equation for a Hertzian nonlinear contact spring with a unilateral restraint:

$$F_n(t) = \begin{cases} \left[\frac{1}{G_{\text{hertz}}} Z_{\text{wrnc}}(t)\right]^{3/2}, & Z_{\text{wrnc}}(t) > 0; \\ 0, & Z_{\text{wrnc}}(t) \leq 0, \end{cases} \tag{24}$$

where $G_{\text{hertz}}$ is the wheel-rail contact constant (m/N$^{2/3}$), which depends on the radii of curvature and the elastic moduli of the wheel and rail, for the given wheel profiles:

$$G_{\text{hertz}} = 3.86 r^{-0.115} \times 10^{-8}, \tag{25}$$

where $r$ is the rolling radius of the wheel. The value of $G_{\text{hertz}}$ changes with the location of the contact point. The $Z_{\text{wrnc}}(t)$ term reflects the amount of normal compression at the wheel-rail contact point, which is defined as an approach between a pair of contact points, one of which belongs to the wheel tread and the other belongs to the rail surface. The condition expressed as $Z_{\text{wrnc}}(t) \leq 0$ reflects the separation between the wheel and the rail, and the condition expressed as $Z_{\text{wrnc}}(t) > 0$ reflects wheel-rail in contact.

The tangential forces of the wheel-rail contact are determined using Kalker's linear creep theory (Kalker 1967) and Shen's model (Shen et al. 1983). First, the wheel-rail creep forces are calculated using Kalker's linear creep theory for small amounts of creep. For large amounts of creep, saturation occurs, resulting in a nonlinear relation that is described using Shen's model (Shen et al. 1983).

## 2.4 Vehicle–Track Excitation Model

The dynamic vehicle–track system used in this study consists of four models (Knothe and Grassie 1993): (1) a stationary load model, (2) a moving-load excitation model, (3) a moving irregularity model, and (4) a moving mass model. A "tracking window" model developed in our previous study (Xiao et al. 2011) was used, which is shown in Fig. 3. In the model, the vehicle remains in a static state with respect to the ground in the longitudinal direction, and the track system moves in the opposite direction of the vehicle motion at the same speed. A detailed description of this vehicle–track model and the derivation of the system equations were presented in (Xiao et al. 2011).

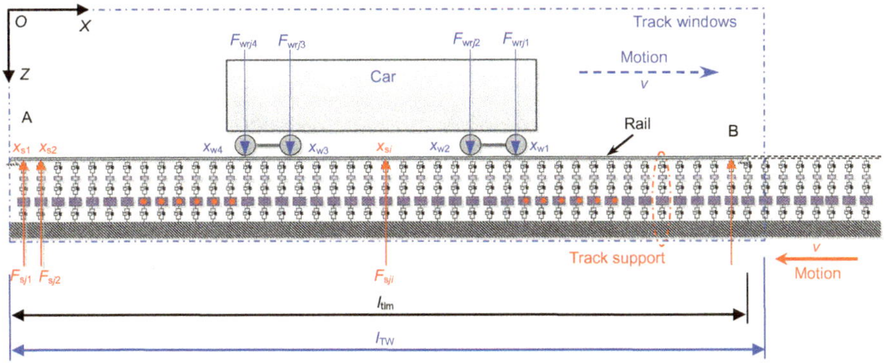


**Fig. 3** Vehicle–track system excitation model

## *2.5 Aerodynamic Forces on the Vehicle*

The aerodynamic forces acting on a railway vehicle subjected to a crosswind can be divided into two parts: steady forces and unsteady forces. Steady crosswind forces are caused by the mean wind speed components of natural wind, and the unsteady wind forces are caused by the fluctuating wind speed components (Xu and Ding 2006). In this study, the crosswind was assumed to be steady, and the mean wind speed was assumed to be in the horizontal direction. The aerodynamic forces due to the crosswinds were modeled as ramp-shaped external forces exerted on the vehicle body. Only aerodynamic forces acting on the car body were taken into account. The crosswind forces $\boldsymbol{F}$ applied to the vehicle body include the side force $F_{wy}$, the lift force $F_{wz}$, the roll moment $M_{wx}$, the pitch moment $M_{wy}$, and the yaw moment $M_{wz}$, as shown in Fig. 4.

Taking into account the transient response of the vehicle in a crosswind scenario, the crosswind forces $\boldsymbol{F}$ can be defined as

$$\boldsymbol{F} = \begin{cases} \frac{\zeta}{L_0}\boldsymbol{F}_0, & 0 \le \zeta \le L_0; \\ \boldsymbol{F}_0, & L_0 < \zeta, \end{cases} \tag{26}$$

where $L_0$ is the vehicle length, and $\zeta$ is the length of the car body immersed in the crosswind scenario. According to the corrected quasi-steady approach, the force vector $\boldsymbol{F}_0 = [F_{wy}, F_{wz}, M_{wx}, M_{wy}, M_{wz}]$ can be expressed as

$$F_{wy}(t) = \frac{1}{2}\rho_{air}A_c C_y(\beta(t))V_{res}^2(t), \tag{27}$$

$$F_{wz}(t) = \frac{1}{2}\rho_{air}A_c C_z(\beta(t))V_{res}^2(t), \tag{28}$$

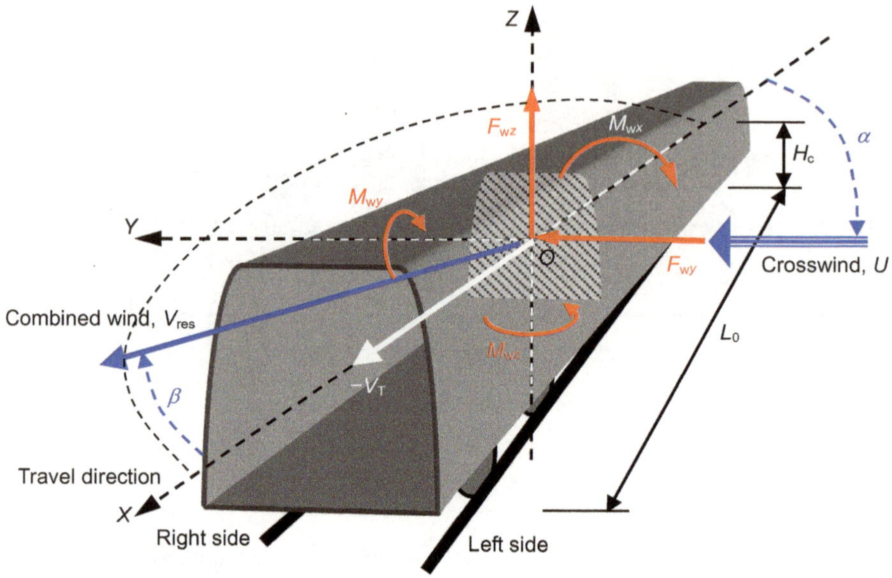


**Fig. 4** Aerodynamic forces on railway vehicle

$$M_{wx}(t) = \frac{1}{2}\rho_{air}A_c H_c C_{mx}(\beta(t))V_{res}^2(t), \tag{29}$$

$$M_{wy}(t) = \frac{1}{2}\rho_{air}A_c H_c C_{my}(\beta(t))V_{res}^2(t), \tag{30}$$

$$M_{wz}(t) = \frac{1}{2}\rho_{air}A_c H_c C_{mz}(\beta(t))V_{res}^2(t), \tag{31}$$

where $\rho_{air}$ is the air density, $A_c$ is the reference area, and $H_c$ is the reference height. A "TSI normalization" with $A_c$ = 10 m$^2$ and $H_c$ = 3 m was adopted in this study (OJEU 2008). The terms $c_y$, $c_z$, $c_{mx}$, $c_{my}$, and $c_{mz}$ correspond to the aerodynamic force coefficients, which depend on the crosswind attack angle $\beta$. The aerodynamic coefficients of the inter city express 2 (ICE2) driving trailer (Orellano and Schober 2003) were used in the calculation of the crosswind forces, as shown in Fig. 5.

The term $V_{res}$ corresponds to the resulting squared wind speed. The terms $\beta$ and $V_{res}$ correspond to spatial averages with respect to the surface of the vehicle. The resulting wind speed $V_{res}(t)$ is defined as

$$V_{res}^2(t) = U^2(t) + V_T^2(t) - 2U(t)V_T(t)\cos(\pi - \alpha_w(t)), \tag{32}$$

and the resultant crosswind attack angle $\beta$ is determined by

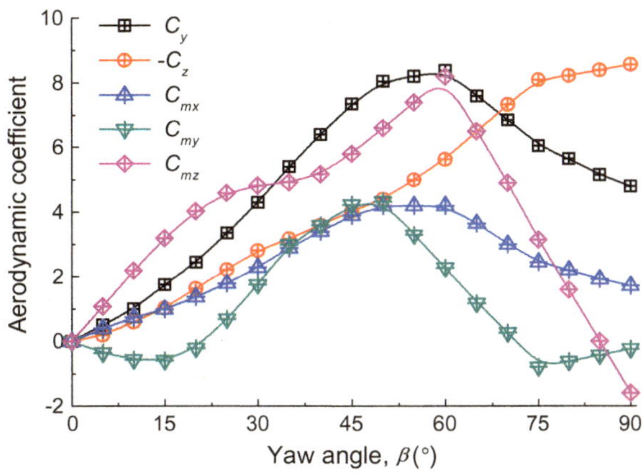


**Fig. 5** Aerodynamic coefficients of extreme forces on the vehicle

$$\beta(t) = \arctan\frac{U(t)\sin\alpha_{\mathrm{w}}(t)}{V_{\mathrm{T}}(t) + U(t)\cos\alpha_{\mathrm{w}}(t)}, \tag{33}$$

where $V_{\mathrm{T}}$ is the vehicle speed, and $U$ is the crosswind velocity. The crosswind attack angle $\alpha_{\mathrm{w}}$ is defined as the relative angle between the direction of the crosswind $U$ and the direction of the vehicle's motion (in the direction of the $x$ axis), as shown in Fig. 4.

## 3 Methods for Safety Assessment of Crosswinds

At present, the derailment criteria for estimating the running safety of trains vary from country to country. Most of the existing criteria consider a single influencing factor or a few influencing factors, and they are regarded as isolated constants in evaluating the operational safety of trains (Ling et al. 2012). The commonly used derailment safety assessment criteria include the following:

(1) Nadal's single-wheel $L/V$ limit criterion ($L/V < 0.8$) (Nadal 1896), where $L$ and $V$ are the wheel-rail lateral and vertical forces, respectively;
(2) Weinstock's axle-sum $L/V$ limit criterion ($(L/V)_{\mathrm{s}} < 1.5$) (Weinstock 1984);
(3) The $L/V$ time duration criterion ($T_{[L/V>0.8]} < 50$ ms) (Japanese National Railways (JNR)) (Yokose 1966);
(4) The $L/V$ distance duration criterion ($\mathrm{Dis}_{[L/V>0.8]} < 1.5$ m) [Federal Railroad Administration (FRA)] (Wu and Wilson 2006);
(5) The bogie-side-sum $L/V$ limit criterion ($(L/V)_{\mathrm{B}} < 0.6$) (Wu and Wilson 2006);
(6) Prudhomme's criterion (transverse axle force) ($F_{ys} < 10 + P_0/3$, where $P_0$ is the static wheelset load) (Wilson et al. 2011);
(7) The wheel load unloading ratio ($\Delta V/V < 0.8$) (Jin et al. 2013);

(8) The vehicle overturning coefficient ($V_D/V_0 < 0.8$) (Jin et al. 2013);
(9) The wheel rise ($Z_{up}$) limit with respect to the rail ($Z_{up} < 28.272$ mm) (Jin et al. 2013);
(10) The lateral coordinate ($y_{con}$) limit of the wheel-rail contact point ($-38.875$ mm $< y_{con} < 57.0$ mm) (Jin et al. 2013).

Criteria (1)–(4) are related to the ratio of the lateral force to the vertical force of a wheel-rail pair. These criteria are applied to assess the climbing derailment safety of railway vehicles. When a high-speed vehicle is in a crosswind scenario, leeward wheel climbing is very likely to occur. Thus, criteria (1)–(4) were used in this study to evaluate the running safety of a high-speed railway vehicle subjected to strong crosswinds. Criterion (5) is usually used to evaluate derailment caused by rail rollover or track gauge widening, and criterion (6) is the track panel shift criterion, which applies in circumstances of strong crosswinds. Criteria (7) and (8) are two useful safety assessment indexes for vehicle overturning. Criteria (1)–(8) are all calculated based on the wheel-rail contact forces. In fact, the separation between wheels and rails occurs quite often. When the wheel loses contact with the rail, the wheel-rail contact forces vanish. In this situation, it is very difficult to determine the status of the vehicle operation using these criteria. We therefore considered two additional derailment criteria based on the wheel-rail contact geometry to evaluate the critical conditions of running safety when high-speed vehicles are subjected to crosswinds. These two criteria are the wheel rise ($Z_{up}$) limit with respect to the rail [criterion (9)] and the lateral coordinate ($y_{con}$) limit of the wheel-rail contact point [criterion (10)]. Based on the dynamic simulation and the derailment safety assessment criteria listed above, the boundaries of the safe operation area, the warning area, and the derailment area were calculated for conditions of strong crosswinds in which high-speed vehicles operate.

## 4 Simulation of High-Speed Vehicle Dynamic Behavior Under Crosswinds

To investigate the effect of crosswinds on the dynamic responses and running safety of high-speed railway vehicles, the coupled vehicle–track dynamic model discussed in Sect. 2 was used to carry out a dynamic analysis in the time domain. The parameters of a Chinese high-speed passenger car and a tangent track were used in the numerical simulation (Xiao et al. 2008). Normal track irregularity was neglected because its effects on the dynamic behavior of the vehicle–track system are very small compared to the effect of crosswind excitation. The dynamic responses of a high-speed vehicle subjected to crosswinds, including the rolling and lateral displacements of the car body and the wheel-rail normal forces, were investigated as described in Sect. 4.1. The safety and overturning risk of the high-speed vehicle were assessed by analyzing the transient values of two derailment criteria: wheel unloading and wheel rise with respect to the rail top. The effects of the crosswind speed, the crosswind attack angle, and the vehicle speed on the running safety of the vehicle were examined in detail, as discussed in Sects. 4.2 and 4.3.

## 4.1 Vehicle Dynamic Responses to Crosswind

First, the dynamic behaviors of the vehicle system as the vehicle enter a crosswind scenario with a constant crosswind attack angle and a constant driving speed were investigated. The vehicle speed $V_T$ was set to 300 km/h, the crosswind speed $U$ was varied from 12 to 24 m/s, and the crosswind attack angle $\alpha_w$ was 90°. The strong crosswind was assumed to blow from the left side to the right side of the vehicle, as shown in Fig. 4.

Figure 6 illustrates the time histories of the lateral and rolling displacements of the car body under the excitation of crosswinds. When the vehicle enters the crosswind scenario, the dynamic response of the car body sharply increases, and there occurs a fierce transient fluctuation of the body. This fluctuation decays periodically with time and returns to a steady-state response. The oscillation period and the amplitude of the transient response of the vehicle system increase as the crosswind speed increases.

As shown in Fig. 6, the crosswind has a great influence on the ride comfort and safety of the passengers. As the crosswind speed increases, the dynamic responses of the car body become very strong. When the crosswind speed reaches 24 m/s, the maximum values of the car body rolling angle and lateral displacement exceed 3° and 50 mm, respectively. In this extreme situation, the high-speed vehicle overturns. Although the amplitudes of the rolling angle and lateral displacement are very large, overturning or derailment does not occur when the crosswind speed is less than 24 m/s. Furthermore, the transient responses of the car body rolling motion are much larger than the steady-state responses, as shown in Fig. 6a. The trends for the lateral displacement are similar, as shown in Fig. 6b. This means that the vehicle can easily overturn or derail during the fierce transient fluctuation period in strong crosswinds.

The derailment criteria most commonly used in the evaluation of the operating safety of a railway vehicle, including the flange climbing derailment coefficient $L/V$ and the wheel unloading ratio $\Delta V/V$, are calculated based on the wheel-rail contact forces. Therefore, the dynamic responses of the wheel-rail contact forces could reflect the derailment or rollover risk when high-speed vehicles operate in crosswinds. Figure 7 shows the time histories of the normal forces of the first wheelset. When the vehicle enters the crosswind scenario, fierce transient fluctuation of the normal forces occurs. The amplitudes of the normal forces on the leeward wheels, i.e., the right wheels, are much larger than those on the left wheels. In other words, the crosswind increases the normal loads on the leeward wheels and reduces the wheel loads on the windward side of the vehicle. When the wind speed is greater than 16 m/s, the minimum value of the normal forces is zero during the first oscillation period, which means that the windward wheels lose contact with the

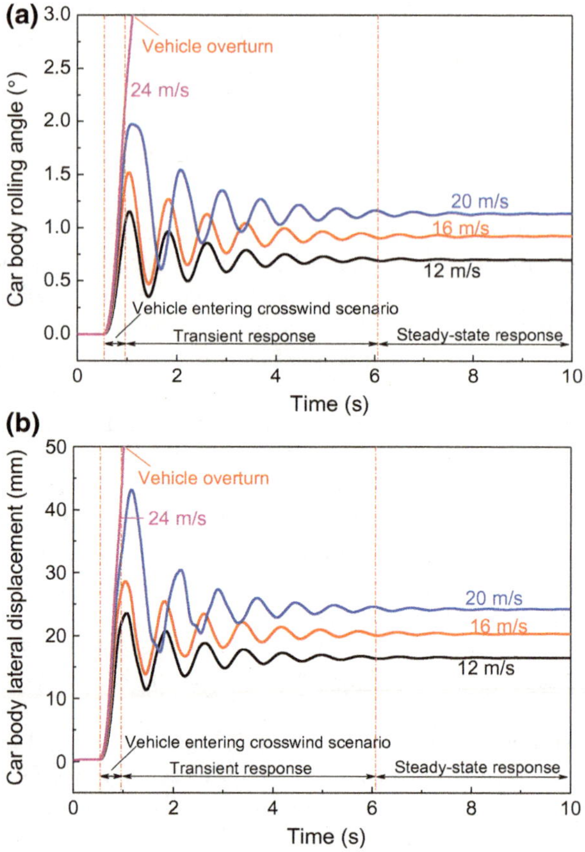


**Fig. 6** Rolling angles (**a**) and lateral displacements (**b**) of car body

left rail, as shown in Fig. 7a. At the same time, the maximum values of the normal forces are greater than 110 kN (Fig. 7b).

In this study, the windward wheels were found to lose contact with the rail and the vehicle overturning was found to take place when the crosswind speed reached 24 m/s (Fig. 7a). During the derailment process, fierce oscillation of the normal forces on the right wheels occurs (Fig. 7b). The maximum values of the normal forces in the course of the vehicle's transient response are much larger than the steady-state values. The variations in the normal forces were found to be similar for all of the wheels considered in the analysis.

The wheel unloading ratio $\Delta V/V$ is an important safety criterion for assessing the overturning risk of railway vehicles subjected to crosswinds. An analysis of $\Delta V/V$ for all wheelsets was therefore carried out for crosswind speeds from 12 to 24 m/s. The results are shown in Fig. 8. The maximum values of the wheel loading reduction occur in the first period, corresponding to the first oscillation period of the normal forces, as shown in Fig. 7. The maximum $\Delta V/V$ values for all of the cases considered in Fig. 8 are greater than 0.8, which is the current limit value of

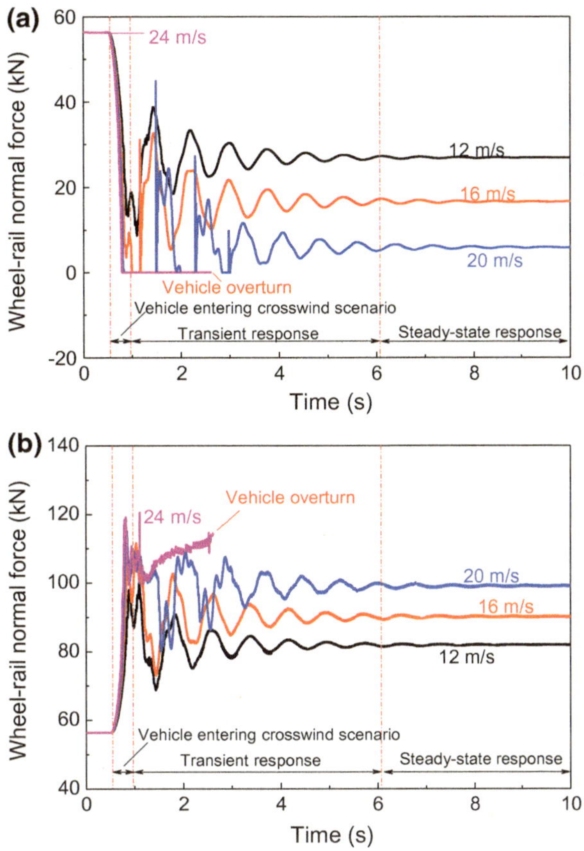


**Fig. 7** Responses of wheel-rail normal forces: **a** windward wheel (*left wheel*) and **b** leeward wheel (*right wheel*)

$\Delta V/V$ for safe operation of high-speed trains in China. For wind speeds in excess of 16 m/s, the peak values of $\Delta V/V$ for all wheelsets are equal to 1.0, which means that wheel-rail separation occurs, as shown in Fig. 8. Figures 7 and 8 show that when the crosswind speed reaches 24 m/s, derailment of the high-speed vehicle occurs.

Figure 8 illustrates another interesting issue that should be considered. Derailment or overturning does not occur when the value of $\Delta V/V$ exceeds 0.8 or even when its value reaches 1.0 (when the wheel-rail separation occurs). This means that a value of 0.8 for the wheel unloading criterion $\Delta V/V$ is somewhat conservative. A value of 0.8 for this criterion is thus not an accurate predictor of when vehicle derailment will occur. A more effective derailment assessment method should be put forward to address this problem.

In this study, the wheel rise $Z_{up}$ was used together with the wheel unloading creation $\Delta V/V$ to evaluate the running safety and derailment mechanism of a high-speed vehicle subjected to strong crosswinds. Figure 9 illustrates the time histories of the wheel rise $Z_{up}$ of the first wheelset, which were obtained from

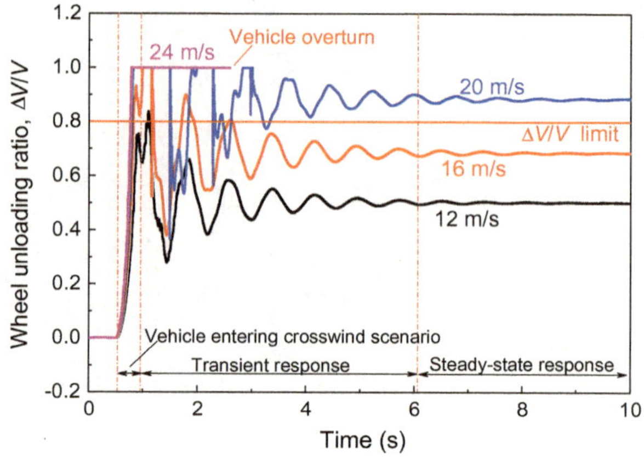


**Fig. 8** Responses of wheel unloading ratio $\Delta V/V$

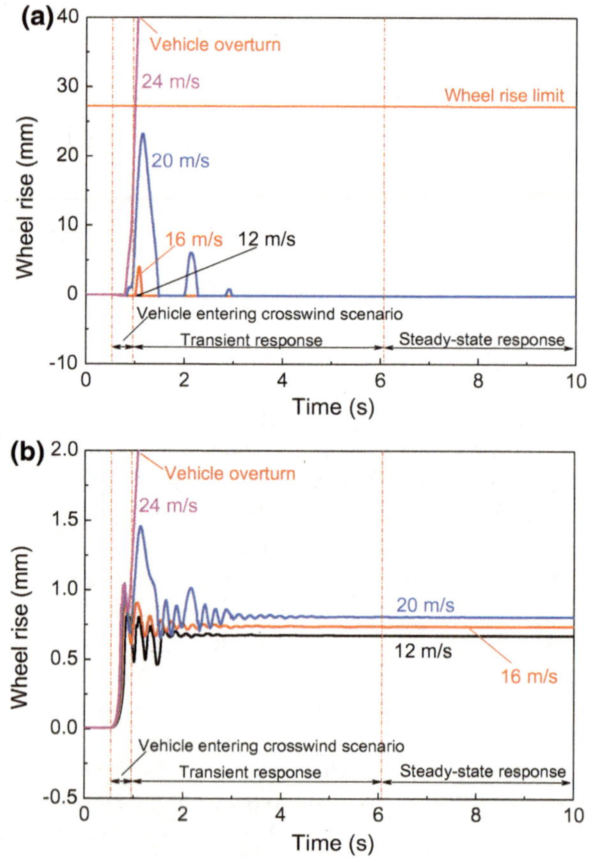


**Fig. 9** Responses of wheel rises: **a** windward wheel (*left wheel*) and **b** leeward wheel (*right wheel*)

calculations of the wheel-rail contact geometry during vehicle operation. The solid horizontal line indicates the wheel rise limit, namely, $\check{Z}_{up} = 28.272$ mm (Fig. 9a).

When the vehicle enters the crosswind scenario, the wheel rises of the leeward wheels increase gradually (Fig. 9b). In this situation, the leeward wheels climb up the right rail top. When the vehicle has entered the wind scenario completely, the crosswind rolling moment increases the vertical load on the climbing leeward wheels, and the wheels stop climbing. As a result, flange climbing derailment is not dominant in railway vehicle derailment caused by strong crosswinds. As the crosswind speed increases, the wheel rises of the windward wheels jump sharply (Fig. 9a). At crosswind speeds less than 20 m/s, the $Z_{up}$ of the windward wheels does not exceed the wheel rise limit $\check{Z}_{up} = 28.272$ mm. At a crosswind speed $U = 24$ m/s, the windward wheels lose the left rail constraint and the vehicle overturns.

## 4.2 *Effect of Crosswind Attack Angle*

In the calculations described above, only the constant crosswind attack angle was considered. The crosswind attack angle $\alpha_w$ can be expected to have a very important effect on the operating safety of the vehicle in crosswinds. Figure 10 illustrates the effects of the crosswind attack angle $\alpha_w$ on the wheel unloading ratio $\Delta V/V$ and the wheel rise $Z_{up}$ at various crosswind speeds. In these calculations, the vehicle operating speed was 300 km/h. The values of the other parameters were the same as those in the analysis described in Sect. 4.1.

As shown in Fig. 10, crosswind attack angles of 75°–90° correspond to the worst-case scenarios. At crosswind attack angles less than 75°, the wheel unloading ratio $\Delta V/V$ and the wheel rise $Z_{up}$ increase gradually as $\alpha_w$ increases. When $\alpha_w$ exceeds 90°, the values of these two derailment criteria decrease as the crosswind attack angle increases. As shown in Fig. 10, at different crosswind speeds considered, the rates of increase in the wheel unloading ratio $\Delta V/V$ are almost the same. However, the influence of the crosswind attack angle on the wheel rise $Z_{up}$ is greater at low crosswind speeds than at high crosswind speeds (Fig. 10b). These results indicate that the crosswind attack angle has a considerable effect on the likelihood of derailment and that the crosswind direction should be taken into account in assessing the safety of high-speed railway vehicles in operation.

## 4.3 *Combined Effects of Vehicle Speed and Crosswind Speed*

This section describes an analysis conducted to assess the combined effects of the vehicle speed and the crosswind speed on the derailment behavior of the vehicle.

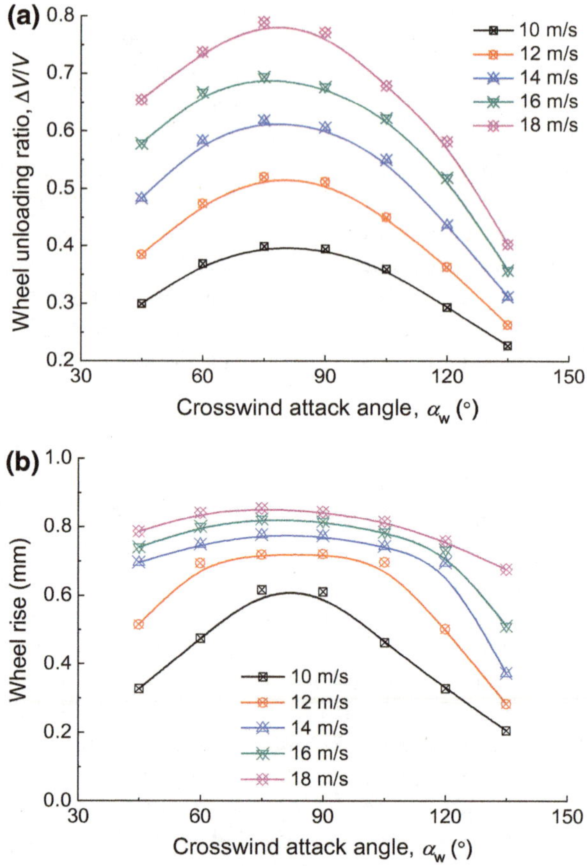


**Fig. 10** Wheel unloading ratio $\Delta V/V$ (**a**) and wheel rise Zup (**b**) versus crosswind attack angle $\alpha_w$

The vehicle speed $V_T$ was varied from 200 to 360 km/h, the crosswind speed $U$ was varied from 10 to 40 m/s, and the crosswind attack angle $\alpha_w$ was held constant at 90°. The other parameter values were the same as in the analysis described in Sect. 4.1.

Figure 11 illustrates the effects of the vehicle speed and crosswind speed on the maximum values of the wheel unloading ratio $\Delta V/V$ and the wheel rise $Z_{up}$ for all of the wheelsets. The bold solid line in Fig. 11a indicates the $\Delta V/V$ limit value of 0.8 that is used in evaluating the safety operations of high-speed trains in China. The flat top of the curved surface in Fig. 11a indicates that when the wheel unloading ratio reaches 1.0, separation of the windward wheels from the rails occurs for the combinations of vehicle speeds and crosswind speeds that fall within this area. The bold solid line in Fig. 11b indicates the $Z_{up}$ limit value of 28.272 mm. In plotting Fig. 11b, 28.272 mm was assigned to $Z_{up}$ when the wheel rise exceeded 28.272 mm. As a result, the top of the curved surface is flat. The variations in the values of $\Delta V/V$ and $Z_{up}$ indicate that both the vehicle speed and the wind speed greatly influence the operating safety of a high-speed train subjected to crosswinds.

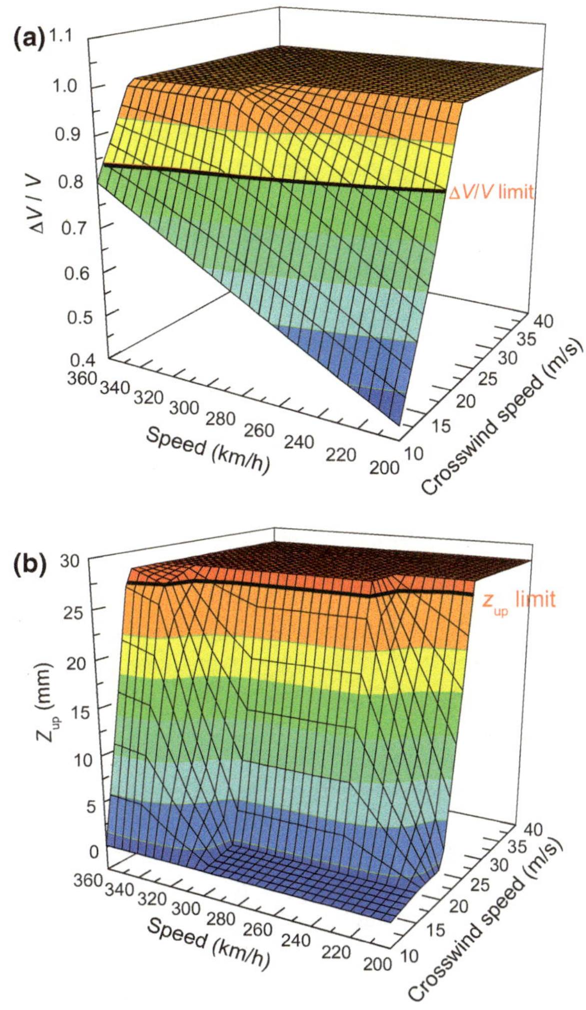


**Fig. 11** Wheel unloading ratio $\Delta V/V$ (**a**) and wheel rise $Z_{up}$ (**b**) versus vehicle speed $V_T$ and crosswind speed $U$

As expected, the crosswind speed $U$ greatly affects the wheel load reduction and wheel rise. As the crosswind speed increases, $\Delta V/V$ increases linearly (Fig. 11a). For vehicle operating speeds $V_T < 300$ km/h and crosswind speeds $U < 15$ m/s, $Z_{up}$ is much less than the limit value of 28.272 mm. Apart from these cases, however, the wheel rise increases rapidly with the mean crosswind velocity. For the range of vehicle speeds considered in this analysis, $Z_{up}$ exceeded the limit at crosswind speeds $U > 25$ m/s (Fig. 11b).

As the vehicle speed increases, the interaction between the vehicle and track increases. Furthermore, increasing the combined wind velocity relative to the vehicle decreases the yaw angle of the combined wind. As shown in Fig. 11, for the range of crosswind speeds considered in this analysis, the rates of increase of the wheel unloading ratio $\Delta V/V$ and the wheel rise $Z_{up}$ were almost the same. It is obvious that the operational speed has a great influence on the vehicle operating safety and that decreasing the operating speed decreases the risk of derailment of a railway vehicle in operation.

## 5 Evaluation of Operational Safety Area for High-Speed Vehicles Under Crosswind Excitations

To estimate the safety surplus of each criterion limit and identify the overturning boundary of high-speed vehicles subjected to crosswinds, the derailment boundaries determined from the dynamic simulation and the operational safety area defined by the safety assessment criteria discussed in Sect. 3 were calculated. The analysis results discussed in Sect. 4 clearly indicate that the crosswind attack angle $\alpha_w$, the vehicle operating speed $V_T$, and the crosswind speed $U$ have a great influence on the operating safety of high-speed vehicles subjected to crosswinds; hence, they were considered in this study to be three key parameters influencing the operating safety of the vehicle.

Figure 12 illustrates the derailment and operational safety boundaries obtained from the results of the dynamic simulation of the vehicle–track coupling for conditions of a tangent track and a steady crosswind. The results shown in Fig. 12a were obtained for crosswinds perpendicular to the direction of the vehicle's motion ($\alpha_w = 90°$) and vehicle speeds $V_T$ from 200 to 400 km/h. The results shown in Fig. 12b were obtained for a fixed vehicle speed of 300 km/h and crosswind attack angles $\alpha_w$ from 45° to 135°. The boundaries determined by the safety assessment criteria $L/V$, $(L/V)_s$, $T_{[L/V>0.8]}$, $\mathrm{Dis}_{[L/V>0.8]}$, $(L/V)_B$, $F_{ys}$, $\Delta V/V$, $V_D/V_0$, $Z_{up}$, and $y_{con}$ were treated as functions of the vehicle operating speed $V_T$ and the crosswind attack angle $\alpha_w$. The calculations were conducted for an $L/V$ limit of 0.8, an $(L/V)_s$ limit of 1.5, a $T_{[L/V>0.8]}$ limit of 50 ms duration, a $\mathrm{Dis}_{[L/V>0.8]}$ limit of 1.5 m, an $F_{ys}$ limit of $10 + P_0/3$, an $(L/V)_B$ limit of 0.6, a $\Delta V/V$ limit of 0.8, a $V_D/V_0$ limit of 0.8, a $Z_{up}$ limit of 28.272 mm, and a $y_{con}$ limit of −38.875 mm. The operational safety boundaries are defined as the separatrices that clearly indicate the safe operation area $A_S$, the warning area $A_W$, and the derailment area $A_D$. The safe operation area $A_S$ and the warning area $A_W$ are divided by the warning boundary $B_W$, which is determined by the boundary of the $\Delta V/V$ limit, as shown in Fig. 12. The boundary separating the derailment area from the warning area is defined as the derailment boundary $B_D$, as indicated by the upper solid curve in Fig. 12. The derailment boundary $B_D$ was determined from the results of the dynamic simulation of the coupled vehicle–track system.

As shown in Fig. 12, the safety boundaries determined by the $\Delta V/V$ limit are the lowest, and the operational safety area surrounded by the boundaries of the $\Delta V/V$ limit is the smallest. That is to say, the critical crosswind speed $U_{cr}$ determined by the $\Delta V/V$ criterion is the lowest. The limit boundaries of the $y_{con}$ criterion are close

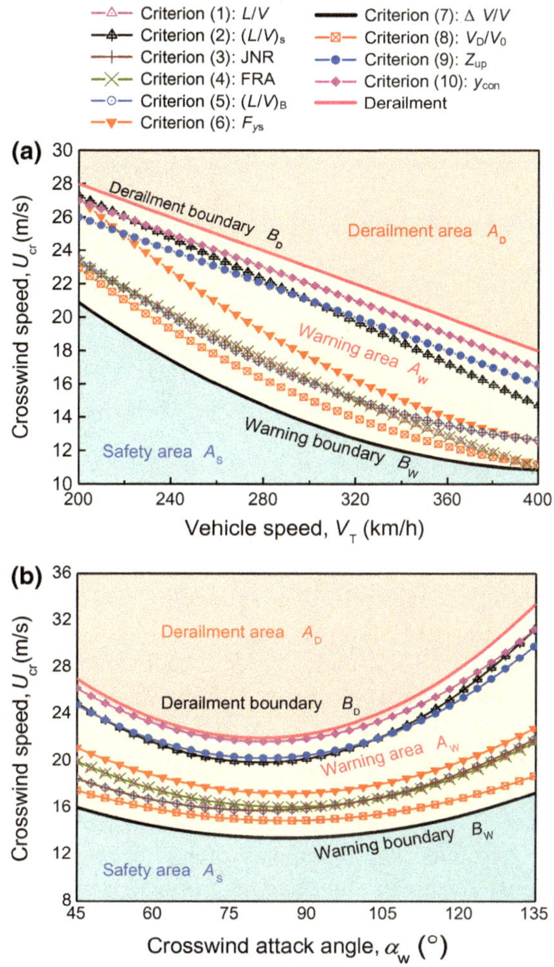


**Fig. 12** Operational safety boundaries as a function of vehicle speed $V_T$ (**a**) and crosswind attack angle $\alpha_w$ (**b**)

to the derailment (vehicle overturning) boundary $B_D$, which means that the critical crosswind speed $U_{cr}$ determined by the $y_{con}$ criterion is the highest. In other words, compared to the other criteria, the $\Delta V/V$ limit is the most conservative or the safest criterion for use in estimating the high-speed operational safety of high-speed railway vehicles in crosswinds, whereas the $y_{con}$ criterion is the least conservative or least safe one. Note that the boundaries determined by the other derailment criterion limits fall between the warning boundary $B_W$ and the derailment boundary $B_D$.

Figure 12 also shows the effects of the vehicle speed $V_T$ and the crosswind attack angle $\alpha_w$ on the safety boundaries and the critical crosswind speeds. The limiting crosswind $U_{cr}$ decreases as the vehicle speed increases, as shown in Fig. 12a. At crosswind attack angles of 75°–90°, the heights of the safety boundaries are the lowest, as shown in Fig. 12b. An increase in the crosswind speed at an attack angle of 75°–90° could easily lead to the overturning of a high-speed vehicle.

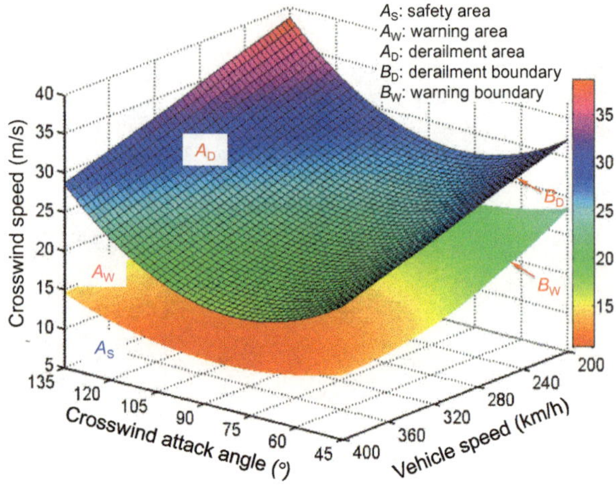


**Fig. 13** Derailment and safe operation areas in crosswinds

A comparison of the limiting values of the crosswind $U_{cr}$ at an attack angle of 45° (the vehicle operating with the wind) and an attack angle of 135° (the vehicle operating against the wind) shows that the vehicle operating against the crosswind is at a lower risk of derailment.

Figure 13 illustrates the derailment and safety areas for high-speed vehicles in crosswinds for vehicle speeds from 200 to 400 km/h, crosswind attack angles from 45° to 135°, and crosswind speeds from 0 to 40 m/s. The upper curved surface corresponds to the derailment boundary $B_D$, and the lower curved surface corresponds to the boundary $B_W$ for the safe operation of high-speed railway vehicles under the given conditions. The boundaries $B_W$ and $B_D$ divide the domain defined by the three key parameters that influence the dynamic behavior of high-speed railway vehicles subjected to crosswinds into three areas. The three areas are the safety area $A_S$, the warning area $A_W$, and the derailment area $A_D$. The three key factors of influence are the vehicle speed, the crosswind attack angle, and the crosswind speed.

The results shown in Fig. 13 can be used in automatic safety control systems installed on high-speed trains. If the sensors of the automatic safety control systems detect that at a crosswind attack angle of 90°, the vehicle and crosswind speeds approach those at the boundary $B_D$ or drop into the warning area $A_W$, the vehicle speed can be reduced rapidly to ensure the safe operation of the high-speed train.

## 6 Conclusion

In this study, a dynamic model for a coupled vehicle–track system was developed to investigate the effect of crosswinds on the operating safety of high-speed railway vehicles. The steady aerodynamic forces caused by crosswinds were modeled as ramp-shaped external forces exerted on the vehicle. Numerical analyses were conducted to investigate the dynamic responses and the dynamic derailment mechanism of a high-speed vehicle in strong crosswind scenarios. The effects of the crosswind speed, crosswind attack angle, and vehicle speed on the operating safety of the vehicle were examined. The operational safety area, warning area, and derailment area and their boundaries were defined and were calculated using the dynamic coupled vehicle–track model and existing criterion limits. The results obtained clearly indicate the operational safety surplus of each derailment criterion for a high-speed train operating in crosswinds, namely, the gap between the criterion limit boundary and the derailment boundary. The following conclusions can be drawn from the numerical results.

1. The crosswind has a great influence on the ride comfort and safety of railway passengers. As the crosswind speed increases, the dynamic responses of the car body and the wheel-rail forces increase linearly. Flange climbing does not play a key role in the likelihood of derailment of high-speed railway vehicles subjected to strong crosswinds. Overturning usually occurs when a vehicle enters into a crosswind scenario.
2. The crosswind attack angle, vehicle speed, and wind speed have a great influence on the operating safety and the likelihood of overturning of a high-speed vehicle operating in crosswinds. As the crosswind speed and vehicle speed increase, the wheel unloading ratio and the wheel rise increase linearly. Crosswind attack angles of 75°–90° correspond to the worst-case scenarios and have the greatest influence on the likelihood of derailment of such vehicles. The crosswind direction should also be taken into account in assessing the safety of high-speed railway vehicles operating in crosswinds.
3. The wheelset unloading ratio $\Delta V/V$ determines the boundary of the common safety area, which is the smallest area defined by the three key factors of influence. This area is considered the safety area for high-speed trains operating in crosswinds. The three key factors of influence are the vehicle speed, the crosswind speed, and the attack angle.

Note that the crosswind scenarios considered involved constant mean wind speeds in this study. In fact, real crosswind scenarios are unsteady and involve fluctuating wind speeds. Unsteady models, such as the "Chinese Hat" wind gust model (CEN 2010) or the "stochastic process" crosswind model (RSSB 2000; Cheli et al. 2006; Xu and Ding 2006), should be considered for use in future research.

It is not common to use a vehicle–track coupling model to evaluate the operating safety of railway vehicles operating in crosswinds. However, the dynamic behavior of vehicles subjected to crosswinds is influenced by many factors, some of which are unknown. Further research should be carried out to assess the sensitivity of the results to the parameters of the dynamic vehicle and track models. The proposed model can be used to assess the most important physical effects that should be modeled in dynamic simulation.

Because the aerodynamic characteristics of different vehicles in the same train may be different, the proposed vehicle–track coupling model needs to be improved to characterize the dynamic behavior of train-track interaction in severe crosswind conditions.

## References


Baker, C., Calleja, F., Jones, J., et al. (2004). Measurements of the cross wind forces on trains. *Journal of Wind Engineering and Industrial Aerodynamics, 92*(7–8), 547–563. doi:10.1016/j.jweia.2004.03.002.

Baker, C., Cheli, F., Orellano, A., et al. (2009). Cross wind effects on road and rail vehicles. *Vehicle System Dynamics, 47*(8), 983–1022. doi:10.1080/00423110903078794.

Carrarini, A. (2006). *Reliability based analysis of the crosswind stability of railway vehicles*. PhD Thesis, Berlin Institute of Technology, Berlin, Germany.

CEN (European Committee for Standardization). (2010). *Railway applications aerodynamics-part 6: Requirements and test procedures for cross wind assessment, EN 14067-6:2010*. Available from http://www.railwayvehiclestandards.com/csn-en-14067-6-railway-applications-aerodynamics-part-6-requirements-and-test-procedures-for-cross-wind-assessment/.

Cheli, F., Belforte, P., Melzi, S., et al. (2006). Numerical-experimental approach for evaluating cross-wind aerodynamic effects on heavy vehicles. *Vehicle System Dynamics, 44*(sup1), 791–804. doi:10.1080/00423110600886689.

Cheli, F., Ripamonti, F., Rocchi, D., et al. (2010). Aerodynamic behaviour investigation of the new EMUV250 train to cross wind using wind tunnel tests and CFD analysis. *Journal of Wind Engineering and Industrial Aerodynamics, 98*(4–5), 189–201. doi:10.1016/j.jweia.2009.10.015.

Chen, G., & Zhai, W. M. (2004). A new wheel/rail spatially dynamic coupling model and its verification. *Vehicle System Dynamics, 41*(4), 301–322. doi:10.1080/00423110412331315178.

Diedrichs, B. (2005). *Computational methods for crosswind stability of railway trains: A literature survey* (pp. 26–36). Stockholm, Sweden: Department of Aeronautical and Vehicle Engineering, Royal Institute of Technology.

Evans, A. W. (2011). Fatal train accidents on Europe's railways: 1980–2009. *Accident Analysis and Prevention, 43*(1), 391–401. doi:10.1016/j.aap.2010.09.009.

Fujii, T., Maeda, T., Ishida, H., et al. (1999). Wind induced accidents of train vehicles and their measures in Japan. *Quarterly Report of RTRI, 40*(1), 50–55. doi:10.2219/rtriqr.40.50.

Gawthorpe, R. G. (1994). Wind effects on ground transportation. *Journal of Wind Engineering and Industrial Aerodynamics, 52*, 73–92. doi:10.1016/0167-6105(94)90040-X.

Jin, X. S., Xiao, X. B., Ling, L., et al. (2013). Study on safety boundary for high-speed trains running in severe environments. *International Journal of Rail Transportation, 1*(1–2), 87–108. doi:10.1080/23248378.2013.790138.

Kalker, J. J. (1967). *On the rolling contact of two elastic bodies in the presence of dry friction*. PhD Thesis, Delft University, the Netherlands.

Knothe, K., & Grassie, S. L. (1993). Modeling of railway track and vehicle/track interaction at high frequencies. *Vehicle System Dynamics, 22*(3–4), 209–262. doi:10.1080/00423119308969027.

Ling, L., Xiao, X. B., & Jin, X. S. (2012). Study on derailment mechanism and safety operation area of high speed trains under earthquake. *Journal of Computational and Nonlinear Dynamics, 7*(4), 041001. doi:10.1115/1.4006727.

Nadal, M. J. (1896). Theorie de stabilit'e des Locomotives, part 2, Mouvement de Lacet. *Annales des Mines, 10*, 232. (in French).

OJEU (Official Journal of the European Union). (2008). *Technical specification for interoperability of high speed rolling stock, TSI/HS-RST-L64-7/3/2008:2008*. Available from http://www.era.europa.eu/Document-Register/Pages/HS-RST-TSI.aspx.

Orellano, A., & Schober, M. (2003). On side-wind stability of high-speed trains. *Vehicle System Dynamics, 40*(sup), 143–160.

RSSB (Rail Safety and Standards Board). (2000). *Resistance of railway vehicles to roll-over in gales, GM/RT2142:2000*. Available from http://www.rgsonline.co.uk/Railway_Group_Standards/RollingStock/RailwayGroupStandards.

Shen, Z. Y., Hedrick, J. K., & Elkins, J. A. (1983). A comparison of alternative creep-force models for rail vehicle dynamic analysis. *Vehicle System Dynamics, 12*(1–3), 79–83. doi:10.1080/00423118308968725.

Silla, A., & Kallberg, V. P. (2012). The development of railway safety in Finland. *Accident Analysis and Prevention, 45,* 737–744. doi:10.1016/j.aap.2011.09.043.

Weinstock, H. (1984). Wheel climb derailment criteria for evaluation of rail vehicle safety. In *Proceedings of the ASME Winter Annual Meeting* (pp. 1–7). New York, USA.

Wilson, N., Fries, R., & Witte, M. (2011). Assessment of safety against derailment using simulations and vehicle acceptance tests: A worldwide comparison of state-of-the-art assessment methods. *Vehicle System Dynamics, 49*(7), 1113–1157. doi:10.1080/00423114.2011.586706.

Wu, H., & Wilson, N. (2006). Railway vehicle derailment and prevention. In S. Inwicki (Ed.), *Handbook of railway vehicle dynamics* (pp. 209–238). London, UK: Taylor & Francis.

Xiao, X. B., Jin, X. S., Deng, Y. Q., et al. (2008). Effect of curved track support failure on vehicle derailment. *Vehicle System Dynamics, 46*(11), 1029–1059. doi:10.1080/00423110701689602.

Xiao, X. B., Jin, X. S., Wen, Z. F., et al. (2011). Effect of tangent track buckle on vehicle derailment. *Multibody System Dynamics, 25*(1), 1–41. doi:10.1007/s11044-010-9210-2.

Xinhua News Agency. (2007). Train overturned by strong wind in NW China. *China Daily*, February 28.

Xu, Y. L., & Ding, Q. S. (2006). Interaction of railway vehicles with track in cross-winds. *Journal of Fluids and Structures, 22*(3), 295–314. doi:10.1016/j.jfluidstructs.2005.11.003.

Yokose, K. (1966). A theory of the derailment of a wheelset. *Quarterly Report of RTRI, 7*(3), 30–34.

Zhai, W. M., Cai, C. B., & Guo, S. Z. (1996). Coupling model of vertical and lateral vehicle/track interactions. *Vehicle System Dynamics, 26*(1), 61–79. doi:10.1080/00423119608969302.

## Author Biography

**Xue-song Jin,** PhD, is a professor at Southwest Jiaotong University, Chengdu, China. He is a leading scholar of wheel-rail interaction in China, and is a State Council Expert for Special Allowance. He is the author of three academic books and over 200 articles. He is an editorial board member of several journals, and has been a member of the Committee of the International Conference on Contact Mechanics and Wear of Rail/Wheel Systems for more than 10 years. He has been a visiting scholar at the University of Missouri-Rolla for 2.5 years. Now his research focuses on wheel-rail interaction, rolling contact mechanics, vehicle system dynamics, vibration and noise.

# Part III
# High-Speed Rail Infrastructure and Material Innovations

# A 2.5D Finite Element Approach for Predicting Ground Vibrations Generated by Vertical Track Irregularities

Xue-cheng Bian, Chang Chao, Wan-feng Jin and Yun-min Chen

**Contents**



## 1 Introduction

High-speed railway lines inevitably pass through densely populated urban areas. Train-induced environmental vibrations have received widely-expressed concerns from residents and railway constructors.

The load acting on railway track generally can be regarded as a combination of the moving quasi-static load and dynamic excitation (Sheng et al. 2003; Auersch 2005; Lombaert and Degrande 2009). The dynamic excitation comes from several sources, such as a parametric excitation due to the discrete supports of the rails, a transient excitation due to the rail joints and wheel flats, and the excitation due to wheel and rail roughness, and track unevenness (Heckl et al. 1996; Sheng et al. 2003).

X. Bian (✉) · C. Chao · W. Jin · Y. Chen
Department of Civil Engineering, MOE Key Laboratory of Soft Soils and Geoenvironmental Engineering, Zhejiang University, Hangzhou 310058, China
e-mail: bianxc@zju.edu.cn

Y. Chen
e-mail: chenyunmin@zju.edu.cn

 
Y. Fang and Y.H. Zhang (eds.), *China's High-Speed Rail Technology*, Advances in High-speed Rail Technology, https://doi.org/10.1007/978-981-10-5610-9_12

Vibrations induced by a moving quasi-static load have been extensively studied. A prediction model for ground vibration, coupling the quasi-static force and the Green function of a half-space, was proposed by Krylov (1995). Vibration induced by a constant or harmonic load moving along a beam resting on layered half-space is presented by Sheng et al. (1999a, b) using an analytical solution. However, the existing model with a moving quasi-static load underestimates the actual response intensities of the railway structure and underlying ground, especially for higher excitation frequencies (Katou et al. 2008). Train wheel axle weights are also not the main factor in generating mid-frequency bands in the far field (Degrande and Lombaert 2001; Takemiya and Bian 2005). The train–track–ground dynamics interaction analysis indicates that the dynamic force generated at the wheel–rail contact cannot be ignored for evaluating ground vibration (Adolfsson et al. 1999; Sheng et al. 2003, 2004). For train travels at speeds below the velocity of wave propagation in the ground, the dynamic excitation is more important than the quasi-static axle loads for the generation of environmental vibration. When train's speed approaches or exceeds the Rayleigh wave velocity in the soil, the quasi-static excitation mechanism dominates the soil response (Galvin et al. 2010a).

Most aforementioned works use analytical solutions to express ground vibration due to moving loads. Although analytical solutions can help us understand the vibration generation mechanism due to traffic loadings, they cannot deal with complex track structures and the railway foundation. Rigueiro et al. (2010) concluded from theoretical analysis and experimental verification that a train–track dynamic interaction model is more reasonable than a moving force model for accessing vibration of ballasted track structure. A 3D multi-body and finite element-boundary element-coupled model was presented by Galvin and Dominguez (2007) to predict vibrations in the time domain, considering the dynamic interaction between train and track.

The literature survey shows that dynamic excitation due to a train running on track with irregularities is essential to the evaluation of train traffic-induced environment vibration.

As an alternative to full 3D finite element models, the so-called 2.5D models have been proposed for the prediction of railway-induced ground vibrations (Yang and Hung 2001; Takemiya 2003; Bian et al. 2008; Galvin et al. 2010b). This method effectively reduces the computational efforts and storage requirements.

In this paper, first, the essential procedures used in establishing the governing equations in terms of 2.5D finite elements are summarized. Then, the one-quarter car model and the harmonic track irregularity are combined to derive the dynamic excitation on track. The vehicle stiffness matrix describing the train motions will be coupled into the 2.5D finite element model for track and ground in the wave-number domain. Some computation results will be presented to discuss the effect of train speed and track irregularities on ground vibration.

## 2 2.5D Finite Element Method

In this section, the major steps to derive a 2.5D finite element model will be briefly introduced. Since the material and geometry of track structure and its supporting ground can be regarded as constant in the direction of the train's movement, the Fourier transform with respect to space coordinate in this direction is applied to simplify the 3D problem. If we assume the train runs in the $x$-direction, the Fourier transform with respect to $x$-coordinate is defined as

$$u^x(\xi_x) = \int_{-\infty}^{+\infty} u(x)\exp(\mathrm{i}\xi_x x)\mathrm{d}x, \tag{1}$$

and its corresponding inverse transform is given by

$$u(x) = \frac{1}{2\pi}\int_{-\infty}^{+\infty} u^x(\xi_x)\exp(-\mathrm{i}\xi_x x)\mathrm{d}\xi_x, \tag{2}$$

where the variable with superscript '$x$' represents the components in the wave-number domain. $\xi_x$ is the $x$-directional wave number.

The 2.5D finite element model of track–ground system is shown in Fig. 1. Double wheel loads $q(t)$ move in track's direction.

The motions of ground with homogeneous and isotropic material assumptions can be described by Navier's equations in the frequency domain:

$$\mu^* u^t_{i,jj} + (\lambda^* + \mu^*)u^t_{j,ji} + \omega^2 \rho u^t_i + f^t_i = 0, \tag{3}$$

where $\omega$ is the excitation frequency, $\rho$ is the material's density, and $\lambda$ and $\mu$ are Lame constants. In this study, the complex Lame constants $\lambda^*$, $\mu^*$ are used to consider the damping effect of wave propagation in ground, $\lambda^* = (1 + 2i\beta)\lambda$, $\mu^* = (1 + 2i\beta)\mu$, where $\beta$ is the damping ratio of ground soil. Variables with superscript '$t$' represent the components in the frequency domain.

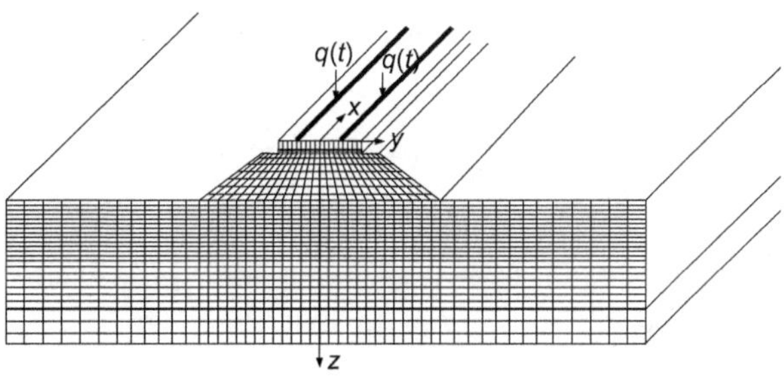


**Fig. 1** 2.5D finite element model of track–ground system

Strain components of an element can be given in the light of the small strain assumption. The predefined Fourier transform in $x$-direction is applied and yields the expressions in the wave-number domain:

$$\begin{aligned} \varepsilon_{xx}^{xt} &= -\mathrm{i}\xi_x \mathrm{u}^{xt}, \quad \gamma_{xy}^{xt} = -\mathrm{i}\xi_x \mathrm{v}^{xt} + \frac{\partial \mathrm{u}^{xt}}{\partial \mathrm{y}}, \\ \varepsilon_{yy}^{xt} &= \frac{\partial v^{xt}}{\partial y}, \quad \gamma_{yz}^{xt} = \frac{\partial w^{xt}}{\partial y} + \frac{\partial v^{xt}}{\partial z}, \\ \varepsilon_{zz}^{xt} &= \frac{\partial w^{xt}}{\partial z}, \quad \gamma_{zx}^{xt} = \frac{\partial u^{xt}}{\partial z} - \mathrm{i}\xi_x \mathrm{w}^{xt}. \end{aligned} \tag{4}$$

Consequently, the strain–displacement relationship in the frequency and wave-number domain can be expressed by

$$\boldsymbol{\varepsilon}^{xt} = \boldsymbol{B}\boldsymbol{u}^{xt}, \tag{5}$$

where $\boldsymbol{\varepsilon}^{xt}$ and $\boldsymbol{u}^{xt}$ are the strain vector and displacement vector of the element, respectively, and their detailed expressions are given as below:

$$\boldsymbol{B} = \begin{bmatrix} -\mathrm{i}\xi_x & 0 & 0 & \frac{\partial}{\partial y} & 0 & \frac{\partial}{\partial z} \\ 0 & \frac{\partial}{\partial y} & 0 & -\mathrm{i}\xi_x & \frac{\partial}{\partial z} & 0 \\ 0 & 0 & \frac{\partial}{\partial z} & 0 & \frac{\partial}{\partial y} & -\mathrm{i}\xi_x \end{bmatrix}^{\mathrm{T}}, \quad \boldsymbol{\varepsilon}^{xt} = \left[\varepsilon_{xx}^{xt} \quad \varepsilon_{yy}^{xt} \quad \varepsilon_{zz}^{xt} \quad \gamma_{xy}^{xt} \quad \gamma_{yz}^{xt} \quad \gamma_{zx}^{xt}\right]^{\mathrm{T}},$$
$$\boldsymbol{u}^{xt} = \left[u^{xt} \quad v^{xt} \quad w^{xt}\right]^{\mathrm{T}}. \tag{6}$$

In addition, the stress—strain relationship can be given by

$$\boldsymbol{\sigma}^{xt} = \boldsymbol{D}\varepsilon^{xt}, \tag{7}$$

where $\boldsymbol{\sigma}^{xt}$ is the stress vector of the element, and $\boldsymbol{D}$ is the elastic matrix:

$$\boldsymbol{D} = \begin{bmatrix} \lambda^* + 2\mu^* & \lambda^* & \lambda^* & 0 & 0 & 0 \\ & \lambda^* + 2\mu^* & \lambda^* & 0 & 0 & 0 \\ & & \lambda^* + 2\mu^* & 0 & 0 & 0 \\ & & & \mu^* & 0 & 0 \\ & \text{sym} & & & \mu^* & 0 \\ & & & & & \mu^* \end{bmatrix}. \tag{8}$$

Since the $\boldsymbol{D}$ matrix is symmetric, only its upper part is given in Eq. (8). The ground is modeled by the quadrilateral element in such a way that the transversal vertical section is meshed by the finite elements, whose nodal displacements are defined in three degrees of freedom. A quadrilateral element with either four or eight nodes can be used to discretize the near field of track and ground. By introducing the shape function $\boldsymbol{N}$, the discretized form of the governing equation in the frequency domain can be derived by the conventional finite element method.

$$\left(\boldsymbol{K}^{xt} - \omega^2\boldsymbol{M}\right)\boldsymbol{U}^{xt} = \boldsymbol{F}^{xt}, \tag{9}$$

where $\boldsymbol{U}^{xt}$ is the displacement vector in frequency and wave-number domain, and $\boldsymbol{M}$, $\boldsymbol{K}^{xt}$, $\boldsymbol{F}^{xt}$ are the mass matrix, stiffness matrix, and equivalent nodal force vector, respectively, and their detailed expressions are

$$\boldsymbol{M} = \sum_e \rho \iint \boldsymbol{N}^{\mathrm{T}}\boldsymbol{N}|\boldsymbol{J}|\mathrm{d}\eta\mathrm{d}\zeta, \tag{10a}$$

$$\boldsymbol{K}^{xt} = \sum_e \rho \iint (\boldsymbol{B}^*\boldsymbol{N})^{\mathrm{T}}\boldsymbol{D}(\boldsymbol{B}\boldsymbol{N})|\boldsymbol{J}|\mathrm{d}\eta\mathrm{d}\zeta, \tag{10b}$$

$$\boldsymbol{F}^{xt} = \sum_e \iint \boldsymbol{N}^{\mathrm{T}}\boldsymbol{f}|\boldsymbol{J}|\mathrm{d}\eta\mathrm{d}\zeta, \tag{10c}$$

where $\eta$ and $\zeta$ are element's local coordinates, and '$e$' represents element-wise integration, $f$ is the external load acting on this element. $\boldsymbol{J}$ is the Jacobi matrix, and $|\boldsymbol{J}|$ is the corresponding determinant, $|\boldsymbol{J}| = \det\boldsymbol{J}$.

The 2.5D finite element method described here has been implemented into the computation code TRAVIB. Since the finite element itself cannot deal with the unbounded soil medium directly, because of the element size, an appropriate boundary must be constructed to prevent wave reflection back into the near field at the edge of the finite element zone. In this study, a frequency-dependent dashpot with viscous components normal and tangent to a given boundary is introduced to simulate the infinity of the ground (Bian et al. 2008).

## 3 Mathematical Model of Train Running on Track with Harmonic Irregularities

Figure 2 shows the one-quarter car model to represent the train. In the model, $k_{\mathrm{p}}$, $c_{\mathrm{p}}$ represent the stiffness and damping of the primary suspension, and $k_{\mathrm{s}}$, $c_{\mathrm{s}}$ represent the stiffness and damping of the secondary suspension. The masses of the car body, the bogie, and the wheel sets are represented by $m_{\mathrm{c}}/4$, $m_{\mathrm{b}}/2$, and $m_{\mathrm{w}}$, respectively. Since the bogie is supported by two wheel axles, one axle shares one half of the total bogie weight. The primary suspension connects the wheels to the bogie, therefore, $k_{\mathrm{p}}$, $c_{\mathrm{p}}$ represent two times the total primary vertical stiffness and viscous damping. The car body is supported by two bogies and every axle carries one quarter of the car body mass, $m_{\mathrm{c}}/4$. $k_{\mathrm{s}}$ and $c_{\mathrm{s}}$ represent one half of the secondary vertical stiffness and viscous damping, respectively. The train is supposed to move

**Fig. 2** One-quarter car model

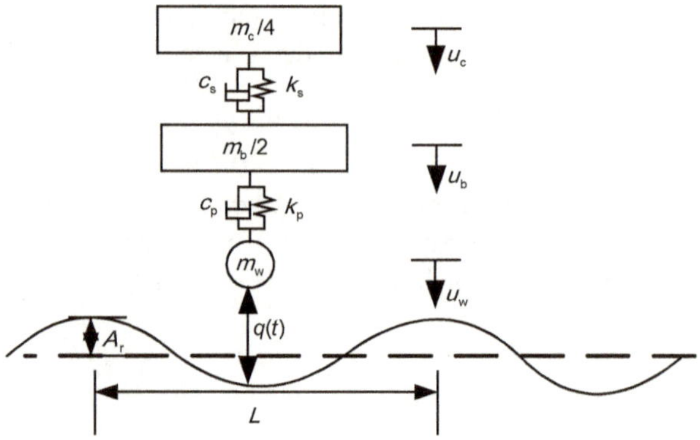


in track's direction at speed $c$. The vertical movements of car body, bogie, and wheels are represented by $u_c$, $u_b$, and $u_w$, respectively. The irregularities at rail surface are defined as a cosine distribution with amplitude $A_r$ and wavelength $L$.

The equilibrium equation of the one-quarter car model can be written as

$$\begin{pmatrix} 0.25m_c & 0 & 0 \\ 0 & 0.5m_b & 0 \\ 0 & 0 & m_w \end{pmatrix}\begin{pmatrix} \ddot{u}_c(t) \\ \ddot{u}_b(t) \\ \ddot{u}_w(t) \end{pmatrix} + \begin{pmatrix} c_s & -c_s & 0 \\ -c_s & c_p + c_s & -c_p \\ 0 & -c_p & c_p \end{pmatrix}\begin{pmatrix} \dot{u}_c(t) \\ \dot{u}_b(t) \\ \dot{u}_w(t) \end{pmatrix}$$
$$+ \begin{pmatrix} k_s & -k_s & 0 \\ -k_s & k_p + k_s & -k_p \\ 0 & -k_p & k_p \end{pmatrix}\begin{pmatrix} u_c(t) \\ u_b(t) \\ u_w(t) \end{pmatrix} = \begin{pmatrix} 0.25m_c g \\ 0.50m_b g \\ m_w g - q(t) \end{pmatrix}. \quad (11)$$

By applying the Fourier transform with respect to time $t$ to Eq. (11), we obtain the interaction forces $q(t)$ in the frequency domain $q(w)$, which can be rearranged as follows:

$$\tilde{q}(\omega) = W_1\delta(\omega) + W_2\delta(\omega - \omega_r) + W_3\delta(\omega + \omega_r), \quad (12)$$

where $\omega_r = 2\pi c/L$, $\omega_1 = 0$, $\omega_2 = \omega_r$, $\omega_3 = -\omega_r$, $W_1$, $W_2$, and $W_3$ can be expressed as

$$W_1 = m_w g + 0.5m_b g + 0.25m_c g,$$
$$W_2 = \frac{A_r}{2} \cdot \frac{-\omega_r^2 L_1 - i\omega_r^3 L_2 + \omega_r^4 L_3 + i\omega_r^5 L_4 - \omega_r^6 L_5}{L_6 + i\omega_r L_7 - \omega_r^2 L_8 - i\omega_r^3 L_9 + \omega_r^4 L_{10}},$$
$$W_3 = \frac{A_r}{2} \cdot \frac{-\omega_r^2 L_1 + i\omega_r^3 L_2 + \omega_r^4 L_3 - i\omega_r^5 L_4 - \omega_r^6 L_5}{L_6 - i\omega_r L_7 - \omega_r^2 L_8 + i\omega_r^3 L_9 + \omega_r^4 L_{10}},$$

where

$$
\begin{aligned}
L_1 &= k_s k_p (m_w g + 0.5 m_b g + 0.25 m_c g),\\
L_2 &= (c_p k_s + k_p c_s)(m_w g + 0.5 m_b g + 0.25 m_c g),\\
L_3 &= c_s c_p (m_w g + 0.5 m_b g + 0.25 m_c g) + k_s (0.25 m_c m_w\\
&\quad + 0.5 m_b m_w) + k_p (0.125 m_c m_b + 0.25 m_c m_w),\\
L_4 &= c_s (0.25 m_c m_w + 0.5 m_b m_w)\\
&\quad + c_p (0.125 m_c m_b + 0.25 m_c m_w),\\
L_5 &= 0.125 m_c m_b m_w,\\
L_6 &= k_s k_p,\\
L_7 &= c_p k_s + k_p c_s,\\
L_8 &= c_s c_p + 0.25 (k_s + k_p) m_c + 0.5 m_b k_s,\\
L_9 &= 0.25 (c_s + c_p) m_c + 0.5 m_b c_s,\\
L_{10} &= 0.125 m_c m_b.
\end{aligned}
$$

For the train traveling at speed $c$ on the track, the interaction forces in the frequency domain can be written as

$$
\begin{aligned}
\tilde{q}(\omega - \xi_x c) &= W_1 \delta(\omega - \xi_x c) + W_2 \delta(\omega - \omega_r - \xi_x c)\\
&\quad + W_3 \delta(\omega + \omega_r - \xi_x c)\\
&= \frac{1}{c} \sum_{i=1}^{3} W_i \delta\left(\xi_x + \frac{\omega - \omega_i}{c}\right).
\end{aligned}
\tag{13}
$$

Equation (13) provides the dynamic loading due to a one-quarter car running at track with irregularities and can be readily coupled into Eq. (11).

## 4 Numerical Results and Discussion

To demonstrate the application of the proposed computation approach for the prediction of train-induced track and ground vibrations, a typical high-speed railway is adopted in the numerical computation. The mass of double rails per unit length is 120 kg/m. The bending stiffness of rail is $1.26 \times 10^7$ N m$^2$, and its loss factor is 0.01. A simplified illustration of the computation model is shown in Fig. 3. Four observation points are indicated in Fig. 4 at the locations of *A*, *B*, *C*, and *D*. Their distances to the track centerline are 0, 6.25, 11.25, and 21.25 m, respectively. Table 1 shows the parameters of the train, and Table 2 shows the parameters of the track structure and ground.

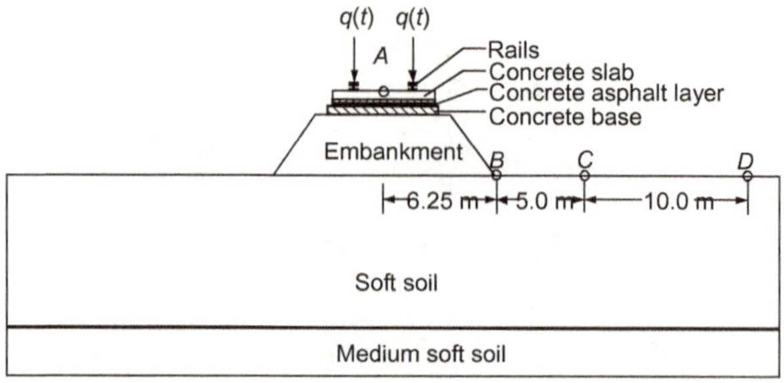


**Fig. 3** Transverse section of the track and ground (not to scale)

## *4.1 Effect of Amplitude of Track Irregularities on Dynamic Responses*

The wavelength of the track vertical irregularity usually lies within the range 0.125–14 m, which can be divided into short wavelength and long wavelength zones. In this study, two typical wavelengths, $L = 0.5$ m and 10.0 m, are chosen to be analyzed. Two typical train speeds are chosen, $c = 50$ m/s and 100 m/s, respectively. In this track–ground model, the shear wave velocity of the upper soft soil is 81.1 m/s, and the train running at high speed can exceed it easily. Three amplitudes of the track irregularities are chosen, $A_r = 0$ mm, 3 mm, and 10 mm. The ground vibration induced by a train passing a track with irregularity wavelength $L = 0.5$ m and 10 m at speed $c = 50$ m/s are presented in Figs. 4 and 5, respectively. For the train running at high-speed $c = 100$ m/s, the computation results presented in Figs. 6 and 7 for wavelength of irregularities at $L = 0.5$ m and 10 m, respectively.

From Figs. 4 and 5, it is seen that dynamic responses both at track and on ground increase monotonically with the irregularity amplitude. Compared to the smooth track, $A_r = 0$ mm, the track irregularities have significant impact on track and ground vibration, especially at a further distance from the track center. The comparison of the results in Figs. 4 and 5 indicates that long wavelength irregularities generate low frequency vibration, which attenuates slowly in more distant ground. Finally, the vibration intensity at the embankment toe is about 10% of that at the track center, which means that the embankment effectively reduces the vibration transmission to adjacent ground.

From Figs. 6 and 7, it is interesting to note that when train's running speed surpasses the critical velocity of the track–ground system (the Rayleigh wave velocity of the upper soft soil layer), track and ground vibrations show different features compared to a slow-speed train. For the track with short wavelength irregularities, track vibrations generated are still dominated by high-frequency responses due to track irregularities, but at a further distance, vibration generated by train's wheel axle weights becomes dominant. In sharp contrast, for the track with long wavelength irregularities, both track and ground vibrations are produced by the movement of train's wheel axle weights. There is very little difference between

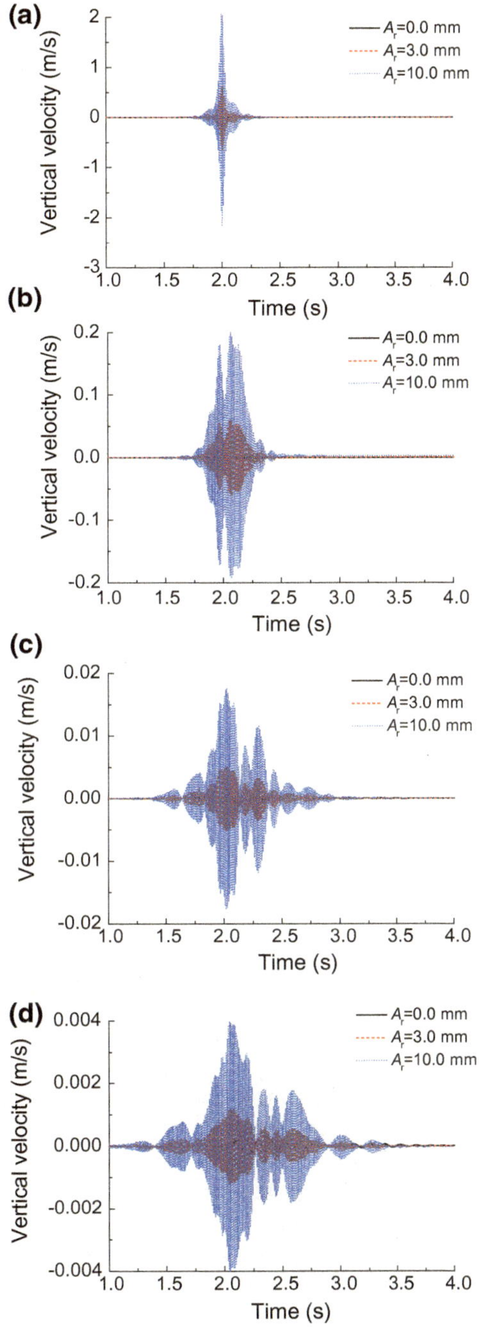


**Fig. 4** Vertical velocity responses for conditions of $L = 0.5$ m and $c = 50$ m/s. **a** Point *A* at track center; **b** point *B* at $y = 6.25$ m; **c** point *C* at $y = 11.25$ m; **d** point *D* at $y = 21.25$ m

**Table 1** Parameters for vehicle model

| Parameter | Value | Parameter | Value |
|---|---|---|---|
| $m_c$ (kg) | 4240 | $c_p$ (N/m) | $5.00 \times 10^3$ |
| $m_b$ (kg) | 3400 | $k_s$ (N/m) | $4.00 \times 10^5$ |
| $m_w$ (kg) | 2200 | $c_s$ (N/m) | $6.00 \times 10^3$ |
| $k_p$ (N/m) | $1.04 \times 10^6$ | | |

**Table 2** Parameters of the track and ground

| Material | Poisson's ratio | Density (kg/m$^3$) | Thickness (m) | Damping ratio | Elastic modulus (MPa) |
|---|---|---|---|---|---|
| Concrete slab | 0.20 | 2500 | 0.20 | 0.05 | 24,000 |
| Concrete asphalt mortar | 0.30 | 1800 | 0.05 | 0.20 | 100 |
| Concrete base | 0.20 | 2400 | 0.30 | 0.05 | 10,000 |
| Embankment | 0.36 | 1400 | 3.00 | 0.05 | 170 |
| Soft soil | 0.47 | 1550 | 15.00 | 0.05 | 30 |
| Medium soft soil | 0.49 | 1750 | 5.00 | 0.05 | 116 |

the vibrations generated by train running on smooth track or track with irregularities.

The comparisons among Figs. 4 and 6, and Figs. 5 and 7 show that, generally, short wavelength irregularities generate more significant ground vibrations than the longer ones.

To explore the vibration attenuation mechanism with the distance from track center, the following definition is adopted:

$$H = 20\lg\left(\frac{V}{V_0}\right), \tag{14}$$

where $V$ is the amplitude of actual vibration velocity, and $V_0$ is the reference value with amplitude of $10^{-8}$ m/s.

As mentioned above, two typical wavelengths are chosen to investigate the influence of track irregularity upon the vibration intensities due to train running. The computation results are presented in Figs. 8 and 9 for train speeds at 50 and 100 m/s, respectively.

Figure 8 shows the variation of vertical vibration intensity with distance from track center for train speed $c$ = 50 m/s. From the computation results, it is found that the irregularity amplitude has sharp impact on the vertical response intensity for low-speed train, both for short wavelength and long wavelength irregularities. However, there are still some differences for these two cases. For short wavelength case, the vibration curves are parallel to each other with the distance increase; however, for long wavelength case, the vibration curves divergence increase gradually with the distance. Especially for the long wavelength case, there is

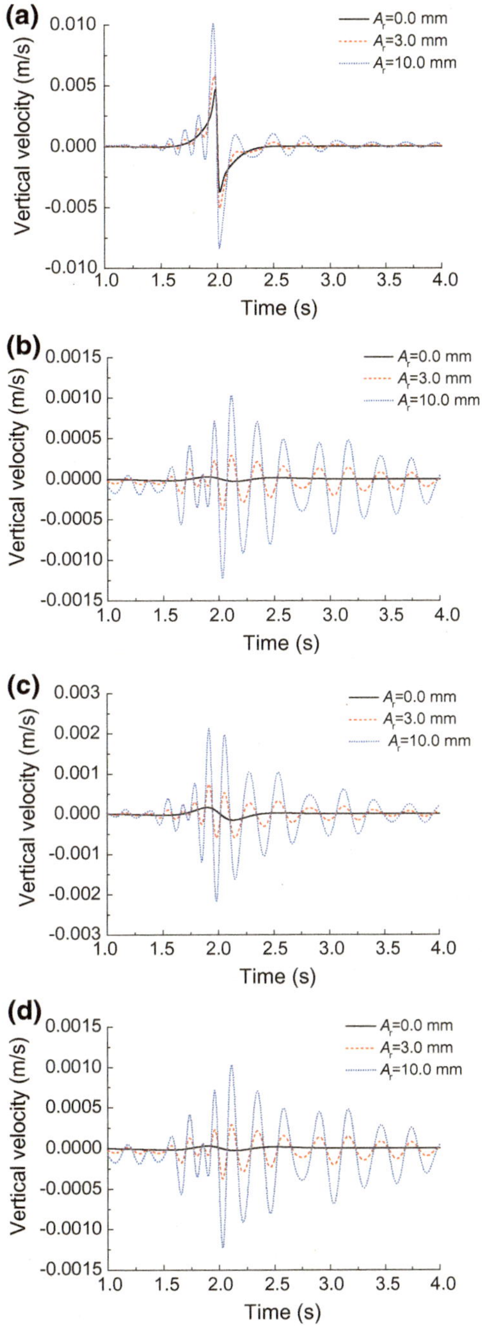


**Fig. 5** Vertical velocity responses for conditions of $L$ = 10.0 m and $c$ = 50 m/s. **a** Point $A$ at track center; **b** point $B$ at $y$ = 6.25 m; **c** point $C$ at $y$ = 11.25 m; **d** point $D$ at $y$ = 21.25 m

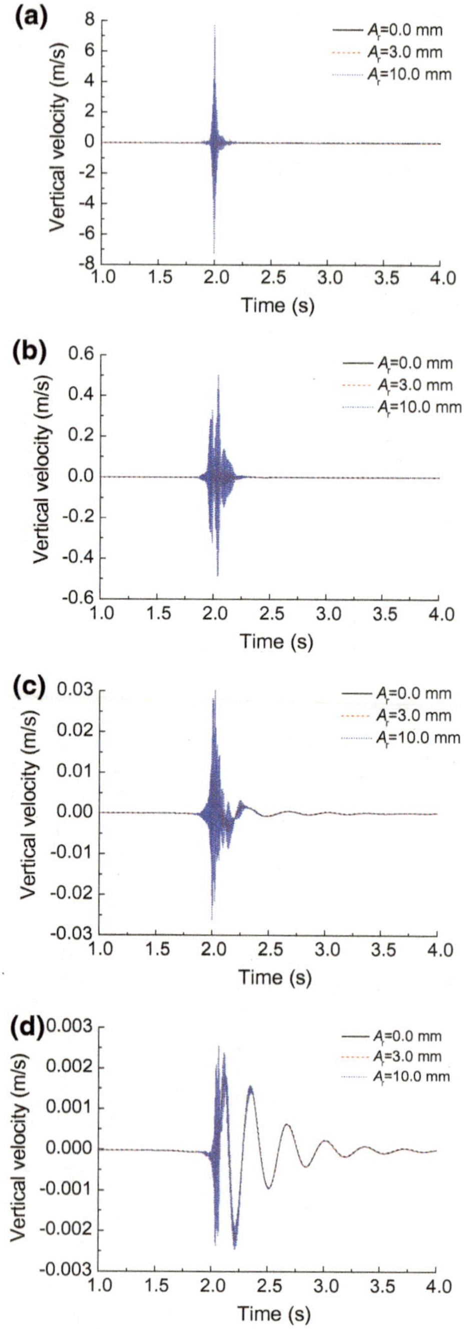


**Fig. 6** Vertical velocity responses for conditions of $L = 0.5$ m and $c = 100$ m/s. **a** Point $A$ at track center; **b** point $B$ at $y = 6.25$ m; **c** point $C$ at $y = 11.25$ m; **d** point $D$ at $y = 21.25$ m

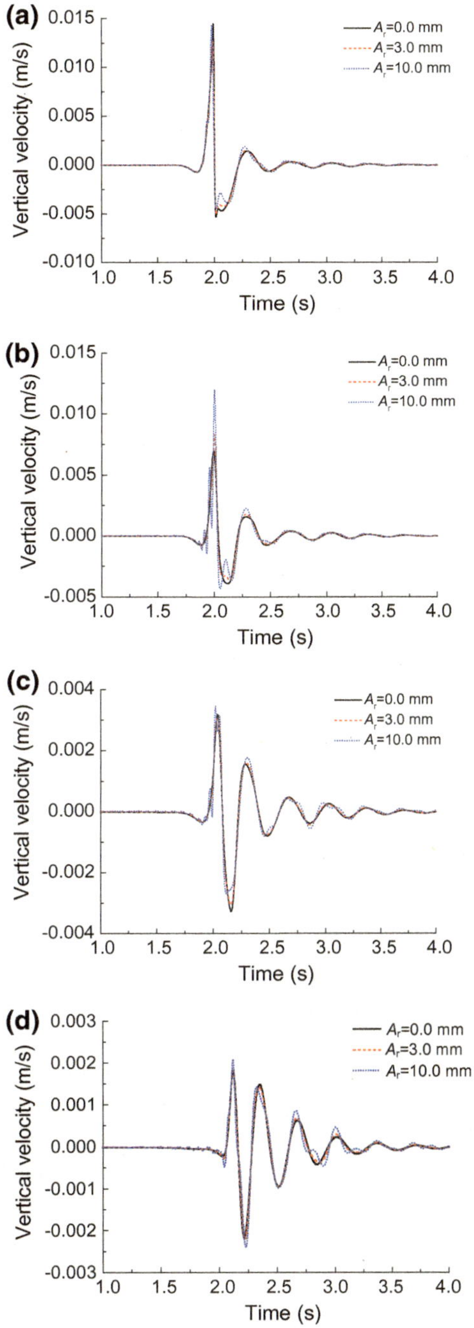


**Fig. 7** Vertical velocity responses for conditions of $L = 10.0$ m and $c = 100$ m/s. a Point $A$ at track center; **b** point $B$ at $y = 6.25$ m; **c** point $C$ at $y = 11.25$ m; **d** point $D$ at $y = 21.25$ m

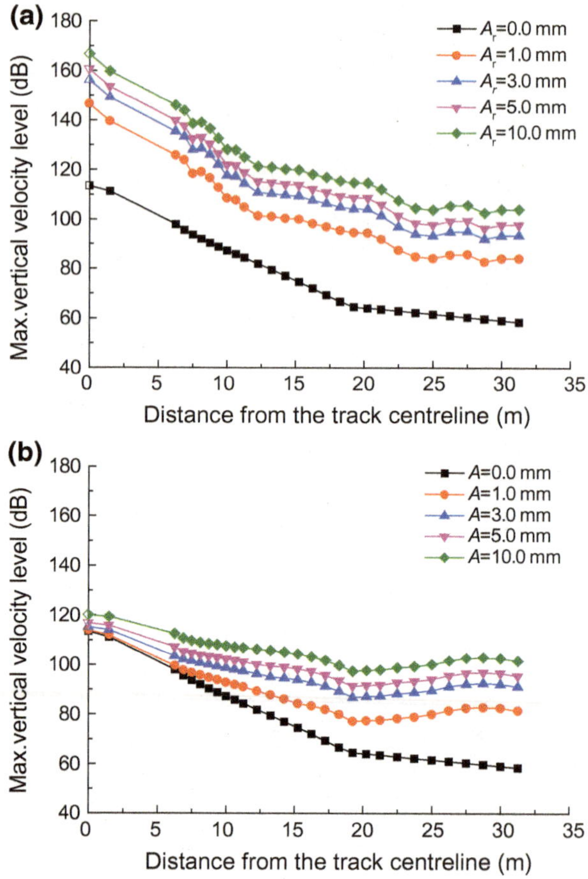


**Fig. 8** Variation of vertical vibration intensity with distance from track center for train speed $c$ = 50 m/s. **a** $L$ = 0.5 m; **b** $L$ = 10 m

vibration amplification at distance about 27 m from the track center, while for the short wavelength case, ground vibration intensity decreases almost monotonically with distance from track center.

Figure 9 shows the variation of vertical vibration intensity with distance from track center for train speed $c$ = 100 m/s. These results show that the irregularity amplitude has a sharp impact on the vertical vibration intensity at ground adjacent to the track for short wavelength case. However, this is not obvious for the long wavelength case and at the ground distant from the track center.

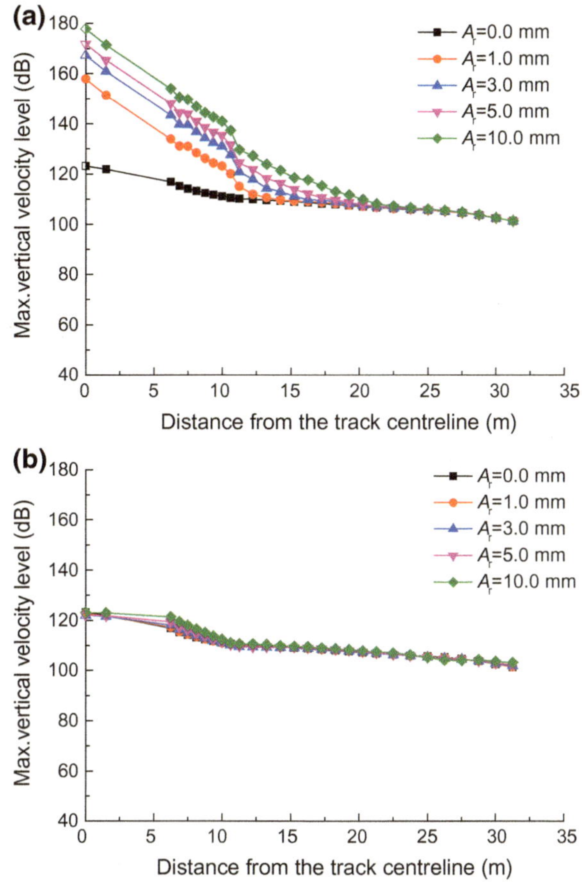


**Fig. 9** Variation of vertical vibration intensity with distance from track center for train speed $c = 100$ m/s. **a** $L = 0.5$ m; **b** $L = 10$ m

## *4.2 Effect of Wavelength of Track Irregularities on Dynamic Responses*

The influence of irregularity wavelength on vibration under $c = 50$ m/s and $c = 100$ m/s are shown in Fig. 10. In this section, the irregularity amplitude is fixed as 3 mm. Figure 10 shows that the track irregularity with shorter wavelength can generate stronger track vibration both for low-speed and high-speed cases. However, when train runs at a low speed, vibration induced by track irregularities can be transmitted over a long distance. When a train runs at high-speed, the wavelength of track irregularities has very little effect on ground vibration at further distances from the track center. Therefore, it can be concluded that the ground vibration further away from the track is dominated by low frequency vibration.

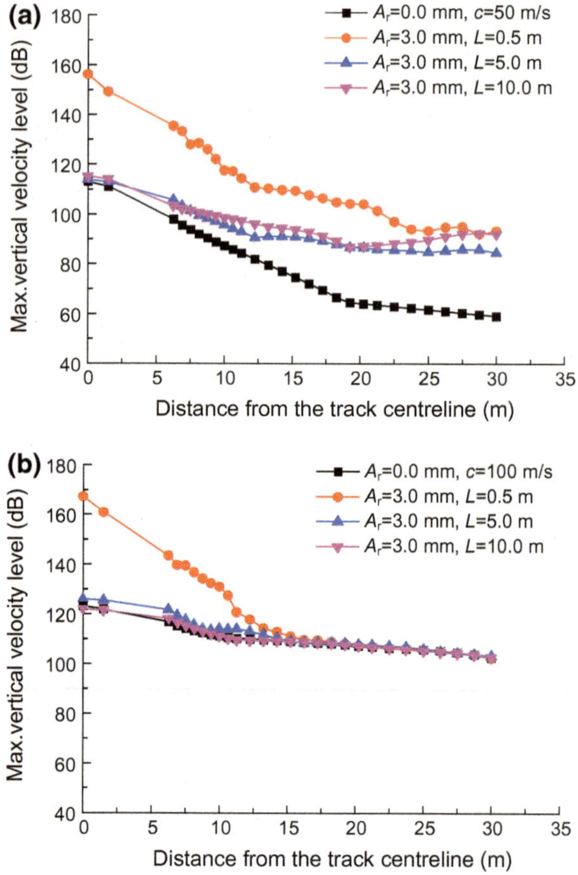


**Fig. 10** Influence of track irregularity wavelength on ground vibration. **a** $c$ = 50 m/s; **b** $c$ = 100 m/s

## 5 Conclusions

In this paper, the track irregularities and one-quarter car model are coupled into the 2.5D finite element model of track and ground. The effects of track irregularities on the track and ground vibrations are discussed. The wave motions at ground surface due to a train running at different speeds are also presented. The following conclusions are made by parametric analyses.

Two main parameters of track irregularities, amplitude and wavelength, have crucial influences on track and ground vibrations, but they operate via different mechanisms. The irregularity amplitude has a direct impact on the vertical response for low-speed trains, both for short wavelength and long wavelength irregularities. Track irregularity with shorter wavelength can generate stronger track vibration both for low-speed and high-speed cases. For low-speed case, vibrations induced by track irregularities dominate far-field responses. For high-speed case, the wavelength of track irregularities has very little effect on ground vibration at distances far from track center.

When train's running speed is close to the critical velocity of the track–ground system, track and ground vibrations show different features. For the short wavelength track irregularities case, track vibrations generated are dominated by high-frequency responses due to track short wavelength irregularities, while at further distances, ground vibration generated by train's wheel axle weights becomes dominant. For the track with long wavelength irregularities, both track and ground vibrations are produced by the movement of the train's wheel axle weights. There is very little difference between the vibrations generated by trains running on a smooth track or a track with irregularities.

## References


Adolfsson, K., Andreasson, B., Bengtson, P. E., Bodare, A., Madshus, C., Massarch, R., et al. (1999). High speed lines on soft ground. Evaluation and analysis of measurements from the west coast line. Technical Report, Banverket, Sweden.

Auersch, L. (2005). The excitation of ground vibration by rail traffic: theory of vehicle-track-soil interaction and measurements on high-speed lines. *Journal of Sound and Vibration, 284*(1–2), 103–132. doi:10.1016/j.jsv.2004.06.017.

Bian, X. C., Chen, Y. M., & Hu, T. (2008). Numerical simulation of high-speed train induced ground vibrations using 2.5D finite element approach. *Science in China Series G: Physics Mechanics and Astronomy, 51*(6), 632–650. doi:10.1007/s11433-008-0060-3.

Degrande, G., & Lombaert, G. (2001). An efficient formulation of Krylov's prediction model for train induced vibrations based on the dynamic reciprocity theorem. *Journal of the Acoustical Society of America, 110*(3), 1379–1390. doi:10.1121/1.1388002.

Galvin, P., & Dominguez, J. (2007). High-speed train-induced ground motion and interaction with structures. *Journal of Sound and Vibration, 307*(3–5), 755–777. doi:10.1016/j.jsv.2007.07.017.

Galvin, P., Françoisa, S., Schevenelsa, M., Bonginic, E., Degrandea, G., & Lombaerta, G. (2010a). A 2.5D coupled FE-BE model for the prediction of railway induced vibrations. *Soil Dynamics and Earthquake Engineering, 30*(12), 1500–1512. doi:10.1016/j.soildyn.2010.07.001.

Galvin, P., Romero, A., & Dominguez, J. (2010b). Fully three-dimensional analysis of high-speed train-track-soil-structure dynamic interaction. *Journal of Sound and Vibration, 329*(24), 5147–5163. doi:10.1016/j.jsv.2010.06.016.

Heckl, M., Hauck, G., & Wettschureck, R. (1996). Structure-borne sound and vibration from rail traffic. *Journal of Sound and Vibration, 193*(1), 175–184. doi:10.1006/jsvi.1996.0257.

Katou, M., Matsuoka, T., Yoshioka, O., Sanada, Y., & Miyoshi, T. (2008). Numerical simulation study of ground vibrations using forces from wheels of a running high-speed train. *Journal of Sound and Vibration, 318*(4–5), 830–849. doi:10.1016/j.jsv.2008.04.053.

Krylov, V. V. (1995). Generation of ground vibration by superfast trains. *Applied Acoustics, 44*(2), 149–164. doi:10.1016/0003-682X(95)91370-I.

Lombaert, G., & Degrande, G. (2009). Ground-borne vibration due to static and dynamic axle loads of InterCity and high-speed trains. *Journal of Sound and Vibration, 319*(3–5), 1036–1066. doi:10.1016/j.jsv.2008.07.003.

Rigueiro, C., Rebelo, C., & Da Silva, L. S. (2010). Influence of ballast models in the dynamic response of railway viaducts. *Journal of Sound and Vibration, 329*(15), 3030–3040. doi:10.1016/j.jsv.2010.02.002.

Sheng, X., Jones, C., & Petyt, M. (1999a). Ground vibration generated by a harmonic load acting on a railway track. *Journal of Sound and Vibration, 225*(1), 3–28. doi:10.1006/jsvi.1999.2232.

Sheng, X., Jones, C., & Petyt, M. (1999b). Ground vibration generated by a load moving along a railway track. *Journal of Sound and Vibration, 228*(1), 129–156. doi:10.1006/jsvi.1999.2406.

Sheng, X., Jones, C., & Thompson, D. J. (2003). A comparison of a theoretical model for quasi-statically and dynamically induced environmental vibration from trains with measurements. *Journal of Sound and Vibration, 267*(3), 621–635. doi:10.1016/S0022-460X(03)00728-4.

Sheng, X., Jones, C., & Thompson, D. J. (2004). A theoretical model for ground vibration from trains generated by vertical track irregularities. *Journal of Sound and Vibration, 272*(3–5), 937–965. doi:10.1016/S0022-460X(03)00782-X.

Takemiya, H. (2003). Simulation of track-ground vibrations due to a high-speed train: The case of X-2000 at Ledsgard. *Journal of Sound and Vibration, 261*(3), 503–526.

Takemiya, H., & Bian, X.C. (2005). Substructure simulation of inhomogeneous track and layered ground dynamic interaction under train passage. *Journal of Engineering Mechanics, 131*(7), 699–711. doi:10.1061/(ASCE)0733-9399(2005)131:7(699).

Yang, B. Y., & Hung, H. H. (2001). A 2.5D finite/infinite element approach for modelling visco-elastic bodies subjected to moving loads. *International Journal for Numerical Methods in Engineering, 240,* 1317–1336. doi:10.1002/nme.208.

## Author Biographies

**Xue-cheng Bian** Ph.D., is a professor of Transpiration Geotechnical Engineering at Zhejiang University, China. He received his PhD degree from Okayama University, Japan, and was a visiting professor in the University of Illinois at Urbana-Champaign, USA, and the University of Edinburgh, UK. He was awarded the Newton Advanced Fellowship by the Royal Society, UK, in 2014, and entitled as Changjiang Young Scholar professor by the Ministry of Education of China in 2015. He also received the NSFC Excellent Young Scientist Award in 2012. He has published over 30 papers in international journals.

**Yun-min Chen** Ph.D., is a professor of Geotechnical Engineering and Dean of the Faculty of Engineering at Zhejiang University, China, and Fellow of the Chinese Academy of Sciences. He received his Ph.D. degree from Zhejiang University in 1989 and worked as PostDoc fellow in IFCO in the Netherlands. His main research interests include soft foundation engineering, geotechnical earthquake engineering, and geoenvironmental engineering. He is involved in consulting projects on the construction of the first airport runway and high-speed railway on soft clay in China, a thermal power plant foundation on liquefiable ground in Indonesia, and the expansion project of the highest municipal.

# Smart Elasto-Magneto-Electric (EME) Sensors for Stress Monitoring of Steel Structures in Railway Infrastructures

**Yuan-feng Duan, Ru Zhang, Yang Zhao, Siu-wing Or, Ke-qing Fan and Zhi-feng Tang**

**Contents**



## 1 Introduction

Steel structures are widely used in railway infrastructures, such as roofs of railway stations, large span cable-stayed bridges, suspension bridges, and steel rails. Structures and facilities under long-term dynamic/static loads deteriorate over time

---


Y.F. Duan (✉) · R. Zhang · Y. Zhao · Z.F. Tang
College of Civil Engineering and Architecture, Zhejiang University, Hangzhou 310058, China
e-mail: ceyfduan@zju.edu.cn

Y. Zhao
e-mail: ceyzhao@zju.edu.cn

S.W. Or
Department of Electrical Engineering, The Hong Kong Polytechnic University, Hong Kong, China

K.Q. Fan
School of Information Engineering, Wuyi University, Jiangmen 529020, China





Y. Fang and Y.H. Zhang (eds.), *China's High-Speed Rail Technology*, Advances in High-speed Rail Technology, https://doi.org/10.1007/978-981-10-5610-9_13

and may become unsafe. Monitoring their stress state is crucial for safety evaluation and in deciding whether to prolong their service life or to retrofit them. The fracture of critical steel components may induce the failure of whole roofs. Damage to steel cables may result in the collapse of whole bridges. The stress distribution in steel rails has a significant influence on the growth rate of cracks and thus affects the occurrence of rail failures (Cannon and Pradier 1996; Sasaki et al. 2008). The stress state of steel rails is the most important determinant for the proper maintenance and for failure prevention of railways, especially for high-speed railways (Cannon et al. 2003; Ekberg and Kabo 2005). However, the non-destructive stress monitoring of in-service steel structures is still a challenging task for civil engineering communities. The current stress monitoring methods, such as those using electric resistance-, vibrating wire-, or optical fiber-strain gauges, or the vibrating frequency method, are unable or unable easily to measure the actual stress (not the relative variation of stress) of in-service steel structures.

An elasto-magnetic (EM) sensor is a promising tool for stress monitoring of steel structures, due to its outstanding superiorities including corrosion resistance, actual-stress measurement, nondestructive monitoring, and long service life. It is composed of a primary coil providing variable flux to the measured steel component, and a secondary (sensing) coil picking up the induced electromotive force that is directly proportional to the change rate of the applied magnetic flux according to Faraday's law of electromagnetic induction. In recent years, several investigators (Kvasnica and Fabo 1996; Wang et al. 1998, 2001; Wang and Wang 2004; Tang et al. 2008) have developed magnetoelastic theory-based systems that are being applied to monitor the stress of civil engineering structures. Kvasnica and Fabo (1996) developed a microcomputer-based instrument as an application of magnetoelastic theory for the investigation of new principles in the non-destructive measurement of large mechanical stress. They found that the change of permeability with tension was linear during the magnetic saturation of the low-carbon steel wires used in the building industry. Wang et al. (1998) introduced the concept of utilizing a novel sensor technology for monitoring structures. They also developed, fabricated, and tested an EM sensor for the direct measurement of stress in steel cables. The sensor was magnetized by a removable C-shaped circuit, rather than by a solenoid (Wang et al. 2001). They later exploited a U-shaped EM sensor (Wang and Wang 2004). Tang et al. (2008) devised a steel strand tension sensor with a different single bypass excitation structure to solve the temperature and installation problems. These techniques, in addition to conventional stress monitoring techniques (Sasada et al. 1986; Seekircher and Hoffmann 1989; Kleinke and Uras 1994; Bartels et al. 1996; Brophy and Brett 1996), are continuously increasing our ability to monitor stress and damage in real time and to obtain an accurate assessment of the actual and future performance of a structure.

Nevertheless, some problems restrict the application of such EM sensors. To magnetize the steel members to magnetic saturation, the primary coil usually has to be large. Precise installation of the secondary coil in accordance with theoretical assumptions and principles is not easy and normally requires skilled technique and a coiling machine. The use of a secondary coil as the signal detection element requires signal integration, which takes a long time and results in a non-real-time monitoring mode. Furthermore, the coils themselves influence the signals due to interference and noise components, resulting in lower accuracy.

This paper presents a novel smart elasto-magneto-electric (EME) sensor for stress monitoring of steel structures, in which the secondary coil is replaced by magneto-electric (ME) laminated composites as the sensing unit. A steel bar was selected as a test specimen. After introducing its working principles, we describe the testing and verification of the performance of the ME sensing unit using a Hall device. A tension test of the selected steel bar was carried out to characterize the developed smart EME sensor.

## 2 Tested Steel Bars

Cylindrical bars, 12 mm in diameter and 800 mm in length, made of steel 45 and processed according to the Chinese National Standard "Quality Carbon Structural Steels" (GB/T699-1999) were used in this study. Their chemical composition and mechanical properties are presented in Tables 1 and 2, respectively. The magnetic properties of this material can be described as follows: coercivity $H_c$ is 592 A/m, remanence $B_r$ is 0.9 T, maximum relative permeability $\mu_{rm}$ is 583 and the corresponding magnetic field strength $H$ is 960 A/m, and the maximum magnetic energy product $(HB)_{max}$ is 0.2 kJ/m$^3$. These data served as a reference to test the reliability of our experiments. The magnetic characteristic curves (Bozorth 1951; Ke et al. 2003), including the fundamental magnetization curve ($B$-$H$), the permeability curve ($\mu$-$H$), the remanence curve ($B_r$-$H$), and the hysteresis loop (the desending

**Table 1** Chemical composition of the steel bars

| Designation | Unified numerical code | Element (%) | | | | | |
|---|---|---|---|---|---|---|---|
| | | C | Si | Mn | Cr | Ni | Cu |
| 45 | U20452 | 0.42–0.50 | 0.17–0.37 | 0.50–0.80 | ≤ 0.25 | ≤ 0.30 | ≤ 0.25 |

**Table 2** Mechanical properties of the steel bars

| Designation | Tensile strength (MPa) | Yield strength (MPa) | Elongation (%) | Percentage reduction of area (%) | Hardness of steel material in delivery state | |
|---|---|---|---|---|---|---|
| | | | | | Non-heat treated steel | Annealed |
| 45 | ≥ 600 | ≥ 355 | ≥ 16 | ≥ 40 | ≤ 229 | ≤ 197 |

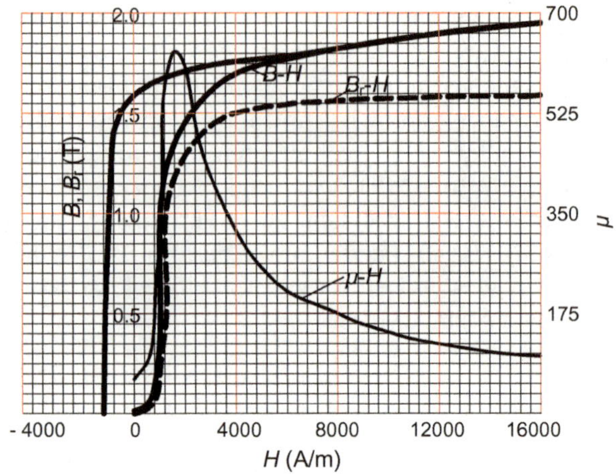


**Fig. 1** Magnetic characteristic curves of the tested steel bars

parts in the first and second quadrants), are shown in Fig. 1 (Bozorth 1951; Ke et al. 2003). All the tests were conducted at room temperature to avoid thermal fluctuation.

# 3 Magneto-Electric Sensing Unit

## 3.1 Working Principle

Figure 2 shows a photograph of the proposed smart sensing unit, which was made of a Terfenol-D alloy/Pb($Zr_{0.52}Ti_{0.48}$)$O_3$ (PZT) ceramic ME laminated composite (Jia et al. 2007; Wang et al. 2008a, 2008b). This is a new form of magnetostrictive/piezoelectric composite material with superior ME effect due to the product effect of the piezoelectric effect and the magnetostrictive effect (Dong et al. 2003). The plates were 12 mm long, 6 mm wide, and 1 mm thick. Under the action of an external magnetic field, mechanical stains arise in the sandwiched Terfenol-D plate due to a magnetostriction effect. These strains are transferred to the PZT plate through the adhesive layer, where they produce an electric signal

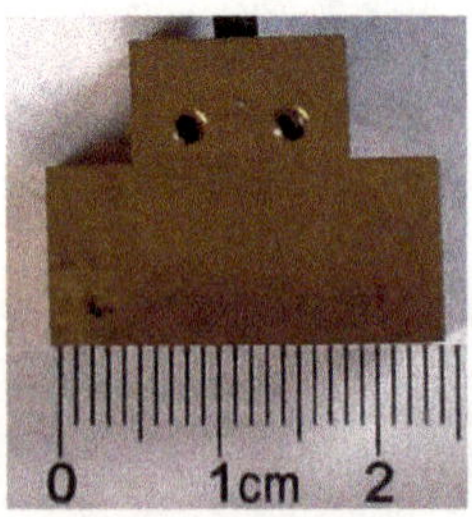


**Fig. 2** ME sensing unit

owing to the piezoelectric effect. The magnetic sensor can be used to measure both direct current (*DC*) and alternating current (*AC*) magnetic fields without an external power supply, either in a 1D or multi-dimensional magnetic field, and can produce a large output voltage in real time, 2000 times higher than the traditional Hall devices. The structure and design of the magnetic sensor are very simple, and each sensing unit is a magnetic component made of smart materials in a packaged solid state. Because of its ultrahigh ME voltage coefficients $\alpha_v$, defined by an induced electrical voltage in response to an applied AC magnetic field $\left(\frac{dV}{dH}\right)$, this magnetic sensor was adopted as the power-free ME sensing unit in our smart EME sensor. Significant advantages such as convenience, low cost, small size, a large magnetic conversion coefficient, fast response, and high sensitivity make these magnetic/piezoelectric laminated materials suitable for the design of the smart EME sensor.

## 3.2 Performance Tests

In practical applications, for a given ME voltage coefficient $\alpha_v$ and the magnetic induction $B_G$, the peak-to-peak value of the ME sensing unit output $V_{ME,pp}$ can be obtained:

$$V_{ME,pp} = \alpha_v \cdot B_G \tag{1}$$

Therefore, performance tests of the ME sensing unit were conducted as shown in Fig. 3. The main instruments included an oscilloscope, a signal generator, and a power amplifier.

By energizing the solenoid with an AC current supply (YE1311 Series Sweep Signal Generator, SINOCERA Piezotronics, Inc.) and a power amplifier (YE5871

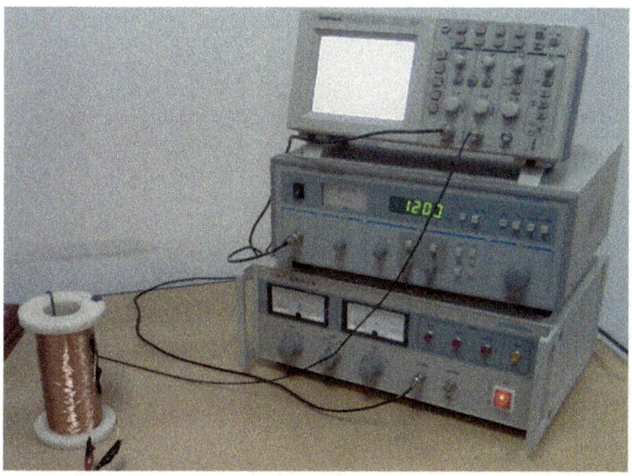

**Fig. 3** Experimental setup for determining the relationship between $V_{ME}$ (mV) and $B$ (mT)

Power Amplifier, SINOCERA Piezotronics, Inc.) at the desired amplitude and frequency, a magnetic field was generated inside the coil. The value of the magnetic induction was reflected by the output of the ME sensing unit $V_{ME}$.

To calibrate and verify the ME sensing unit, a Hall probe connected to a Gaussmeter (Model 410) was also used to measure the magnetic induction $B_G$. $V_{ME,pp}$ was obtained from the oscilloscope. From the slope of $B_G - V_{ME,pp}$ plot, $\alpha_v$ was determined. The results of the performance tests of the ME sensing unit are shown in Figs. 4 and 5.

Figure 4a illustrates the waveforms of the measured output ($V_{ME}$) of the ME sensing unit due to an applied AC voltage $V_{in}$ with the peak-to-peak value of 40.4 V and the frequency of 120.0 Hz. It is clear that $V_{ME}$ follows $V_{in}$ steadily and has a maximum peak-to-peak amplitude of 211.6 mV. The fact that $V_{ME}$ and $V_{in}$ are in opposite phase can be explained by the negative sign of the piezoelectric coefficient in the expression of $\alpha_v$ (Jia et al. 2007).

Figure 4b plots $V_{ME,pp}$ and $B_G$ as a function of input peak voltage $V_{in,pp}$ at a frequency of 120.0 Hz. The linear regression equation of $V_{ME,pp}$ and $V_{in,pp}$ is $y_1 = 10.26x - 1.815$, with the correlation coefficient $R = 0.998$, and the linear regression equation of $B_G$ and $V_{in,pp}$ is $y_2 = 0.040x - 0.0860$, with the correlation coefficient $R = 0.999$, indicating good linearity between $V_{ME,pp}$ and $B_G$.

**Fig. 4** Measured time-history of $V_{ME}$ due to an applied AC voltage $V_{in}$ (**a**) and $V_{ME,pp}$ and $B_G$ as a function of $V_{in,pp}$ (**b**) at the frequency of 120.0 Hz

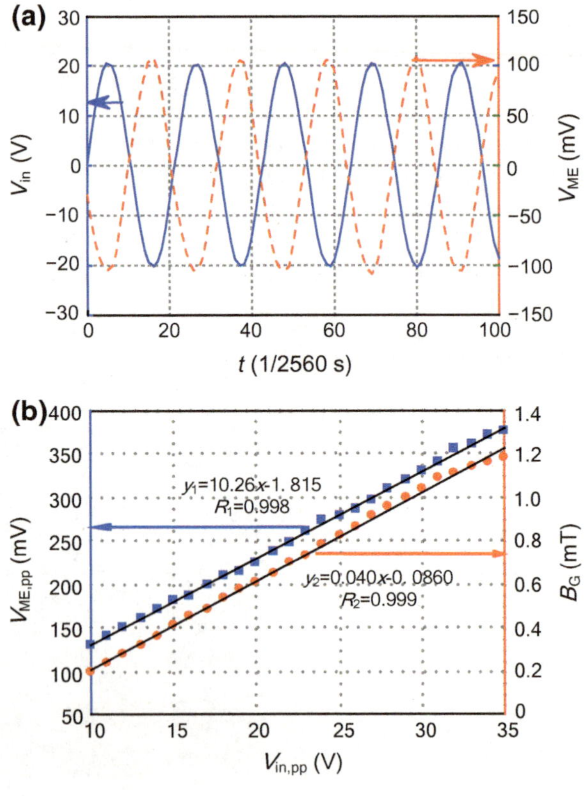

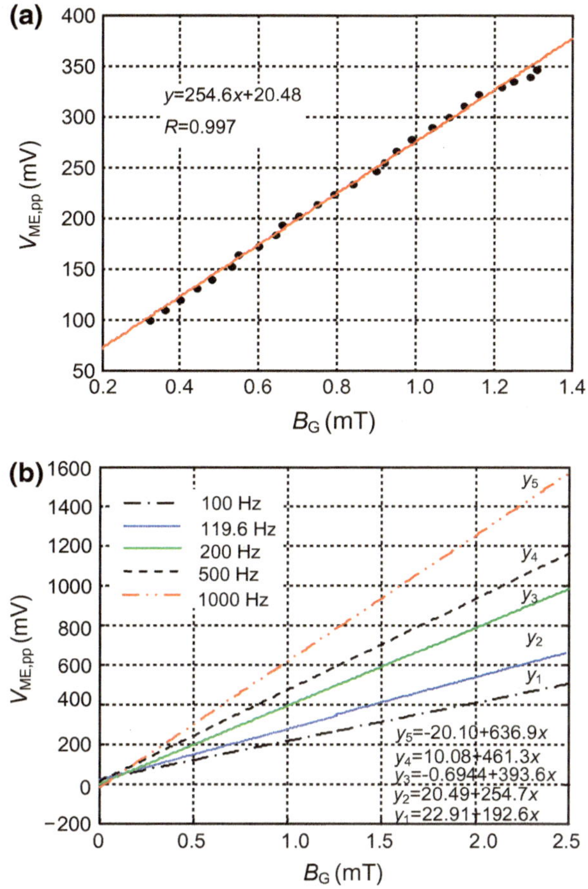


**Fig. 5** Relationship between $V_{ME,pp}$ and $B_G$ with the applied AC voltage at the frequency of 120.0 Hz (**a**) and at various frequencies (**b**)

Figure 5a shows the relationship between $V_{ME}$ and $B_G$ with the applied AC voltage at the frequency of 120.0 Hz. Its linear regression equation $y = 254.6x + 20.48$ and correlation coefficient $R = 0.997$, which suggests that there is good linearity between the output signals of the smart sensing unit and the magnetic induction $B_G$, and thus can be used to measure the magnetic induction. Similarly, $V_{ME,pp}$ as a function of the measured $B_G$ and the linear regression equations for some other excitation frequencies are plotted in Fig. 5b. $V_{ME,pp}$ has good linear responses to the magnetic induction in the measured ranges, and $\alpha_v$ at each certain excitation frequency can be determined from the slopes of the $B_G - V_{ME,pp}$ plot.

Note that the ME sensing unit exhibits good linearity for a certain excitation frequency in the range of 100–1000 Hz. To obtain a high and stable conversion factor, this specific range of excitation frequency should be utilized in the design and operation of the sensing unit and its associated devices.

## 4 Smart EME Sensor and Tension Tests

The smart EME sensor is composed of the energizing apparatus providing the necessary magnetic field and the ME sensing unit measuring the magnetic field under various stresses.

The tests were conducted using the setup as shown in Fig. 6. Axial stress below the yield strength was applied through the tension testing machine (CSS5200, SANS), which was operated in accordance with the pre-set procedures, including the loading magnitude and speed. The proportional limit in the stress of this wire is 353 MPa (the limit tension of the tested steel bar is 40 kN). The tests were conducted at room temperature and with the tension in the range of 0–20 kN at 2 kN intervals. The peak of input voltage $V_{in,pp}$ was 41.0 V and the frequency was 120.0 Hz. Dynamic signal acquisition and processing in the experiment was performed by the data acquisition card and signal conditioning instruments (Fig. 7).

The sinusoidal input voltages to the solenoid and the output signals from the sensing unit were recorded by the data acquisition and analysis system. Each data point was obtained by averaging the results of ten tests, with deviation no larger than 1% (Fig. 8). The linear regression equation of $V_{ME,pp}$ and $T$ (tension in kN) was $y_1 = 217.8 + 1.042x$, with a correlation coefficient of $R = 0.998$, for the loading process. For the unloading process, the linear regression equation of $V_{ME,pp}$ and $T$ was $y_2 = 216.4 + 1.108x$, with a correlation coefficient of $R = 0.986$. Therefore, a good linearity between $V_{ME,pp}$ and $T$ is indicated. The error was below 1% and the repeatability was good, as observed in the tests.

Several factors may have contributed to the differences between the loading and unloading results including: the existence of residual deformation and initial imperfection, the inconsistent force-displacement curves (Fig. 9), the

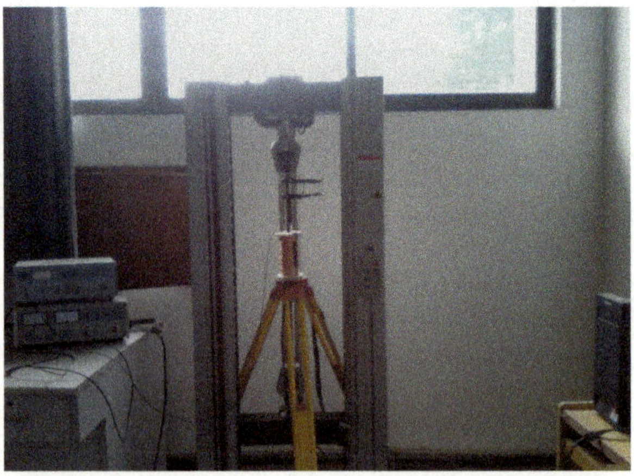

**Fig. 6** Experimental setup for tension tests

**Fig. 7** Signal conditioning and data acquisition instruments

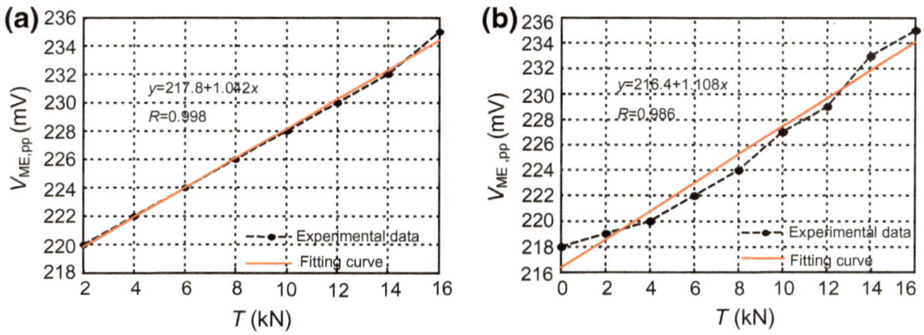


**Fig. 8** Relationship between $V_{ME,pp}$ and $T$ under loading (**a**) and unloading (**b**) processes

non-uniformity of the material, and the impact of loading conditions (both ends are clamped) resulting in distortion of the steel bars. So the results for the loading process should be more reliable and were selected for analysis. The measurement result was in accordance with the theoretical analysis of the sensor. The correlation coefficient of linear regression exceeded 0.99 and the repeating error of the sensor was less than 0.15% (Fig. 10).

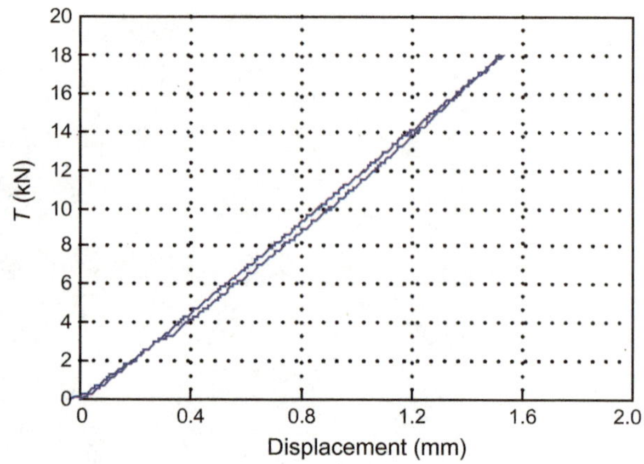


**Fig. 9** Force-displacement curve during the loading and unloading processes

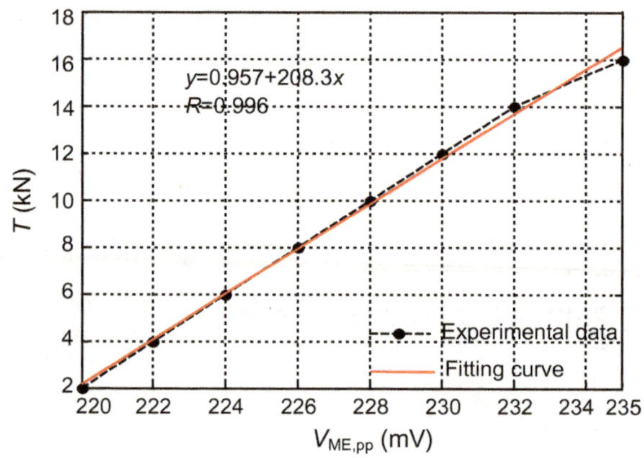


**Fig. 10** Measured results of the tension tests

## 5 Conclusions

A novel smart EME sensor has been developed, fabricated, and tested. Compared to conventional coil-wound EM sensors, our smart EME sensor has distinct advantages due to the replacement of the sensing coil by an ME sensing unit. The new EME sensor with an ME sensing unit is small, lightweight, and easy to install. Its high precision and good repeatability are demonstrated by our test results. Thus, it is suitable for stress monitoring, with increased sensitivity in real-time of steel structures in railway and other civil infrastructures, such as the roofs of railway stations, large span cable-stayed bridges and suspension bridges, and steel rails.

## References


Bartels, K. A., Kwun, H., & Hanley, J. J. (1996). Magnetostrictive sensors for the characterization of corrosion in rebars and prestressing strands. *Proceedings of SPIE, 2946,* 40–50. doi:10.1117/12.259151.

Bozorth, R. M. (1951). *Ferromagnetism*. New York, USA: IEEE Press.

Brophy, J. W., & Brett, C. R. (1996). Guided UT wave inspection of insulated feedwater piping using magnetostrictive sensors. *Proceedings of SPIE, 2947,* 205–209. doi:10.1117/12.259168.

Cannon, D. F., & Pradier, H. (1996). Rail rolling contact fatigue research by the European Rail Research Institute. *Wear, 191*(1–2), 1–13. doi:10.1016/0043-1648(95)06650-0.

Cannon, D. F., Edel, K. O., Grassie, S. L., & Sawley, K. (2003). Rail defects: An overview. *Fatigue and Fracture of Engineering Materials and Structures, 26*(10), 865–886. doi:10.1046/j.1460-2695.2003.00693.x.

Dong, S. X., Li, J. F., & Viehland, D. (2003). Ultrahigh magnetic field sensitivity in laminates of TERFENOL-D and $Pb(Mg_{1/3}Nb_{2/3})O_3$-$PbTiO_3$ crystals. *Applied Physics Letters, 83*(11), 2265–2267. doi:10.1063/1.1611276.

Ekberg, A., & Kabo, E. (2005). Fatigue of railway wheels and rails under rolling contact and thermal loading—An overview. *Wear, 258*(7–8), 1228–1300. doi:10.1016/j.wear.2004.03.039.

GB/T 699-1999. Quality carbon structural steels. National Standard of People's Republic of China (in Chinese).

Jia, Y. M., Or, S. W., Wang, J., Chan, H. L. W., Zhao, X. Y., & Luo, H. S. (2007). High magnetoelectric effect in laminated composites of giant magnetostrictive alloy and lead-free piezoelectric ceramic. *Journal of Applied Physics, 101*(10), 104103. doi:10.1063/1.2732420.

Ke, S., Ye, D. P., Zhang, G. J., & Su, L. G. (2003). *Quick manual for magnetic characteristic curves of comman steels*. Beijing, China (in Chinese): China Machine Press.

Kleinke, D. K., & Uras, H. M. (1994). A magnetostrictive force sensor. *Review of Scientific Instruments, 65*(5), 1699–1710. doi:10.1063/1.1144863.

Kvasnica, B., & Fabo, P. (1996). Highly precise non-contact instrumentation for magnetic measurement of mechanical stress in low-carbon steel wires. *Measurement Science & Technology, 7*(5), 763–767. doi:10.1088/0957-0233/7/5/007.

Sasada, I., Uramoto, S., & Harada, K. (1986). Noncontact torque sensors using magnetic heads and a magnetostrictive layer on the shaft surface-application of plasma jet spraying process. *IEEE Transactions on Magnetics, 22*(5), 406–408. doi:10.1109/TMAG.1986.1064383.

Sasaki, T., Takahashi, S., Kanematsu, Y., Satoh, Y., Iwafuchi, K., Ishida, M., et al. (2008). Measurement of residual stresses in rails by neutron diffraction. *Wear, 265*(9–10), 1402–1407. doi:10.1016/j.wear.2008.04.047.

Seekircher, J., & Hoffmann, B. (1989). New magnetoelastic force sensor using amorphous alloys. *Sensors and Actuators, A: Physical, 22*(1–3), 401–405. doi:10.1016/0924-4247(89)80002-0.

Tang, D. D., Huang, S. L., Chen, W. M., & Jiang, J. S. (2008). Study of a steel strand tension sensor with difference single bypass excitation structure based on the magneto-elastic effect. *Smart Materials and Structures, 17*(2), 025019. doi:10.1088/0964-1726/17/2/025019.

Wang, G. D., & Wang, M. L. (2004). The utilities of U-shape EM sensor in stress monitoring. *Journal of Structural Engineering and Mechanics, 17*(3–4), 291–302.

Wang, M. L., Koontz, S., & Jarosevic, A. (1998). Monitoring of cable forces using magneto-elastic sensors. In *Proceedings of 2nd US-China Symposium Workshop on Recent Developments and Future Trends of Computational Mechanics in Structural Engineering*, Dalian, China (pp. 337–349).

Wang, M. L., Lloyd, G. M., & Hovorka, O. (2001). Development of a remote coil magnetoelastic stress sensor for steel cables. *Proceedings of SPIE, 4337,* 122–128. doi:10.1117/12.435584.

Wang, Y. J., Or, S. W., Chan, H. L. W., Zhao, X. Y., & Luo, H. S. (2008a). Magnetoelectric effect from mechanically mediated torsional magnetic force effect in NdFeB magnets and shear piezoelectric effect in $0.7Pb(Mg_{1/3}Nb_{2/3})O_3$-$0.3PbTiO_3$ single crystal. *Applied Physics Letters, 92*(12), 123510. doi:10.1063/1.2901162.

Wang, Y. J., Cheung, K. F., Or, S. W., Chan, H. L. W., & Luo, H. S. (2008b). PMN-PT single crystal and Terfenol-D alloy magnetoelectric laminated composites for electromagnetic device applications. *Journal of the Ceramic Society of Japan, 116*(1352), 540–544. doi:10.2109/jcersj2.116.540.

## Author Biography

**Yuan-feng Duan** Ph.D., is a professor of Structural and Bridge Engineering, College of Civil Engineering and Architecture, Zhejiang University, China. His research interests include structural health monitoring and vibration control, vector mechanics and structural dynamics. He is a recipient of the National Natural Science Grant for Excellent Young Scientist, the Zhejiang Provincial Grant for Distinguished Young Scientist, and a Fok Ying Tung Grant. He has published one monograph and 80 technical papers, including 20 SCI-indexed and 20 EI-indexed papers. He has been awarded five international patents. His research outputs have been applied to several large-scale structures, such as the Dongting Lake Bridge, the Quanzhou Bay Bridge, the 2nd Jiaojiang Bridge, and the Chenglang Bridge.

# Recent Research on the Track-Subgrade of High-Speed Railways

**Ren-peng Chen, Jin-miao Chen and Han-lin Wang**

**Contents**



## 1 Background

In recent years, the rapid development of high-speed railways in China has surprised the world with the so-called China speed. The total mileage of high-speed railway will reach $1.6 \times 10^4$ km by 2020. As a result, "High-speed Railway Diplomacy" has become a national strategy. In the world, many countries are making their high-speed railway plans. The safety and comfort of high-speed trains raises the strict demands on the performance of the track-subgrade system during the service life over 100 years, for example, strict post-construction settlement at millimeter level, appropriate dynamic stiffness, and long-term durability. Under extreme climatic conditions, such as heavy rainfall, persistent drought, and extreme low or high temperatures, long-term dynamic loading on the track-subgrade will cause many engineering problems, including excessive settlement, mud pumping, cracks in the slab, large voids under the slab, erosion of the reinforced concrete structure. Those engineering problems have been found in many operational high-speed railways. The research into the problems is still insufficient.

R. Chen (✉) · J. Chen · H. Wang
Department of Civil Engineering, Zhejiang University, Hangzhou 310058, China
e-mail: chenrp@zju.edu.cn



Y. Fang and Y.H. Zhang (eds.), *China's High-Speed Rail Technology*,
Advances in High-speed Rail Technology, https://doi.org/10.1007/978-981-10-5610-9_14

## 2 Dynamic Response of Track-Subgrade

The research on the dynamic response of train-track-subgrade has made many achievements. Zhai et al. (2009; 2013a, b) have established a robust 35-degree-of-freedom vehicle-track coupled dynamics model. Dynamic responses of the carriages can be analyzed by this model in the case of carriages passing over curved tracks. Much research on the dynamic response of the subgrade has been conducted using analytical methods (Metrikine 2004), 3D dynamic finite element method (Hall 2003), 2.5D finite element method (Bian et al. 2008), field monitoring (Mishra et al. 2012; Verbraken et al. 2012; Cui et al. 2014), and model tests (Chen et al. 2013, 2014a, b). Those studies have shown that the track-subgrade vibration is significantly associated with the train speed. There exists a critical train speed close to the Rayleigh wave velocity of the track-subgrade (Bian et al. 2008). When the train speed is slower than the critical speed, the vibration level increases almost linearly with the train speed. When the train speed equals the critical speed, the resonance of the track-subgrade causes a great increase in the vibration level. In practice, due to the high quality of fill material used in the subgrade construction, the critical speed is always higher than the train speed. Hence, the resonance can hardly be observed through field monitoring and model tests.

The dynamic stress on the subgrade surface is an important design load which is used to design the subgrade and the soil improvement measures. There are many factors that influence the dynamic stress on the subgrade, including carriage type, train speed, track type, and environmental factors. From field measurement, it can be found that the dynamic stress on the subgrade surface ranges from 15 to 20 kPa for ballastless track, and 50–100 kPa for ballast track. In the Chinese Code for Design of High Speed Railway TB 10621 (MRC 2009), the dynamic load magnification factor (DLF) for the subgrade is 3.0 for a train speed of 300 km/h and 2.5 for a train speed of 250 km/h. In the German Railway Standard Rail 836 (2008), for a train speed of 300 km/h, the DLF of the slab is 1.7–2.1 and 1.24–1.5 for the subgrade. The influence of environmental factors, such as the unevenness of the rail caused by the settlement of the subgrade, the degradation of the subgrade stiffness on the dynamic stress needs further study. *In situ* measurement or full-scale model tests can be used to study the distribution of dynamic stress in the subgrade.

It is regarded that the influence depth of the dynamic stress on the performance of the subgrade is only 3 m for ballastless track. Beneath 3 m, the dynamic stress is so small that it can be neglected. The distribution of the dynamic stress in the subgrade with the depth can be derived from Boussinesq's solution, though the shape of the subgrade and the ground soil is not an exact elastic half-space. Most of the research on the dynamic stress was conducted without the consideration of soil improvement piles. In China, the soft or loose soils under the embankment will be improved with piles. The method for calculating the dynamic stress in the subgrade with Boussinesq's solution should be further verified for a piled embankment. It is found from a full-scale model test for a low embankment improved with piles that the soil arching (Fig. 1) changes the distribution of dynamic stress in the subgrade

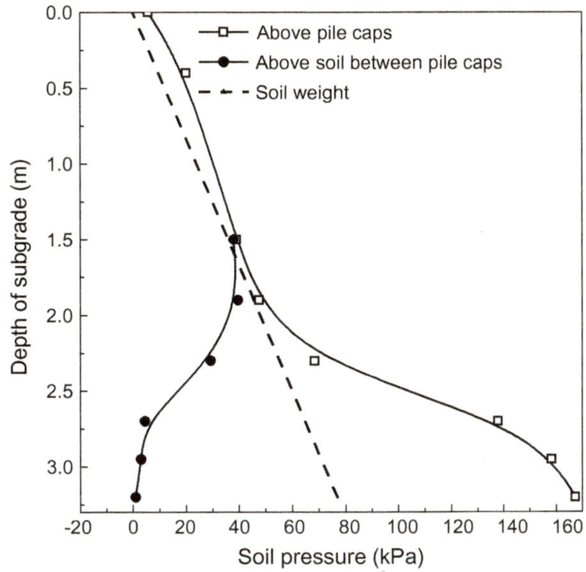


**Fig. 1** Static soil pressure in the subgrade. Reprinted from Chen et al. (2015), with permission from *Journal of the China Railway Society*

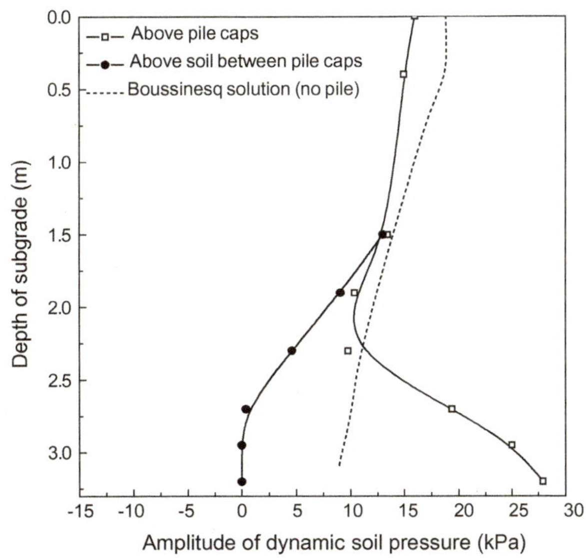


**Fig. 2** Dynamic soil pressure in the subgrade. Reprinted from Chen et al. (2015), with permission from *Journal of the China Railway Society*

(Chen et al. 2015): the dynamic stress above the pile caps is enhanced greatly; on the contrary, the dynamic stress above the soil between the caps decreases greatly (Fig. 2). The large number of cycles of dynamic loading on the pile cap will cause the accumulative settlement of the pile. It is found that when (SLR + CLR) < 0.5 (SLR: static loading ratio; CLR: cyclic loading ratio), there is no accumulative settlement of the pile in silt soils (Chen et al. 2011). The study on the accumulative settlement of the pile under a large number of cycles of dynamic loading should be made in future studies.

Future research will focus on the ground and structure vibration due to the train passing, and protection methods will be studied for vibration sensitive structures. Studies on certain kinds of railway sections, such as the inhomogeneous soil profile in longitudinal and lateral directions, and transition sections, should be proposed.

## 3 Post-construction Settlement of the Track-Subgrade

The post-construction settlement of the track consists of the consolidation settlement of foundation soils, accumulative settlement of the subgrade under train loading, and the accumulative settlement of foundation soils (or piles) under train loading. Generally, the post-construction settlement of the subgrade and foundation under static load belongs to a traditional soil mechanics problem. The post-construction settlement of piled embankments for highways is rarely considered due to the limited total settlement of soil improved with piles. But the post-construction settlement of a piled embankment for high-speed railways becomes important due to the strict demands of the settlement. Zhou et al. (2012) developed a semi-solution for solving the consolidation of a piled embankment in soft ground. Puppala and Chittoori (2012) evaluated the effectiveness of deep soil mixing (DSM) columns for stabilizing soils in arresting the distress posed to the pavements. Accumulative settlement of the subgrade under train loadings is a hot research topic. The commonly used research methods are a full-scale triaxial test (Suiker et al. 2005; Ishikawa et al. 2011; Mishra et al. 2013), full-scale model tests (Chen et al. 2014a, b, 2015), and discrete element method (Huang and Chrismer 2013). It shows that these are three types of the plastic strain development model: plastic shakedown, plastic creep, and incremental collapse (Werkmeister et al. 2005). The development model of plastic strain depends on the ratio between the dynamic stress and initial confining pressure. The greater the ratio is, the more likely the soil will fail. Experimental studies have taken the stress level, number of cycles, particle size distribution, and degree of compaction into consideration.

Although some approaches have been developed to calculate the accumulative settlement, the accuracy of those approaches is low. This is mainly due to the very small accumulative settlement and the uncertainties of the soil parameters. Hence, it is better to propose some threshold values related to the dynamic soil stress to control the development of the accumulative settlement. Further studies should focus on the influence factors on the threshold values, such as saturation, compaction coefficient, gradation. Discrete element method (DEM) and computer tomography (CT) are two useful measures to study the behavior of the subgrade. With the use of pile supported low embankment in soft soil area, pile settlement in the dynamic and static load combinations has become an important part of subgrade

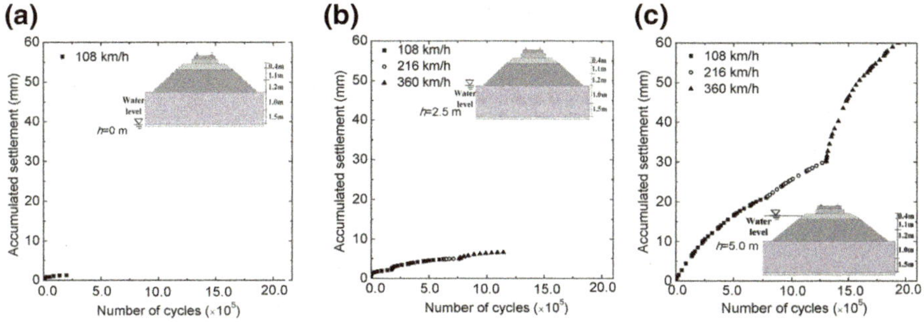


**Fig. 3** Accumulative subgrade settlement under different water levels: **a** water level is at the bottom of the foundation; **b** water level is at the top of the foundation; **c** water level is at the top of the subgrade. Reprinted from Chen et al. (2014a) by permission of Taylor & Francis Ltd.

settlement. Currently, research in this area is not sufficient, and high-speed rail design specifications have no clear rules and calculation methods.

## 4 Long-Term Serviceability of Subgrade

During the service life of high-speed railways, the track-subgrade will suffer drying–wetting cycles, dynamic stress cycles, and temperature cycles. For example, the variation of temperature in north China will be 60 °C. Nurmikolu (2012) thought it was crucial to understand and take into account the frost action mechanism in cold climate, especially where seasonal frost occurs. The water content in the subgrade will also change greatly from the optimum water content during the construction to the saturation status during the wet season. The change of the water content in the subgrade has a significant impact on the subgrade settlement (Fig. 3, Chen et al. 2014a). Future research tendencies will concentrate on the following aspects: (1) the soil–water characteristic curve and permeability of coarse-grained subgrades; (2) the water movement in the subgrade under heavy rainfall. Then, the engineering parameters of the high-speed railway subgrade will be proposed, and advice about fine particles content, grading, and structural design will also be given.

## 5 Summary

The high-speed train makes strict demands on the long-term performance of the track-subgrade. The key scientific point of research on the performance of high-speed railway subgrade is the mechanical and hydraulic properties of the subgrade under the coupling of dynamic cycles, dry–wet cycles, and temperature cycles. Much further research work should be done for the maintenance of existing high-speed railways and the new construction of high-speed railways.

## References

Bian, X. C., Chen, Y. M., & Hu, T. (2008). Numerical simulation of high-speed train induced ground vibrations using 2.5D finite element approach. *Science in China Series G: Physics, Mechanics and Astronomy, 51*(6), 632–650. doi:10.1007/s11433-008-0060-3.

Chen, R. P., Ren, Y., & Chen, Y. M. (2011). Experimental investigation on single stiff pile subjected long-term axial cyclic loading. *Chinese Journal of Geotechnical Engineering, 33* (12), 1926–1933. (in Chinese).

Chen, R. P., Zhao, X., Wang, Z. Z., et al. (2013). Experimental study on dynamic load magnification factor for ballastless track-subgrade of high-speed railway. *Journal of Rock Mechanics and Geotechnical Engineering, 5*(4), 306–311. doi:10.1016/j.jrmge.2013.04.004.

Chen, R. P., Chen, J. M., Zhao, X., et al. (2014a). Accumulative settlement of track subgrade in high-speed railway under varying water levels. *International Journal of Rail Transportation, 2* (4), 205–220. doi:10.1080/23248378.2014.959083.

Chen, R. P., Zhao, X., Jiang, H. G., et al. (2014b). Model test on deformation characteristics of slab track-subgrade under changes of water level. *Journal of the China Railway Society, 36*(3), 87–93. (in Chinese).

Chen, R. P., Wang, Y. W., Chen, J. M., et al. (2015). Experimental study on soil arching effect in pile-supported reinforced embankment under dynamic train loads with larger cycle number of vibration cycles. *Journal of the China Railway Society, 37*(9), 107–113. (in Chinese).

Cui, Y. J., Lamas-Lopez, F., Trinh, V. N., et al. (2014). Investigation of interlayer soil behavior by field monitoring. *Transportation Geotechnics, 1*(3), 91–105. doi:10.1016/j.trgeo.2014.04.002.

German Railway Standard Rail 836. (2008). Erdbauwerkeplanen, bauen und instandhalten (in German).

Hall, L. (2003). Simulations and analyses of train-induced ground vibrations in finite element models. *Soil Dynamics and Earthquake Engineering, 23*(5), 403–413. doi:10.1016/S0267-7261(02)00209-9.

Huang, H., & Chrismer, S. (2013). Discrete element modeling of ballast settlement under trains moving at "Critical Speeds". *Construction and Building Materials, 38*, 994–1000. doi:10.1016/j.conbuildmat.2012.09.007.

Ishikawa, T., Sekine, E., & Miura, S. (2011). Cyclic deformation of granular material subjected to moving-wheel loads. *Canadian Geotechnical Journal, 48*(5), 691–703. doi:10.1139/t10-099.

Metrikine, A. V. (2004). Steady state response of an infinite string on a non-linear visco-elastic foundation to moving point loads. *Journal of Sound and Vibration, 272*(3–5), 1033–1046. doi:10.1016/j.jsv.2003.04.001.

Mishra, D., Tutumluer, E., Stark, T. D., et al. (2012). Investigation of differential movement at railroad bridge approaches through geotechnical instrumentation. *Journal of Zhejiang University-SCIENCE A (Applied Physics & Engineering), 13*(11), 814–824. doi:10.1631/jzus.A12ISGT7.

Mishra, D., Kazmee, H., Tutumluer, E., et al. (2013). Characterization of railroad ballast behavior under repeated loading. *Transportation Research Record: Journal of the Transportation Research Board, 2374*(1), 169–179. doi:10.3141/2374-20.

MRC (Ministry of Railways of China). (2009). Chinese code for design of high speed railway TB 10621, MRC (in Chinese).

Nurmikolu, A. (2012). Key aspects on the behaviour of the ballast and substructure of a modern railway track: research-based practical observations in Finland. *Journal of Zhejiang University-SCIENCE A (Applied Physics & Engineering), 13*(11), 825–835. doi:10.1631/jzus.A12ISGT1.

Puppala, A. J., & Chittoori, B. C. (2012). Transportation infrastructure settlement and heave distress: challenges and solutions. *Journal of Zhejiang University-SCIENCE A (Applied Physics & Engineering), 13*(11), 850–857. doi:10.1631/jzus.A12ISGT9.

Suiker, A. S. J., Selig, E. T., & Frenkel, R. (2005). Static and cyclic triaxial testing of ballast and sub-ballast. *Journal of Geotechnical and Geoenvironmental Engineering ASCE, 131*(6), 771–782. doi:10.1061/(ASCE)1090-0241(2005)131:6(771).

Verbraken, H., Lombaert, G., & Degrande, G. (2012). Experimental and numerical prediction of railway induced vibration. *Journal of Zhejiang University-SCIENCE A (Applied Physics & Engineering), 13*(11), 802–813. doi:10.1631/jzus.A12ISGT8.

Werkmeister, S., Dawson, A. R., & Wellner, F. (2005). Permanent deformation behavior of granular materials. *Road Materials and Pavement Design, 6*(1), 31–51. doi:10.1080/14680629.2005.9689998.

Zhai, W. M., Wang, K. Y., & Cai, C. B. (2009). Fundamentals of vehicle-track coupled dynamics. *Vehicle System Dynamics, 47*(11), 1349–1376. doi:10.1080/00423110802621561.

Zhai, W. M., Xia, H., Cai, C. B., et al. (2013a). High-speed train-track-bridge dynamic interactions–Part I: theoretical model and numerical simulation. *International Journal of Rail Transportation, 1*(1–2), 3–24. doi:10.1080/23248378.2013.791498.

Zhai, W. M., Xia, H., Cai, C. B., et al. (2013b). High-speed train-track-bridge dynamic interactions–Part II: experimental validation and engineering application. *International Journal of Rail Transportation, 1*(1–2), 25–41. doi:10.1080/23248378.2013.791497.

Zhou, W. H., Chen, R. P., Zhao, L. S., et al. (2012). A semi-analytical method for the analysis of pile-supported embankments. *Journal of Zhejiang University-SCIENCE A (Applied Physics & Engineering), 13*(11), 888–894. doi:10.1631/jzus.A12ISGT4.


## Author Biography

**Ren-peng Chen,** obtained his Master's and Ph.D. degrees in Civil Engineering from Zhejiang University in 1997 and 2001, respectively, following a BS degree in Civil Engineering from Hunan University in 1994. His areas of research include soil improvement, road and railway engineering, and pile foundation. He was selected for the National Youth Talents for Innovation of Science and Technology in 2013. Currently, he is a Professor and Dean of School of Civil Engineering, Hunan University. Prior to joining Hunan University, he was a professor in the Department of Civil Engineering, Zhejiang University. He is the recipient of numerous grants and the author of over 100 peer-reviewed articles.

# Microstructure and Properties of Cold Drawing Cu-2.5% Fe-0.2% Cr and Cu-6% Fe Alloys

**Guo-huan Bao, Yi Chen, Ji-en Ma, You-tong Fang, Liang Meng, Shu-min Zhao, Xin Wang and Jia-bin Liu**

**Contents**



## 1 Introduction

Cu-based in situ filamentary alloys prepared by heavy cold processing, such as using Cu–Nb, Cu–Cr, Cu–Fe, and Cu–Ag, have been studied for decades due to their high strength and high conductivity (Mattissen et al. 1999; Tian et al. 2011; Jia et al. 2013; Raju et al. 2013; Deng et al. 2014; Pantsyrny et al. 2014; Wang et al. 2014). These high strength and high conductivity alloys are expected to be used as conductor materials in high-field magnet and recently developing high-speed railways.

G. Bao · Y. Chen · L. Meng · J. Liu (✉)
Department of Materials Science and Engineering, Zhejiang University, Hangzhou 310027, China
e-mail: liujiabin@zju.edu.cn

J. Ma · Y. Fang
College of Electrical Engineering, Zhejiang University, Hangzhou 310027, China

S. Zhao · X. Wang
College of Materials and Environmental Engineering, Hangzhou Dianzi University, Hangzhou 310018, China

 
Y. Fang and Y.H. Zhang (eds.), *China's High-Speed Rail Technology*, Advances in High-speed Rail Technology, https://doi.org/10.1007/978-981-10-5610-9_15

In these alloys, the second elements always have limited solubility in Cu at room temperature. The microstructure of these alloys contains a Cu matrix and second phase. During the cold drawing, both the Cu matrix and second phase are elongated and evolved into filamentary structures at high drawing strains. The density of the phase interface increases as the diameter and interval of the second phase decreases. The strength of the heavy drawn alloys significantly exceeds the predicted strength by the rule of mixture for high drawing strains. The abundant phase interface is thought to play a major role in the strengthening behavior of these alloys. The production of abundant phase interface relies on the co-deformation of the Cu matrix and second phase. Much attention was focused on the deformation behavior of the alloys, especially the second phase, during cold drawing. The co-deformation of Cu–Ag alloys has been well documented in our previous study (Liu et al. 2011a). Since both Cu and Ag phases belong to the face centered cubic (FCC) phase, and they have a cube-on-cube orientation relationship, the two phases could keep an almost synchronous co-deformation during cold drawing. However, the co-deformation of the Cu-body centered cubic (BCC) (Cu–Nb, Cu–Cr, or Cu–Fe) system is very complicated (Spitzig et al. 1987; Biselli and Morris 1996; Sinclair et al. 1999). The BCC phase has a different dislocation slip system in association with a Cu matrix. Moreover, the BCC phase always has different strain hardening characteristics in association with a Cu matrix. Some general criteria for the co-deformation of FCC-BCC alloys were stated by Sinclair et al. (1999), including the resolved shear stress, the angle between the incident and activated systems, and the resulting configuration at the phase interface. The original Nb dendrites in Cu–Nb alloys always have a scale of tens of micrometers (Spitzig et al. 1987). The Nb dendrites gradually evolved into Nb ribbons with a scale of nanometers at high drawing strains. When included in Cu–Cr alloys, original Cr dendrites and Cr–Cu eutectic always have a scale of several micrometers. Both the Cr dendrites and Cr–Cu eutectic gradually evolved into ribbons at high strains with a large scale of nanometers, because Cr appears in two microstructure morphologies (Raabe et al. 2000).

Similar situations occur with the Fe primary dendrites in Cu–Fe alloys (Biselli and Morris 1994; Xie et al. 2011). In our previous studies, the strain of Fe dendrite linearly increased with an increase in the drawing strain up to 6 and deviated from the linear relationship when the drawing strain was higher than 6 in Cu-12% Fe (in weight) (Lu et al. 2014). The thickness, width, and spacing of Fe ribbons in the filamentary structure exponentially decreased with an increase in the drawing strain. The density of the interface between Cu and Fe phases exponentially increased with an increase in the aspect ratio of Fe ribbons. In addition, for the Fe primary dendrites, plenty of Fe precipitate particles were produced in the solution and aging of the Cu–Fe alloys (Monzen and Kita 2002; Watanabe et al. 2008). These Fe precipitate particles have an initial scale of only several nanometers. Whether the deformation behavior of these nano-particles is the same with that of the micro-dendrites in Cu–Fe alloys still remains unclear. In this study, Cu-2.5% Fe-0.2% Cr and Cu-6% Fe (in weight) alloys were used to prepare Fe nano-particles and Fe micro-dendrites. The deformation behavior between Fe nano-particles and

Fe micro-dendrites was compared and analyzed. The evolution of the mechanical and electrical properties of both alloys was investigated based on the microstructure analysis.

## 2 Materials and Methods

Cu-2.5% Fe-0.2% Cr and Cu-6% Fe alloys were melted in a vacuum induction furnace and cast into cylindrical ingots of 22 mm in diameter in a copper mold. According to the Cu–Fe phase diagram, the maximum solubility of Fe in Cu is 4.1% at 1098 °C. In this study, we chose Cu-6% Fe to obtain primary Fe dendrites and chose Cu-2.5% Fe-0.2% Cr to obtain precipitate Fe particles. Very little Cr addition in Cu–Fe alloys was found to be able to promote the precipitation of Fe and to strengthen the Fe precipitate (Hong and Song 2001; Song et al. 2001). The Cu-2.5% Fe-0.2% Cr ingots were homogenized at 900 °C for 8 h, and solution treated at 1050 °C for 4 h followed by water quenching, then aged at 700 °C for 33 h. The Cu-6% Fe ingots were homogenized at 900 °C for 8 h, and solution treated at 1050 °C for 4 h followed by water quenching. The surface layer of the ingots was turned off to remove the surface oxides and defects. The ingots were cold worked by rolling and drawing to a drawing strain of $\eta = 6.6$, where $\eta = \ln(A_0/A)$, and $A_0$ and $A$ are the original and final transverse section areas, respectively. No intermediate annealing was applied to the drawn wires.

Ultimate tensile strength was determined at ambient temperature using an electronic tensile testing machine at a strain rate of $2 \times 10^{-3}\ s^{-1}$. Electrical resistivity was measured at room temperature using a standard four-point technique. The microstructure was observed by optical microscopy (OM), scanning electron microscopy (SEM, Hitachi S4800, Japan), and transmission electron microscopy (TEM, FEI F20, USA). TEM observation was carried out in the TEM operating at 200 kV. The TEM samples were prepared by mechanical thinning to about 40 μm and then ion milled at 4 kV with an incidence angle of 8°. The energy-dispersive spectrometer (EDS, Oxford INCA, UK) equipped in F20 was used to perform composition analysis.

## 3 Results

The microstructure of the solution treated Cu-6% Fe alloy consists of Fe primary particles and Cu grains with an average diameter of 75.7 μm (Fig. 1a). The Fe primary particles with an average diameter of 4.8 μm are dispersed in the Cu grains or near the grain boundaries. The microstructure of aged Cu-2.5% Fe-0.2% Cr alloy consists of Cu grains with an average diameter of 97.3 μm (Fig. 1b). No Fe primary particle was observed in the microstructure of Cu-2.5% Fe-0.2% Cr, which is different with the microstructure of Cu-6% Fe. TEM results show that there are

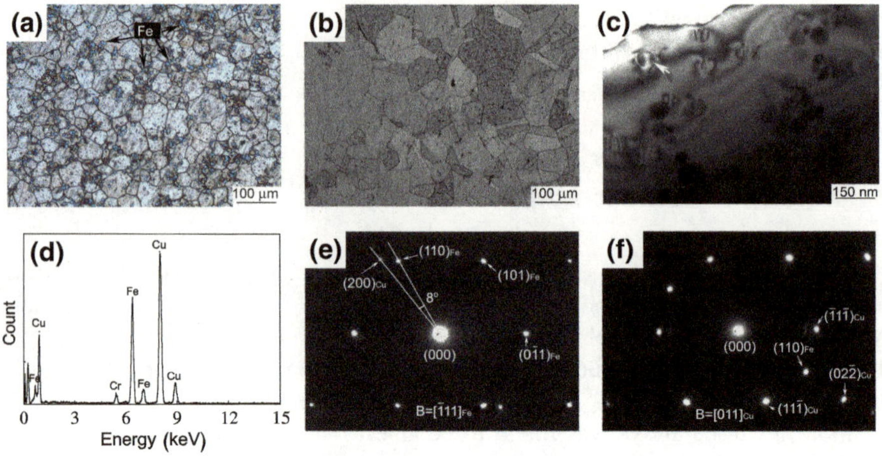


**Fig. 1** Optical images of solution treated Cu-6% Fe (**a**) and aged Cu-2.5% Fe-0.2% Cr (**b**), TEM image of aged Cu-2.5% Fe-0.2% Cr (**c**), EDS spectra of a precipitate particle pointed by *white arrow* in Fig. 1c (**d**), and SAED patterns of Cu matrix and Fe particles when the electron beam is almost parallel with $[\bar{1}11]_{Fe}$ (**e**) and $[011]_{Cu}$ (**f**)

plenty of Fe-rich precipitate particles with an average diameter of 52.4 nm in the Cu matrix (Fig. 1c). EDS results of these particles indicate that the particles mainly contain Fe elements and some Cr elements (Fig. 1d). The signal of the Cu element should come from the surrounding Cu matrix. Selected area electron diffraction (SAED) patterns are shown in Fig. 1e, f. When the electron beam is almost parallel with $[\bar{1}11]_{Fe}$, only a pair of spots of $\{200\}_{Cu}$ are observed. There is a misfit angle of 8° between $(200)_{Cu}$ and $(110)_{Fe}$. When the electron beam is almost parallel with $[011]_{Cu}$, only a pair of spots of $\{110\}_{Fe}$ are observed and are nearly parallel with the $\{220\}_{Cu}$. In a word, the SAED investigation indicates that there should be no special orientation relationship between the Cu matrix and Fe particles in either Cu-2.5% Fe-0.2% Cr or Cu-6% Fe alloys.

After heavy drawing, the original equal-axial Cu grains in the Cu-6% Fe alloy were elongated along the drawing direction (Fig. 2a). The Fe primary particles were also elongated and evolved into parallel Fe ribbons. At the transversal section of the drawn specimens, the Fe ribbons exhibit a curved morphology with an average thickness of 68.4 nm (Fig. 2b). The formation of the curved ribbon-like morphology has been widely observed in Cu–Fe and Cu–Nb alloys (Spitzig et al. 1987; Biselli and Morris, 1996; Liu et al. 2013; Deng et al. 2014). The composites were reported to be highly textured: Nb had a ⟨110⟩ orientation whereas Cu had ⟨111⟩ and ⟨100⟩ textures (Raabe et al. 1992, 1995a). The reason was fully explained that the BCC phase was forced to curl and fold due to the constraint of the surrounding FCC matrix, which is able to accommodate axially symmetric flow (Bevk et al. 1978). The microstructure of the Cu-2.5% Fe-0.2% Cr alloy on the longitudinal section is similar to that of the Cu-6% Fe alloy, while there are only Cu filaments in

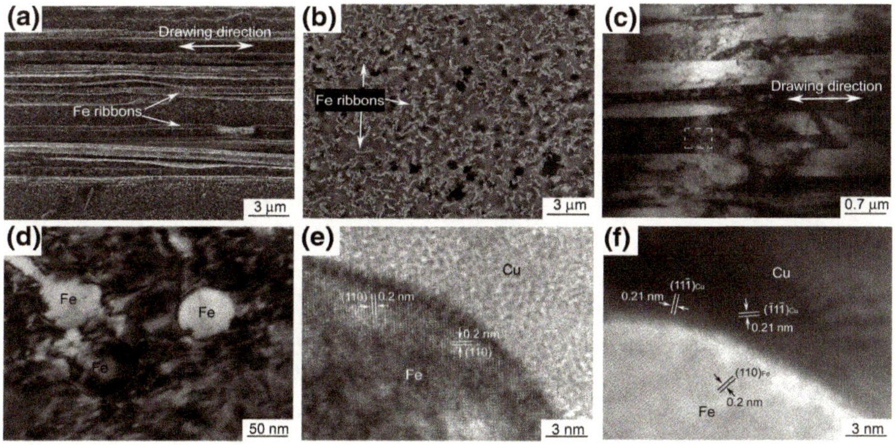


**Fig. 2** SEM images of Cu-6% Fe at $\eta$=6 longitudinal section (**a**) and transversal section (**b**), TEM images of Cu-2.5% Fe-0.2% Cr at $\eta$=6 low magnification of longitudinal section (**c**) and high magnification of Fe precipitate particles in the *white rectangle* in Fig. 2c (**d**), and HRTEM images of a Fe precipitate particle embedded in Cu matrix when the electron beam is almost parallel with $[100]_{Fe}$ (**e**) and $[011]_{Cu}$ (**f**)

the Cu-2.5% Fe-0.2% Cr alloy (Fig. 2c). High density of dislocation was observed inside the Cu filaments. Further investigation of the zone with high density of dislocation found many dislocations surrounding the Fe precipitate particles (Fig. 2d). The Fe precipitate particles still keep most of their spherical morphology, which indicates that they undergo little deformation during the drawing. The average diameter of the Fe precipitate particles is 62.2 nm in the Cu-2.5% Fe-0.2% Cr alloy at $\eta$=6, which is almost the same with that of the aged Cu-2.5% Fe-0.2% Cr alloy. High resolution transmission electron microscopy (HRTEM) investigation was employed to analyze the interface structure between the Cu matrix and Fe particles. When the electron beam is almost parallel with $[100]_{Fe}$, no low-index of Cu matrix is parallel with the electron beam. Both planes of $(110)_{Fe}$ and $(1\bar{1}0)_{Fe}$ are clearly observed while nearly no lattice plane of the Cu matrix is visible as shown in Fig. 2e. When the electron beam is almost parallel with $[011]_{Cu}$, both planes of $(11\bar{1})_{Cu}$ and $(\bar{1}1\bar{1})_{Cu}$ are clearly observed as shown in Fig. 2f. At the same time, the planes of $(110)_{Fe}$ are visible but are not parallel with neither $(11\bar{1})_{Cu}$ nor $(\bar{1}1\bar{1})_{Cu}$. The HRTEM investigation indicates that there should be no special orientation relationship between Fe precipitate particles and the Cu matrix, which agrees very well with the results of the SAED patterns. The interface structure of Fe precipitate particles and the Cu matrix also shows typical incoherent phase interface characteristics.

Figure 3 gives the mechanical and electrical properties of Cu-2.5% Fe-0.2% Cr and Cu-6% Fe alloys at various drawing strains. For comparison, the properties of pure Cu are also given in Fig. 3. The strength of the tested alloys increases with the increasing drawing strain. The strength increasing rates of Cu-2.5% Fe-0.2% Cr and

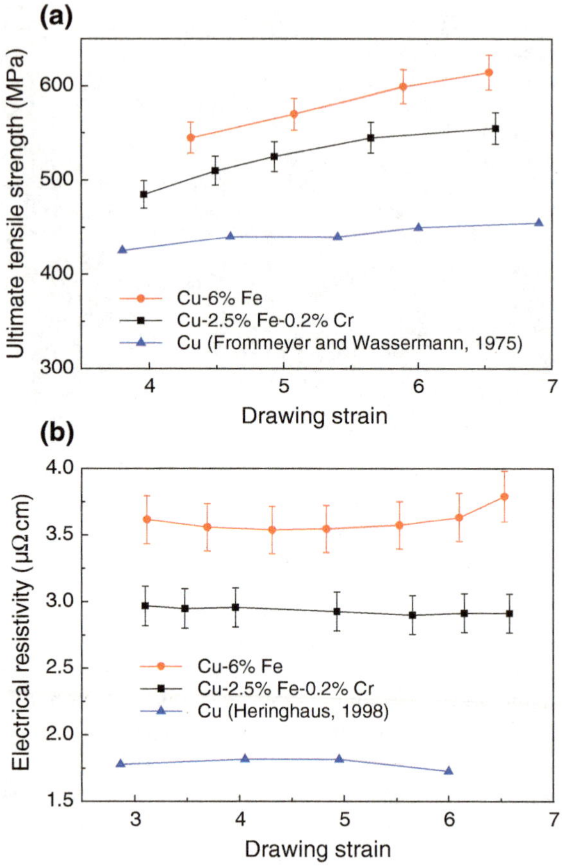


**Fig. 3** Ultimate tensile strength (**a**) and electrical resistivity (**b**) of Cu-2.5% Fe-0.2% Cr, Cu-6% Fe, and pure Cu (Frommeyer and Wassermann 1975; Heringhaus 1998) dependent on the drawing strain

Cu-6% Fe alloys are higher than that of pure Cu. The electrical resistivity of the tested alloys changes slightly with the increasing drawing strain. The electrical resistivity of Cu-2.5% Fe-0.2% Cr and Cu-6% Fe alloys are obviously higher than that of pure Cu.

## 4 Discussion

Fe primary particles with a size of several micrometers were drawn into nano-ribbons by heavy drawing. However, Fe precipitate particles with a size of tens of nanometers were hardly deformed. Size effect should play a major role in different deformation behaviors of Fe primary particles and Fe precipitate particles. Studies on mechanical property of small size samples suggest that the flow stress of the samples increases with the size decreasing, following a Hall-Petch like equation (Hangen and Raabe 1995; Legros et al. 2000):

$$\sigma = \sigma_0 + kd^{-n}, \tag{1}$$

where $\sigma$ and $\sigma_0$ are the flow stress and the lattice friction stress, respectively, $d$ is the size of the sample, $k$ is constant, and $n$=0.5–1.0. Dunstan and Bushby (2013) indicated that the strength should be inversely proportional to the size of the samples, and suggested that this effect could be explained by the room of the sample to activate the dislocation source. Namely, the smaller sample is stronger due to its harder activation of the dislocation source. According to these results, the deformation of Fe precipitate particles in Cu-2.5% Fe-0.2% Cr alloy should be much harder than Fe primary particles in Cu-6% Fe alloy.

Since Fe primary particles or Fe precipitate particles in tested alloys are embedded in the Cu matrix, taking into account the interaction between Cu and Fe phases during cold drawing, the plastic deformation behavior should be more complex. Studies of Cu–Ag alloys reveal that Ag precipitates produced by aging treatment are rod-like and are dozens of nanometers in diameter (Liu et al. 2011b; Sitarama Raju et al. 2013). These nano-scaled Ag precipitates could be evolved into extreme fine nanofibers by cold drawing. However, the nano-scaled Fe precipitate particles kept almost spherical morphology and underwent little deformation in this study. The main reason may lie in the different dislocation slip situations between Cu–Ag and Cu–Fe systems. For the Cu matrix and Ag precipitates, it has been confirmed that Ag precipitates have a cube-on-cube orientation relationship with the Cu matrix, that is $\langle 110\rangle_{Ag}//\langle 110\rangle_{Cu}$ and $\{111\}_{Ag}//\{111\}_{Cu}$ (Han et al. 2003; Tian and Zhang, 2009). This orientation relationship can be well maintained even when the Cu–Ag alloys are heavily cold drawn (Liu et al. 2011b). The cube-on-cube orientation relationship could guarantee Ag precipitates have the same dislocation slip system with the Cu matrix, which produces a greater benefit for the co-deformation of both phases. For the Cu matrix and Fe precipitate particles, the Fe precipitate particles are spherical and have no special orientation relationship. Little or even no favorite slip system of the Cu matrix is parallel with that of the Fe precipitate particles, which provides little help for the deformation of the Fe precipitate particles.

The evolution of Fe primary particles into Fe ribbons greatly increases the density of the Cu/Fe phase interface, which should be beneficial for the strength of the Cu-6% Fe alloy. It has been documented that the phase interface plays a major strengthening role in the heavily drawn Cu–$X$ ($X$=Nb, Fe, Ag, Cr) alloys (Biselli and Morris 1996; Wu et al. 2009; Badinier et al. 2014). As a result, the Cu-6% Fe alloy shows a much higher strength than pure Cu. The strengthening situation in the Cu-2.5% Fe-0.2% Cr is notably different with that in Cu-6% Fe. Since the Fe precipitate particles are hardly deformed, the strengthening effect mainly results from the precipitation hardening by the well-known Orowan mechanism.

The resistivity of composites can be partitioned into the contribution of four scattering mechanisms: photon scattering, dislocation scattering, interface scattering, and impurity scattering (Raabe et al. 1995b; Hong and Hill 1999). The photon scattering component can be ignored at room temperature. The impurity scattering component is proportional to the impurity concentration, which is almost constant during the drawing process. The contribution from the dislocation scattering component is about 0.2 μΩ cm, which is relatively low (Karasek and Bevk 1981). Therefore, the interface scattering component plays a main role. The density of the Cu/Fe phase interface in the Cu-6% Fe alloy increases with the increase of the drawing strain, which should enhance the electron scattering in the Cu-6% Fe alloy. As a result, the electrical resistivity of the Cu-6% Fe alloy slightly increases at $\eta < 6$ and greatly enhances at $\eta > 6$ at which stage the Fe particles are fully evolved into ribbons. However, the electrical resistivity of the Cu-2.5% Fe-0.2% Cr alloy keeps almost constant even at high drawing strains. Since the Fe precipitate particles are hardly deformed, the density of the Cu/Fe phase interface should be kept nearly constant, and therefore the electrical resistivity has little change.

## 5 Conclusions

Cu-2.5% Fe-0.2% Cr and Cu-6% Fe alloys were designed to compare the deformation behavior of the Fe phase with different scales. Fe primary particles with a size of about 5 μm were produced in the Cu-6% Fe alloy by solution treatment. Fe precipitate particles with a size of about 50 nm were produced in the Cu-2.5% Fe-0.2% Cr alloy by a solution and aging treatment. During the cold drawing, Fe primary particles were elongated and evolved into nano-scaled ribbons. The density of the Cu/Fe phase interface gradually increases with the drawing strain. As a result, the strength and the electrical resistivity of the Cu-6% Fe alloy also increase with the drawing strain. In contrast, the Fe precipitate particles were hardly deformed and kept their spherical morphology even at $\eta = 6$. High density of dislocation surrounds the Fe precipitate particles. The strength of the Cu-2.5% Fe-0.2% Cr alloy increases with the increase of the drawing strain and can be described by the Orowan mechanism. The electrical resistivity of the Cu-2.5% Fe-0.2% Cr alloy keeps almost constant since the density of the Cu/Fe phase interface shows hardly any change during cold drawing. The size effect and the incoherent interface of the Fe precipitate particles and Cu matrix play primary roles in the unchanging of Fe precipitate particles during cold drawing.

## References

Badinier, G., Sinclair, C. W., Allain, S., et al. (2014). The Bauschinger effect in drawn and annealed nanocomposite Cu–Nb wires. *Materials Science and Engineering A, 597,* 10–19. doi:10.1016/j.msea.2013.12.031.

Bevk, J., Harbison, J. P., & Bell, J. L. (1978). Anomalous increase in strength of in situ formed Cu–Nb multifilamentary composites. *Journal of Applied Physics, 49*(12), 6031–6038. doi:10.1063/1.324573.

Biselli, C., & Morris, D. G. (1994). Microstructure and strength of Cu–Fe in-situ composites obtained from prealloyed Cu–Fe powders. *Acta Metallurgica et Materialia, 42*(1), 163–176. doi:10.1016/0956-7151(94)90059-0.

Biselli, C., & Morris, D. G. (1996). Microstructure and strength of Cu–Fe in situ composites after very high drawing strains. *Acta Materialia, 44*(2), 493–504. doi:10.1016/1359-6454(95)00212-X.

Deng, L. P., Han, K., Hartwig, K. T., et al. (2014). Hardness, electrical resistivity, and modeling of in situ Cu–Nb microcomposites. *Journal of Alloys and Compounds, 602,* 331–338. doi:10.1016/j.jallcom.2014.03.021.

Dunstan, D. J., & Bushby, A. J. (2013). The scaling exponent in the size effect of small scale plastic deformation. *International Journal of Plasticity, 40,* 152–162. doi:10.1016/j.ijplas.2012.08.002.

Frommeyer, G., & Wassermann, G. (1975). Microstructure and anomalous mechanical properties of in situ-produced silver-copper composite wires. *Acta Metallurgica, 23*(11), 1353–1360. doi:10.1016/0001-6160(75)90144-3.

Han, K., Vasquez, A. A., Xin, Y., et al. (2003). Microstructure and tensile properties of nanostructured Cu-25 wt%Ag. *Acta Materialia, 51*(3), 767–780. doi:10.1016/S1359-6454(02)00468-8.

Hangen, U., & Raabe, D. (1995). Modeling of the yield strength of a heavily wire drawn Cu-20-percent-Nb composite by use of a modified linear rule of mixtures. *Acta Metallurgica et Materialia, 43*(11), 4075–4082. doi:10.1016/0956-7151(95)00079-B.

Heringhaus, F. (1998). *Quantitative analysis of the influence of the microstructure on strength, resistivity, and magnetoresistance of eutectic silver-copper.* PhD Thesis, Institut fur Metallkunde und Metallphysik, RWTH Aachen, Germany; National High Magnetic Field Laboratory Tallahassee, USA.

Hong, S. I., & Hill, M. A. (1999). Mechanical stability and electrical conductivity of Cu–Ag filamentary microcomposites. *Materials Science and Engineering A, 264*(1–2), 151–158. doi:10.1016/S0921-5093(98)01097-1.

Hong, S. I., & Song, J. S. (2001). Strength and conductivity of Cu-9Fe-1.2*X* (*X*=Ag or Cr) filamentary microcomposite wires. *Metallurgical and Materials Transaction A, 32*(4), 985–991. doi:10.1007/s11661-001-0356-7.

Jia, N., Roters, F., Eisenlohr, P., et al. (2013). Simulation of shear banding in heterophase co-deformation: Example of plane strain compressed Cu–Ag and Cu–Nb metal matrix composites. *Acta Materialia, 61*(12), 4591–4606. doi:10.1016/j.actamat.2013.04.029.

Karasek, K. R., & Bevk, J. (1981). Normal-state resistivity of in situ-formed ultrafine filamentary Cu–Nb composites. *Journal of Applied Physics, 52*(3), 1370–1375. doi:10.1063/1.329767.

Legros, M., Elliott, B. R., Rittner, M. N., et al. (2000). Microsample tensile testing of nanocrystalline metals. *Philosophical Magazine A: Physics of Condensed Matter, Structure, Defects and Mechanical Properties, 80*(4), 1017–1026. doi:10.1080/01418610008212096.

Liu, J. B., Zhang, L., & Meng, L. (2011a). Codeformation in Cu-6wt.% Ag nanocomposites. *Scripta Materialia, 64*(7), 665–668. doi:10.1016/j.scriptamat.2010.12.015.

Liu, J. B., Zhang, L., Yao, D. W., et al. (2011b). Microstructure evolution of Cu/Ag interface in the Cu–6 wt.% Ag filamentary nanocomposite. *Acta Materialia, 59*(3), 1191–1197. doi:10.1016/j.actamat.2010.10.052.

Liu, K. M., Lu, D. P., Zhou, H. T., et al. (2013). Influence of a high magnetic field on the microstructure and properties of a Cu–Fe–Ag in situ composite. *Materials Science and Engineering A, 584,* 114–120. doi:10.1016/j.msea.2013.07.016.

Lu, X. P., Yao, D. W., Chen, Y., et al. (2014). Microstructure and hardness of Cu-12% Fe composite at different drawing strains. *Journal of Zhejiang University-SCIENCE A (Applied Physics & Engineering), 15*(2), 149–156. doi:10.1631/jzus.A1300164.

Mattissen, D., Rabbe, D., & Heringhaus, F. (1999). Experimental investigation and modeling of the influence of microstructure on the resistive conductivity of a Cu–Ag–Nb in situ composite. *Acta Materialia, 47*(5), 1627–1634. doi:10.1016/S1359-6454(99)00026-9.

Monzen, R., & Kita, K. (2002). Ostwald ripening of spherical Fe particles in Cu-Fe alloys. *Philosophical Magazine Letters, 82*(7), 373–382. doi:10.1080/09500830210137399.

Pantsyrny, V. I., Khlebova, N. E., Sudyev, S. V., et al. (2014). Thermal stability of the high strength high conductivity Cu–Nb, Cu–V, and Cu–Fe nanostructured microcomposite wires. *IEEE Transaction of Applied Superconductor, 24*(3), 0502804.

Raabe, D., Ball, J., & Gottstein, G. (1992). Rolling textures of a Cu-20-percent-Nb composite. *Scripta Metallurgica et Materialia, 27*(2), 211–216. doi:10.1016/0956-716X(92)90115-U.

Raabe, D., Heringhaus, F., Hangen, U., et al. (1995a). Investigation of a Cu-20 mass-percent Nb in-situ composite. 1. Fabrication, microstructure and mechanical properties. *Zeitschrift Fur Metallkunde, 86,* 405–415.

Raabe, D., Heringhaus, F., Hangen, U., et al. (1995b). Investigation of a Cu-20 mass-percent Nb in-situ composite. 2. Electromagnetic properties and application. *Zeitschrift Fur Metallkunde, 86,* 416–422.

Raabe, D., Miyake, K., & Takahara, H. (2000). Processing, microstructure, and properties of ternary high-strength Cu–Cr–Ag in situ composites. *Materials Science and Engineering A, 291* (1–2), 186–197. doi:10.1016/S0921-5093(00)00981-3.

Raju, K. S., Sarma, V. S., Kauffmann, A., et al. (2013). High strength and ductile ultrafine-grained Cu–Ag alloy through bimodal grain size, dislocation density and solute distribution. *Acta Materialia, 61*(1), 228–238. doi:10.1016/j.actamat.2012.09.053.

Sinclair, C. W., Embury, J. D., & Weatherly, G. C. (1999). Basic aspects of the co-deformation of BCC/FCC materials. *Materials Science and Engineering A, 272*(1), 90–98. doi:10.1016/S0921-5093(99)00477-3.

Sitarama Raju, K., Subramanya Sarma, V., Kauffmann, A., et al. (2013). High strength and ductile ultrafine-grained Cu–Ag alloy through bimodal grain size, dislocation density and solute distribution. *Acta Materialia, 61*(1), 228–238. doi:10.1016/j.actamat.2012.09.053.

Song, J. S., Hong, S. I., & Kim, H. S. (2001). Heavily drawn Cu–Fe–Ag and Cu–Fe–Cr microcomposites. *Journal of Materials Processing Technology, 113*(1–3), 610–616. doi:10.1016/S0924-0136(01)00665-3.

Spitzig, W. A., Pelton, A. R., & Laabs, F. C. (1987). Characterization of the strength and microstructure of heavily cold worked Cu–Nb composites. *Acta Metallurgica, 35*(10), 2427–2442. doi:10.1016/0001-6160(87)90140-4.

Tian, Y. Z., & Zhang, Z. F. (2009). Microstructures and tensile deformation behavior of Cu–16wt.%Ag binary alloy. *Materials Science and Engineering A, 508*(1–2), 209–213. doi:10.1016/j.msea.2008.12.050.

Tian, Y. Z., Wu, S. D., Zhang, Z. F., et al. (2011). Comparison of microstructures and mechanical properties of a Cu–Ag alloy processed using different severe plastic deformation modes. *Materials Science and Engineering A, 528*(13–14), 4331–4336. doi:10.1016/j.msea.2011.01.057.

Wang, Y. F., Gao, H. Y., Wang, J., et al. (2014). First-principles calculations of Ag addition on the diffusion mechanisms of Cu–Fe alloys. *Solid State Communications, 183,* 60–63. doi:10.1016/j.ssc.2013.11.025.

Watanabe, D., Watanabe, C., & Monzen, R. (2008). Effect of coherency on coarsening of second-phase precipitates in Cu-base alloys. *Journal of Materials Science, 43*(11), 3946–3953. doi:10.1007/s10853-007-2373-4.

Wu, Z. W., Chen, Y., & Meng, L. (2009). Microstructure and properties of Cu–Fe microcomposites with prior homogenizing treatments. *Journal of Alloys and Compounds, 481*(1–2), 236–240. doi:10.1016/j.jallcom.2009.03.078.

Xie, Z., Gao, H., Wang, J., et al. (2011). Effect of homogenization treatment on microstructure and properties for Cu–Fe–Ag in situ composites. *Materials Science and Engineering A, 529,* 388–392. doi:10.1016/j.msea.2011.09.047.

## Author Biographies

**Ji-en Ma** Ph.D., is Associate Professor in Zhejiang University, Hangzhou, China. She received a PhD degree in mechatronics from Zhejiang University in 2009. Then she did postdoctoral work in the College of Electrical Engineering, Zhejiang University, China. Her recent work is on electrical machines and drives. Her research interests include PM machines and drives for traction applications, and mechatronics machines such as the magneto fluid bearing.

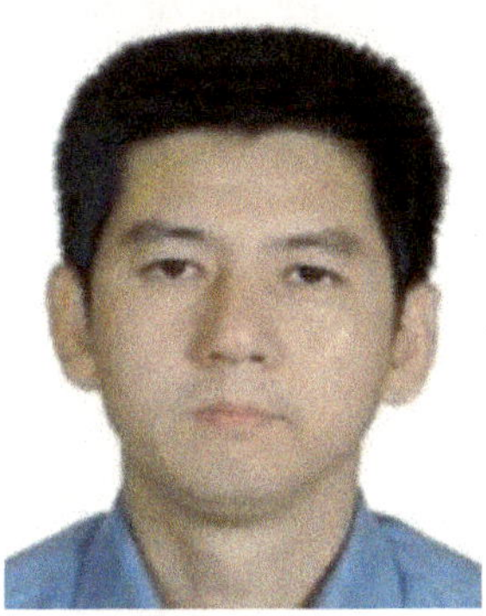

**You-tong Fang** is Professor at Zhejiang University. He is chairman of the High Speed Rail Research Center in Zhejiang University, deputy director of the National Intelligent Train Research Center, on the committee of the China High Speed Rail Innovation Plan, and expert of the National 863 Program (China High Technology Development Program) in modern transportation and advanced carrying technology. He is also director of three projects of the National Natural Science Foundation of China (NSFC) and more than 10 National 863 projects and Technology Support Programs. Recent work has been on electrical machines and drives. His research interests include permanent magnet (PM) machines and drives for traction applications.

**Liang Meng** Ph.D., is Professor in School of Materials Science and Engineering at Zhejiang University, China. He was an advanced visiting scholar at Northwestern University, IL, USA and a guest professor at the Ecole Nationale Supérieure d'Arts et Métiers, French. He has published more than 200 papers in internationally recognized journals. His research interests focus on metal and function film materials. He has managed and completed a national science and technology support project on high strength and high conductivity contact line used in the high-speed electric train.

**Jia-bin Liu** Ph.D., is Associate Professor in the School of Materials Science and Engineering at Zhejiang University, China. He earned his Bachelor and Ph.D. degrees in Materials Science and Engineering from Zhejiang University in 2004 and 2009, respectively. In 2013–2014 he was a visiting scholar at the City University of Hong Kong. He is the recipient of several grants and the author of over 50 articles.

# Microstructure and Hardness of Cu-12% Fe Composite at Different Drawing Strains

**Xiao-pei Lu, Da-wei Yao, Yi Chen, Li-tian Wang, An-ping Dong, Liang Meng and Jia-bin Liu**

## Contents



## 1 Introduction

Some Cu-based in situ composites, such as Cu–Fe, Cu–Nb, and Cu–Cr, containing body-centered cubic (bcc) filaments have attracted considerable attention because of their strong mechanical properties and electrical conductivity (Jin et al. 1997; Hong and Hill 2001; Gao et al. 2005, 2007; Wu et al. 2009a). The filamentary composite structure in those alloys was produced generally by heavily drawing the

X. Lu · D. Yao · Y. Chen · L. Meng · J. Liu (✉)
Department of Materials Science and Engineering,
Zhejiang University, Hangzhou 310027, China
e-mail: liujiabin@zju.edu.cn

L. Wang · A. Dong
China Railway Construction Electrification
Bureau Group Co., Ltd., Beijing 100036, China

J. Liu
College of Electrical Engineering,
Zhejiang University, Hangzhou 310027, China

 
Y. Fang and Y.H. Zhang (eds.), *China's High-Speed Rail Technology*,
Advances in High-speed Rail Technology, https://doi.org/10.1007/978-981-10-5610-9_16

as-cast double-phase structure and showed that the practical strength was much higher than that predicted by the linear rule of mixtures (Funkenbusch and Courtney 1985; Go and Spitzig 1991; Biselli and Morris 1994, 1996; Sauvage et al. 2005).

In these in situ composites, the Cu–Fe system has attracted particular attention because of the relatively low cost of the Fe constituent (Jeong et al. 2009; Qu et al. 2011; Morris and Muñoz-Morris 2011). Cu–Fe alloy could produce a high deformation without breakage during drawing strain at room temperature because of the excellent plasticity of Fe and Cu phases (Funkenbusch and Courtney 1981). During the deformation of Cu–Fe alloys, the Cu matrix mainly presented a ⟨111⟩ texture and the Fe phase mainly a ⟨110⟩ texture (Brokmeier et al. 2000). The Cu matrix and Fe phase evolved into ribbon-like filaments and the filamentary scale decreased with an increase in the deformation degree (Biselli and Morris 1994, 1996). The fine microstructure of Cu–Fe composite can benefit the mechanical properties since the high density of the phase boundary can effectively provide a strengthening benefit (Stepanov et al. 2013). Cu–Fe composite with fine microstructure produced from rapid solidification and heavy drawing deformation results in higher strength than that predicted by the rule of mixture averages (Biselli and Morris 1994; He et al. 2000). The mechanical properties depend obviously on the filamentary spacing in the drawn microstructure. For example, the strength of Cu-20% Fe in situ composites increased with the reduction in the spacing between Fe filaments and obeyed a Hall–Petch relationship (Go and Spitzig 1991).

In this study, Cu-12% Fe (in weight) in situ filamentary composite was prepared by casting, pretreating, and cold drawing to different strain levels. The microstructure evolution was observed, and the Vickers hardness was determined. The relationship between the preferred orientation and the drawing strain was investigated. The effect of the Fe filament spacing on the hardness at different drawing strains was discussed.

## 2 Materials and Methods

Cu-12% Fe alloy was melted in a vacuum induction furnace and cast into a copper mold in a $1.0 \times 10^4$ Pa Ar-shielded atmosphere to obtain cylindrical ingots of 21.0 mm in diameter. The inner and outer diameters of the copper mold were 21.0 mm and 60.0 mm, respectively. Electrolytic Cu with 99.99% purity and master alloy of Cu-50% Fe were used as starting materials. To promote the precipitation of secondary Fe particles, the as-cast ingots were pretreated at 1000 °C for 1 h, quenched in water and then aged at 550 °C for 4 h. The surface layer about 1.0 mm in thickness was turned off to remove the surface oxides and defects. The ingots were heavily drawn into fine wires by multiple drawing performed at ambient temperature with a straight wire drawing machine (LZ100, Hangzhou Drawing Factory, China), which has a power of 5 kW and a working length of 6.5 m. The level of drawing reduction was evaluated by $\eta = \ln(A_0/A)$ and referred to as the drawing strain, where $A_0$ and $A$ are the original and final transverse section areas of the drawing specimens, respectively. The drawing reduction in per pass was $\Delta\eta < 0.2$, and wires with about 50 mm in length were cut as test specimens at $\eta$ = 1.0, 2.0, 3.0, 4.0, 5.0, 6.0, 7.0, 8.0 and 8.6. The total drawing reduction was $\eta$ = 8.6 while the final length of the wire reached about 6 m.

The microstructure was observed by scanning electron microscopy (SEM) (Hitachi S4800, Japan). The measured values of average width, thickness and spacing of Fe grains at different drawing strains were taken from the arithmetical averages at least 50 measured points on the transverse section of the specimens. The average length of the interface between Cu and Fe phases per unit area was also measured from the microstructure on the transverse section of the specimens. The phase structure was identified by X-ray diffraction (XRD) using Cu-K$\alpha$ radiation from 30° to 100°. The operating voltage and current were 40 kV and 100 mA, respectively. The hardness was determined on the polished transverse section of the specimens using a Vickers hardness tester (MH-5, Laizhou Weiyi Experiment Machine Manufacturing Co., Ltd, China) with a load of 100 g and a dwell time of 15 s. Each value was taken from the arithmetical mean measured from more than ten indentations.

# 3 Results

## 3.1 Microstructure

Figure 1a shows the microstructure of pretreated Cu-12% Fe. Coarse primary Fe dendrites and some secondary Fe particles are observed in the pretreated alloy. Both Cu and Fe phases are elongated along the drawing direction during cold drawing. The Fe phase evolves into pencils at small drawing strains and even into nano-fibers at high drawing strains on the longitudinal section (Fig. 1b, c). The length of the Fe filaments continuously increases and the interval of the Fe filaments decreases with the increase in the drawing strains.

Figure 2 shows the microstructure on the transverse section of Cu-12% Fe at different drawing strains. Fe dendrites are curled into a ribbon-like morphology, which is the result of co-deformation between Fe and Cu phases. It has been well

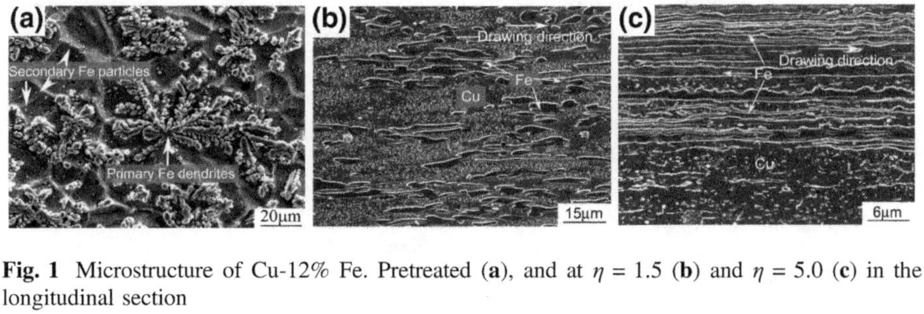


**Fig. 1** Microstructure of Cu-12% Fe. Pretreated (**a**), and at $\eta$ = 1.5 (**b**) and $\eta$ = 5.0 (**c**) in the longitudinal section

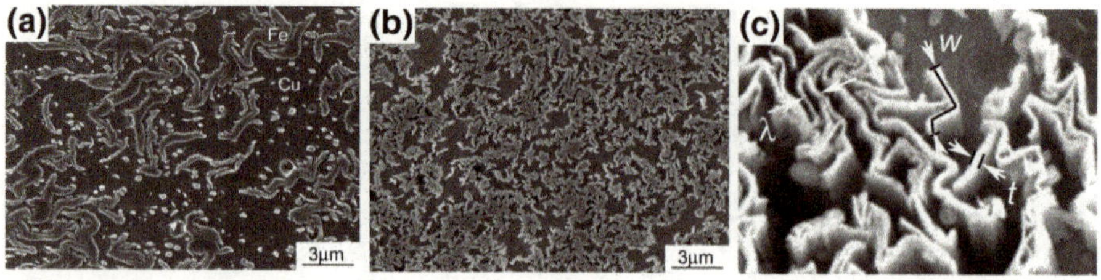


**Fig. 2** Microstructure on the transverse section of the Cu-12% Fe. **a** $\eta = 5.0$, **b** $\eta = 7.0$ and **c** schematic illustration of the measurement of thickness ($t$), width ($w$), and spacing ($\lambda$) of Fe ribbons

**Fig. 3** Dependence of the average thickness ($t$), width ($w$), and spacing ($\lambda$) of Fe ribbons on the drawing strain for Cu-12% Fe

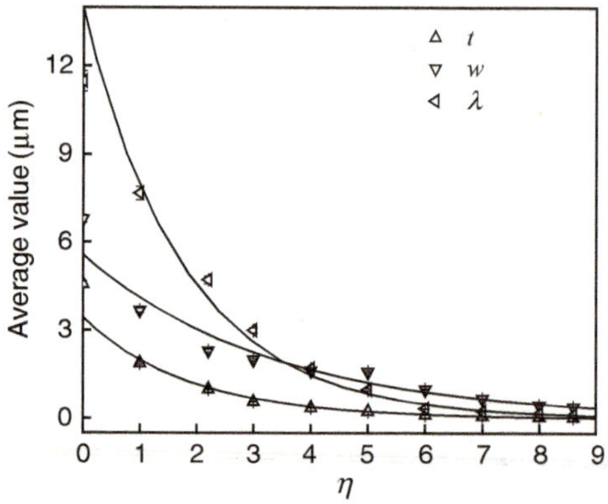


documented that the bcc phase is forced to curl and fold due to the constraint of the surrounding face-centered cubic matrix, which is able to accommodate axially symmetric flow (Bevk et al. 1978). The ribbon-like morphology has been widely observed in Cu–Fe and Cu–Nb alloys (Bevk et al. 1978; Raabe et al. 2009; Wu et al. 2009b). Increasing the drawing strain to $\eta = 7.0$ produces a more uniform dense distribution of Fe ribbons in the Cu matrix.

The thickness ($t$), width ($w$) and spacing ($\lambda$) of Fe ribbons could be measured on the SEM images, as schematically shown in Fig. 2c. The change in the average thickness, width and spacing of Fe ribbons with various drawing strains is given in Fig. 3. These values decrease with an increase in the drawing strain. In particular, the reduction in the spacing is more obvious than that in the ribbon width or thickness of Fe ribbons. Some exponential relationships between the microstructure scale and drawing strain can be fitted from the obtained data as

$$t = 3.40\mathrm{e}^{-0.54\eta}, \tag{1}$$

$$w = 5.54\mathrm{e}^{-0.30\eta}, \tag{2}$$

$$\lambda = 13.85\mathrm{e}^{-0.56\eta}. \tag{3}$$

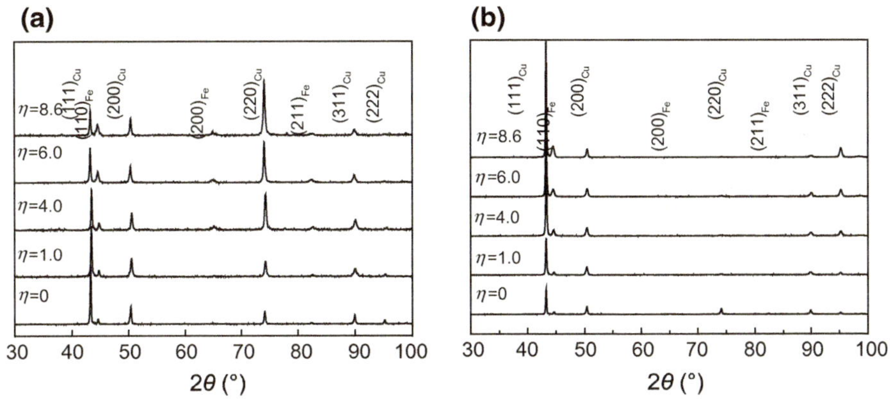


**Fig. 4** XRD patterns of Longitudinal section (**a**) and transverse section (**b**) of Cu-12% Fe at different drawing strains

## 3.2 *Structure Orientation*

Figure 4 shows the XRD patterns of Cu-12% Fe at different drawing strains. The intensities of (*hkl*) diffractions are given in Table 1. The alloy consists of Cu and α-Fe phases, but the relative intensity between different diffraction peaks changes with the drawing strain. For example, from the detection on the longitudinal section, the intensity of $(111)_{Cu}$ is higher than that of $(220)_{Cu}$ at lower drawing strains, but the intensity of $(111)_{Cu}$ becomes lower than that of $(220)_{Cu}$ at higher drawing strains. From the detection on the transverse section, the intensities of $(111)_{Cu}$ and $(110)_{Fe}$ increase with the drawing strain but that of $(220)_{Cu}$ almost disappears at high drawing strains. These results imply that the drawing deformation results in a change in crystal preferred orientation or texture distribution.

The degree of preferred orientation in the alloy can be described by the Lotgering factor (Lotgering 1959)

$$L_{(hkl)} = \frac{P_{(hkl)} - P^0_{(hkl)}}{1 - P^0_{(hkl)}}, \tag{4}$$

where

$$P_{(hkl)} = I_{(hkl)} / \sum_j I_j, \tag{5}$$

**Table 1** Related XRD data and experimental counts to $I_j$, $I_j^0$, and $P_{(hkl)}$ on the longitudinal section of the specimens

| $\eta$ | Intensity of related (*hkl*) diffractions (counts per second) | | | | | | | $P_{(hkl)}$ | | | | | |
|---|---|---|---|---|---|---|---|---|---|---|---|---|---|
| | $(111)_{Cu}$ | $(110)_{Fe}$ | $(200)_{Cu}$ | $(220)_{Cu}$ | $(200)_{Fe}$ | $(311)_{Cu}$ | $I_j^0$ | $(111)_{Cu}$ | $(110)_{Fe}$ | $(200)_{Cu}$ | $(220)_{Cu}$ | $(200)_{Fe}$ | $(311)_{Cu}$ |
| 0 | 12,209.4 | 460.5 | 3532.8 | 104.1 | 2685.6 | 1879.5 | 21,786.0 | 0.56 | 0.02 | 0.16 | 0.005 | 0.12 | 0.04 |
| 1.0 | 14,590.0 | 11,450.5 | 13,287.5 | 10,088.0 | 11,933.5 | 11,563.0 | 79,199.5 | 0.18 | 1.24 | 0.17 | 0.13 | 0.15 | 0.13 |
| 4.0 | 21,847.8 | 21,400.5 | 22,225.2 | 20,448.6 | 27,003.1 | 21,768.4 | 153,711.3 | 0.14 | 0.14 | 0.14 | 0.14 | 0.18 | 0.14 |
| 6.0 | 36,452.4 | 31,444.2 | 32,509.2 | 30,295.2 | 38,502.6 | 31,468.2 | 224,973.6 | 0.16 | 0.16 | 0.17 | 0.16 | 0.17 | 0.14 |
| 8.6 | 45,290.0 | 460.5 | 43,422.0 | 40,279.0 | 51,686.0 | 41,244.0 | 257,678.5 | 0.18 | 0.02 | 0.17 | 0.16 | 0.20 | 0.16 |

$$P^0_{(hkl)} = I^0_{(hkl)} / \sum I^0_j, \tag{6}$$

where $I_{(hkl)}$ and $I^0$ $(hkl)$ are the intensities of $(hkl)$ diffraction in the deformed and undeformed specimens, and $I_j$ and $I^0_j$ are the intensities of any diffractions in the deformed and undeformed specimens, respectively. It is obvious that there should be a random orientation of $(hkl)$ if $L_{(hkl)} = 0$ or there should be a perfect preferred orientation of $(hkl)$ if $L_{(hkl)} = 1$. In general, the presented probability of the preferred orientation of $(hkl)$ must decrease with the reduction in $L_{(hkl)}$. Based on Eqs. (4)–(6) and the experimental XRD data, related experimental counts to $I_j$, $I^0_j$, and $P_{(hkl)}$ are given in Tables 1 and 2.

Figure 5 shows the Lotgering factors of different crystal planes from the longitudinal and transverse sections of the alloy at different drawing strains. From the longitudinal section of Cu-12% Fe, $L_{(220)}$ of Cu obviously increases and $L_{(200)}$ of Fe slightly increases with an increase in the drawing strain. $L_{(111)}$ of Cu and $L_{(110)}$ of Fe decrease while $L_{(200)}$ of Cu hardly changes with an increase in the drawing strain. From the transverse section, $L_{(111)}$ of Cu and $L_{(110)}$ of Fe increase with the drawing strain. $L_{(200)}$ and $L_{(220)}$ of Cu and $L_{(200)}$ of Fe maintain basically a constant or slightly decrease with an increase in the drawing strain. Those results indicate that the drawing strain can produce the preferred orientations of $\langle 110 \rangle$ in Cu filaments and $\langle 100 \rangle$ in Fe filaments on the longitudinal section, and the drawing strain can produce the preferred orientations of $\langle 111 \rangle$ in Cu filaments and $< 110 >$ in Fe filaments on the transverse section.

### 3.3 Vickers Hardness

Figure 6 shows the change of the hardness of Cu-12% Fe with the drawing strain. The hardness obviously increases in the initial drawing process and at $\eta > 5.0$, and slowly increases at $\eta = 2.2$–$5.0$.

## 4 Discussion

### 4.1 Microstructure Evolution

During drawing, Fe dendrites were evolved into aligned filaments on the longitudinal section and into curled ribbons on the transverse section. The aspect ratio $k$ is defined to describe the morphological evolution by determining the width $w$ and thickness $t$ of Fe ribbons on the transverse section

**Table 2** Related XRD data and experimental counts to $I_j$, $I_j^0$, and $P_{(hkl)}$ on the transverse section of the specimens

| $\eta$ | Intensity of related (*hkl*) diffractions (counts per second) | | | | | | | $P_{(hkl)}$ | | | | | |
|---|---|---|---|---|---|---|---|---|---|---|---|---|---|
| | $(111)_{Cu}$ | $(110)_{Fe}$ | $(200)_{Cu}$ | $(220)_{Cu}$ | $(200)_{Fe}$ | $(311)_{Cu}$ | $I_j^0$ | $(111)_{Cu}$ | $(110)_{Fe}$ | $(200)_{Cu}$ | $(220)_{Cu}$ | $(200)_{Fe}$ | $(311)_{Cu}$ |
| 0 | 6,446.6 | 570.4 | 1960.4 | 81.4 | 1418.0 | 992.4 | 12,032.1 | 0.54 | 0.05 | 0.13 | 0.01 | 0.11 | 0.08 |
| 1.0 | 19,511.0 | 10,707.0 | 12,006.0 | 10,068.0 | 10,249.0 | 10,613.0 | 93,744.0 | 0.26 | 0.14 | 0.16 | 0.13 | 0.14 | 0.14 |
| 4.0 | 35,784.0 | 21,464.0 | 22,038.4 | 20,056.0 | 20,204.8 | 20,856.0 | 181,590.4 | 0.25 | 0.15 | 0.15 | 0.14 | 0.14 | 0.14 |
| 6.0 | 44,679.0 | 31,526.4 | 31,863.0 | 30,041.4 | 30,203.4 | 30,811.8 | 260,746.8 | 0.21 | 0.15 | 0.15 | 0.15 | 0.14 | 0.14 |
| 8.6 | 67,311.4 | 42,376.0 | 41,769.4 | 40,025.2 | 40,127.8 | 40,412.2 | 354,277.4 | 0.23 | 0.15 | 0.15 | 0.14 | 0.14 | 0.14 |

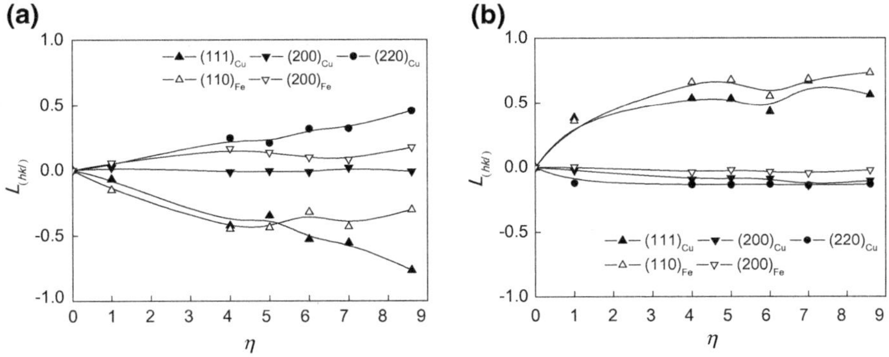


**Fig. 5** Lotgering factors of different crystal planes from the longitudinal section (**a**) and transverse section (**b**) of Cu-12% Fe at different drawing strains

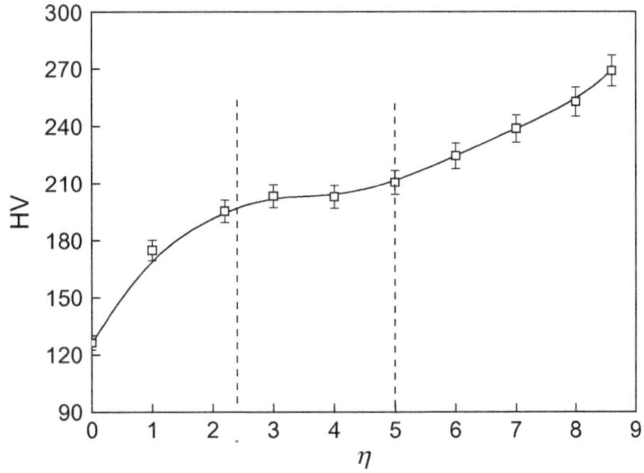


**Fig. 6** Dependence of the hardness on longitudinal section of Cu-12% Fe with the drawing strain

$$k = w/t. \tag{7}$$

It is certain that a high drawing strain must produce a high aspect ratio. The true strain of Fe grains can be expressed as

$$\varepsilon_{\mathrm{Fe}} = \ln(k_{\mathrm{Fe}}/k_{\mathrm{Fe}}^{0}), \tag{8}$$

where $k_{\mathrm{Fe}}$ and $k_{\mathrm{Fe}}^{0}$ are the average aspect ratios of Fe grains in deformed and undeformed specimens, respectively.

The relationship between the strain of Fe grains and the drawing strain of the wires is shown in Fig. 7. At $\eta$ = 0–6.0, the strain of Fe grains increases almost linearly with the drawing strain. This implies that Fe grains undergo a homogeneous strain with the reduction in the wire section or both strains of Fe and Cu phases being practically isochronous. At $\eta$ > 6.0, the increase in the strain of Fe

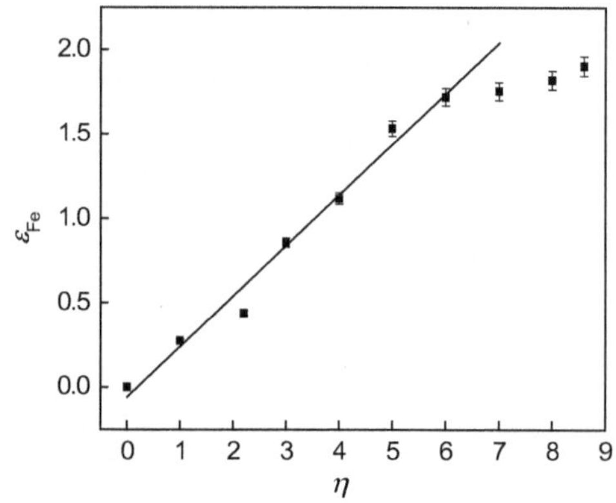


**Fig. 7** Dependence of the strain of Fe grains on the drawing strain for Cu-12% Fe

grains becomes slow and deviates from the linear relation. This implies that the strain degree of Fe grains is lower than the strain of the wire or that the Cu matrix endures more strain than Fe ribbons at higher drawing strains. In this case, the Fe ribbons may need more curl to fit the strain of the Cu matrix, which is similar to the situation of Nb ribbons (Raabe et al. 2009; Hao et al. 2013).

The microstructure evolution during the drawing strain must result in the change in the interface density and shows the relationship between the interface density of Cu and Fe phases and the aspect ratio of Fe ribbons. The interface density obviously increases with the aspect ratio of Fe ribbons and can be expressed by the exponential relationship as shown in Fig. 8,

$$S = 0.14e^{0.44k}. \tag{9}$$

## 4.2 *Hardness*

The hardness as a function of the Fe filament spacing is plotted in Fig. 9. It is obvious that the relationship between the hardness and Fe filament spacing can be expressed by the well-known Hall–Petch equation similarly as

$$\mathrm{HV} = \mathrm{HV}_0 + k_{\mathrm{H}}\lambda^{-1/2}, \tag{10}$$

where $k_{\mathrm{H}}$ is the Hall–Petch coefficient reflecting the change in hardness with filament spacing, and $\mathrm{HV}_0$ is the intrinsic hardness of the specimen with a rather coarse filament.

The change in strain hardening with filament spacing shows the behavior in two stages. In the initial strain stage of $\eta < 3.0$, the multiplication of dislocations or the refinement of subcells has a high rate in elongated Cu and Fe grains due to low dislocation density in the coarse initial microstructure, which results in a higher

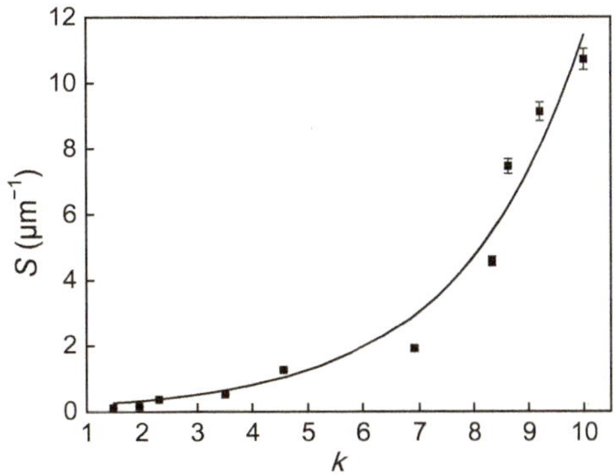


**Fig. 8** Dependence of the interface density of Cu and Fe phases on the aspect ratio for Cu-12% Fe

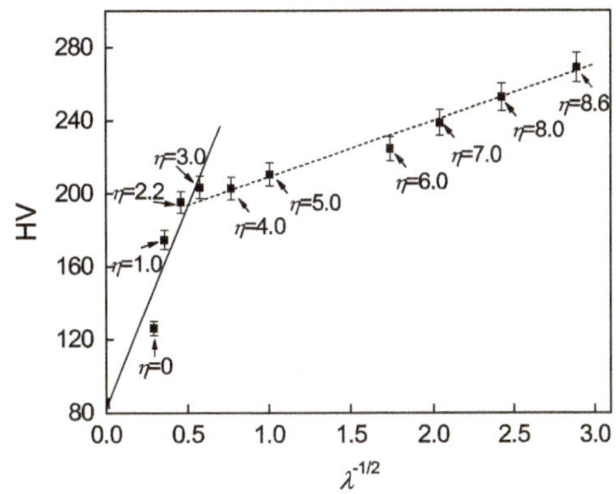


**Fig. 9** Change in the hardness on the longitudinal section of Cu-12% Fe wire with the Fe ribbon spacing

hardening rate or larger Hall–Petch coefficient from the microstructure refining. In the high strain stage of $\eta > 3.0$, the dynamic recovery may occur because high deformation heat sufficiently releases the strain storage energy, which results in near equilibrium between dislocation generation and annihilation. The dynamic recovery has been well observed and discussed in cold drawn pure Cu and Cu–Nb alloys (Cairns et al. 1971; Spitzig et al. 1987). Moreover, at large drawing strains, the interface spacing approaches the minimum size of stable dislocation cells or the subcells have failed to be refined further in the subsequent drawing strain (Spitzig 1991; Biselli and Morris 1996; Zheng et al. 2013). Both mechanisms can be responsible for the reduced hardening rate or reduced Hall–Petch coefficient due to microstructure refining in the stage of a heavy drawing strain at $\eta > 3.0$.

## 5 Conclusions

The Cu matrix and Fe dendrites in Cu-12% Fe evolved into the composite filamentary structure during the drawing process. The preferred orientations of $\langle 110 \rangle$ in Cu filaments and $\langle 100 \rangle$ in Fe filaments tend to be formed on the longitudinal section of the wires. The preferred orientations of ⟨111⟩ in Cu filaments and ⟨110⟩ in Fe filaments tend to be formed on the transverse section.

The thickness, width, and spacing of Fe ribbons in the filamentary structure exponentially decrease with an increase in the drawing strain. The density of the interface between Cu and Fe phases exponentially increases with an increase in the aspect ratio of Fe ribbons. With an increase in the drawing strain, the strain of Fe grains linearly increases at drawing strains lower than 6.0 and deviates from the linear relationship at drawing strains higher than 6.0.

Heavy drawing deformation results in high strain hardening. There is a similar Hall–Petch relationship between the hardness and Fe filament spacing. The hardening from the microstructure refining is more obvious at drawing strains below 3.0 than over 3.0.

## References

Bevk, J., Harbison, J. P., & Bell, J. L. (1978). Anomalous increase in strength of in situ formed Cu-Nb multifilamentary composites. *Journal of Applied Physics, 49*, 6031–6038.

Biselli, C., & Morris, D. G. (1994). Microstructure and strength of Cu–Fe in situ composites obtained from prealloyed Cu–Fe powders. *Acta Metallurgica et Materialia, 42*(1), 163–176. doi:10.1016/0956-7151(94)90059-0.

Biselli, C., & Morris, D. G. (1996). Microstructure and strength of Cu–Fe in situ composites after very high drawing strains. *Acta Materialia, 44*(2), 493–504. doi:10.1016/1359-6454(95)00212-X.

Brokmeier, H. G., Bolmaro, R. E., Signorelli, J. A., et al. (2000). Texture development of wire drawn Cu–Fe composites. *Physica B: Condensed Matter, 276–278,* 888–889. doi:10.1016/S0921-4526(99)01540-9.

Cairns, J. H., Clough, J., Dewey, M. A. P., & Nutting, J. (1971). The structure and mechanical properties of heavily deformed copper. *Journal of the Institute of Metals, 99,* 93–97.

Funkenbusch, P. D., & Courtney, T. H. (1981). Microstructural strengthening in cold worked in situ Cu-14.8 Vol.% Fe composites. *Scripta Metallurgica, 15*(12), 1349–1354. doi:10.1016/0036-9748(81)90096-X.

Funkenbusch, P. D., & Courtney, T. H. (1985). On the strength of heavily cold worked in situ composites. *Acta Metallurgica, 33*(5), 913–922. doi:10.1016/0001-6160(85)90116-6.

Gao, H. Y., Wang, J., Shu, D., et al. (2005). Effect of Ag on the microstructure and properties of Cu–Fe *in situ* composites. *Scripta Materialia, 53*(10), 1105–1109. doi:10.1016/j.scriptamat.2005.07.028.

Gao, H. Y., Wang, J., Shu, D., et al. (2007). Microstructure and strength of Cu–Fe–Ag in situ composites. *Materials Science and Engineering A, 452–453,* 367–373. doi:10.1016/j.msea.2006.10.111.

Go, Y. S., & Spitzig, W. A. (1991). Strengthening in deformation-processed Cu-20% Fe composites. *Journal of Materials Science, 26*(1), 163–171. doi:10.1007/BF00576047.

Hao, S. J., Cui, L. S., Jiang, D. Q., et al. (2013). A transforming metal nanocomposite with large elastic strain, low modulus, and high strength. *Science, 339*(6124), 1191–1194. doi:10.1126/science.1228602.

He, L., Allard, L. F., & Ma, E. (2000). Fe-Cu two-phase nanocomposites: Application of a modified rule of mixtures. *Scripta Materialia, 42*(5), 517–523. doi:10.1016/S1359-6462(99)00300-0.

Hong, S. I., & Hill, M. A. (2001). Microstructure and conductivity of Cu-Nb microcomposites fabricated by the bundling and drawing process. *Scripta Materialia, 44*(10), 2509–2515. doi:10.1016/S1359-6462(01)00665-0.

Jeong, E., Han, S., Goto, M., et al. (2009). Effects of thermo-mechanical processing and trace amount of carbon addition on tensile properties of Cu–2.5Fe–0.1P alloys. *Materials Science and Engineering A, 520*(1–2), 66–74. doi:10.1016/j.msea.2009.05.021.

Jin, Y., Adachi, K., Takeuchi, T., et al. (1997). Correlation between the electrical conductivity and aging treatment for a Cu-15 wt% Cr alloy composite formed in-situ. *Materials Letters, 32*(5–6), 307–311. doi:10.1016/S0167-577X(97)00053-0.

Lotgering, F. K. (1959). Topotactical reactions with ferrimagnetic oxides having hexagonal crystal structures—I. *Journal of Inorganic and Nuclear Chemistry, 9*(2), 113–123. doi:10.1016/0022-1902(59)80070-1.

Morris, D. G., & Muñoz-Morris, M. A. (2011). The effectiveness of equal channel angular pressing and rod rolling for refining microstructures and obtaining high strength in a Cu–Fe composite. *Materials Science and Engineering A, 528*(19–20), 6293–6302. doi:10.1016/j.msea.2011.04.076.

Qu, L., Wang, E. G., Zuo, X. W., et al. (2011). Experiment and simulation on the thermal instability of a heavily deformed Cu–Fe composite. *Materials Science and Engineering A, 528*(6), 2532–2537. doi:10.1016/j.msea.2010.12.015.

Raabe, D., Ohsaki, S., & Hono, K. (2009). Mechanical alloying and amorphization in Cu-Nb-Ag *in situ* composite wires studied by transmission electron microscopy and atom probe tomography. *Acta Materialia, 57*(17), 5254–5263. doi:10.1016/j.actamat.2009.07.028.

Sauvage, X., Wetscher, F., & Pareige, P. (2005). Mechanical alloying of Cu and Fe induced by severe plastic deformation of a Cu–Fe composite. *Acta Materialia, 53*(7), 2127–2135. doi:10.1016/j.actamat.2005.01.024.

Spitzig, W. A. (1991). Strengthening in heavily deformation processed Cu-20%Nb. *Acta Metallurgica et Materialia, 39*(6), 1085–1090. doi:10.1016/0956-7151(91)90195-7.

Spitzig, W. A., Pelton, A. R., & Laabs, F. C. (1987). Characterization of the strength and microstructure of heavily cold worked Cu-Nb composites. *Acta Metallurgica, 35*(10), 2427–2442. doi:10.1016/0001-6160(87)90140-4.

Stepanov, N. D., Kuznetsov, A. V., Salishchev, G. A., et al. (2013). Evolution of microstructure and mechanical properties in Cu–14%Fe alloy during severe cold rolling. *Materials Science and Engineering A, 564,* 264–272. doi:10.1016/j.msea.2012.11.121.

Wu, Z. W., Chen, Y., & Meng, L. (2009a). Effects of rare earth elements on annealing characteristics of Cu-6 wt.% Fe composites. *Journal of Alloys and Compounds, 477*(1–2), 198–204. doi:10.1016/j.jallcom.2008.10.047.

Wu, Z. W., Liu, J. J., Chen, Y., & Meng, L. (2009b). Microstructure, mechanical properties and electrical conductivity of Cu–12wt% Fe microcomposite annealed at different temperatures. *Journal of Alloys and Compounds, 467*(1–2), 213–218. doi:10.1016/j.jallcom.2007.12.020.

Zheng, S. J., Beyerlein, I. J., Carpenter, J. S., et al. (2013). High-strength and thermally stable bulk nanolayered composites due to twin-induced interfaces. *Nature Communications, 4,* 1696. doi:10.1038/ncomms2651.

## Author Biographies

**Liang Meng, Ph.D.,** is a professor in School of Materials Science and Engineering at Zhejiang University, China. He was an advanced visiting scholar at Northwestern University, IL, USA, and a guest professor at the Ecole Nationale Supérieure d'Arts et Métiers, French. He has published more than 200 papers in internationally recognized journals. His research interests focus on metal and function film materials. He has managed and completed a national science and technology support project on high strength and high conductivity contact line used in the high-speed electric train.

**Jia-bin Liu, Ph.D.,** is the associate professor in the School of Materials Science and Engineering at Zhejiang University, China. He earned his Bachelor and Ph.D. degrees in Materials Science and Engineering from Zhejiang University in 2004 and 2009, respectively. In 2013–2014 he was a visiting scholar at the City University of Hong Kong. He is the recipient of several grants and the author of over 50 articles.

# Part IV
# High-Speed Rail Wheel/Rail Dynamics

# Modeling of High-Speed Wheel–Rail Rolling Contact on a Corrugated Rail and Corrugation Development

**Xin Zhao, Ze-feng Wen, Heng-yu Wang, Xue-song Jin and Min-hao Zhu**

## Contents



## 1 Introduction

Rail corrugation is a long-standing problem observed worldwide on many kinds of railway tracks, including tram, metro, traditional railway, heavy haul, and high-speed tracks. Once present, corrugation can worsen the wheel–rail and

X. Zhao (✉) · Z. Wen · H. Wang · X. Jin · M. Zhu
State Key Laboratory of Traction Power,
Southwest Jiaotong University, Chengdu 610031, China
e-mail: xinzhao@home.swjtu.edu.cn

 
Y. Fang and Y.H. Zhang (eds.), *China's High-Speed Rail Technology*,
Advances in High-speed Rail Technology, https://doi.org/10.1007/978-981-10-5610-9_17

**Fig. 1** Corrugated rail section observed on a Chinese high-speed line in 2011 (Its wavelength of about 80 mm can be estimated from the sleeper span of 0.65 m. The running speed on this section is about 300 km/h, currently the maximum commercial speed in China.)

vehicle–track interactions, leading to poor ride quality and an exacerbated rate of deterioration of the system, e.g., rail support failure (Zhou and Shen 2013). Recently, rail corrugation, particularly short-pitch rail corrugation (hereinafter referred to as corrugation for short), was observed on a recently opened high-speed line in China, causing great concern in the industry. Figure 1 shows a corrugated rail section on the high-speed line.

It is well known that corrugation should be a consequence of the accumulation of irregular wear and/or irregular plastic deformation (i.e., the material damage mechanism). Such irregular material damage is likely to be caused by certain eigenmodes of the vehicle–track system excited by some imperfections in the system (i.e., the wavelength fixing mechanism), and the unstable wheel–rail rolling contact resulting from those eigenmodes is the immediate cause. Considering different material damage and wavelength fixing mechanisms, Grassie and Kalousek (1993) and Grassie (2005) classified corrugation into six different categories based on engineering experience, which significantly increased the understanding of corrugation.

The corrugation mechanisms proposed by Grassie and Kalousek (1993), Grassie (2005), and Knothe and Groβ-Thebing (2008) imply that the key to understanding and predicting corrugation initiation and development is to solve the dynamic vehicle–track interaction and the transient wheel–rail rolling contact. The present work employs a 3D transient rolling contact finite element (FE) model to solve the high-speed wheel–rail rolling contact and the vehicle–track interaction on a corrugated rail in the time domain. This FE model is valid for a rolling speed of up to 500 km/h. A Chinese high-speed railway system is considered. The emphasis of analysis is placed on detailed contact solutions. On the basis of numerical results and field measurements, a better understanding of the mechanism of corrugation development is achieved.

Traditionally, the vehicle–track interaction on corrugation sites was treated without detailed wheel–rail contact modeling. A simplified Hertz spring was usually employed to represent the contact, together with beams for rails and lumped masses for wheels, respectively (Knothe and Grassie 1993; Hiensch et al. 2002; Wu and Thompson 2004; Knothe and Groβ-thebing 2008; Nielsen 2008; Xie and Iwnicki 2008; Iwnicki et al. 2009; Li et al. 2009; Xiao et al. 2010; Zhai et al. 2013). Moreover, most traditional models considered only the normal wheel–rail interaction (Knothe and Grassie 1993; Hiensch et al. 2002; Wu and Thompson 2004; Nielsen 2008; Xie and Iwnicki 2008; Iwnicki et al. 2009; Li et al. 2009). Nevertheless, the importance of the tangential interaction can be seen from the fact that both the plastic deformation and wear of rails are dominated by the tangential contact load on many corrugation sites, such as on corrugated curves. Taking into account the tangential interaction, Clark et al. (1988) proposed a mechanism of slip-stick vibrations to explain the occurrence of corrugation. Knothe and Groβ-Thebing (2008) and Groβ-Thebing et al. (1992) treated the tangential interaction by using a viscous damper to simulate the steady rolling contact and the combination of a viscous damper and a spring for the non-steady case. Kalker (1990)'s creep coefficient was employed to determine the characteristics of the viscous damper for the steady rolling contact, and a frequency-dependent creep coefficient (Gross-Thebing 1989) for the corresponding parameters of the non-steady case. In addition, approaches have also been developed (Iwnicki et al. 2009) to derive the dynamic tangential contact force from the normal contact force and the geometric and material properties of the wheel and rail (Afshari and Shabana 2010).

Recent studies have shown that the structural flexibility of the wheelset has a significant influence on the vehicle–track interaction (Ripke and Knothe 1995; Chaar and Berg 2006) and even the rotation of the wheel might also play an important role in high-frequency vehicle–track dynamics under certain conditions (Baeza et al. 2008). Wen et al. (2005), Pang and Dhanasekar (2006), and Pletz et al. (2009) considered the detailed geometries of the wheel and rail to take their flexibility into account, but only the normal wheel–rail interaction was solved for cases at joints or in crossings.

The FE modeling approach employed in this study origins from that published in (Li et al. 2008). This approach has been validated by Li et al. (2008, 2011) and Molodova et al. (2011) for the high-frequency vehicle–track interaction at squats (in the frequency range between a few hundred and about 2000 Hz), and by Zhao and Li (2011) for the normal and tangential contact solutions. Li et al. (2012) employed the FE modeling approach to study the wheel–rail rolling contact on corrugation and the resulting wear pattern at a speed of 108 km/h, for which a ballasted track was considered. However, the transient rolling contact of a wheel over a corrugated rail at high speeds has not been studied yet.

Following the same modeling approach, a 3D transient rolling contact FE model is developed for a slab track system of the Chinese high-speed line shown in Fig. 1 to study the corrugation phenomenon. The main advance of the model is that a rolling speed of up to 500 km/h can be simulated, while the maximum rolling speed considered in previous studies was 140 km/h on a ballasted track. In the model, the actual geometries of a wheelset and a rail are included by a mesh of solid elements, based on which a detailed surface-to-surface contact algorithm is employed to solve the transient rolling contact in the time domain. The flexibility of the vehicle and track subsystems and the wheel–rail continua are both included, and the rolling-sliding behavior of the wheel on the rail is simulated. The contact filter effect, which eliminates the corrugation components with a wavelength close to or less than the width of the contact patch (Knothe and Groβ-Thebing 2008), is considered inherently. Hence, transient contact stresses, including both normal and tangential stresses, and their derivatives are obtained through the numerical simulation, together with the resultant contact forces. These ensure the applicability of the FE model to high-frequency vehicle–track interaction on corrugation sites. Furthermore, idealized corrugation models are applied and simulated to better understand the fundamentals of the corrugation phenomenon, even though measured corrugation profiles can be introduced. For clarity and ease of explanation, the case of a smooth rail (without any irregularities) is analyzed before considering corrugations.

## 2 Model Descriptions

## *2.1 FE Model*

### 2.1.1 An Overview

Figure 2 illustrates a 3D transient rolling contact FE model developed with ANSYS/LS-DYNA, which considers a high-speed vehicle and a typical slab track on a Chinese high-speed line. The modeling approach for the vehicle is the same as that used by Li et al. (2008) and Zhao and Li (2011). The track is composed of the rail, fastenings, slabs, and the mortar layer. For the investigated high-speed line, the minimum radius of curvature is 7000 m, on which the lateral movement of the wheelset is very well controlled, as predicted by multi-body simulations performed in the State Key Laboratory of Traction Power in Southwest Jiaotong University, China. Further considering that the rolled distance of the wheelset in a typical simulation in this study is less than 3.5 m, the lateral movement of the wheelset becomes negligible. Therefore, only a half wheelset and a half straight track are modeled in view of the symmetry of the system (Fig. 2b), and the lateral movement of the wheelset and the track is ignored. The simulated track is 15.2 m long and includes 24 fastenings.

A Lagrangian mesh of solid elements is applied to the wheelset and the rail. The minimum element size is 1.1 mm, used in the contact surface of the solution zone (*BC* in Fig. 2a), where irregularities such as corrugation are applied by modifying the coordinates of the nodes involved. The wheel profile is of the type $LM_A$, and the rail is the standard CN60 with an inclination of 1:40. A penalty method-based surface-to-surface contact algorithm is employed to solve the wheel–rail rolling

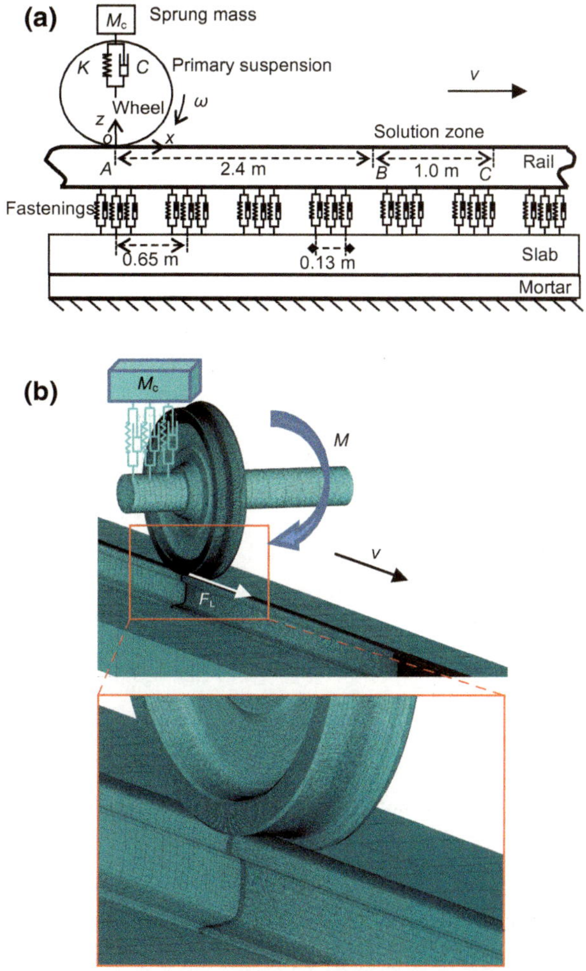


**Fig. 2** Three-dimensional transient FE model for wheel–rail rolling contact. **a** A schematic diagram; **b** The mesh. The slab layer is composed of prefabricated slabs of 6.5 m long and the gaps of 50 mm in between filled with concrete. Reprinted from Zhao et al. (2014) by permission of Taylor & Francis Ltd.

contact, in which Coulomb's law of friction is used. A mesh of solid elements is also applied to the slabs and the mortar layer. A fastening system is simulated by 12 groups of parallel springs and dampers (three columns in the longitudinal and four rows in the lateral directions). In total, there are $1.46 \times 10^6$ elements and $1.29 \times 10^6$ nodes.

For solution, boundary conditions are applied as follows: Symmetric boundary conditions are applied to the axle ends of the wheelset and to the rail ends; the bottom of the mortar layer is fixed; the fastenings, the slabs, and the mortar layer can only move vertically. Table 1 lists the values of the parameters involved. To simulate the worst-case scenario, the sprung mass $M_c$ is determined by considering the weight of the loaded coach, the non-uniform distribution of the weight on different wheels, and the dynamic loads at low frequencies. A 3D right-handed

**Table 1** Values of parameters involved in this study

| Parameter | | Value |
|---|---|---|
| Coefficient of friction, $f$ | | 0.5 |
| Lumped sprung mass, $M_c$ (kg) | | 8000 |
| Wheel diameter, $\phi$ (m) | | 0.86 |
| Wheelset mass, $M_w$ (kg) | | 586 |
| Unsprung mass attached to wheelset, $M_a$ (kg) | | 340 |
| Stiffness of primary suspension, $K$ (kN/m) | | 880 |
| Damping of primary suspension, $C$ (kN·s/m) | | 4 |
| Stiffness of fastenings, $K_f$ (MN/m) | | 22 |
| Damping of fastenings, $C_f$ (kN·s/m) | | 200 |
| Wheel and rail material | Young's modulus, $E$ (GPa) | 205.9 |
| | Poisson's ratio, $\upsilon$ | 0.3 |
| | Density, $\rho$ (kg/m$^3$) | 7790 |
| | Damping constant, $\beta$ (s) | $1.0 \times 10^{-4}$ |
| Material of prefabricated slabs | Young's modulus, $E_s$ (GPa) | 34.5 |
| | Poisson's ratio, $\upsilon_s$ | 0.25 |
| | Density, $\rho_s$ (kg/m$^3$) | 2400 |
| Material filled in slab gaps | Young's modulus, $E_g$ (GPa) | 29.5 |
| | Poisson's ratio, $\upsilon_g$ | 0.25 |
| | Density, $\rho_g$ (kg/m$^3$) | 2400 |
| Mortar material | Young's modulus, $E_m$ (GPa) | 8 |
| | Poisson's ratio, $\upsilon_m$ | 0.2 |
| | Density, $\rho_m$ (kg/m$^3$) | 1600 |

Reprinted from Zhao et al. (2014) by permission of Taylor & Francis Ltd.

Cartesian coordinate system ($Oxyz$) is defined, of which the origin $O$ is located in the initial position of the contact patch center (position $A$ in Fig. 2a, hereinafter referred to as the initial position of the wheel). The $x$-axis is defined along the rolling direction (i.e., the longitudinal direction), and the $z$-axis is in the vertical direction. Note that the difference between the vertical direction and the normal direction of the contact is negligible in this study since no lateral movement of the wheelset is considered. Such an FE model, developed specially for investigations into high-frequency dynamics, is not suitable for low-frequency vibrations such as vehicle hunting.

### 2.1.2 A Typical Process of Numerical Simulation and the Explicit Time Integration

The simulation is composed of two steps. Step 1: The static equilibrium state of the system under gravity is first solved by an implicit solver in the initial position of the wheel (position $A$ in Fig. 2a). Step 2: An explicit solver is employed to simulate

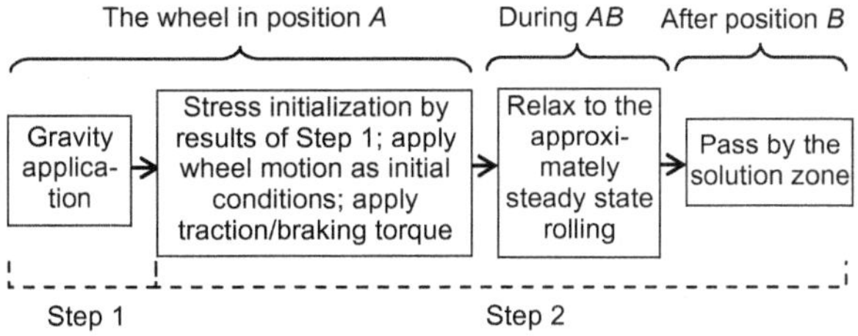


**Fig. 3** Schematic diagram of the simulation process

the transient rolling contact in the presence of friction, for which the displacement field obtained in Step 1 is used for stress initialization (at $t = 0$), the predefined rotational and forward speeds of the wheelset are also applied at $t = 0$ as initial conditions, and a specified acceleration or deceleration is further modeled by applying the time-dependent traction or braking torque to the wheel axle ($M$ in Fig. 2b). The distance before the solution zone (i.e., $AB$ in Fig. 2a) is designed to ensure that the wheelset achieves the steady state rolling approximately before entering the solution zone. When the wheelset passes by the solution zone, the transient results on forces, stresses, and strains are obtained. Such a process is sketched in Fig. 3. The determination of the distance $AB$ is presented in Sect. 3.1. Note that the speed increase/decrease in the simulation caused by the torque $M$ is negligible in comparison with the simulated speed because of the short time period simulated (typically 0.04 s) and, therefore, is ignored in later explanations.

A central difference method-based explicit scheme is employed to treat the time integration in Step 2. A very small time step (e.g., $8.9 \times 10^{-8}$ s for the model in this study) is required to meet the Courant stability condition of the explicit solver or to ensure the convergence of the method. Because of such a tiny time step, the high-frequency dynamic behavior of the vehicle–track system and the transient rolling contact phenomenon can be captured effectively. Due to the high computational costs, only one wheel passage is simulated at present.

### 2.1.3 Traction and Creepage

The coefficient of friction (COF, $f$), defined by Eq. (1), is reported to be typically 0.4–0.65 for dry-clean wheel–rail contact and less than 0.3 in the presence of a thin film of water or oil (Cann 2006). In this study, the COF is taken as 0.5 (Table 1) to simulate the dry-clean condition.

$$f = F_{\mathrm{T}}/F_{\mathrm{N}}, \tag{1}$$

where $F_{\mathrm{T}}$ and $F_{\mathrm{N}}$ are the tangential and the normal (vertical) contact forces, respectively, transmitted through a full-sliding contact. Obviously, the maximum tangential load that can be transmitted through a wheel–rail contact is $fF_{\mathrm{N}}$, i.e., in full-sliding contact, while a smaller load is transmitted in the case of rolling-sliding contact. Hereinafter, the tangential load actually transmitted is measured by a traction coefficient ($\mu$) defined by

$$\mu = F_{\mathrm{L}}/F_{\mathrm{N}} \leq f, \tag{2}$$

where $F_{\mathrm{L}}$ is the longitudinal component of the tangential wheel–rail contact force, being the traction force for a driving wheelset in acceleration or the braking force for a wheelset in braking. Different traction/braking loads or different friction exploitation levels are simulated by specifying the corresponding driving torque $M$ in the model. The torque $M$ is assumed first to increase linearly from zero to its maximum and then remain constant (Fig. 4). Note that $F_{\mathrm{L}}$ is equivalent to $F_{\mathrm{T}}$ in this study because the tangential plane coincides approximately with the horizontal plane and no lateral friction force is considered.

Referring to Fig. 5, the longitudinal creepage $\xi$ corresponding to a specified traction load is calculated using

$$\xi = \frac{\omega R - v}{(\omega R + v)/2}, \tag{3}$$

where $\omega$, $R$, and $v$ are the angular speed around the axis, the radius of the wheel, and the translating speed, respectively, and ωR is the linear speed of a node in the wheel contact surface. Note that, traditionally, "creepage" is a concept defined by

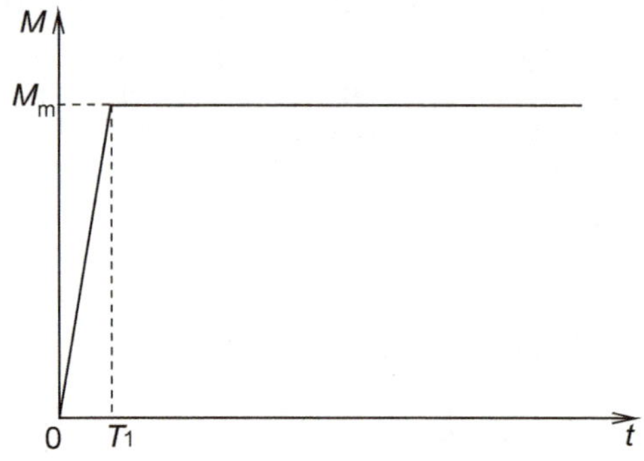


**Fig. 4** Variation of the driving torque $M$ with time in the simulation

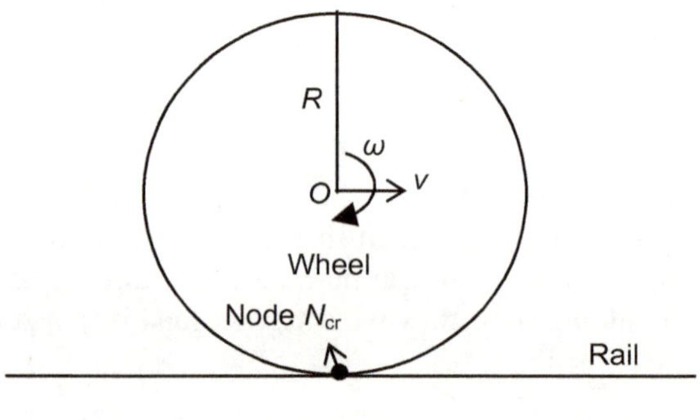


**Fig. 5** Wheel rolls on a rail when only the longitudinal creepage exists. Reprinted from Zhao et al. (2014) by permission of Taylor & Francis Ltd.

rigid motion, whereas elastic deformation is considered in this study. In Eq. (3), the linear speed of a node in the contact surface is taken as the linear speed of the wheel, through which the continuum vibrations excited by the contact load are considered in the calculated creepage.

### 2.1.4 Material Model

A linear elastic material model is used for the wheel and the rail for the following reasons: (1) For steels, known as Hookean solids, the assumption of linear elastic behavior is valid in many cases (Meyers and Chawla 1999); (2) shakedown is expected for most cases because of the generation of residual stresses and work hardening, which ensures the relatively long service lives of wheels and rails; (3) even at locations where plastic deformation occurs, its magnitude must be very small in each wheel passage, leading to a continuous increase of the rail surface hardness during the first two years after installation (Olofsson and Telliskivi 2003). More accurate material models, once ready and if necessary, can easily be introduced. The slab and mortar materials are also assumed to be linear elastic.

Material damping is considered by Rayleigh damping ($C_{\mathrm{m}}$):

$$C_{\mathrm{m}} = \alpha m + \beta K, \tag{4}$$

where $\alpha$ and $\beta$ are the mass ($m$) and stiffness ($K$) proportional damping constants, respectively. Trial simulations confirmed that the mass proportional damping is more effective for low-frequency vibrations and also damps out rigid body motions, as stated in the keyword user's manual of LS-DYNA. Therefore, the mass proportional damping is not suitable for this work and only the stiffness proportional damping is employed. The value of $\beta$ is chosen based on the estimate made by Kazymyrovych et al. (2010), in which $\beta$ for an FE model was determined by matching the measured stress level during testing with the level calculated using the FE method.

## 2.2 *Frictional Work and Wear Prediction*

The frictional work at a point of the rail contact surface ($W_{\mathrm{f}}$) is calculated as

$$W_{\mathrm{f}} = \int_0^t \tau s \mathrm{d}t = \sum_{i=1}^{n_{\mathrm{t}}} \tau_i s_i \Delta t_{\mathrm{W}}, \tag{5}$$

where $\tau$ and $s$ are the surface shear stress and micro-slip (i.e., the relative speed between contacting particles) at the point, respectively, being functions of time. $\tau_i$ and $s_i$ correspond to the instant $i\Delta t_{\mathrm{W}}$, the time step $\Delta t_{\mathrm{W}}$ is taken as $1 \times 10^{-5}$ s (the wheel translates 0.83 mm within $\Delta t_{\mathrm{W}}$ at 300 km/h), and $n_t$ is the number of calculated time steps. Material wear, if assumed to be proportional to the frictional work (Clark et al. 1988), can directly be scaled from the frictional work. More complicated/realistic wear behavior is beyond the scope of this work.

## 2.3 Corrugation Model

Figure 6 shows an example of the 3D corrugation models. The distribution of the corrugation depth (*d*) is assumed to be sinusoidal in the vertical-longitudinal section and parabolic along the lateral direction, as determined by

$$d_{\mathrm{C}} = -0.5 d_{\mathrm{m}} (1 - \sin[2\pi(x - x_{\mathrm{s}})/L + \pi/2]), \tag{6}$$

$$d = d_{\mathrm{C}}[1 - (y/W)^2], \tag{7}$$

where $d_{\mathrm{m}}$, *L*, and *W* are the maximum depth, wavelength, and width of the corrugation, being 0.14, 80, and 30 mm for corrugation *D* (default), respectively; $d_{\mathrm{C}}$ is the maximum depth in the lateral direction and is located in the middle of the corrugation; $x_{\mathrm{s}}$ is the longitudinal coordinate at the starting point of the corrugation and is equal to 2.4 m. To maximize the influence of the corrugation for better analysis, $d_{\mathrm{C}}$ is applied to the position where the maximum contact pressure occurs in a case with smooth rail surfaces, i.e., in the vertical-longitudinal section of

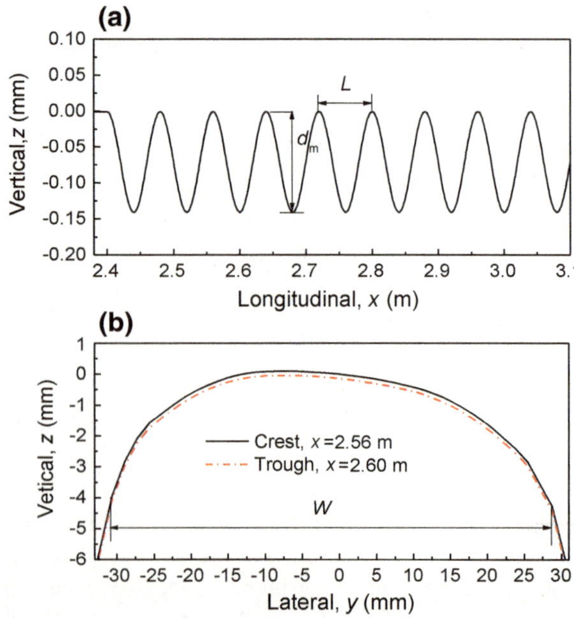


**Fig. 6** Geometry of corrugation *D* (default). **a** In the vertical-longitudinal section of *y* = −2.8 mm (the deepest); **b** Along the lateral direction

$y = -2.8$ mm (Fig. 17b). Note that the inclination of the rail is included in Fig. 6 and different scales are applied in Fig. 6a, b for clarity. The smooth case serves as a trial simulation for corrugation applications.

# 3 Results of Smooth Contact Surface

## 3.1 Dynamic Relaxation

It is mentioned above that the static solution obtained by the implicit solver is employed to initialize the transient simulation. Figure 7 shows that the vertical displacement field of the rail surface is symmetric in the vicinity of the contact patch in the static case, while it becomes asymmetric in the transient analysis due to the moving load. Hence, when the wheel rolls forward in a transient analysis, the displacement field gradually changes from the static to the transient case. Such a process inevitably introduces an oscillation (referred to later as the initial oscillation), as observed from the vertical (contact) forces plotted in Fig. 8. Note that Fig. 7 shows the results only along a longitudinal line where the maximum pressure is, i.e., the longitudinal axis of the contact patch when the contact patch is an ellipse. More studies have shown that the influence of the rolling speed on the pressure distribution is negligible in the case with a smooth rail contact surface. This is because the resultant speed at the contact point (assuming a rigid contact, the contact patch reduces to a point) is zero, being independent of the rolling speed, i.e., the instantaneous center of the wheel. Hereinafter, the longitudinal line is referred to as the longitudinal axis, although the contact patch is not necessarily an ellipse, e.g., on corrugated rails. Results shown later are all taken from the rail side.

From Fig. 8, it is seen that *AB* (Fig. 2a), designed for a dynamic relaxation, should be at least 2.4 m long for the system to relax to an acceptable oscillation level (10% of static load) in position *B* at 500 km/h. This is how *AB* was determined. For a speed under 300 km/h, the vertical force becomes very stable after a rolled distance of 2.4 m. Note that the abscissa of Fig. 8 is set to be the rolled distance for comparison.

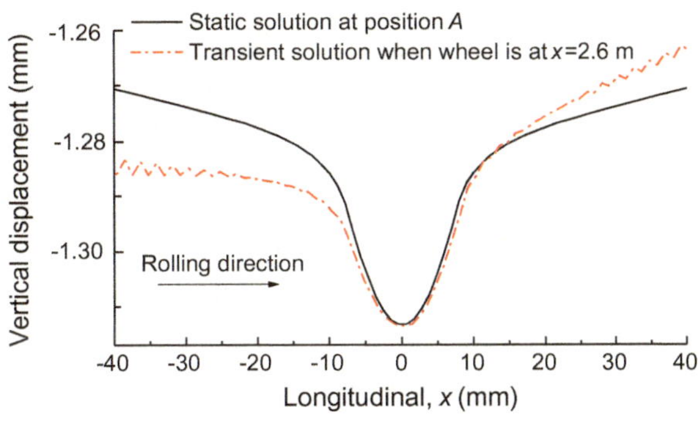


**Fig. 7** Vertical displacement fields in the static and transient cases ($v = 300$ km/h) (Only the results in the rail surface along the longitudinal axis are given and a longitudinal shift is applied to the transient solution for comparison)

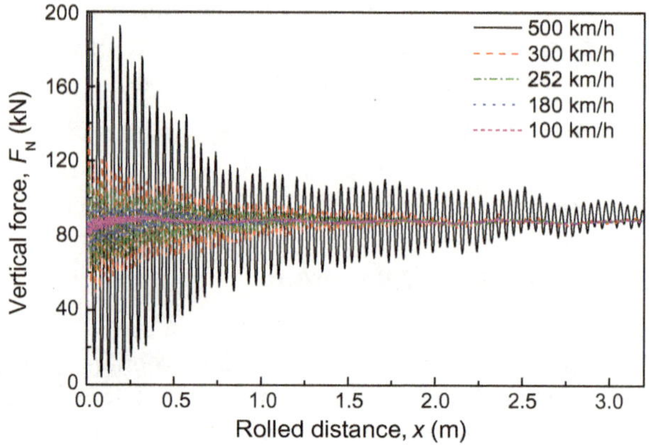


**Fig. 8** Vertical force variations at different rolling speeds

Note that the displacement difference in Fig. 7 may not be the only reason behind the initial oscillation. Other factors, probably related to the initial conditions, may also play certain roles. Nevertheless, this is not discussed further because this study focuses on the contact solution after achieving an approximate rolling state.

## 3.2 Longitudinal and Vertical Forces

The initial increase rate of the driving torque is varied by setting $T_1$ (Fig. 4) at different values, from zero to sufficiently large. Figure 9 shows the contact force results of a smooth case when $T_1$ is varied, for which a rolling speed of 300 km/h and a traction coefficient of 0.3 are assumed. For clarity, only representative results corresponding to three values of $T_1$, namely 0.0001, 0.005, and 0.01 s, are plotted. As expected, the vertical force does not change considerably with $T_1$ (a difference of much less than 1%). Hence, only one vertical force result scaled by the COF (i.e., $fF_N$) is plotted in the figures to illustrate the limit of the longitudinal force.

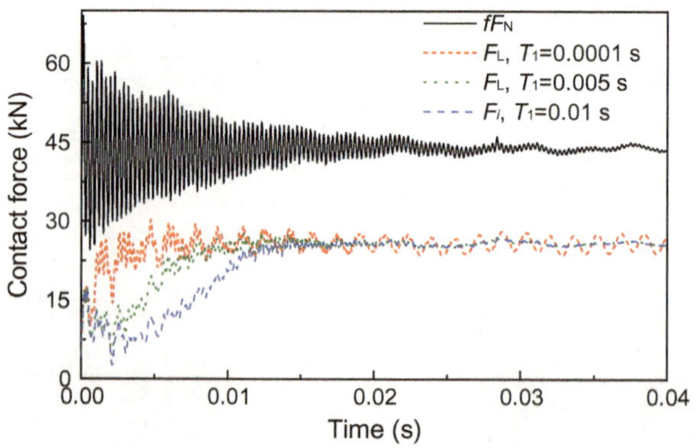


**Fig. 9** Vertical and longitudinal forces as $T_1$ varies ($\mu = 0.3$, $v = 300$ km/h)

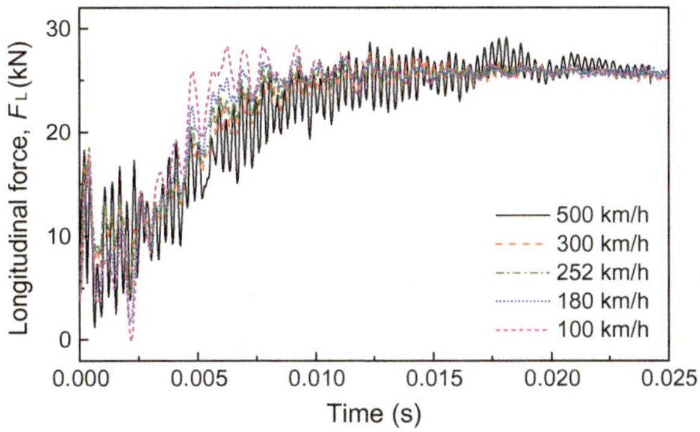


**Fig. 10** Longitudinal forces at different rolling speeds ($\mu$ = 0.3)

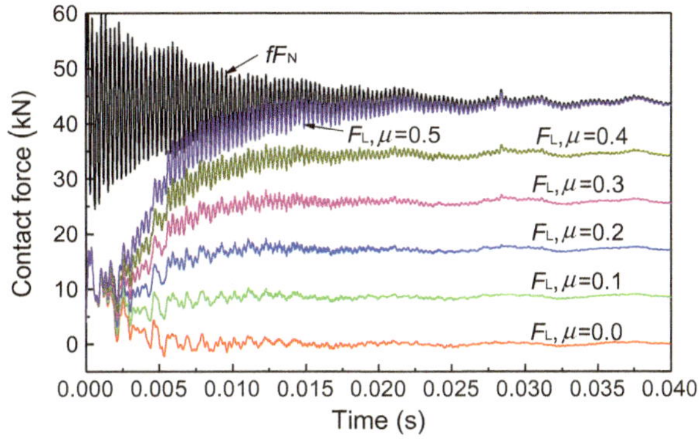


**Fig. 11** Vertical and longitudinal forces at different traction coefficients ($v$ = 300 km/h)

The initial oscillation also influences the longitudinal (contact) forces considerably (Fig. 9). After the dynamic relaxation ($t > 0.028$ s), the longitudinal force becomes very stable, like the vertical force, when $T_1$ is 0.005 or 0.01 s, but not in the case when $T_1$ = 0.0001 s. Moreover, the longitudinal force does not reach the specified value at $T_1$ but has a delay of about 0.005 s. This delay is the response time of the material in contact to the torque applied to the axle. Hereinafter, cases with $T_1$ = 0.005 s are used for analyses. Variations of the longitudinal force in the case with $T_1$ = 0.0001 s are not further studied here.

Contact force results at different rolling speeds and with different traction coefficients are shown in Figs. 10 and 11, respectively. A stable rolling contact is achieved after the dynamic relaxation in all cases.

## 3.3 Contact Stresses and Frictional Work

For the case of $v$ = 300 km/h and $\mu$ = 0.3, distributions of the maximum contact stresses along the longitudinal axis are shown in Fig. 12 together with the

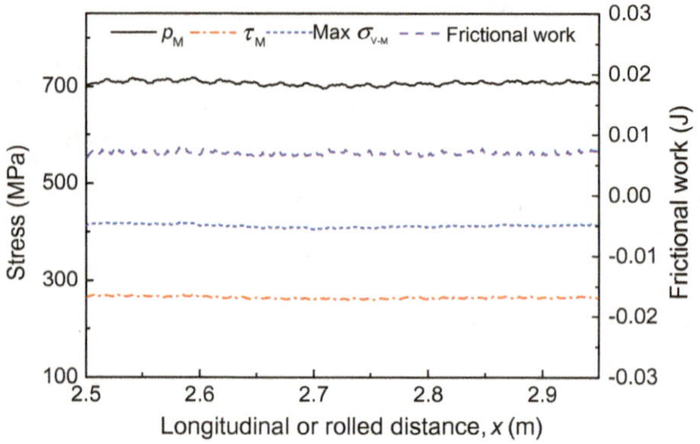


**Fig. 12** Distributions of maximum pressure ($P_m$), maximum surface shear stress ($\tau_m$), maximum V-M stress (max $\sigma_{V\text{-}M}$), and frictional work along the longitudinal axis ($\mu = 0.3$, $v = 300$ km/h, $x = -0.00278$ m)

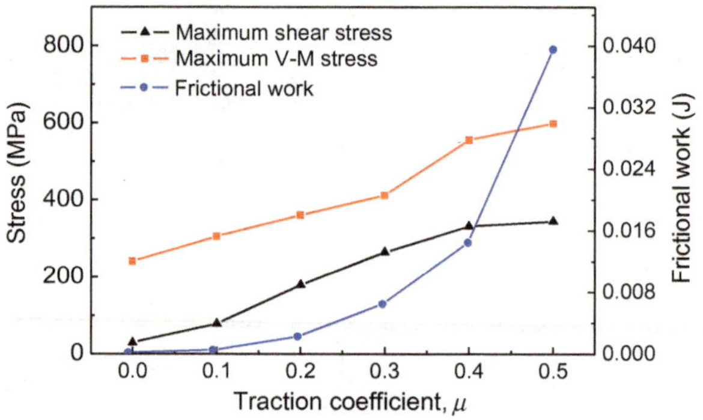


**Fig. 13** Stresses and frictional work versus the traction coefficient at $v = 300$ km/h

corresponding frictional work. A maximum stress at a location means the maximum value reached at that location during the simulated wheel passage. The frictional work given in the figure indicates the calculated result in a longitudinal strip with a width of 1.1 mm (the width of the elements there). These results confirm that approximately steady rolling is achieved in the simulation.

As the traction coefficient becomes larger, the magnitudes of the stresses and frictional work shown in Fig. 13 all increase, as expected. Note that the Von Mises (V-M) stresses presented in this study are of the surface layer of elements (at their centers) because constant stress elements are employed. Pressure is not plotted in Fig. 13 because it is independent of the traction level.

# 4 Corrugation

## *4.1 Measurements on a High-Speed Line*

Corrugation occurred on a Chinese high-speed line was measured using a corrugation analysis trolley (CAT). Figure 14a shows roughness level spectra in 1/3 octave bands measured at two sites where the running speeds were about 300 km/h. Two typical wavelengths, namely around 65 and 125 mm, are observed, and the first one dominates. Note that the wavelength of 80 mm shown in Fig. 1 (a picture taken on site 2 illustrated in Fig. 14) is not obvious in the roughness level spectra. This may be because those spectra were obtained from measurements of rails of several hundred meters (about 300 and 1000 m at sites 1 and 2, respectively), in which the wavelength of the corrugation varied around 65 mm and the wavelength of 80 mm existed at only some locations. The physical essence behind such a wavelength variation is likely to be the inevitable variations of the vehicle–track parameters along the measured site. In addition, the discrete frequencies taken in the

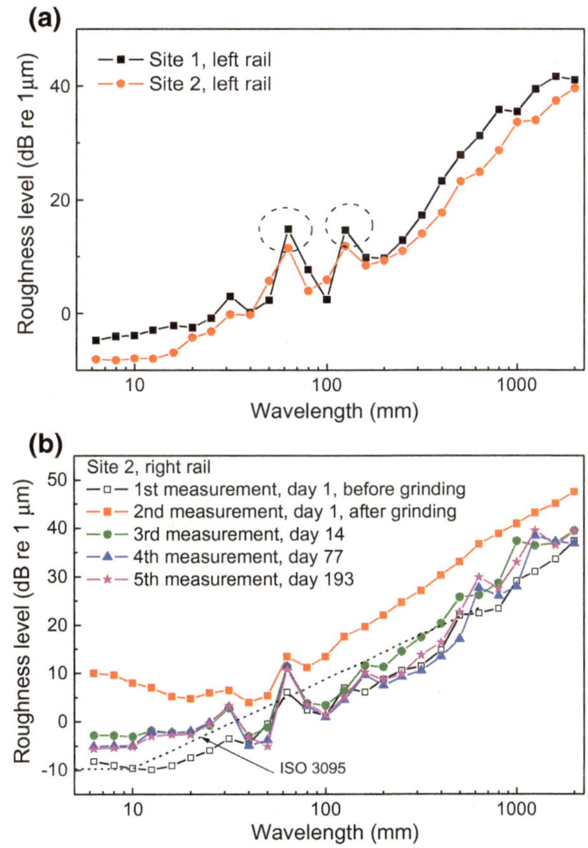


**Fig. 14** Roughness level spectra in 1/3 octave bands measured on a Chinese high-speed line. **a** Typical measurements at two corrugation sites; **b** Corrugation development with time. Two measurements were conducted before and after a rail grinding on day 1

1/3 octave bands may also cause some errors in the wavelength estimation. Figure 14b shows five CAT measurements at site 2 conducted during a grinding cycle. It is seen that the corrugation re-occurred after grinding and gradually became stabilized in a short period.

In this study, corrugation with a wavelength range of 65–95 mm is modeled to study its influence on transient wheel–rail interaction at high speeds. All corrugation models are applied in the same way as corrugation *D* (with the default wavelength of 80 mm, Fig. 6), i.e., with the same location and phase in the longitudinal direction, and the same depth distribution in the lateral direction. Other corrugation models are later referred to by their wavelengths and maximum depths, without presenting their geometry in figures.

## 4.2 Transient Wheel–Rail Interaction

The transient wheel–rail interaction on a corrugated rail is analyzed in this section, for which corrugation *D* is considered with $v$ = 300 km/h and $\mu = 0.3$. Figure 15 shows variations of the contact forces together with the geometry of corrugation *D*, in which the vertical force is scaled by the COF ($f$) for comparison. As expected, both the vertical and longitudinal forces significantly fluctuate at the corrugated site. The maximum vertical force ($F_N$) is 193.19 kN, and the minimum is as low as 3.02 kN, i.e., the wheel and rail almost separate from each other. At several corrugation troughs, $fF_N$ coincides with $F_L$, i.e., full sliding occurs. The phenomenon that the force peaks do not occur exactly at the corresponding corrugation crests (referred to as the longitudinal shift hereinafter) will be discussed later.

Furthermore, the vertical force is larger when the wheel is above the fastenings than in between, demonstrating the considerable influence of the discrete supports of the rail. Such a result is in line with authors' observations that some corrugations are deeper in positions above fastenings than in between, which has also been reported in (Clayton and Allery 1982; Jin et al. 2008). Note that the influence of the

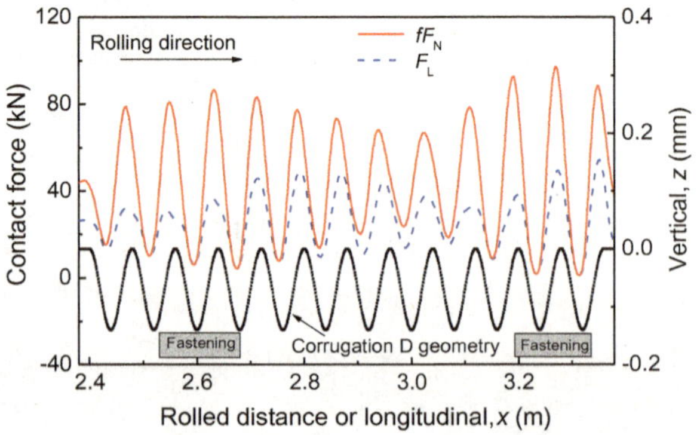


**Fig. 15** Dynamic forces excited by corrugation *D* ($v$ = 300 km/h)

discrete supports on the longitudinal force is different from that on the vertical one. The relationships between the normal and the tangential wheel–rail interactions will be discussed later.

Pressure and surface shear stress distributions at 10 typical instants (i.e., *t*1–*t*10) are plotted in Fig. 16 to show the transient effects. As expected, the contact stresses vary greatly on the corrugated rail. Within the selected instants, the rolling contact changes from the rolling-sliding state to full sliding (at *t*4), and then back to rolling sliding. It should be specified that the pressure reaches local maximums (within a wavelength) at *t*1 and *t*7, and local minimums at *t*4 and *t*10. At *t*1, *t*7, and *t*4, the surface shear stress also reaches local maximums or minimums, but another local minimum occurs at *t*10′, being slightly different from *t*10. In a word, the typical instants are not completely the same in Fig. 16a, b for better illustration.

It is further shown from Fig. 16a that local maximums of pressure occur before the contact patch center reaches a corrugation crest. This is because the pressure is determined by both the contact geometry and the dynamic vertical force (as mentioned above, a longitudinal shift exists). Figure 17a shows the 3D distribution of the maximum pressure in the same section as shown in Fig. 16. Comparing Fig. 17a with the corresponding results on the smooth rail shown in Fig. 17b, great influences of corrugation can be observed clearly.

Figure 18 presents the creepage variation caused by corrugation *D*. From the creepage results calculated at node $N_{cr}$ (located in the contact patch at $t = 0$, Fig. 5), it is seen that the maximum creepage reached on the corrugated rail is close to double the stabilized value on the smooth contact surface (before the wheel enters the corrugation, about 0.276%). Fluctuation of the creepage becomes obviously fiercer at $x = 2.69$ m because the node $N_{cr}$ comes into contact again after a rolled distance of 2.69 m (a cycle), i.e., continuum (local) vibrations are excited by the contact load. To filter out the influence of the local vibrations, the calculated creepages at 28 selected nodes (distributed over the whole circumference of the wheel) are averaged and also plotted in Fig. 18. Little further explanation of the creepage results is given hereinafter because: (1) the transient contact stresses and the resulting irregular frictional work and/or irregular V-M stress, not the irregular

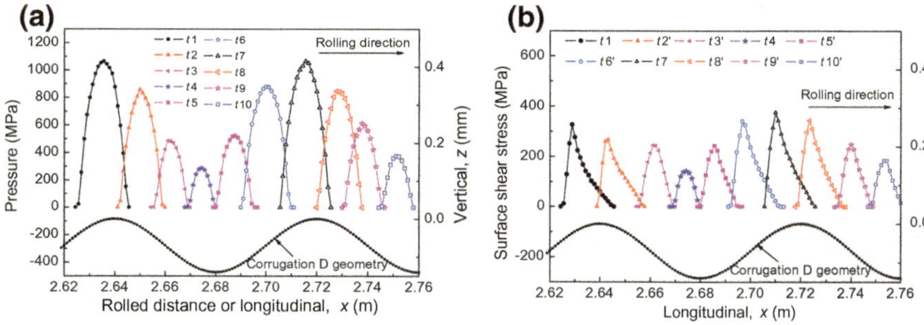


**Fig. 16** Transient contact stress distributions along the longitudinal axis. **a** Pressure; **b** Surface shear stress. The symbols represent nodes

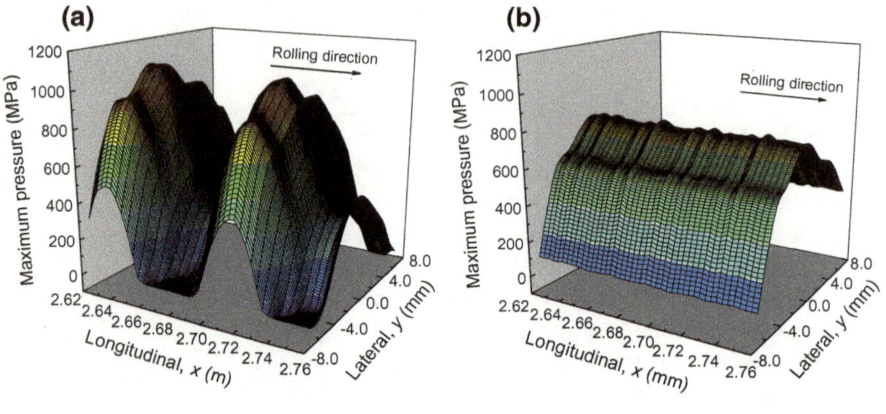


**Fig. 17** Three-dimensional distributions of the maximum pressure on corrugated rail (corrugation *D*) (**a**) and smooth rail (**b**)

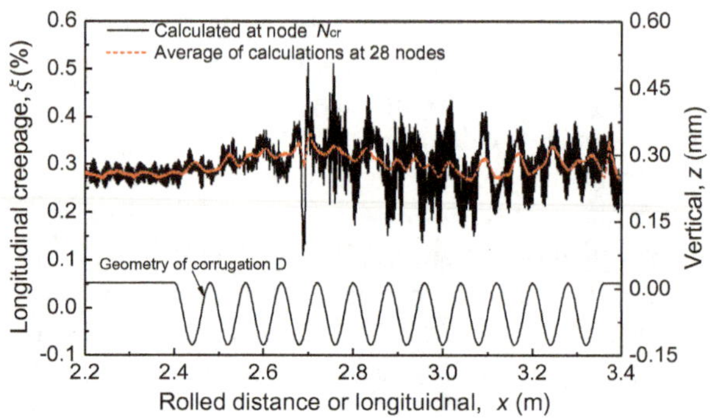


**Fig. 18** The creepage variation caused by corrugation *D*

creepage, are the immediate causes of corrugation development; (2) according to rolling contact theories, higher creepage usually means larger tangential contact force, higher frictional work, and larger V-M stress, which is valid for cases simulated in this study.

Figure 19 shows distributions of the maximum pressure, maximum surface shear stress, maximum V-M stress, and frictional work along the longitudinal axis in the corrugated section. It is observed that the stresses all vary following the pattern of corrugation, but with slightly different longitudinal phases, and the relative positions of the stress peaks with respect to the corresponding crests change slightly at different waves of corrugation. Moreover, the patterns of the maximum V-M stress and the frictional work are closer to those of the maximum surface shear stress than to those of the maximum pressure, as expected.

Fig. 19 Distributions of the maximum stresses and frictional work along the longitudinal axis

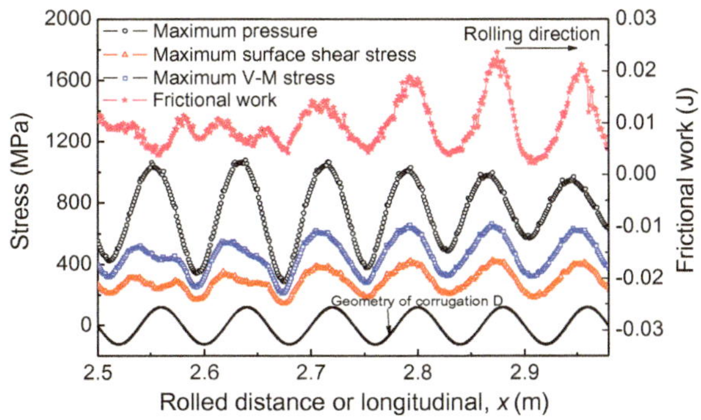


## 4.3 Different Wavelengths and Depths

Keeping $v$ = 300 km/h and $\mu$ = 0.3, the wavelength ($L$) and depth ($d_m$) of corrugation are varied separately to study their influences. Considering the wavelength range reported in Sect. 4.1, three wavelengths, namely 65, 80, and 95 mm, are considered. Figure 20 shows that among the simulated cases, the maximum vertical and longitudinal forces both occur in the case of $L$ = 80 mm. For $L$ = 65 mm, the longitudinal force does not follow the pattern of the simulated corrugation any more, but shows a shorter characteristic wavelength (Fig. 20b), and its magnitude is much smaller than in the other two cases. This may be explained as follows: The excitation frequency of corrugation ($f_{ex} = v/L$, i.e., the passing frequency) monotonically decreases with the corrugation wavelength; for a wavelength of 80 mm, $f_{ex}$ is closest (among the three cases) to an eigenfrequency of the system related to corrugations, leading to the strongest response.

The mean value of the maximum V-M stress in the corrugated section is approximately constant for different wavelengths, while its fluctuation range is the largest at $L$ = 80 mm (Fig. 21). This is in line with the longitudinal force results in Fig. 20b. In contrast, the largest fluctuation of the frictional work occurs at $L$ = 95 mm (being 0.016, 0.023, and 0.024 J at $L$ = 65, 80, and 95 mm,

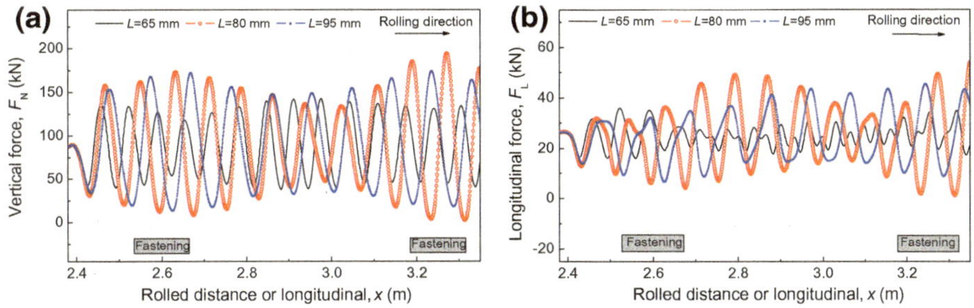


**Fig. 20** Dynamic forces caused by corrugations of different wavelengths. **a** Vertical force; **b** longitudinal force. The depth remains constant, $d_m$=0.14 mm

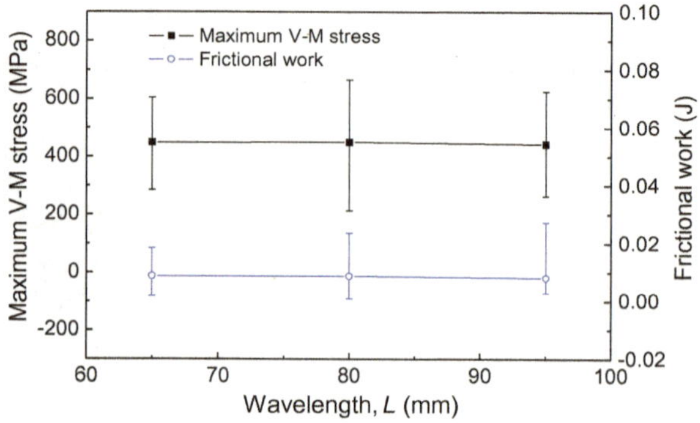


**Fig. 21** Mean values (*symbols*) and fluctuation ranges (*error bars*) of the maximum V-M stress and the frictional work over the section in Fig. 19, as the wavelength varies

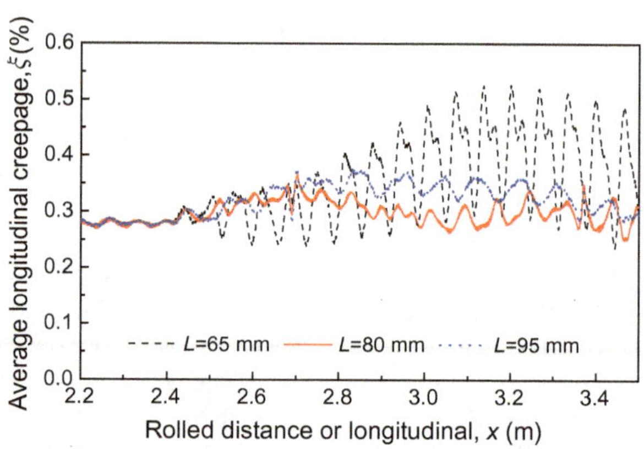


**Fig. 22** Creepage variations caused by corrugations of different wavelengths. Average of calculations at 28 nodes, $d_m = 0.14$ mm

respectively), because the frictional work is determined not only by the contact stresses (or the contact forces) but also by the micro-slip. Significant influences of the wavelength on the micro-slip in the corrugated section can be seen from the creepage (integration of the micro-slip over the contact patch) variations given in Fig. 22. No more detailed distributions of the maximum V-M stress and the frictional work in corrugated sections are given hereinafter, considering that their patterns are similar to those of the longitudinal force, as mentioned above.

When the corrugation depth increases, fluctuation ranges of the dynamic forces, the maximum V-M stress, and the frictional work all increase as expected (Fig. 23). Detailed variations of the dynamic forces are not illustrated here because their patterns remain constant for different depths.

## 4.4 Different Traction Coefficients

The traction coefficient is varied from 0.0 to 0.5 to examine its influence on the tangential wheel–rail interaction, for which corrugation *D* is considered and the

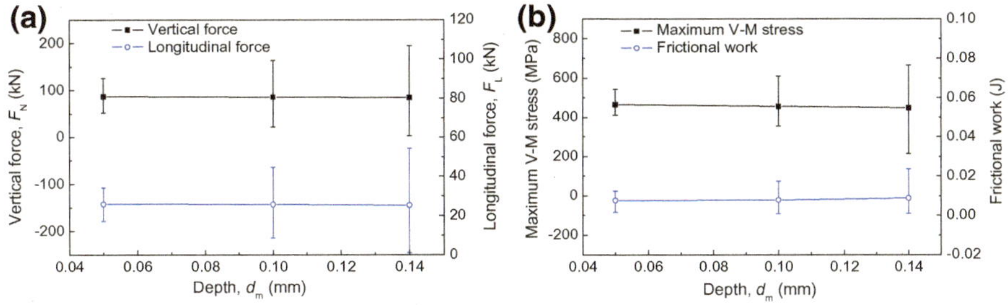


**Fig. 23** Influences of the corrugation depth on mean values (*symbols*) and fluctuation ranges (*error bars*). **a** Dynamic forces, over the section in Fig. 15; **b** maximum V-M stress and frictional work, over the section in Fig. 19

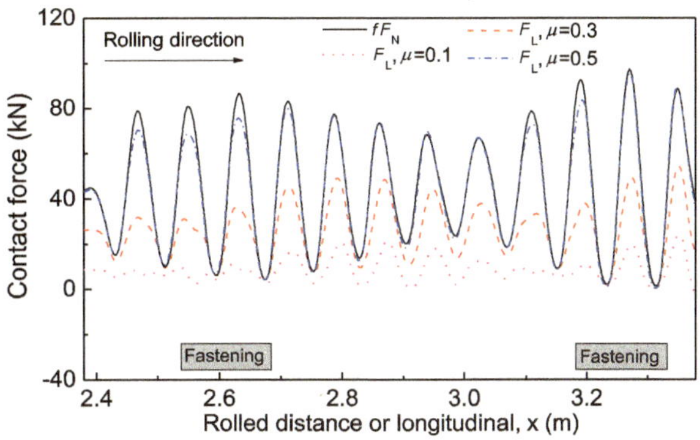


**Fig. 24** Dynamic forces excited by corrugation *D* at different traction levels. The vertical force does not vary with the traction coefficient

rolling speed is kept at 300 km/h. From the dynamic forces shown in Fig. 24, it is found that the longitudinal force gradually becomes in phase with the vertical force scaled by the COF as the traction coefficient increases. This is determined by Coulomb's law of friction employed in the model, i.e., the surface shear stress distribution approaches the scaled pressure distribution with the increase of traction coefficient. When the traction coefficient is very low, variation of the pressure distribution (i.e., variation of the upper limit of the surface shear stress distribution) has little influence on the surface shear stress distribution since the slip area in the contact patch (where the upper limit is reached) is very small. Obviously, such an influence becomes more significant as the slip area enlarges, i.e., as the traction coefficient increases. This is why the fluctuation range of the longitudinal force increases with the traction coefficient (Fig. 24). Figure 25 presents the mean values and fluctuation ranges of the maximum V-M stress and the frictional work over the section in Fig. 19. It is seen that the mean values and the fluctuation ranges all increase with the traction coefficient, showing the same trend as the longitudinal force.

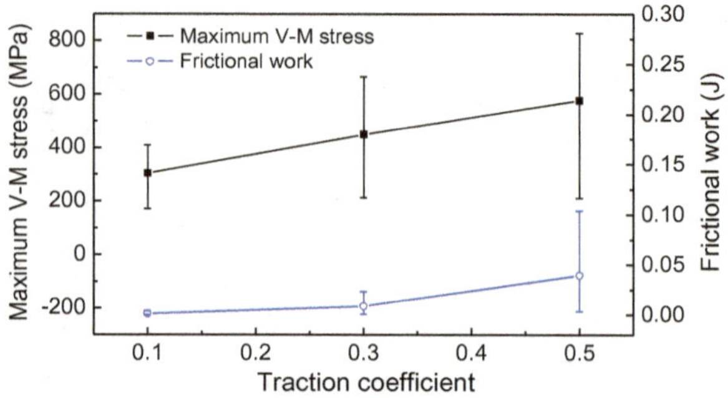


**Fig. 25** Mean values (*symbols*) and fluctuation ranges (*error bars*) of the maximum V-M stress and the frictional work at different traction levels, over the section of corrugation *D* in Fig. 19

## 4.5 Different Rolling Speeds

The dynamic forces caused by corrugation *D* at different rolling speeds are illustrated in Fig. 26, in which the traction coefficient remains 0.3. The maximum vertical and longitudinal forces first increase with the rolling speed when the speed is less than 300 km/h, and then decrease (i.e., the maximums at 500 km/h are lower than those at 300 km/h). This is a consequence of the following phenomena: (1) The excitation frequency of the corrugation changes with the rolling speed, leading to the strongest response at a certain speed (the same reason as that mentioned in Sect. 4.3); (2) among the simulated speeds, the influence of the discrete supports of the rail is the greatest at 300 km/h (Fig. 26).

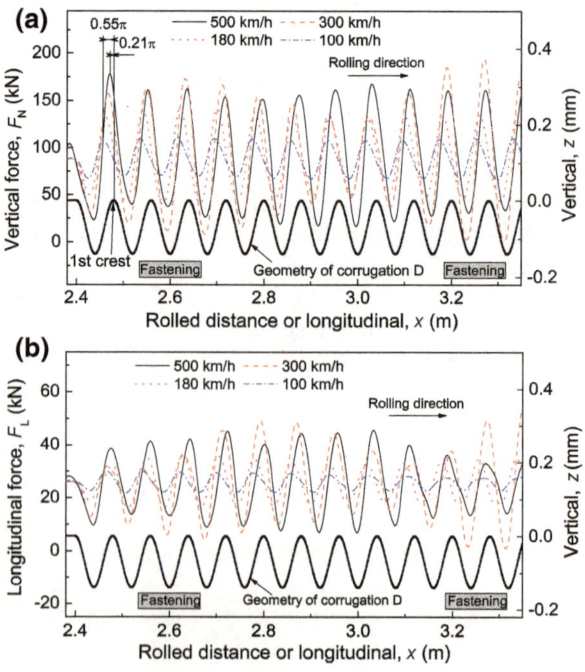


**Fig. 26** Dynamic forces caused by corrugation *D* at different rolling speeds. **a** Vertical force; **b** longitudinal force

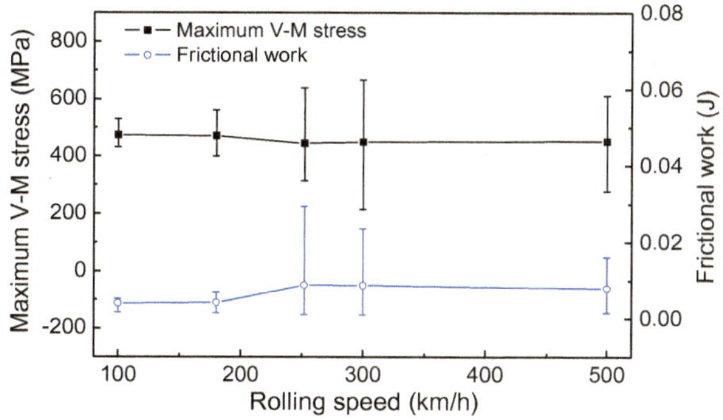


**Fig. 27** Mean values (*symbols*) and fluctuation ranges (*error bars*) of the maximum V-M stress and the frictional work at different rolling speeds, over the section of corrugation *D* in Fig. 19

Furthermore, the longitudinal shift of the vertical force decreases with the rolling speed. For example, as indicated in Fig. 26a, it varies from $0.55\pi$ at 100 km/h to $0.21\pi$ at 500 km/h at the first corrugation crest (a shift of $2\pi$ corresponds to a corrugation wavelength). Note that the relative positions of the force peaks with respect to the corresponding crests are not constant at different waves due to the transient effects.

Influences of the speed on the longitudinal shift of the longitudinal force are more complicated than those of the vertical force (comparing Fig. 26a and b). This is because the tangential interaction has its own eigenmodes and in the meantime is limited by the normal one. As mentioned in Sect. 4.4, the tangential force will follow exactly the vertical force scaled by the COF in a full-sliding case due to the application of Coulomb's law of friction.

From Fig. 27, it is further seen that the fluctuation range of the maximum V-M stress caused by corrugation *D* is the largest at 300 km/h, while at 250 km/h, the fluctuation range of the frictional work reaches its maximum. From the results in Figs. 26 and 27, it could be concluded that, for the simulated speed range, the responses of the system to corrugation *D* reach the maximum at a speed between 250 and 300 km/h. Moreover, Fig. 27 shows that the mean values vary considerably when the rolling speed changes from 180 to 250 km/h.

## 4.6 Comparison with the Multi-body Approach

The vertical force obtained from the transient FE simulation is compared to that of the traditional multi-body approach (Jin et al. 2005, 2008; Xiao et al. 2014) in Fig. 28, for which a corrugation with a wavelength of 80 mm and a depth of $d_m = 0.18$ mm is considered at 500 km/h. It is seen that the transient FE result is significantly lower. Contact loss corresponding to the vertical force of zero is predicted by the multi-body approach, but not by the transient FE simulation. Such a difference can be explained as follows: In the multi-body approach, the whole wheel is lumped into one mass particle, the rail is represented by Euler beams, and a

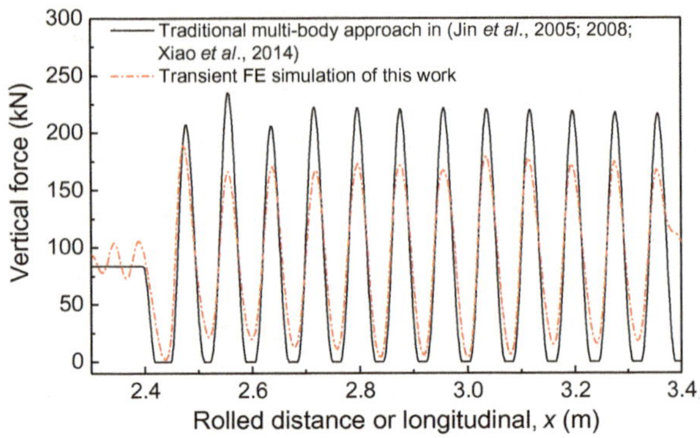


**Fig. 28** Comparison between the vertical forces obtained by the traditional multi-body approach and the transient FE model at 500 km/h ($L$ = 80 mm, $d_m$ = 0.18 mm)

Hertz spring is used to model the wheel–rail contact, which significantly exaggerates the contact stiffness and assumes the contact patch is infinitesimal. In other words, the results from the transient FE model should be more reasonable and the multi-body approach overestimates the dynamic wheel–rail interaction under the excitation of corrugations. Note that the longitudinal force is not compared here because it cannot be obtained by the traditional multi-body approach.

It should be specified that with a modified Hertz spring, the traditional multi-body approach may still provide accurate predictions of the contact forces caused by corrugations. Such an approach is appealing due to the low computational costs of multi-body simulations. To this end, the transient FE model developed in this study provides a potential calibration and validation tool for the suitable contact stiffness. Once calibrated and validated, more accurate contact force predictions may be realized without increasing the computational costs.

## 5 Discussion

Parameter variation analyses show that fluctuation ranges of the maximum V-M stress and the frictional work on a corrugated rail increase with the traction coefficient. This means that for the same excitation (e.g., a corrugation in this study), the irregular material response, namely the irregular plastic deformation and wear, probably becomes more severe as the friction exploitation level increases. Such results may explain why corrugations are more often observed on curves where the transmitted friction force is relatively high.

Moreover, for the simulated system, fluctuation ranges of the contact forces, the V-M stress, and the frictional work caused by a corrugation are found to reach their maximums at a speed between 250 and 300 km/h. As the wavelength varies, the maximum fluctuation ranges of the contact forces and the V-M stress all occur at the wavelength of 80 mm, whereas the frictional work fluctuation at the wavelength of 80 mm is slightly less than that at the wavelength of 95 mm, but significantly larger than that at the wavelength of 65 mm. Considering that components of the random rail roughness leading to higher dynamic responses than others may gradually develop, the above-mentioned results seem to explain why a corrugation with a wavelength of about 80 mm occurred in the rail section shown in Fig. 1 where the running speed was about 300 km/h.

The simulations also show that the longitudinal force variation excited by the corrugation with a wavelength of 65 mm becomes different from the corrugation in pattern and has a small magnitude, demonstrating the low possibility of occurrence of such a corrugation and of those with shorter wavelengths. This is in agreement with observations that a lower bound always exists for the corrugation on a track. Note that values of the system parameters, such as the wheel diameter and stiffness of the rail fastenings, are all nominal and kept constant in simulations, for which new wheels, new rails, and many designed values are considered. In reality, however, the wheels and rails are constantly worn and regularly re-profiled or ground, and track characteristics vary from section to section and with time. These factors should be born in mind when interpreting the numerical results. For example, the above-shown results suggest that the corrugation wavelength on the monitored Chinese high-speed line should be around 80 mm, while according to the CAT measurement, it is around 65 mm (Sect. 4.1). In other words, the nominal values of the parameters may represent the track section shown in Fig. 1, but not every section of the monitored track, which should be studied further in the future.

Results in Sect. 3 show that the calculated material response would be regular along the rail if no corrugation was applied. This means that the initiation mechanism of corrugation (growing up from smooth rails) is not included in the FE model. The irregular material response corresponding to the results in Sect. 4 is the consequence of the existing corrugation or caused by the geometric variation at the corrugation. So, what is the relationship between the initiation mechanism (the wavelength fixing mechanism) and the existing corrugation's consequence? The authors here propose a theory, explained in the following paragraph, to answer this question.

The irregular material response caused by the initiation mechanism, hereinafter referred to as the irregular response I, should be larger at corrugation valleys and smaller at crests, i.e., leading to the occurrence of a corrugation. This is the dominant mechanism in the relatively early stages of a corrugation. Once the corrugation comes into being, another irregular material response also starts to act due to the geometric variation at the corrugation, referred hereinafter to as the irregular response II. The irregular response II is larger at crests than at valleys (Fig. 19), being very different from the irregular response I, especially at high speeds (smaller longitudinal shifts at higher speeds, Fig. 26a). Obviously, the irregular response II will alleviate the irregular response I due to their different patterns. Further considering that the irregular response II becomes more irregular with the corrugation depth (Fig. 23), at a certain depth the combined material response (irregular response I + II) starts to become regular, or the initiation mechanism and the consequence of the existing corrugation become balanced, i.e., the corrugation stabilizes. The phenomenon of corrugation stabilization has been

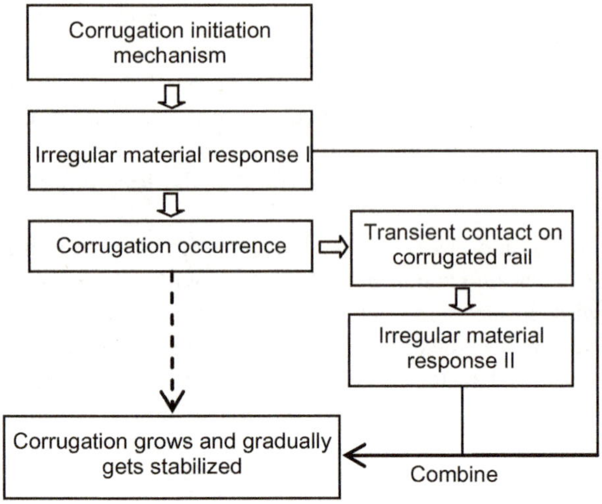


**Fig. 29** Schematic diagram of corrugation occurrence

observed in the field, e.g., the high-speed corrugations shown in Fig. 14b. Figure 29 shows a schematic diagram to help understand the mechanism explained above. Note that other factors such as material work hardening and residual stresses could also play certain roles in the stabilization of corrugation.

By changing the driving torque definition, braking cases can also be simulated by the 3D transient FE model, although only traction cases are presented in this paper due to space limitations. To authors' experience, for the same friction exploitation level, the results of the braking and the traction cases indeed show some differences in contact stresses. In the future work, the 3D transient FE model can be employed to further study the material damage mechanisms on corrugated rails. Finally, it might be worthwhile to note that all discussions are mostly based on the numerical results shown above, in which influences of the lateral movement of the wheelset and the initiation mechanism of corrugation are not included.

## 6 Summary and Conclusions

On the basis of field measurements and observations of (short-pitch) corrugations occurred on a recently opened Chinese high-speed line, a 3D transient rolling contact FE model is developed by an explicit FE approach to analyze the high-speed vehicle–track interaction in the presence of rail corrugations. The vehicle and the track subsystems are considered to ensure that the vehicle–track interaction is solved accurately in both the vertical and longitudinal directions. Detailed contact solutions on corrugated rails, including both the normal and the tangential solutions, are examined at different traction levels and at rolling speeds of up to 500 km/h. A summary of the results shown above and some conclusions are as follows.

1. The vertical and longitudinal (contact) forces, the pressure, and the surface shear stress all vary following the pattern of the corrugation, but with slightly different longitudinal phases. The patterns of the V-M stress and the frictional work are closer to those of the surface shear stress than to those of the pressure.
2. The discrete supports of the rail have considerable influences on the vehicle–track interaction on the corrugated rail at certain rolling speeds.
3. At certain friction exploitation levels, the state of rolling contact may oscillate between rolling sliding and full sliding at the passing frequency of corrugation, leading to a significantly higher creepage than on smooth rails.
4. Fluctuation ranges of the V-M stress and the frictional work caused by a corrugation increase with the traction coefficient. This may explain why corrugation is more often observed on curves where the transmitted friction force is relatively high.
5. According to simulations by the nominal parameters, the wavelength of the corrugation occurring on the monitored Chinese high-speed line is most probably around 80 mm and a speed between 250 and 300 km/h is most detrimental, corresponding well to the corrugation shown in Fig. 1. Inevitable variations of the parameters along the track and their changes from the nominal values probably explain why the dominant corrugation wavelength found by CAT measurements is about 65 mm on sections where the running speed is about 300 km/h.
6. The traditional multi-body approach overestimates the dynamic wheel–rail interaction on corrugated rails, whereas results from the transient FE model should be more reasonable.
7. A theory is proposed to explain the observed phenomenon that the corrugation gradually stabilizes.

## References


Afshari, A., & Shabana, A. A. (2010). Directions of the tangential creep forces in railroad vehicle dynamics. *Journal of Computational and Nonlinear Dynamics, 5*(2), 021006. doi:10.1115/1.4000796.

Baeza, L., Fayos, J., Roda, A., et al. (2008). High frequency railway vehicle-track dynamics through flexible rotating wheelsets. *Vehicle System Dynamics, 46*(7), 647–662. doi:10.1080/00423110701656148.

Cann, P. M. (2006). The "leaves on the line" problem—a study of leaf residue film formation and lubricity under laboratory test conditions. *Tribology Letters, 24*(2), 151–158. doi:10.1007/s11249-006-9152-2.

Chaar, N., & Berg, M. (2006). Simulation of vehicle-track interaction with flexible wheelsets, moving track models and field tests. *Vehicle System Dynamics, 44*(S1), 921–931. doi:10.1080/00423110600907667.

Clark, R. A., Scott, G. A., & Poole, W. (1988). Short wave corrugations—an explanation based on slip-stick vibrations. *Applied Mechanics Rail Transportation Symposium, 96,* 141–148.

Clayton, P., & Allery, M. B. P. (1982). Metallurgical aspects of surface damage problems in rails. *Canadian Metallurgical Quarterly, 21*(1), 31–46. doi:10.1179/cmq.1982.21.1.31.

Grassie, S. L. (2005). Rail corrugation: Advances in measurement, understanding and treatment. *Wear, 258*(7–8), 1224–1234. doi:10.1016/j.wear.2004.03.066.

Grassie, S. L., & Kalousek, J. (1993). Rail corrugation: Characteristics, causes and treatments. *Proceedings of the Institution of Mechanical Engineers, Part F: Journal of Rail and Rapid Transit, 207*(16), 57–68. doi:10.1243/PIME_PROC_1993_207_227_02.

Gross-Thebing, A. (1989). Frequency-dependent creep coefficients for three-dimensional rolling contact problem. *Vehicle System Dynamics, 18*(6), 359–374. doi:10.1080/00423118908968927.

Groβ-Thebing, A., Knothe, K., & Hempelmann, K. (1992). Wheel-rail contact mechanics for short wavelengths rail irregularities. *Vehicle System Dynamics, 20*(S1), 210–224. doi:10.1080/00423119208969399.

Hiensch, M., Nielsen, J. C. O., & Verheijen, E. (2002). Rail corrugation in the Netherlands—measurements and simulations. *Wear, 253*(1–2), 140–149. doi:10.1016/S0043-1648(02)00093-5.

Iwnicki, S., Bezin, Y., Xie, G., et al. (2009). Advances in vehicle-track interaction tools. *Railway Gazette International, 165*(9), 47–52.

Jin, X. S., Wang, K. Y., Wen, Z. F., et al. (2005). Effect of rail corrugation on vertical dynamics of rail vehicle coupled with a track. *Acta Mechanica Sinica, 21*(1), 95–102. doi:10.1007/s10409-004-0010-x.

Jin, X. S., Xiao, X. B., Wen, Z. F., et al. (2008). Effect of sleeper pitch on rail corrugation at the tangent track in vehicle hunting. *Wear, 265*(9–10), 1163–1175. doi:10.1016/j.wear.2008.01.028.

Kalker, J. J. (1990). *Three-dimensional Elastic Bodies in Rolling Contact*. Dordrecht, the Netherlands: Kluwer Academic Publishers. doi:10.1007/978-94-015-7889-9.

Kazymyrovych, V., Bergström, J., & Thuvander, F. (2010). Local stresses and material damping in very high cycle fatigue. *International Journal of Fatigue, 32*(10), 1669–1674. doi:10.1016/j.ijfatigue.2010.03.007.

Knothe, K. L., & Grassie, S. L. (1993). Modelling of railway track and vehicle/track interaction at high frequencies. *Vehicle System Dynamics, 22*(3–4), 209–262. doi:10.1080/00423119308969027.

Knothe, K. L., & Groβ-Thebing, A. (2008). Short wavelength rail corrugation and non-steady-state contact mechanics. *Vehicle System Dynamics, 46*(1–2), 49–66. doi:10.1080/00423110701590180.

Li, M. X. D., Berggren, E. G., & Berg, M. (2009). Assessment of vertical track geometry quality based on simulations of dynamic track-vehicle interaction. *Proceedings of the Institution of Mechanical Engineers, Part F: Journal of Rail and Rapid Transit, 223*(2), 131–139. doi:10.1243/09544097JRRT220.

Li, Z. L., Dollevoet, R., Molodova, M., et al. (2011). Squat growth—some observations and the validation of numerical predictions. *Wear, 271*(1–2), 148–157. doi:10.1016/j.wear.2010.10.051.

Li, S. G., Li, Z. L., Dollevoet, R. (2012). Wear study of short pitch corrugation using an integrated 3D FE train-track interaction model. In the 9th International Conference on Contact Mechanics and Wear of Rail/Wheel Systems, Chengdu, China, pp. 216–222.

Li, Z. L., Zhao, X., Esveld, C., et al. (2008). An investigation into the causes of squats: correlation analysis and numerical modeling. *Wear, 265*(9–10), 1349–1355. doi:10.1016/j.wear.2008.02.037.

Meyers, M. A., & Chawla, K. K. (1999). *Mechanical Behavior of Materials*. Upper Saddle River, USA: Prentice Hall.

Molodova, M., Li, Z. L., & Dollevoet, R. (2011). Axle box acceleration: Measurement and simulation for detection of short track defects. *Wear, 271*(1–2), 349–356. doi:10.1016/j.wear.2010.10.003.

Nielsen, J. C. O. (2008). High-frequency vertical wheel-rail contact forces—validation of a prediction model by field testing. *Wear, 265*(9–10), 1465–1471. doi:10.1016/j.wear.2008.02.038.

Olofsson, U., & Telliskivi, T. (2003). Wear, plastic deformation and friction of two rail steels—a full-scale test and a laboratory study. *Wear, 254*(1–2), 80–93. doi:10.1016/S0043-1648(02)00291-0.

Pang, T., Dhanasekar, M. (2006). Dynamic finite element analysis of the wheel-rail interaction adjacent to the insulated joints. In *7th International Conference on Contact Mechanics and Wear of Rail/Wheel Systems* (pp. 509–516). Brisbane, Australia.

Pletz, M., Daves, W., Fischer, F. D., et al. (2009). A dynamic wheel set-crossing model regarding impact, sliding and deformation. In *8th International Conference on Contact Mechanics and Wear of Rail/Wheel Systems* (pp. 801–808). Florence, Italy.

Ripke, B., & Knothe, K. (1995). Simulation of high frequency vehicle-track interactions. *Vehicle System Dynamics, 24*(S1), 72–85. doi:10.1080/00423119508969616.

Wen, Z. F., Jin, X. S., & Zhang, W. H. (2005). Contact-impact stress analysis of rail joint region using the dynamic finite element method. *Wear, 258*(7–8), 1301–1309. doi:10.1016/j.wear.2004.03.040.

Wu, T. X., & Thompson, D. J. (2004). The effects of track non-linearity on wheel/rail impact. *Proceedings of the Institution of Mechanical Engineers, Part F: Journal of Rail and Rapid Transit, 218*(1), 1–15. doi:10.1243/095440904322804394.

Xiao, X. B., Ling, L., Xiong, J. Y. et al. (2014). Study on the safety of operating high-speed railway vehicles subjected to crosswinds. *Journal of Zhejiang University-SCIENCE A (Applied Physics & Engineering), 15*(9), 694–710. [doi:10.1631/jzus.A1400062].

Xiao, G. W., Xiao, X. B., Guo, J., et al. (2010). Track dynamic behavior at rail welds at high speed. *Acta Mechanica Sinica, 26*(3), 449–465. doi:10.1007/s10409-009-0332-9.

Xie, G., & Iwnicki, S. D. (2008). Simulation of wear on a rough rail using a time-domain wheel-track interaction model. *Wear, 265*(11–12), 1572–1583. doi:10.1016/j.wear.2008.03.016.

Zhai, W. M., Xia, H., Cai, C. B., et al. (2013). High-speed train-track-bridge dynamic interactions-Part I: theoretical model and numerical simulation. *International Journal of Rail Transportation, 1*(1–2), 3–24. doi:10.1080/23248378.2013.791498.

Zhao, X., & Li, Z. L. (2011). The solution of frictional wheel-rail rolling contact with a 3-D transient finite element model: Validation and error analysis. *Wear, 271*(1–2), 444–452. doi:10.1016/j.wear.2010.10.007.

Zhao, X., Wen, Z., Zhu, M., et al. (2014). A study on highspeed rolling contact between a wheel and a contaminated rail. *Vehicle System Dynamics*, *52*(10):1270–1287. doi:10.1080/00423114.2014.934845.

Zhou, L., Shen, Z. Y. (2013). Dynamic analysis of a high-speed train operating on a curved track with failed fasteners. *Journal of Zhejiang University-SCIENCE A (Applied Physics & Engineering), 14*(6):447–458. [doi:10.1631/jzus.A1200321].

## Author Biographies

**Xin Zhao** Ph.D., is Associate Professor in vehicle application engineering at Southwest Jiaotong University, Chengdu, China. He is a member of the Committee of the International Conference on Contact Mechanics and Wear of Rail/Wheel Systems. Since 2003, he has been carrying out research on the phenomenon of wheel–rail rolling contact and the problems related to it such as high-frequency vehicle–track interactions, rolling contact fatigue, and thermal fatigue. In recent years, he has developed a numerical approach to solve transient rolling contact problems, this being a breakthrough in contact mechanics. He received a PhD in 2012 from Delft University of Technology, the Netherlands.

**Xue-song Jin** Ph.D., is a Professor at Southwest Jiaotong University, Chengdu, China. He is a leading scholar of wheel–rail interaction in China and is a State Council Expert for Special Allowance. He is the author of three academic books and over 200 articles. He is an editorial board member of several journals and has been a member of the Committee of the International Conference on Contact Mechanics and Wear of Rail/Wheel Systems for more than 10 years. He has been a visiting scholar at the University of Missouri, Rolla, for 2.5 years. Now, his research focuses on wheel–rail interaction, rolling contact mechanics, vehicle system dynamics, vibration, and noise.

# A 3D Model for Coupling Dynamics Analysis of High-Speed Train/Track System

**Liang Ling, Xin-biao Xiao, Jia-yang Xiong, Li Zhou, Ze-feng Wen and Xue-song Jin**

**Contents**



## 1 Introduction

High-speedrailways are developing rapidly in many countries around the world. The mileage of commercial high-speed railway in China now exceeds 6000 km. The operating speed of high-speed trains ranges from 200 to 350 km/h. China plans to construct 16,000–18,000 km of passenger dedicated lines by 2020, with operating speeds exceeding 200 km/h (Zhang 2009). Nowadays, more and more people consider high-speed trains as a comfortable, safe, low-emission, and clean energy

L. Ling · X. Xiao · J. Xiong · L. Zhou · Z. Wen · X. Jin (✉)
State Key Laboratory of Traction Power, Southwest Jiaotong University,
Chengdu 610031, China
e-mail: xsjin@home.swjtu.edu.cn

 
Y. Fang and Y.H. Zhang (eds.), *China's High-Speed Rail Technology*,
Advances in High-speed Rail Technology, https://doi.org/10.1007/978-981-10-5610-9_18

consumption transportation tool with a high on-schedule rate. But increasing the speed posts very high requirements in service performance, running safety, and vibration control in environments which are all closely related to the dynamic performance of the train/track coupling system. Therefore, the following studies on train/track system dynamics are very important for designing well-matched high-speed trains/tracks and ensuring the safe operation of high-speed trains.

Railway system dynamics studies have been performed for almost a century, resulting in thousands of papers and theoretical models being published (Knothe and Grassie 1993; Popp et al. 1999; Evans and Berg 2009; Zhai et al. 2009; Arnold et al. 2011; Xiao et al. 2014). Throughout previous studies, there are mainly two types of simulation models: single-vehicle/track coupling models and models for multi-vehicles (or trains) coupled with a rigid or nearly rigid track.

In traditional railway vehicle dynamics simulations and track modeling using commercial software, such as SIMPACK, NUCARS, GENSYS, and VAMPIRE, the railway track is often assumed to be a rigid or nearly rigid structure. However, many studies have pointed out that track flexibility has a significant influence on wheel–rail contact behavior and vehicle/track dynamics. Neglecting track dynamic behavior may lead to a significant overestimation of railway vehicle dynamics performance, including hunting stability, wheel–rail contact forces, and other vehicle system dynamical behaviors that are involved (Jin et al. 2002; Zhai et al. 2009; Di Gialleonardo et al. 2012). In addition, the classical vehicle dynamics study using a simplified rigid track model cannot solve the dynamic problems caused by the failure of a track component and other severe conditions, such as the running safety of railway vehicles passing over unsupported tracks, broken rails, and buckled tracks. These models, of course, cannot characterize the dynamic behavior of track components or the ground vibration induced by high-speed trains in operation. Another important factor to consider is that a train running on a track is a large-scale coupling system and that the dynamic behaviors of the train and the track, and the neighboring vehicles significantly affect each other. Thus, it is necessary to develop a 3D dynamic model of a high-speed train coupled with a flexible track to allow a deeper investigation into the dynamic behavior of high-speed trains under various conditions. That is the purpose of the present study.

The widely used coupled single-vehicle/track models (VTMs) can simulate the basic phenomena of a vehicle coupled with a track. An overview of single-vehicle/track modeling and its interaction analysis can be seen in (Knothe and Grassie 1993; Popp et al. 1999; Zhai et al. 2009). Most of the existing models were used to deal with single-vehicle/track vertical interaction problems (Nielsen and Igeland 1995; Fröhling 1998; Oscarsson and Dahlberg 1998; Sun and Dhanasekar 2002; Lei and Mao 2004; Cai et al. 2008), and a few were used to analyze lateral and vertical dynamical behavior (Zhai et al. 1996; Sun et al. 2003; Jin et al. 2006; Baeza and Ouyang 2011; Xiao et al. 2011; Zhou and Shen 2013). In addition, a few models for train/railway structure interactions were developed to investigate railway system dynamics (Yang and Wu 2002; Xia et al. 2003; Tanabe et al. 2008; Ju and Li 2011).

Although coupled VTMs can solve many scientific problems effectively, there are some issues which these models cannot deal with. The most prominent one is that they cannot consider the effect of inter-vehicle connections on the dynamic behavior of the train/track system. Most modern high-speed trains are equipped with tight-lock inter-vehicle connections, such as tight-lock couplers and inter-vehicle dampers. When high-speed trains run in complex operating environments, such as a derailment occurring due to strong crosswinds, earthquakes, or serious track buckling, the mutual influence between the adjacent vehicles on the system's dynamic behavior should not be neglected in a dynamic behavior analysis (Evans and Berg 2009; Zhang 2009; Jin et al. 2013). In these environments, any VTM cannot characterize the behavior of the vehicle and track accurately and reliably. To the authors' knowledge, no previous research results have been published regarding the difference between the dynamic behaviors calculated by using a VTM and an entire-train/track model (TTM) even when a train operates under normal conditions. In the present investigation, the differences in several key dynamic behaviors between these two types of dynamic models are clarified.

To meet the challenges of the various complex dynamics problems of high-speed trains coupled with tracks, the existing models need to be further improved in two ways: the space scale of trains coupled with tracks and the modeling of their key components. In this study, a 3D dynamic model of a high-speed train coupled with a ballast track is developed, which extends the single-vehicle/track vertical–lateral coupling model to a multi-vehicle/track vertical–lateral–longitudinal coupling model. In the 3D coupled train/track model, each vehicle is modeled as a 42 degrees of freedom (DOFs) multi-body system, which considers the nonlinear dynamic characteristics of the suspension systems and the longitudinal motion of the vehicle components. To simulate the interaction between adjacent vehicles, a detailed inter-vehicle connection model is developed, which considers nonlinear couplers, nonlinear inter-vehicle dampers, and a linear tight-lock vestibule diaphragm. The track is a flexible 3-layer model consisting of rails, sleepers, and ballast. The dynamic behavior and elastic structure of the track components are considered. An improved wheel/rail contact geometry model is introduced to take the effect of the profiles and the instant deformation of the wheel and the rail into account (Chen and Zhai 2004; Xiao et al. 2011). The modified model is also able to deal with separation occurring between the wheels and the rails. A moving sleeper support track model is adopted to simulate train/track excitation caused by the discrete sleepers (Xiao et al. 2011). The reliability of the 3D coupled train/track model is then validated through a detailed numerical comparison with the commercial software, SIMPACK, and the contrast caused by different track modeling methods is analyzed. Also, the differences are investigated between the dynamical behaviors obtained by VTM and TTM, with the results calculated using the proposed TTM being more reasonable. The investigated dynamic behaviors include vibration frequency components, ride comfort, and curving performance, which are important in estimating the operational qualities and dynamic characteristics of trains and tracks.

## 2 3D Modeling of High-Speed Train/Track System

A 3D dynamic model of a high-speed train coupled with a ballast track is developed in this study (Fig. 1). The coupled train/track dynamic model consists of four subsystems: the vehicle, the inter-vehicle connection, the track, and the wheel/rail contact. The interaction of the vehicles and the track is characterized through the wheels/rails in rolling contact, and the interaction between adjacent vehicles is transferred via the inter-vehicle connection. They are described in Sects. 2.1–2.4 in detail.

### *2.1 Modeling Vehicle Subsystem*

A new generic Chinese high-speed train, named CRH380A, is selected to be modeled in this study. The train consists of six power vehicles and two trailing vehicles, and its highest operating speed reaches 380 km/h. The calculation model of a high-speed vehicle coupled with a ballast track is shown in Fig. 2. In the coupled dynamic model, each power vehicle or each trailing vehicle is modeled as a 42 DOFs nonlinear multi-body system, which includes seven rigid components: a car body, two bogies, and four wheelsets.

In Fig. 2, the coordinate system $x$–$y$–$z$ is a Cartesian system and the initial one. Axis $x$ is in the moving direction of the high-speed train, axis $z$ is in the vertical direction, and axis $y$ is in the lateral direction of the track. For convenience, the front bogie and the rear bogie are numbered 1 and 2, respectively; the leading wheelset and the trailing wheelset of the front bogie are numbered as 1 and 2, respectively; and the corresponding wheelsets of the rear bogie are indicated by 3 and 4, respectively. The subscript $j$ ($j$ = $L$ or $R$) refers to the left or right side when looking in the direction of movement of the train. Each component of the vehicle has six DOFs: the longitudinal displacement $X$, the lateral displacement $Y$, the vertical displacement $Z$, the roll angle $\phi$, the pitch angle $\beta$, and the yaw angle $\psi$.

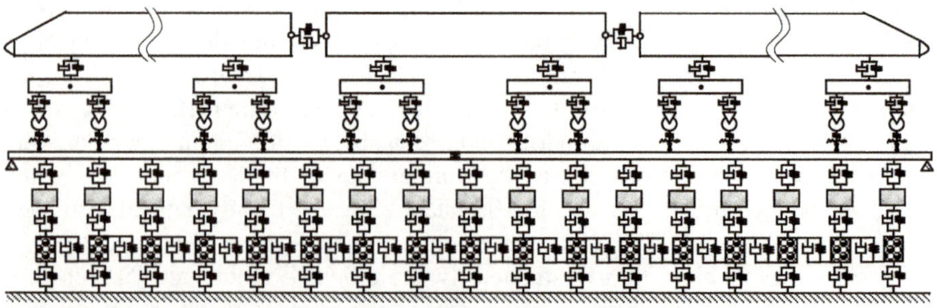

**Fig. 1** High-speed train/track coupling model

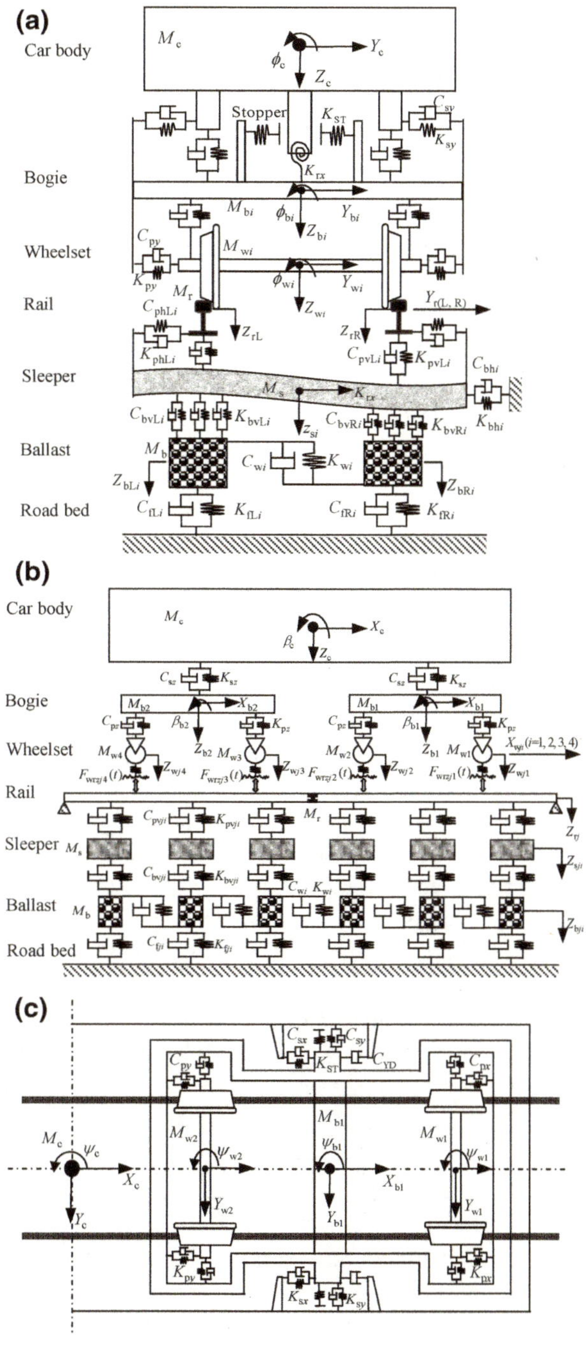


**Fig. 2** 3D views of the vehicle and track model: **a** elevation; **b** side elevation; **c** planform

In Fig. 2, the notations $C$ and $K$ with subscripts stand for the coefficients of the equivalent dampers and the stiffness coefficients of the equivalent springs, respectively. The equivalent dampers and springs are used to replace the connections between the components of the high-speed vehicle and the ballast track.

The equations of motion of the car body in the longitudinal, lateral, vertical, rolling, pitching, and yawing directions are

$$\begin{cases} M_c\ddot{X}_c = -F_{xs1} - F_{xs2} - F_{xcf} - F_{xcb}, \\ M_c\ddot{Y}_c = F_{ys1} + F_{ys2} - F_{ycf} - F_{ycb} + M_c g\phi_{sec} + F_{ycc}, \\ M_c\ddot{Z}_c = -F_{zs1} - F_{zs2} - F_{zcf} - F_{zcb} + M_c g + F_{zcc}, \\ I_{cx}\ddot{\phi}_c = -M_{xs1} - M_{xs2} + M_{xcf} + M_{xcb} + M_{xcc}, \\ I_{cy}\ddot{\beta}_c = -M_{ys1} - M_{ys2} + M_{ycf} + M_{ycb}, \\ I_{cz}\ddot{\psi}_c = -M_{zs1} - M_{zs2} + M_{zcf} + M_{zcb} + M_{zcc}, \end{cases} \tag{1}$$

where $M_c$ is the mass of the car body; $I_{cx}$, $I_{cy}$, $I_{cz}$ are the rolling, pitching, and yawing moments of inertia, respectively; $\ddot{X}_c$, $\ddot{Y}_c$, $\ddot{Z}_c$, $\ddot{\phi}_c$, $\ddot{\beta}_c$, and $\ddot{\psi}_c$ are the accelerations of the car body center in the longitudinal, lateral, vertical, rolling, pitching, and yawing directions, respectively; $\phi_{sec}$ is the angular deflection of the car body rolling caused by the cant of the high rail; $F_{xsi}$, $F_{ysi}$, $F_{zsi}$, $M_{xsi}$, $M_{ysi}$, and $M_{zsi}$ ($i$ = 1, 2) denote the mutual forces and moments between car body and bogie frames in the $x$, $y$, and $z$ directions; subscripts 1 and 2 indicate the front and rear bogies; $F_{xci}$, $F_{yci}$, $F_{zci}$, $M_{xci}$, $M_{yci}$, and $M_{zci}$ ($i$ = $f$ or $b$) denote the inter-vehicle forces and moments caused by inter-vehicle connections between the adjacent car bodies in the $x$, $y$, and $z$ directions; and subscripts f and b indicate the front and end of each car body. Detailed expressions of the inter-vehicle forces between the adjacent vehicles will be given in Sect. 2.2. $F_{ycc}$, $F_{zcc}$, $M_{xcc}$, and $M_{zcc}$ denote the external forces on the car bodies resulting from the centripetal acceleration when a train is negotiating a curved track. Lastly, $g$ is the gravitational acceleration.

The equations of motion of the bogie $i$ ($i$ = 1, 2) in the longitudinal, lateral, vertical, rolling, pitching, and yawing directions are

$$\begin{cases} M_b\ddot{X}_{bi} = F_{xsi} - F_{xf(2i-1)} - F_{xf(2i)}, \\ M_b\ddot{Y}_{bi} = F_{yf(2i-1)} + F_{yf(2i)} - F_{ysi} + M_b g\phi_{sebi} + F_{ycbi}, \\ M_b\ddot{Z}_{bi} = F_{zsi} - F_{zf(2i-1)} - F_{zf(2i)} + M_b g + F_{zcbi}, \\ I_{bx}\ddot{\phi}_{bi} = -M_{xf(2i-1)} - M_{xf(2i)} + M_{xsi} + M_{xcbi}, \\ I_{by}\ddot{\beta}_{bi} = -M_{yf(2i-1)} - M_{yf(2i)} + M_{ysi}, \\ I_{bz}\ddot{\psi}_{bi} = -M_{zf(2i-1)} - M_{zf(2i)} + M_{zsi} + M_{zcbi}, \end{cases} \tag{2}$$

where $M_b$ is the mass of the bogie; $I_{bx}$, $I_{by}$, and $I_{bz}$ are the moments of inertia of the bogie in rolling, pitching, and yawing motions; $\ddot{X}_b$, $\ddot{Y}_b$, $\ddot{Z}_b$, $\ddot{\phi}_b$, $\ddot{\beta}_b$, and $\ddot{\psi}_b$ are the accelerations of the bogie center in the longitudinal, lateral, vertical, rolling, pitching, and yawing directions, respectively; $\phi_{seb}$ is the angular deflection of the bogie rolling caused by the cant of the high rail; $F_{xfi}$, $F_{yfi}$, $F_{zfi}$, $M_{xfi}$, $M_{yfi}$, and $M_{zfi}$

($i$ = 1, 2, 3, 4) denote the mutual forces and moments between bogie frames and wheelsets in the $x$, $y$, and $z$ directions; subscripts 1, 2, 3, 4 indicate the four wheelsets of the vehicle, respectively; and $F_{\mathrm{ycb}i}$, $F_{\mathrm{zcb}i}$, $M_{\mathrm{xcb}i}$, and $M_{\mathrm{zcb}i}$ ($i$ = 1, 2) denote the external forces on bogies resulting from the centripetal acceleration when the vehicle is negotiating curved track.

The equations of motion of the wheelset $i$ ($i$ = 1, 2, 3, 4) in the longitudinal, lateral, vertical, rolling, pitching, and yawing directions are

$$\begin{cases} M_{\mathrm{w}}\ddot{X}_{\mathrm{w}i} = F_{\mathrm{xf}i} + F_{\mathrm{wrx}i}, \\ M_{\mathrm{w}}\ddot{Y}_{\mathrm{w}i} = -F_{\mathrm{yf}i} + F_{\mathrm{wry}i} + M_{\mathrm{w}}g\phi_{\mathrm{sew}i} + F_{\mathrm{ycw}i}, \\ M_{\mathrm{w}}\ddot{Z}_{\mathrm{w}i} = F_{\mathrm{zf}i} - F_{\mathrm{wrz}i} + M_{\mathrm{w}}g + F_{\mathrm{zcw}i}, \\ I_{\mathrm{wx}}\ddot{\phi}_{\mathrm{w}i} = M_{\mathrm{xf}i} - M_{\mathrm{wrz}i} + M_{\mathrm{xcw}i}, \\ I_{\mathrm{wy}}\ddot{\beta}_{\mathrm{w}i} = M_{\mathrm{wry}i} + M_{\mathrm{TB}i}, \\ I_{\mathrm{wz}}\ddot{\psi}_{\mathrm{w}i} = M_{\mathrm{zf}i} + M_{\mathrm{wrz}i} + M_{\mathrm{zcw}i}, \end{cases} \tag{3}$$

where $M_{\mathrm{w}}$ is the mass of the wheelset; $I_{\mathrm{wx}}$, $I_{\mathrm{wy}}$, and $I_{\mathrm{wz}}$ are the moments of inertia of the wheelset in rolling, pitching, and yawing motions, respectively; $\ddot{X}_{\mathrm{w}}$, $\ddot{Y}_{\mathrm{w}}$, and $\ddot{Z}_{\mathrm{w}}$ are the accelerations of the wheelset in the longitudinal, lateral, and vertical directions, respectively; $\ddot{\phi}_{\mathrm{w}}$, $\ddot{\beta}_{\mathrm{w}}$, and $\ddot{\psi}_{\mathrm{w}}$ are the angular accelerations in rolling, spin, and yawing directions, respectively; $\phi_{\mathrm{sew}}$ is the angular deflection of the wheelset rolling caused by the cant of the high rail; $F_{\mathrm{wrx}i}$, $F_{\mathrm{wry}i}$, $F_{\mathrm{wrz}i}$, $M_{\mathrm{wrx}i}$, $M_{\mathrm{wry}i}$, and $M_{\mathrm{wrz}i}$ ($i$ = 1, 2, 3, 4) denote the contact forces and moments between the wheels and the rails in the $x$, $y$, and $z$ directions, respectively; $F_{\mathrm{ycw}i}$, $F_{\mathrm{zcw}i}$, $M_{\mathrm{xcw}i}$, and $M_{\mathrm{zcw}i}$ ($i$ = 1, 2, 3, 4) denote the external forces on the wheelsets resulting from the centripetal acceleration when the train is negotiating curved track; and $M_{\mathrm{TB}i}$ is the traction or braking moment acting on the wheelsets when the train is accelerating or decelerating. In this study, a constant traveling speed of the train is assumed. Thus, $M_{\mathrm{TB}i}$ equals zero here.

In the present train/track model, each bogie is equipped with double suspension systems. The wheelsets and the bogies are connected by the primary suspensions, while the car body is supported on the bogies through the secondary suspensions. The primary and secondary suspension systems were represented using 3D spring–damper elements, and the nonlinear dynamic characteristics of the suspension systems were considered. The nonlinear suspension elements include the yaw and lateral dampers and the bump-stops installed on the secondary suspension, and the vertical dampers installed on the primary suspension, as illustrated in Fig. 3. In the model developed in this study, the nonlinear behavior of the suspension system components was modeled using bilinear spring and damping elements, as shown in Fig. 4.

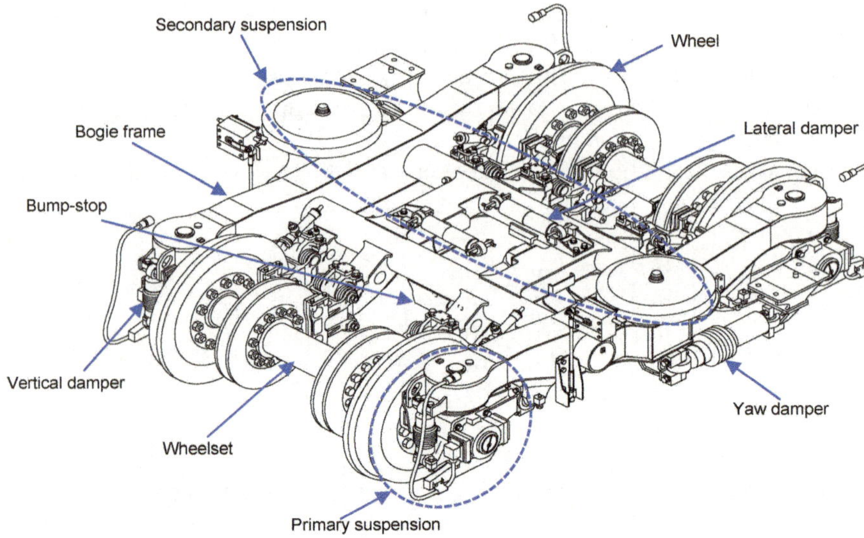


**Fig. 3** Bogie of a Chinese high-speed train

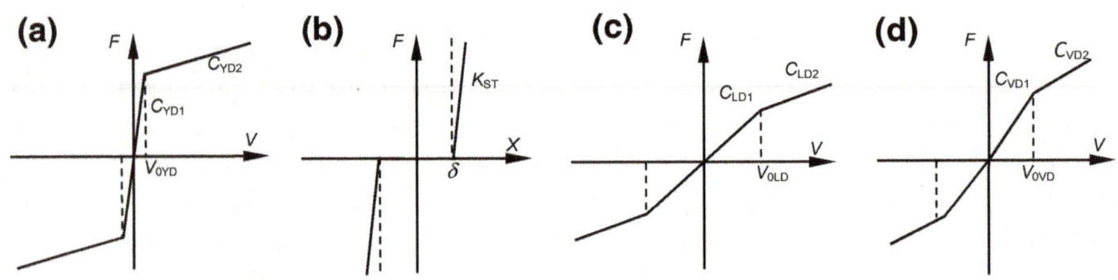


**Fig. 4** Nonlinear characteristics of the vehicle suspensions. **a** Yaw damper; **b** bump-stop; **c** lateral damper; **d** vertical damper

According to the bilinear postulation, the forces between the bogies and the car body or the wheelsets are

$$F_{x\mathrm{YD}} = \begin{cases} C_{\mathrm{YD1}}\Delta\dot{x}_{\mathrm{YD}}, & |\Delta\dot{x}_{\mathrm{YD}}| < V_{0\mathrm{YD}}, \\ \mathrm{sign}(\Delta\dot{x}_{\mathrm{YD}})[C_{\mathrm{YD1}}V_{0\mathrm{YD}} + C_{\mathrm{YD2}}(|\Delta\dot{x}_{\mathrm{YD}}| - V_{0\mathrm{YD}})], & |\Delta\dot{x}_{\mathrm{YD}}| \geq V_{0\mathrm{YD}}, \end{cases} \tag{4}$$

$$F_{y\mathrm{ST}} = \begin{cases} 0, & |\Delta y_{\mathrm{ST}}| < \delta, \\ K_{\mathrm{ST}}(|\Delta y_{\mathrm{ST}}| - \delta), & |\Delta y_{\mathrm{ST}}| \geq \delta, \end{cases} \tag{5}$$

$$F_{y\mathrm{LD}} = \begin{cases} C_{\mathrm{LD1}}\Delta\dot{y}_{\mathrm{LD}}, & |\Delta\dot{y}_{\mathrm{LD}}| < V_{0\mathrm{LD}}, \\ \mathrm{sign}(\Delta\dot{y}_{\mathrm{LD}})[C_{\mathrm{LD1}}V_{0\mathrm{LD}} + C_{\mathrm{LD2}}(|\Delta\dot{y}_{\mathrm{LD}}| - V_{0\mathrm{LD}})], & |\Delta\dot{y}_{\mathrm{LD}}| \geq V_{0\mathrm{LD}}, \end{cases} \tag{6}$$

$$F_{z\mathrm{VD}} = \begin{cases} C_{\mathrm{VD1}}\Delta\dot{z}_{\mathrm{VD}}, & |\Delta\dot{z}_{\mathrm{VD}}| < V_{0\mathrm{VD}}, \\ \mathrm{sign}(\Delta\dot{z}_{\mathrm{VD}})[C_{\mathrm{VD1}}V_{0\mathrm{VD}} + C_{\mathrm{VD2}}(|\Delta\dot{z}_{\mathrm{VD}}| - V_{0\mathrm{VD}})], & |\Delta\dot{z}_{\mathrm{VD}}| \geq V_{0\mathrm{VD}}, \end{cases} \tag{7}$$

where $C_{\mathrm{YD}}$, $C_{\mathrm{LD}}$, and $C_{\mathrm{VD}}$ stand for the equivalent coefficients of the yaw dampers, the lateral dampers, and the vertical dampers, respectively; $K_{\mathrm{ST}}$ is the contact stiffness when the car body is in contact with the bump-stops; $V_{0\mathrm{YD}}$, $V_{0\mathrm{LD}}$, and $V_{0\mathrm{VD}}$ are the load-off velocities of the yaw dampers, the lateral dampers, and the vertical dampers, respectively; $\delta$ is the lateral clearance between the car body and the bump-stops on the bogie frames; $\Delta\dot{x}_{\mathrm{YD}}$ is the longitudinal relative velocity between the car body and the side frame; $\Delta y_{\mathrm{ST}}$ is the lateral relative displacement between the bottom of the car body and the bogies; $\Delta\dot{y}_{\mathrm{LD}}$ is the lateral relative velocity between the bottom of the car body and the bogies; $\Delta\dot{z}_{\mathrm{VD}}$ is the vertical relative velocity between the axle and the side frame; $F_{x\mathrm{YD}}$, $F_{y\mathrm{ST}}$, and $F_{y\mathrm{LD}}$ are the forces of the yaw dampers, the bump-stops, and the lateral dampers between bogies and car body, respectively; and $F_{z\mathrm{VD}}$ is the force of the vertical dampers between bogies and wheelsets.

## 2.2 Modeling the Inter-vehicle Connection Subsystem

The design of the inter-vehicle connection is very important for a high-speed train because it has to include mechanical and electrical connections between adjacent vehicles. In addition, it should provide passengers with a comfortable and safe passage. Among the inter-vehicle suspensions of a high-speed train, three devices have a significant influence on the dynamics of the train/track system: couplers, inter-vehicle dampers, and tight-lock vestibule diaphragms. In the present model, the nonlinear couplers and inter-vehicle dampers are replaced with nonlinear spring–damper elements and are retractable only along the axial direction. The tight-lock vestibule diaphragm is simplified as a linear 3D spring element, which can restrain the adjacent vehicles in the longitudinal, lateral, vertical, rolling, pitching, and yawing directions. Therefore, the inter-vehicle forces can be calculated based on the deformation of the connectors and the relative angles between connectors and car bodies.

To improve running stability and ride comfort during acceleration or deceleration, tight-lock couplers are installed comprehensively on modern high-speed trains. A type of tight-lock coupler system used on the Chinese high-speed trains is modeled in this study, as shown in Fig. 5a. In this type of tight-lock coupler, the couplers installed on adjacent vehicles are fixed by the coupler connection, and the slackless is very small. The couplers can rotate by a certain angle around the coupler yoke in the horizontal and vertical directions. The coupler body is approximately rigid, and the inter-vehicle contact stiffness is offered by the draft gear. In this model, the nonlinear stiffness characteristic of the draft gear is considered, and the draft gear is modeled by a bilinear spring element, as shown in Fig. 5b. According to the bilinear assumption, the coupler forces are

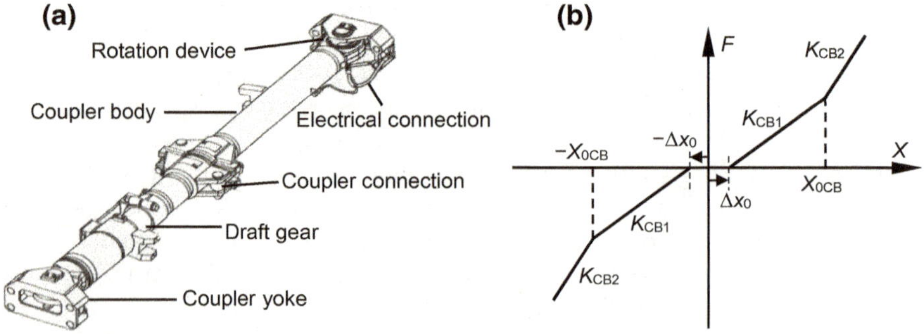


**Fig. 5** Nonlinear coupler model (**a**) and nonlinear characteristic of coupler system (**b**)

$$F_{cg} = \begin{cases} 0, & |\Delta x| < \Delta x_0, \\ K_{CB1}(\Delta x - \Delta x_0), & \Delta x_0 \leq |\Delta x| \leq X_{0CB}, \\ K_{CB1}(X_{0CB} - \Delta x_0) + K_{CB2}(\Delta x - X_{0CB}), & \Delta x| > X_{0CB}, \end{cases} \quad (8)$$

where $\Delta x$ is the relative displacement between the two ends of the couplers connecting the adjacent vehicles in the axial direction, $\Delta x_0$ is the slackless of the coupler, $X_{0CB}$ is the initial length of the coupler, and $K_{CB}$ is its equivalent stiffness coefficient.

According to the dynamic responses of the vehicles and the geometric relationship between couplers and car body ends, the lateral and vertical angles between the coupler and the adjacent vehicles can be calculated (Garg and Dukkipati 1984). The longitudinal, lateral, and vertical components of the coupler forces applied to the adjacent vehicles near to the coupler are then obtained.

Inter-vehicle dampers are widely used on high-speed trains, such as the German ICE, the French TGV, the Japanese Shinkansen train sets, and the Chinese CRH. In the present model, a type of longitudinal inter-vehicle damper used on Chinese high-speed trains is introduced, as shown in Fig. 6a. Field tests and numerical studies (Zhang 2009) highlight that this kind of damper can reduce the longitudinal impacts between the vehicles and improve the lateral stability and ride comfort of high-speed trains. The inter-vehicle dampers are also replaced with bilinear spring–damper elements, and their damping and stiffness are considered, as shown in Fig. 6b, c. Based on Fig. 6, the forces on the inter-vehicle dampers are

$$F_{CDL,R} = \begin{cases} C_{CD1}\Delta V_{CDL,R}, & |\Delta V_{CDL,R}| < V_{0CD}, |\Delta X_{CDL,R}| \leq X_{0CD}, \\ \mathrm{sign}(\Delta V_{CDL,R})[C_{CD1}V_{0CD} \\ + C_{CD2}(|\Delta V_{CDL,R}| - V_{0CD})], & |\Delta V_{CDL,R}| \geq V_{0CD}, |\Delta X_{CDL,R}| \leq X_{0CD}, \\ \mathrm{sign}(\Delta X_{CDL,R})K_{CD}(|\Delta X_{CDL,R}| - X_{0CD}), & |\Delta V_{CDL,R}| \geq V_{0CD}, |\Delta X_{CDL,R}| > X_{0CD}, \end{cases} \quad (9)$$

where $F_{CDi}$ ($i = L, R$) are the interaction forces of the longitudinal inter-vehicle dampers; $C_{CD}$ and $K_{CD}$ stand for the coefficients of the equivalent damper and the

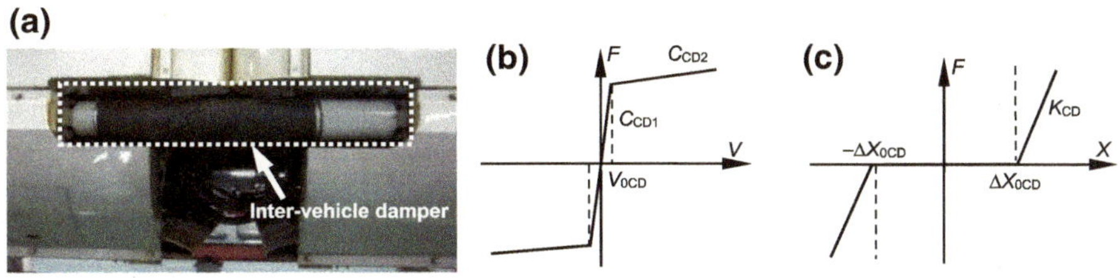


**Fig. 6** Inter-vehicle damper model. **a** Inter-vehicle damper; **b** Nonlinear damping; **c** Nonlinear stiffness

equivalent spring, respectively; $V_{0\mathrm{CD}}$ is the load-off velocity of the inter-vehicle dampers; $X_{0\mathrm{CD}}$ is the initial length of the inter-vehicle damper; $\Delta V_{\mathrm{CD}i}$ and $\Delta X_{\mathrm{CD}i}$ ($i = L$, $R$) are the relative velocity and displacement between two ends of the inter-vehicle dampers connecting adjacent vehicles in the axial direction, respectively; and the subscript $i$ ($i = L$, $R$) refers to the left or right longitudinal inter-vehicle damper. Using the same process as in the coupler angle calculation, the relative angles between the dampers and the car body ends are then calculated. Thus, the forces caused by the inter-vehicle dampers in $x$, $y$, and $z$ directions can be obtained.

The tight-lock vestibule diaphragm also has an impact on the dynamics of a high-speed train. For simplicity, it is replaced with 3D linear spring elements in the present model, which can supply the car body with restraining stiffness in the longitudinal, lateral, vertical, rolling, pitching, and yawing directions.

## 2.3 Modeling the Track Subsystem

The model of the track is a flexible one consisting of rails, sleepers, and ballasts, as shown in Fig. 2. In the track model, rails are assumed to be Timoshenko beams supported by discrete sleepers, and the effects of vertical and lateral motions and rail roll on wheel/rail creepage are taken into account. Each sleeper is treated as an Euler beam supported by a uniformly distributed stiffness and damping in its vertical direction, and a lumped mass is used to replace the sleeper in its lateral direction. The ballast bed is replaced by equivalent rigid ballast bodies in the calculation model, taking into account only the vertical motion of each ballast body. The motion of the roadbed is neglected. The equivalent springs and dampers are used as the connections between rails and sleepers, between sleepers and ballast blocks, and between ballast blocks and the roadbed.

The bending deformations of the rails are described by the Timoshenko beam theory. Using the modal synthesis method and normalized shape functions of a Timoshenko beam, the fourth-order partial differential equations of the rails are converted into second-order ordinary differential equations as follows:

For the lateral bending motion:

$$\begin{cases} \ddot{q}_{\mathrm{ry}k}(t) + \frac{\kappa_{\mathrm{ry}} G_{\mathrm{r}} A_{\mathrm{r}}}{\rho_{\mathrm{r}} A_{\mathrm{r}}} \left(\frac{k\pi}{l_{\mathrm{r}}}\right)^2 q_{\mathrm{ry}k}(t) - \kappa_{\mathrm{ry}} G_{\mathrm{r}} A_{\mathrm{r}} \frac{k\pi}{l_{\mathrm{r}}} \sqrt{\frac{1}{m_{\mathrm{r}} \rho_{\mathrm{r}} I_{\mathrm{rz}}}} w_{\mathrm{ry}k}(t) \\ \quad = -\sum_{i=1}^{N_{\mathrm{s}}} R_{yi}(t) Y_{\mathrm{r}k}(x_{\mathrm{s}i}) + \sum_{j=1}^{N_{\mathrm{w}}} F_{\mathrm{wry}j}(t) Y_{\mathrm{r}k}(x_{\mathrm{w}j}), \\ \ddot{w}_{\mathrm{ry}k}(t) + \left[\frac{\kappa_{\mathrm{ry}} G_{\mathrm{r}} A_{\mathrm{r}}}{\rho_{\mathrm{r}} I_{\mathrm{rz}}} + \frac{E_{\mathrm{r}} I_{\mathrm{rz}}}{\rho_{\mathrm{r}} I_{\mathrm{rz}}} \left(\frac{k\pi}{l_{\mathrm{r}}}\right)^2\right] w_{\mathrm{ry}k}(t) \\ - \kappa_{\mathrm{ry}} G_{\mathrm{r}} A_{\mathrm{r}} \frac{k\pi}{l_{\mathrm{r}}} \sqrt{\frac{1}{m_{\mathrm{r}} \rho_{\mathrm{r}} I_{\mathrm{rz}}}} q_{\mathrm{ry}k}(t) = 0, \quad k = 1, 2, \ldots, \mathrm{NMY}; \end{cases} \tag{10}$$

For the vertical bending motion:

$$\begin{cases} \ddot{q}_{\mathrm{rz}k}(t) + \frac{\kappa_{\mathrm{rz}} G_{\mathrm{r}} A_{\mathrm{r}}}{\rho_{\mathrm{r}} A_{\mathrm{r}}} \left(\frac{k\pi}{l_{\mathrm{r}}}\right)^2 q_{\mathrm{rz}k}(t) - \kappa_{\mathrm{rz}} G_{\mathrm{r}} A_{\mathrm{r}} \frac{k\pi}{l_{\mathrm{r}}} \sqrt{\frac{1}{m_{\mathrm{r}} \rho_{\mathrm{r}} I_{\mathrm{ry}}}} w_{\mathrm{rz}k}(t) \\ \quad = -\sum_{i=1}^{N_{\mathrm{s}}} R_{zi}(t) Z_{\mathrm{r}k}(x_{\mathrm{s}i}) + \sum_{j=1}^{N_{\mathrm{w}}} F_{\mathrm{wrz}j}(t) Z_{\mathrm{r}k}(x_{\mathrm{w}j}), \\ \ddot{w}_{\mathrm{rz}k}(t) + \left[\frac{\kappa_{\mathrm{rz}} G_{\mathrm{r}} A_{\mathrm{r}}}{\rho_{\mathrm{r}} I_{\mathrm{ry}}} + \frac{E_{\mathrm{r}} I_{\mathrm{ry}}}{\rho_{\mathrm{r}} I_{\mathrm{ry}}} \left(\frac{k\pi}{l_{\mathrm{r}}}\right)^2\right] w_{\mathrm{rz}k}(t) \\ - \kappa_{\mathrm{rz}} G_{\mathrm{r}} A_{\mathrm{r}} \frac{k\pi}{l_{\mathrm{r}}} \sqrt{\frac{1}{m_{\mathrm{r}} \rho_{\mathrm{r}} I_{\mathrm{ry}}}} q_{\mathrm{rz}k}(t) = 0, \quad k = 1, 2, \ldots, \mathrm{NMZ}; \end{cases} \tag{11}$$

For the torsional motion:

$$\begin{aligned} \ddot{q}_{\mathrm{rT}k}(t) + \frac{G_{\mathrm{r}} K_{\mathrm{r}}}{\rho_{\mathrm{r}} I_{\mathrm{r}0}} \left(\frac{k\pi}{l_{\mathrm{r}}}\right)^2 q_{\mathrm{rT}k}(t) &= -\sum_{i=1}^{N_{\mathrm{s}}} M_{\mathrm{s}i}(t) \Phi_{\mathrm{r}k}(x_{\mathrm{s}i}) \\ + \sum_{j=1}^{N_{\mathrm{w}}} M_{\mathrm{G}j}(t) \Phi_{\mathrm{r}k}(x_{\mathrm{w}j}), &\quad k = 1, 2, \cdots, \mathrm{NMT}. \end{aligned} \tag{12}$$

In Eqs. (10)–(12), $q_{\mathrm{ry}k}(t)$, $q_{\mathrm{rz}k}(t)$, and $q_{\mathrm{rT}k}(t)$ are the generalized coordinates of the lateral, vertical, and rotational deflection of the rail, respectively, while $w_{\mathrm{ry}k}(t)$ and $w_{\mathrm{rz}k}(t)$ are the generalized coordinates of the deflection curve of the rail with respect to the $z$-axis and the $y$-axis. The material properties of the rail are indicated by the density $\rho_{\mathrm{r}}$, the shear modulus $G_{\mathrm{r}}$, and Young's modulus $E_{\mathrm{r}}$. $m_{\mathrm{r}}$ is the mass per unit longitudinal length. The geometry of the cross section of the rail is represented by the area $A_{\mathrm{r}}$, the second moments of area $I_{\mathrm{ry}}$ and $I_{\mathrm{rz}}$ around the $y$-axis and the $z$-axis, respectively, and the polar moment of inertia $I_{\mathrm{r}0}$. The shear coefficients $\kappa_{\mathrm{ry}} = 0.4057$ and $\kappa_{\mathrm{rz}} = 0.5329$ for the lateral and the vertical bending and the shear coefficient $K_{\mathrm{r}} = 2.473{,}346 \times 10^{-6}$ are obtained through a finite element analysis of the rail profile of Chinese CN 60 using the software package ANSYS. The calculation length of the beam is denoted by $l_{\mathrm{r}}$, the value of which was set at 420 m when considering an eight-vehicle train running on the calculated track. In this case, 1000 vibration modes of the rail were considered, and the frequency of the highest mode was approximately 1.2 kHz. $R_{yi}$ and $R_{zi}$ are the lateral and vertical forces between the rail and sleeper $i$, respectively. The wheel/rail forces at the wheel $j$ in the lateral and vertical directions are represented by $F_{wryj}$ and $F_{wrzj}$, respectively. $M_{si}$ and $M_{Gj}$ denote the equivalent moments acting on the rail. $x_{si}$ and $x_{wj}$ denote the longitudinal positions of the sleeper $i$ and the wheel $j$, respectively, and $N_w$ and $N_s$ are the number of wheelsets and sleepers within the analyzed rail, respectively. The subscript $i$ indicates sleeper $i$ and $j$ for wheel $j$. NMY, NMZ, and NMT are the total numbers of the shape functions, and $Y_{rk}(x)$, $Z_{rk}(x)$, and $\Phi_{rk}(x)$ are the $k$th shape functions, which are given by

$$Y_{rk}(x) = \sqrt{\frac{2}{\rho_r A_r l_r}} \sin\left(\frac{k\pi}{l_r} x\right), \tag{13}$$

$$Z_{rk}(x) = \sqrt{\frac{2}{\rho_r A_r l_r}} \sin\left(\frac{k\pi}{l_r} x\right), \tag{14}$$

$$\Phi_{rk}(x) = \sqrt{\frac{2}{\rho_r I_{r0} l_r}} \sin\left(\frac{k\pi}{l_r} x\right). \tag{15}$$

The sleeper in the present model is treated as an Euler–Bernoulli beam with free-free ends in the vertical direction, while a lumped mass is used to replace it for its lateral motion. The longitudinal rigid motion and rotating motion of each sleeper are neglected, as shown in Fig. 2. Using the modal synthesis method and the normalized shape functions of the Euler beam, the fourth-order partial differential equations of its vertical vibration can be simplified as a second-order ordinary differential equation as follows:

$$\begin{aligned} \ddot{q}_{szk}(t) + \frac{E_s I_s}{m_s}\left(\frac{k\pi}{l_s}\right)^4 q_{szk}(t) &= -\sum_{i=1}^{N_b} F_{bzi}(t) Z_{sk}(y_{bi}) \\ &+ \sum_{j=1}^{N_r} R_{zj}(t) Z_{sk}(y_{rj}), \quad k = 1, 2, \ldots, \text{NMS}, \end{aligned} \tag{16}$$

where $q_{szk}(t)$ are the generalized coordinates of the sleeper vertical deflection, $E_s$ is Young's modulus, $I_s$ is the second moment of area of the sleeper cross section about the $y$-axis, $m_s$ is the mass per unit longitudinal length, $l_s$ is the length of the sleeper, $N_b$ and $N_r$ are the number of ballast and rails within the analyzed sleeper, respectively, $F_{bzi}$ is the force between the sleeper and the ballast body in the action spot $i$, $R_{zj}$ is the force between the sleeper and the rail in the action location $j$, NMS is the total number of the shape functions, and $Z_{sk}(y)$ is the $k$th modal function, which is given by

$$Z_{sk}(y) = \begin{cases} \sqrt{1/m_s}, & k = 1, \\ \sqrt{3/m_s}(1 - 2y/l_s), & k = 2, \\ \sqrt{1/m_s}(\cosh(\alpha_k y) + \cos(\alpha_k y)) & \\ \quad - C_k(\sinh(\alpha_k y) + \sin(\alpha_k y))], & k = 3, 4, \ldots, \text{NMS}, \end{cases} \tag{17}$$

where $\alpha_k$ and $C_k$ are the frequency coefficient and the function coefficient of a beam with free-free boundary conditions, respectively.

The equation of the lateral rigid motion of the sleeper is

$$M_s \ddot{Y}_{si} = F_{yLi} + F_{yRi} - F_{byi}, \tag{18}$$

where $F_{yLi}$ and $F_{yRi}$ are the lateral forces between sleeper $i$ and the left and right rails, and $F_{byi}$ is the equivalent lateral support force by the ballast body. The longitudinal rigid motion and rotating motion of each sleeper are neglected.

The ballast bed is replaced by equivalent rigid ballast blocks in this calculation model, while only the vertical motion of each ballast body is taken into account. The vertical equations of motion of the ballast body $i$ are

$$M_{bs} \ddot{Z}_{bLi} = F_{bzLi} + F_{zrLi} + F_{zLRi} - F_{zgLi} - F_{zfLi}, \tag{19}$$

$$M_{bs} \ddot{Z}_{bRi} = F_{bzRi} + F_{zrRi} - F_{zLRi} - F_{zgRi} - F_{zfRi}, \tag{20}$$

where $F_{zfLi}$, $F_{zrLi}$, $F_{zfRi}$, $F_{zrRi}$, and $F_{zLRi}$ are the vertical shear forces between neighboring ballast bodies, $F_{zgLi}$ and $F_{zgRi}$ are the vertical forces between ballast bodies and the roadbed, and $M_{bs}$ is the mass of each ballast body. Such a ballast model can represent the in-phase and out-of-phase motions of two vertical rigid modes in the vertical–lateral plane of the track. For brevity, the detailed derivation of track system equations, which can be seen in (Zhai et al. 2009; Xiao et al. 2011), is omitted here. Note that it is easy to develop the present track model in the case of a slab track or other ballastless tracks. The results for a slab track are not given here. A detailed description of the slab track model can be seen in (Xiao et al. 2012).

## 2.4 Modeling the Wheel/Rail Contact Subsystem

The wheel/rail contact is an essential element that couples the vehicle subsystem with the track subsystem. The wheel/rail contact model includes two basic issues: the geometric relationship and the contact forces between the wheel and the rail. The wheel/rail contact geometry calculation is necessary to acquire the location of the contact point on the wheel and rail surfaces and the wheel/rail interaction forces. In this study, an improved geometric calculation model of the wheel/rail contact based on the method discussed in (Jin et al. 2005) is introduced. The modified spatial wheel/rail geometric contact model is able to take the instant motion and deformation of the rails into account and to deal with the separation of wheel and rail (Chen and Zhai 2004; Xiao et al. 2011).

In this study, calculating the wheel/rail normal force uses the Hertzian nonlinear contact spring model, and the creep force calculation uses Shen et al. (1983)'s model based on Kalker (1967)'s linear creep theory. These two models are based on the assumptions of Hertzian contact theory. The contact points were previously calculated in the wheel/rail force calculation. The detailed contact point calculation is described as follows.

The wheel/rail contact points vary with the lateral displacement $y_{\mathrm{w}}$, yawing angle $\psi_{\mathrm{w}}$, and rolling angle $\phi_{\mathrm{w}}$ of the wheelset; the lateral displacements $Y_{r\mathrm{L,R}}$, vertical displacements $Z_{r\mathrm{L,R}}$, and torsion angles $\phi_{r\mathrm{L,R}}$ of the rail obtained through the dynamics calculation; and the given profiles of the wheel and rail. The profiles of the rails and wheels are expressed with the discrete datum, which is described in coordinate systems *OXYZ* and *o'x'y'z'*, respectively, as shown in Fig. 7. The origin of *o'x'y'z'* is fixed at the center of the wheelset, and its axis *y'* coincides with the axle of the wheelset. By solving the vehicle and track system equations, the instant motions of the wheelset and the two rails and the positions of the rails at any given moment in a fixed reference configuration *OXYZ* are calculated, as shown in Fig. 7. In the contact geometry calculation, the height $Z_{\mathrm{w0}}$ of the wheelset in *OXYZ* is then set high enough to ensure no penetration occurred between the wheels and the rails. Using the wheel/rail contact point trace method (Wang 1984), the minimum vertical distances between the wheels and the rails are calculated on both of the left and right sides. Hence, the two points on the wheel and rail treads with the smallest distance for each side wheel/rail are obtained, respectively. These two points constitute a pair of contact points $C_{\mathrm{L,R}}$ between the wheelset and the two rails before their deformation.

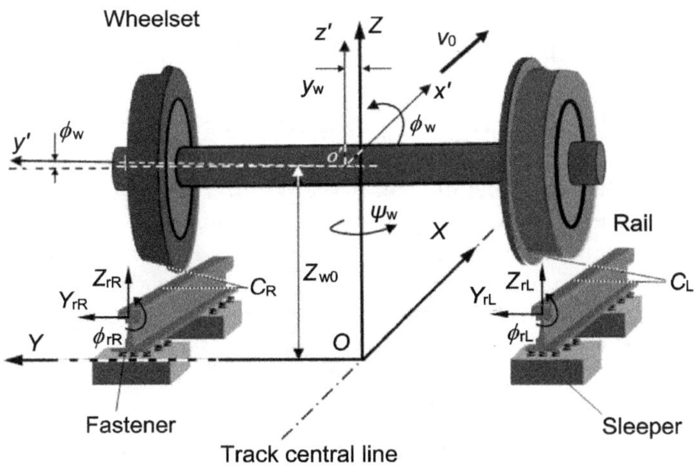


**Fig. 7** Wheel/rail contact geometry calculation model

Using the known locations of the contact points, one obtains the curvature radii of the wheels/rails at their contact points according to the prescribed wheel/rail profiles. Using the radii and the static wheel normal load, one calculates the semi-axle lengths of the wheel/rail contact patches and the initial wheel/rail normal approach by means of Hertzian contact theory; then, the Kalker (1967)'s creep coefficients can be found from his creep coefficient table. So far the calculation of the wheel/rail forces (normal and tangent) can be carried out by using the Hertzian nonlinear contact spring model and Shen et al. (1983)'s model.

The calculation model of the wheel/rail normal force, which characterizes the relationship law of the normal load and deformation between the wheel and rail, is described by a Hertzian nonlinear contact spring with a unilateral restraint and reads

$$F_{\mathrm{n}}(t) = \begin{cases} \left[\frac{1}{G} Z_{\mathrm{wrnc}}(t)\right]^{3/2} & Z_{\mathrm{wrnc}}(t) > 0, \\ 0, & Z_{\mathrm{wrnc}}(t) \leq 0, \end{cases} \tag{21}$$

where $G$ is the wheel/rail contact constant (m/N$^{2/3}$), which can be obtained using the Hertzian contact theory. $Z_{\mathrm{wrnc}}(t)$ is the normal compressing amount (or the normal approach) at the wheel/rail contact point. $Z_{\mathrm{wrnc}}(t)$ is strictly defined as an approach between the two far points, one belonging to the wheel and the other belonging to the rail. It can be determined by solving the system of equations and calculating the contact geometry of the wheelset and the rails discussed above. In Eq. (21), $Z_{\mathrm{wrnc}}(t) > 0$ indicates the wheel/rail in contact, and $Z_{\mathrm{wrnc}}(t) \leq 0$ stands for their separation. The creep force calculation employs Shen et al. (1983)'s model, which is based on Kalker (1967)'s linear creep theory. Kalker (1967)'s linear creep theory is only available for small creepages. When large creepages are generated as, for example, in the case of wheel/rail flange contact, the creep force saturates, and then, the creep forces vary nonlinearly with the creepages.

## 2.5 *Train/Track Excitation Model*

In the train/track dynamics calculation, there are four existing models (Popp et al., 1999): (1) the stationary load model; (2) the moving load model; (3) the moving irregularity model; and (4) the moving mass model. The most realistic one is the so-called moving mass model. However, it is very difficult to carry out numerical implementation using such a model because of the continuously updated track under the running train. For simplicity, a moving track support model (Xiao et al., 2007) developed by the authors is used to simulate the effect of the discrete periodic track support between the interaction of a high-speed vehicle and a track when high-speed trains run at constant speeds. The model of a half vehicle (one bogie) coupled with a track was extended to consider a whole vehicle (two bogies) in (Xiao et al. 2011). In this study, the model of Xiao et al. (2011) is further extended

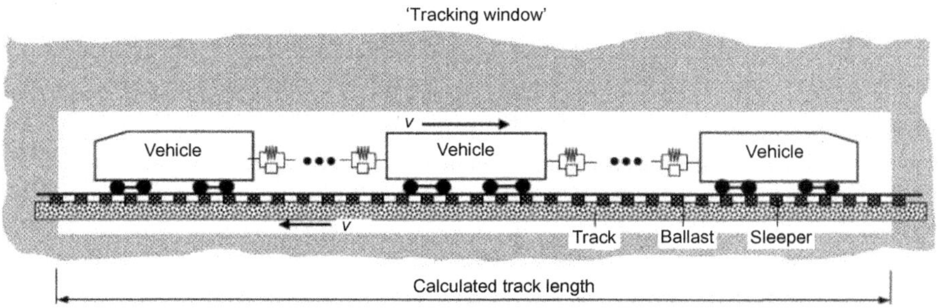


**Fig. 8** Train/track excitation model: 'Tracking window'

to consider the multi-vehicles of a train or the whole train coupled with a track, as shown in Fig. 8.

The model is seen as if one watches the behavior of a vehicle of the train running along the track through a window of $l_{tim}$ width. The window moves forward at the speed of the moving train. It is assumed that the vehicle always vibrates in the window. The track passes through the window in the inverse direction at the speed of the train, as shown in Fig. 8. The advantage of this model is that it allows rapid calculation of the train/track interaction of a train running on an infinitely long flexible track.

## 2.6 *Initial and Boundary Conditions of the Coupled Train/Track System*

Before solving the equations of the dynamic system, the initial and boundary conditions should be prescribed. Both ends of the Timoshenko beam modeling the rails are hinged, and the deflections and the bending moments at the hinged beam ends are assumed to be zero. The vertical motion of the ballast bodies at both ends of the calculation track is assumed to be always zero, and the static state of the systems is regarded as the original point of reference. The initial displacements and velocities of all components of the track are set to zero. The initial displacements and the initial vertical and lateral velocities of all components of the high-speed train are also set to zero, and the initial longitudinal velocity is the running speed of the train, which is a constant.

It is obvious that the equations of coupled train/track model form a large-scaled nonlinear system. The stability, calculation speed, and accuracy of the numerical method for the equations are very important. A numerical method developed by Zhai (1996), termed as 'new fast numerical integration for dynamics analysis of large systems,' is used to analyze the equations in a time step of $1.4 \times 10^{-5}$ s in this study.

## 3 Verification of the Train/Track Model

Based on the mathematical model described in Sect. 2, a computer simulation program, named high-speed train/track system dynamics (HSTTSD), was developed to analyze the dynamics of the coupled train/track system. To verify the 3D coupled train/track model, the dynamic results calculated by the present model are compared with those obtained by the commercial software SIMPACK. In this section, the vehicle/track dynamic interactions in the vertical and lateral directions are analyzed, by comparing the system responses obtained through HSTTSD and SIMPACK, under the excitation of vertical and lateral track irregularity on the tangent track. In the calculation, the vehicle parameters and the fastening parameters used are the same, and the vehicle speed is 300 km/h. The track irregularities are artificially generated sine wave defects with a length of 20 m and an amplitude of 10 mm.

Figure 9a, b is the wheelset vertical displacements and wheel/rail vertical forces, respectively, calculated by SIMPACK and HSTTSD. From Fig. 9, it is clear that the vertical displacements of the wheelsets are very close. Strictly speaking, the vertical displacement calculated by HSTTSD is a little larger than that obtained by SIMPACK, which is not clearly shown in Fig. 9a. The vertical force calculated by HSTTSD is also a little larger than that calculated by SIMPACK.

The lateral interaction of the wheel/rail system has a great influence on running safety against derailment of a train and wear of the wheels and rails. Figure 10a, b indicates the wheelset lateral displacements and wheel/rail lateral forces,

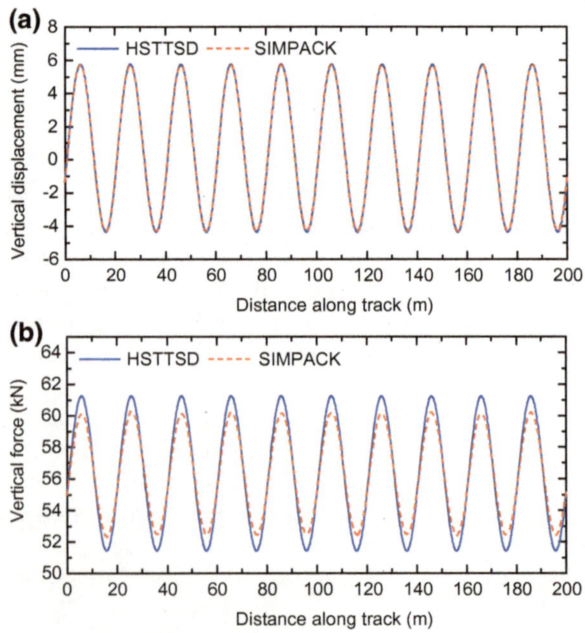


**Fig. 9** Comparison of vertical dynamic responses. **a** Wheelset vertical displacement; **b** wheel/rail vertical force

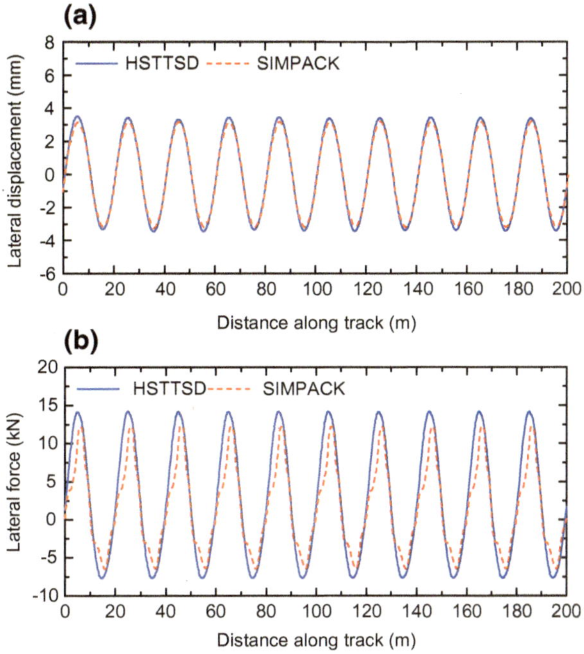


**Fig. 10** Comparison of lateral dynamic responses. **a** Wheelset lateral displacement; **b** Wheel/rail lateral force

respectively, achieved by SIMPACK and HSTTSD. It is obvious that the lateral displacements and forces calculated by HSTTSD are larger than those obtained by SIMPACK, which is similar to the phenomena that occurred in the results relating to the vertical interaction of the vehicle and the track, as described in Fig. 9.

The reason for the above phenomenon is that the track model in HSTTSD is different from that in SIMPACK. The track model in HSTTSD considers a flexible three-layer infrastructure consisting of rails, sleepers, and ballast bed. The connections between rails and sleepers, between sleepers and ballast blocks, and between ballast blocks and roadbed are replaced with the equivalent dampers and springs. The structure deformations of rails and sleepers are taken into account. Thus, the vertical (lateral) stiffness of the track characterized by HSTTSD is lower than that characterized by SIMPACK, which leads to the vertical (lateral) displacement calculated by HSTTSD being slightly larger than that obtained by SIMPACK, as shown in Figs. 9a and 10a.

Figures 9b and 10b show that the difference between wheel/rail forces calculated by HSTTSD and by SIMPACK is significant, i.e., the relative errors are approximately 10%. Compared to the simplified track model in SIMPACK, the flexible track model in HSTTSD also considers the longitudinal propagating vibration waves induced in the rails and the periodical excitation caused by discrete sleepers. The structure deformation of rails, wave reflection from the adjacent wheels, and the moving track excitation may result in larger wheel/rail contact forces, and their corresponding contribution onto these differences needs to be examined in future work. However, the differences between the calculated results by the two models can be accepted in practice. Through the results discussed above, the proposed vehicle/track model is verified to be reliable, and it can be extended to a 3D coupled train/track model, as discussed in Sect. 2.

Through the comparisons, it can be concluded that the track model in HSTTSD is more reasonable than that of SIMPACK, because HSTTSD considers the flexibility and the dynamic behavior of the track components. But when we simulated a high-speed vehicle running over a 1000-m-length straight track at a speed of 350 km/h by using the Windows operating system on a 2.79 GHz CPU DELL Studio XPS (which has one node with eight processors), the computational time required for HSTTSD and SIMPACK was 470 and 121 s, respectively. This means the computation speed for SIMPACK is approximately three times faster than that for HSTTSD. In other words, we should try to optimize the numerical algorithm to improve the calculation efficiency of the current model in the future.

## 4 Comparison of Dynamic Performances Obtained by TTM and VTM

Traditional dynamics studies of railway vehicle/track systems were mainly based on the coupled VTM, while the cross-influence between the adjacent vehicles and the effect of the vehicle location in a train were neglected. However, the interaction of the neighboring vehicles has a great influence on the dynamic performance of the train/track system due to the tight-lock inter-vehicle connections installed on modern high-speed trains. In this situation, the difference in dynamic performance obtained by TTM and VTM should be taken into account. To obtain more accurate and reliable results from the dynamics simulation, the differences between the two types of dynamic models should first be pointed out.

In this section, several key dynamics performances, including vibration frequency response, ride comfort, and curving performance, obtained by TTM and VTM, are compared, which will be discussed in Sects. 4.1–4.3. In the calculation, the TTM used a Chinese high-speed train comprised of eight vehicles coupled with the ballast track. For simplicity, the parameters of the vehicle and the track used in the two dynamic models are the same. The measured track irregularities of a Chinese high-speed line from Beijing to Tianjin were used in this calculation.

### *4.1 Comparison of Vibration Frequency Components*

To make clear the differences in the dynamics performances obtained by TTM and VTM, the random responses of the car bodies and the wheel/rail forces were firstly compared. In this simulation, the 3D high-speed train/track model described in

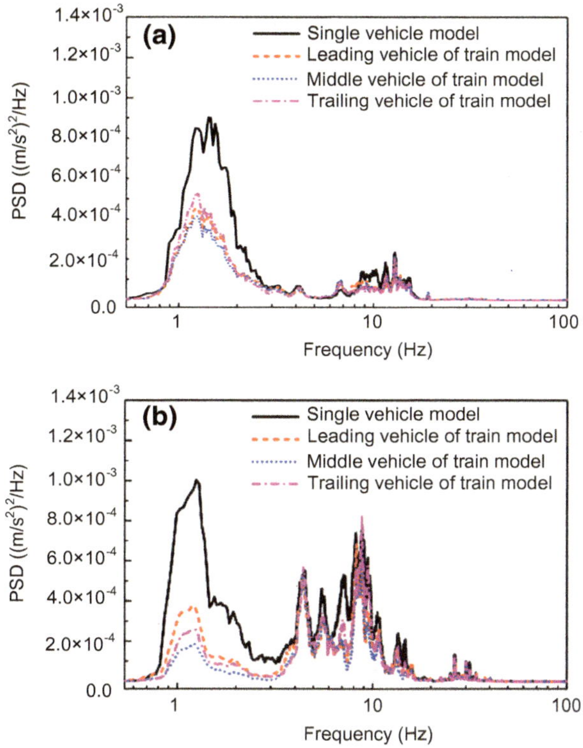


**Fig. 11** Comparison of car body vibration frequency components: **a** car body vertical acceleration PSD; **b** car body lateral acceleration PSD

Sect. 2 was used, a tangent track was considered, and the operating speed was 350 km/h. The power spectral densities (PSDs) of the vertical and lateral car body accelerations calculated by VTM and TTM are shown in Fig. 11, and the PSDs of the vertical and lateral wheel/rail forces are shown in Fig. 12. In these figures, the leading and trailing vehicles mean the 1st and 8th vehicles of the train, respectively, and the 4th vehicle is taken as the middle vehicle.

Figure 11 shows significant difference occurs on vertical accelerations of the car body center upper the bogie for frequencies below 3 Hz, while 4 Hz for lateral accelerations, calculated by the two types of dynamic models, whereas the difference is small at higher frequencies due to the dominant low-frequency vibration of the rigid car body model. From Fig. 11, it can be found that the car body PSD responses obtained by VTM are much higher than those obtained by TTM, especially in the frequency range of 1–3 Hz. The reason for this phenomenon is that the tight-lock inter-vehicle connections between the adjacent vehicles of the train effectively restrain the relative motion of the neighboring vehicle ends, including the vertical, lateral, pitching, and yawing motions of the vehicles. The role of the tight-lock inter-vehicle connections can be characterized in TTM. But in VTM, the two ends of the car body are considered to be free. In this situation, the motions at the ends of the vehicle calculated by VTM are larger than those calculated by TTM,

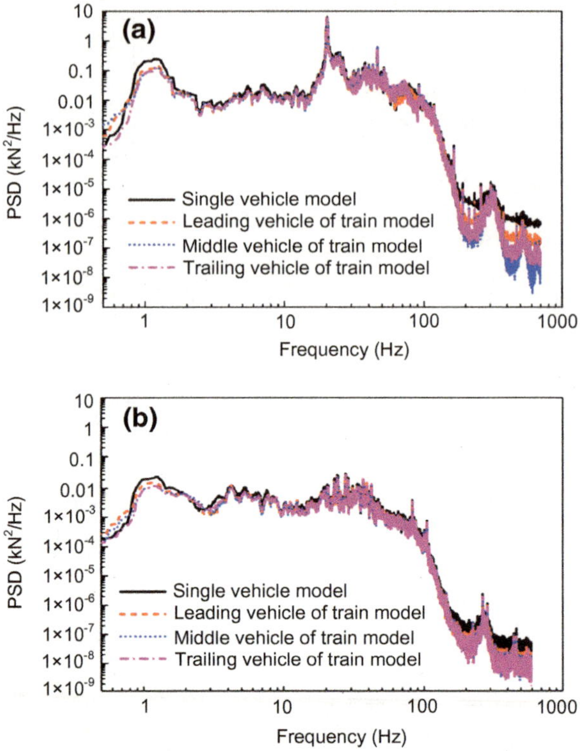


**Fig. 12** Comparison of wheel/rail force vibration frequency components: **a** wheel/rail vertical force PSD; **b** wheel/rail lateral force PSD

especially at low frequencies. From Fig. 11, it can also be seen that the PSD of the middle car is lower than that of the leading car and trailing car, especially at 1–3 Hz, as shown in Fig. 11b. For vertical car body acceleration, the peak response quite often occurs in the trailing car, while the greatest lateral acceleration of the car body is found in the leading car.

Figure 12 indicates the PSDs of the vertical and lateral wheel/rail forces of the first left wheel achieved by VTM and TTM. From Fig. 12, it can be seen that there is a little difference between the wheel/rail vertical and lateral forces calculated by the two models in the frequency range below 100 Hz, but there is a significant difference at higher frequencies. The wheel/rail force PSD obtained from VTM is larger than that obtained from TTM in the high-frequency range. These differences are caused by the wave reflections between the wheels. Wu and Thompson (2002) pointed out that there is a big difference between the wheel/rail contact forces in the frequency region of 550–1200 Hz obtained by a multiple-wheel/rail interaction model and a single-wheel/rail interaction model due to the effect of wave reflections between the wheels.

This explanation is also appropriate for the results of Fig. 12. The first wheelset of VTM receives wave reflections from three other wheelsets, while the leading wheelset of TTM receives reflections from 31 other wheelsets. These wave reflections between wheels would make the responses of wheel/rail interaction calculated by VTM and TTM differently. The wheel/rail force PSD of the leading car is larger than that of the middle car and trailing car in the frequency range of more than 100 Hz. The vertical wheel/rail PSD of the middle car is the smallest, compared to those of the leading and trailing cars.

The comparison shown in Figs. 11 and 12 clearly indicates that there is a significant difference in the dynamic behavior characteristics of the vehicles characterized by VTM and TTM. The vehicle location also has an important influence on the dynamic behavior. It is important to consider the vehicle location and the cross-influence of adjacent vehicles in the analysis of vertical and lateral car body accelerations in the frequency range below 20 Hz and the wheel/rail force variations at high frequencies.

## *4.2 Comparison of Ride Comfort*

The ride comfort, one of the key dynamic performance targets of high-speed trains, is closely related to the vibration characteristics of the car body in the low-frequency range. The analysis in Sect. 4.1 indicates that the vibration frequency components of the car body in the frequency range below 20 Hz obtained by VTM and TTM are very different, which means the ride comforts calculated by the two types of dynamic models are different. To clarify this difference, a comparison of ride comfort performance is carried out in this section. In this calculation, the tangent ballast track was used, and the operating speed ranged from 200 to 400 km/h. Other parameters were the same as those used in Sect. 4.1. The comparison results of the lateral and vertical Sperling's comfort indices are shown in Fig. 13.

From Fig. 13a, it can be clearly seen that the lateral Sperling's comfort index calculated by VTM is larger than that calculated by TTM in all speed ranges. The maximum difference in the results between the single-vehicle model and the middle vehicle and the leading vehicle reach 0.25 and 0.11, respectively. The difference between the two types of dynamic models increases with increasing train speed. When the running speed reaches 400 km/h, the maximum lateral Sperling's comfort indices of the leading vehicle, middle vehicle, and the trailing vehicle, calculated by TTM, are 2.42, 2.28, and 2.32, respectively. However, the maximum lateral Sperling's comfort index of VTM reaches 2.53, which is greater than the comfort index limit value of the 'Excellent grade' used in Chinese Railways (SAC 1985). It means that the lateral comfort of high-speed trains would be overestimated by VTM in practical engineering application. Thus, when the lateral comfort of high-speed trains is investigated though numerical simulation, using TTM is more reasonable. The vehicle location also has a great influence on the ride comfort. Among the three vehicles compared, the lateral comfort index of the middle car is the smallest in the speed range, and the ride comfort of the leading car is the worst.

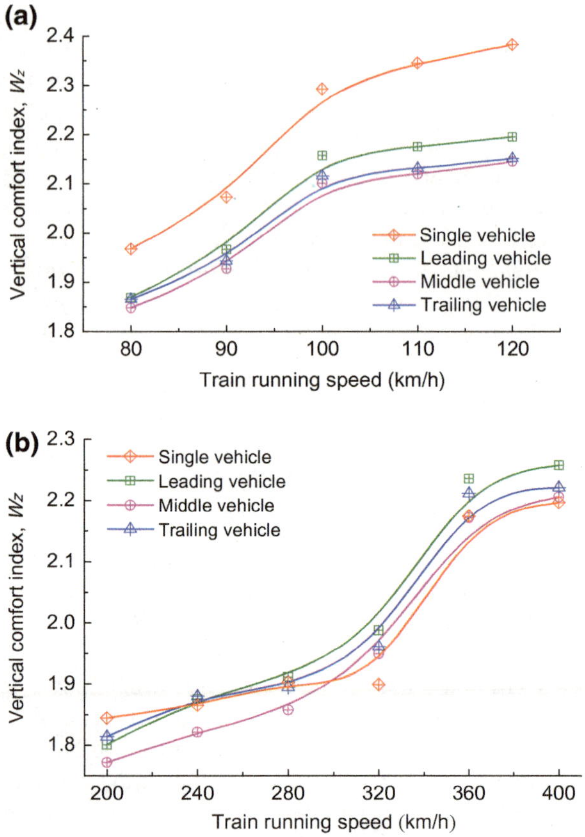


**Fig. 13** Comparison of ride comfort: **a** lateral Sperling's comfort index; **b** vertical Sperling's comfort index

Compared to the obvious difference of the lateral comfort indexes calculated by the two models, the difference of the vertical comfort indexes is not so significant, as shown in Fig. 13b. From the comparison results of the vertical comfort index, it can be concluded that VTM is appropriate for analyzing the vertical comfort index of the vehicles when a long high-speed train operates on a tangent track without serious irregularities, such as corrugated rails, rail welding dips, and track subsidence. However, it can be expected that if the track irregularity is severe, the difference of the vertical ride comfort when using these two models would be large. Furthermore, the operating speed has a great influence on the ride comfort. With increasing speed, the differences in the lateral and vertical Sperling's comfort indices calculated by the two models increase rapidly.

### 4.3 Comparison of Curving Performance

When a high-speed train negotiates a curved track, large lateral forces are generated between the wheels and rails. These large lateral forces, in combination with the small vertical forces, may cause wheel climbing and rail rollover as the train negotiates the curve. Therefore, curving performance is very important for evaluating the running safety of high-speed trains. In this section, the curving performances obtained by TTM and VTM are compared. The curved track had a circle curve radius of 9000 m, a transition curve length of 490 m, a circle curve length of 400 m, and a superelevation of 125 mm. The running speed of the train ranged from 200 to 400 km/h. The track irregularities and other concerned parameters are the same as in Sect. 4.1.

To evaluate curving performance, two safety criteria used in Chinese Railway were selected. One is the derailment coefficient (or Nadal coefficient) (SAC 1985) defined as the ratio of the lateral force to the total vertical force on the same wheel. The other is the wheel load reduction, which is defined as the ratio of the reduction in the vertical dynamical forces on both wheels of a wheelset to the total vertical wheelset loading. The total vertical force is the sum of the static wheel load and the dynamic vertical force on the same wheel. The safety limit values of both derailment coefficient and wheel load reduction are 0.8 in the evaluation of the operating safety of high-speed trains in China. TTM and VTM are used to calculate the two safety criteria when the train passes over curved track at different speeds. The calculated results are compared and discussed as follows.

Figure 14 shows the maximum values of the dynamic derailment coefficient and wheel loading reduction of all the wheelsets calculated by VTM and TTM. As expected, the derailment coefficient and wheel loading reduction increase as the train speed increases. When the train speed is greater than 350 km/h, the maximum wheel loading reduction is greater than its safety limit value, 0.8. This means that the running speed of the high-speed train should be limited when it is negotiating a curved track.

From Fig. 14, it can also be seen that the interaction of neighboring vehicles and the vehicle location have a large effect on the derailment coefficient, but their effects on wheel unloading are not significant due to the large radius of the curved track. Figure 14a illustrates the great difference of derailment coefficients calculated by the two models. The derailment coefficient calculated by VTM is much larger than that calculated by TTM in all the analyzed speed ranges. Specifically, the derailment coefficient calculated using VTM is larger than reality when the train passes over the curved track. The maximum difference occurs between the middle-vehicle and the leading-vehicle models, which are calculated by VTM and TTM, respectively. Compared to the results of the leading and trailing vehicles of the same train, the derailment coefficient of the middle vehicle is the smallest. Note that the difference of the results obtained with the two models increases with increasing operating speed. On the other hand, Fig. 14b shows a good agreement between the wheel load reductions calculated by the two models in all the analyzed speed ranges

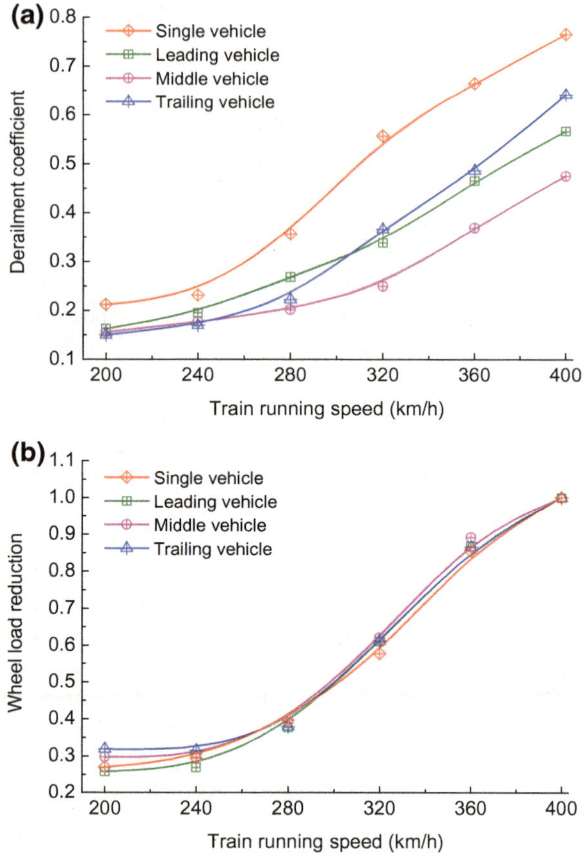


**Fig. 14** Comparison of dynamic performances on a large radius curved track: **a** derailment coefficient; **b** wheel load reduction

under the present curved track conditions. However, it can be predicted that if the radius of the curved track is small, the difference of wheel load reductions calculated by the two models would be large.

The above results discussed show that the vertical comfort indexes on the tangent track and the wheel load reduction on large radius curved tracks calculated by VTM and TTM are close. However, if the operating environment is bad or the radius of the curved track is small, how much would be the difference between the two types of dynamic models? To measure that difference, a comparison of the dynamic responses on a small radius curved track obtained by TTM and VTM is carried out. The curved track has a circle curve radius of 600 m, a transition curve length of 100 m, a circle curve length of 280 m, and a cant of 100 mm. The operating speed of the train ranges from 80 to 120 km/h, and other concerned parameters used in this numerical simulation are the same as in Sect. 4.1. Figure 15 shows the results of vertical comfort indexes and wheel load reductions calculated by VTM and TTM, respectively.

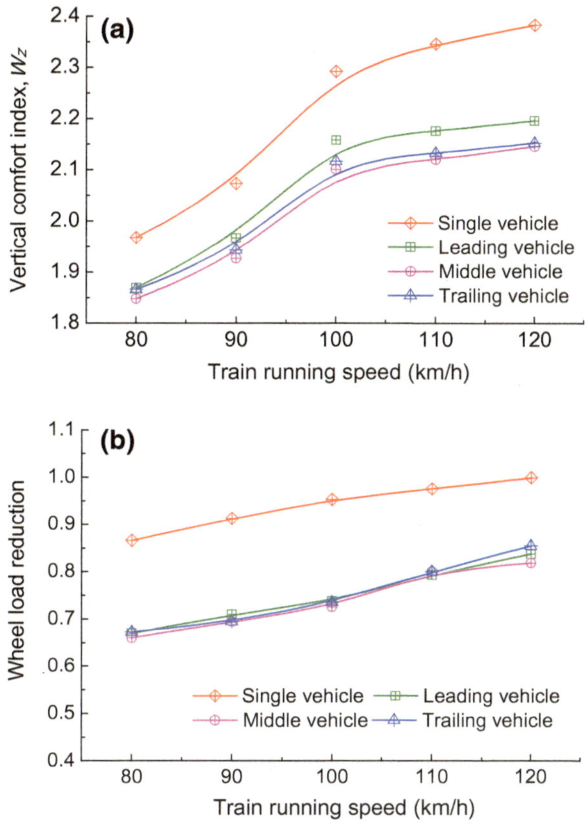


**Fig. 15** Comparison of dynamic performance on a small radius curve track: **a** vertical Sperling's comfort indices; **b** wheel load reductions

The difference in the dynamical behavior calculated by the two models is evident for a train operating on a curved track with a relatively small radius. The dynamical behavior of the different vehicles of the same train calculated by TTM is also different under the same operating conditions. From Fig. 15a, the differences of vertical comfort indexes of these vehicles increase with increasing operating speed. Figure 15b shows that the wheelset load reductions of the vehicles approach to 1 with increasing operating speed. This is because the speed increase causes the normal load between the wheels and the low rail reduces to zero; that is to say, the wheels lose contact with the low rail.

Through the detailed comparisons of the results obtained by VTM and TTM, it is noticeable that the dynamical behavior of the vehicle/track system calculated by VTM will be overstated, and it is more reasonable that TTM is used to calculate the dynamic behavior of the train and the track, especially in the situation of trains with strong lateral and vertical vibrations. Since the neighboring vehicles of a train influence each other and each vehicle has different boundary conditions, the dynamic behavior of each vehicle is different from the others in the same train. Therefore, it is necessary that a 3D dynamic model of a train coupled with a flexible track is carried forward to estimate the dynamical behavior of the train and the track in high-speed operations.

## 5 Conclusions

A 3D dynamic model of a nonlinear high-speed train coupled with a flexible ballast track is put forward. The advantages of this model are as follows: (1) The mutual influence of the adjacent vehicles on the dynamic behavior of high-speed vehicles and the track is considered; (2) it is possible to carry out fast dynamics calculations on a long train running on an infinitely long flexible track. The reliability of the 3D coupled train/track model was verified through a detailed numerical comparison with the commercial software SIMPACK, and the difference caused by the track modeling was then analyzed. Several key dynamic performances, including vibration frequency components, ride comfort, and curving behavior, obtained by TTM and VTM, are compared and discussed. Subsequently, the following conclusions were reached:

1. There is a distinct difference in the vibration frequency components calculated between VTM and TTM. The inter-vehicle connections of a train have an important influence on the dynamic behavior of a car body in the frequency range below 20 Hz and the wheel/rail forces at high frequencies.
2. The lateral comfort index calculated by VTM is greater than that calculated by TTM, which can be predicted. Therefore, in practical engineering applications, using TTM is more reasonable. The vertical comfort indexes obtained by the two models are close when the train operates on a curved track of large radius, but the difference is very large when the train operates on a small radius curved track.
3. The difference of derailment coefficients obtained by the two models is very large when the train negotiates curved tracks with large radii. It is obvious that the derailment coefficient is overestimated using VTM, and using TTM is more reasonable in practical engineering applications. The wheel load reductions obtained by the two models have a good agreement when the train operates on a curved track with a large radius. If the radius of the curved track is small, the difference is obvious.
4. The difference in lateral dynamic behavior is relatively large when looking at different vehicle locations in a high-speed train, but the difference in vertical dynamic performance is relatively small when a high-speed train operates on a usually tangent track. Among the vehicles of a long train, the results calculated by TTM show that the ride comfort and curving performance of the intermediate vehicles are better than those of the leading and trailing vehicles because the two ends of the intermediate vehicles are restrained by their neighbors.

## References


Arnold, M., Burgermeister, B., Führer, C., et al. (2011). Numerical methods in vehicle system dynamics: State of the art and current developments. *Vehicle System Dynamics, 49*(7), 1159–1207. doi:10.1080/00423114.2011.582953.

Baeza, L., & Ouyang, H. (2011). A railway track dynamics model based on modal substructuring and a cyclic boundary condition. *Journal of Sound and Vibration, 330*(1), 75–86. doi:10.1016/j.jsv.2010.07.023.

Cai, Y., Sun, H., & Xu, C. (2008). Response of railway track system on poroelastic half-space soil medium subjected to a moving train load. *International Journal of Solids and Structures, 45*(18–19), 5015–5034. doi:10.1016/j.ijsolstr.2008.05.002.

Chen, G., & Zhai, W. M. (2004). A new wheel/rail spatially dynamic coupling model and its verification. *Vehicle System Dynamics, 41*(4), 301–322. doi:10.1080/00423110412331315178.

Di Gialleonardo, E., Braghin, F., & Bruni, S. (2012). The influence of track modelling options on the simulation of rail vehicle dynamics. *Journal of Sound and Vibration, 331*(19), 4246–4258. doi:10.1016/j.jsv.2012.04.024.

Evans, J., & Berg, M. (2009). Challenges in simulation of rail vehicle dynamics. *Vehicle System Dynamics, 47*(8), 1023–1048. doi:10.1080/00423110903071674.

Fröhling, R. D. (1998). Low frequency dynamic vehicle/track interaction: Modelling and simulation. *Vehicle System Dynamics, 29*(S1), 30–46. doi:10.1080/00423119808969550.

Garg, V. K., & Dukkipati, R. V. (1984). *Dynamics of railway vehicle systems*. Canada: Academic Press.

Jin, X. S., Wu, P. B., & Wen, Z. F. (2002). Effects of structure elastic deformations of wheelset and track on creep forces of wheel/rail in rolling contact. *Wear, 253*(1–2), 247–256. doi:10.1016/S0043-1648(02)00108-4.

Jin, X. S., Wen, Z. F., Zhang, W. H., et al. (2005). Numerical simulation of rail corrugation on curved track. *Computers & Structures, 83*(25–26), 2052–2065. doi:10.1016/j.compstruc.2005.03.012.

Jin, X. S., Wen, Z. F., Wang, K. W., et al. (2006). Three-dimensional train-track model for study of rail corrugation. *Journal of Sound and Vibration, 293*(3–5), 830–855. doi:10.1016/j.jsv.2005.12.013.

Jin, X. S., Xiao, X. B., Ling, L., et al. (2013). Study on safety boundary for high-speed trains running in severe environments. *International Journal of Rail Transportation, 1*(1–2), 87–108. doi:10.1080/23248378.2013.790138.

Ju, S. H., & Li, H. C. (2011). Dynamic interaction analysis of trains moving on embankments during earthquakes. *Journal of Sound and Vibration, 330*(22), 5322–5332. doi:10.1016/j.jsv.2011.05.032.

Kalker, J. J. (1967). *On the rolling contact of two elastic bodies in the presence of dry friction*. Ph. D. thesis. The Netherlands: Delft University.

Knothe, K., & Grassie, S. L. (1993). Modeling of railway track and vehicle/track interaction at high frequencies. *Vehicle System Dynamics, 22*(3–4), 209–262. doi:10.1080/00423119308969027.

Lei, X. Y., & Mao, L. J. (2004). Dynamic response analyses of vehicle and track coupled system on track transition of conventional high speed railway. *Journal of Sound and Vibration, 271*(3–5), 1133–1146. doi:10.1016/S0022-460X(03)00570-4.

Nielsen, J. C., & Igeland, A. (1995). Vertical dynamic interaction between train and track-influence of wheel and track imperfections. *Journal of Sound and Vibration, 187*(5), 825–839. doi:10.1006/jsvi.1995.0566.

Oscarsson, J., & Dahlberg, T. (1998). Dynamic train/track/ballast interaction—Computer models and full-scale experiments. *Vehicle System Dynamics, 29*(S1), 73–84. doi:10.1080/00423119808969553.

Popp, K., Kruse, H., & Kaiser, I. (1999). Vehicle-track dynamics in the mid-frequency range. *Vehicle System Dynamics, 31*(5–6), 423–464. doi:10.1076/vesd.31.5.423.8363.

SAC (Standardization Administration of the People's Republic of China). (1985). *Railway vehicles—Specification for evaluation the dynamic performance and accreditation test, GB/T 5599-85*. China: SAC. (in Chinese).

Shen, Z. Y., Hedrick, J. K., & Elkins, J. A. (1983). A comparison of alternative creep-force models for rail vehicle dynamic analysis. *Vehicle System Dynamics, 12*(1–3), 79–83. doi:10.1080/00423118308968725.

Sun, Y. Q., & Dhanasekar, M. (2002). A dynamic model for the vertical interaction of the rail track and wagon system. *International Journal of Solids and Structures, 39*(5), 1337–1359. doi:10.1016/S0020-7683(01)00224-4.

Sun, Y. Q., Dhanasekar, M., & Roach, D. (2003). A three-dimensional model for the lateral and vertical dynamics of wagon-track systems. *Proceedings of the Institution of Mechanical Engineers, Part F: Journal of Rail and Rapid Transit, 217*(1), 31–45. doi:10.1243/095440903762727339.

Tanabe, M., Matsumoto, N., Wakui, H., et al. (2008). A simple and efficient numerical method for dynamic interaction analysis of a high-speed train and railway structure during an earthquake. *Journal of Computational and Nonlinear Dynamics, 3*(4), 041002. doi:10.1115/1.2960482.

Wang, K. (1984). The track of wheel contact points and the calculation of wheel/rail geometric contact parameters. *Journal of Southwest Jiaotong University, 19*(1), 88–99. (in Chinese).

Wu, T. X., & Thompson, D. J. (2002). Behaviour of the normal contact force under multiple wheel/rail interaction. *Vehicle System Dynamics, 37*(3), 157–174. doi:10.1076/vesd.37.3.157.3533.

Xia, H., Zhang, N., & Roeck, G. D. (2003). Dynamic analysis of high-speed railway bridge under articulated trains. *Computers & Structures, 81*(26–27), 2467–2478. doi:10.1016/S0045-7949(03)00309-2.

Xiao, X. B., Jin, X. S., & Wen, Z. F. (2007). Effect of disabled fastening systems and ballast on vehicle derailment. *Journal of Vibration and Acoustics, 129*(2), 217–229. doi:10.1115/1.2424978.

Xiao, X. B., Jin, X. S., Wen, Z. F., et al. (2011). Effect of tangent track buckle on vehicle derailment. *Multibody System Dynamics, 25*(1), 1–41. doi:10.1007/s11044-010-9210-2.

Xiao, X. B., Ling, L., & Jin, X. S. (2012). A study of the derailment mechanism of a high speed train due to an earthquake. *Vehicle System Dynamics, 50*(3), 449–470. doi:10.1080/00423114.2011.597508.

Xiao, X. B., Ling, L., Xiong, J. Y., et al. (2014). Study on the safety of operating high-speed railway vehicles subjected to crosswinds. *Journal of Zhejiang University-Science A (Applied Physics & Engineering), 15*(9), 694–710. doi:10.1631/jzus.A1400062.

Yang, Y. B., & Wu, Y. S. (2002). Dynamic stability of trains moving over bridges shaken by earthquakes. *Journal of Sound and Vibration, 258*(1), 65–94. doi:10.1006/jsvi.2002.5089.

Zhai, W. M. (1996). Two simple fast integration methods for large-scale dynamic problems in engineering. *International Journal for Numerical Methods in Engineering, 39*(24), 4199–4214. doi:10.1002/(SICI)1097-0207(19961230)39:24<4199:AID-NME39>3.3.CO;2-P.

Zhai, W. M., Cai, C. B., & Guo, S. Z. (1996). Coupling model of vertical and lateral vehicle/track interactions. *Vehicle System Dynamics, 26*(1), 61–79. doi:10.1080/00423119608969302.

Zhai, W. M., Wang, K. Y., & Cai, C. B. (2009). Fundamentals of vehicle-track coupled dynamics. *Vehicle System Dynamics, 47*(11), 1349–1376. doi:10.1080/00423110802621561.

Zhang, S. G. (2009). *Design method of high-speed train*. Beijing, China: Chinese Railway Press. (in Chinese).

Zhou, L., & Shen, Z. Y. (2013). Dynamic analysis of a high-speed train operating on a curved track with failed fasteners. *Journal of Zhejiang University-Science A (Applied Physics & Engineering), 14*(6), 447–458. doi:10.1631/jzus.A1200321.

## Author Biography

**Xue-song Jin** Ph.D., is a professor at Southwest Jiaotong University, Chengdu, China. He is a leading scholar of wheel–rail interaction in China and is a State Council Expert for Special Allowance. He is the author of three academic books and over 200 articles. He is an editorial board member of several journals and has been a member of the Committee of the International Conference on Contact Mechanics and Wear of Rail/Wheel Systems for more than 10 years. He has been a visiting scholar at the University of Missouri-Rolla for 2.5 years. Now, his research focuses on wheel–rail interaction, rolling contact mechanics, vehicle system dynamics, vibration, and noise.

# Effect of the First Two Wheelset Bending Modes on Wheel–Rail Contact Behavior

**Shuo-qiao Zhong, Jia-yang Xiong, Xin-biao Xiao, Ze-feng Wen and Xue-song Jin**

## Contents



## 1 Introduction

High-speed railways are currently popular globally. However, there are some problems including passenger riding comfort, noise pollution, and even operational safety (Jin et al. 2013). Rail corrugation, rail welding irregularity, wheel burning, and wheel out-of-roundness (OOR) generate high-frequency components of the dynamic wheel–rail contact forces that contribute significantly to the total wheel–rail contact forces (Nielsen et al. 2003), and reduce the life of the components of track and vehicle, such as wheels, rails, and fasteners. Rail grinding and wheel reprofiling are the most common measures that have been proved to be effective in controlling rail irregularities and wheel OOR. However, these measures lead to notably high maintenance costs. A lot of measurements at the sites and coupling vehicle–track

S. Zhong (✉) · J. Xiong · X. Xiao · Z. Wen · X. Jin
State Key Laboratory of Traction Power, Southwest Jiaotong University, Chengdu 610031, China
e-mail: zhongsq1234@163.com

X. Jin
e-mail: xsjin@home.swjtu.edu.cn



Y. Fang and Y.H. Zhang (eds.), *China's High-Speed Rail Technology*, Advances in High-speed Rail Technology, https://doi.org/10.1007/978-981-10-5610-9_19

dynamics modeling have been carried out to investigate the mechanism and development of these phenomena. In the vehicle–track dynamics modeling, a rigid multi-body system is often adopted to simulate railway vehicles, based on several commercial codes available for the low-frequency domain, such as GENSYS, NUCARS, SIMPACK, and VAMPIRE. These computer programs are generally used to analyze railway vehicle dynamics responses at frequencies below 20 Hz, where the influence of rigid motions of the vehicle on wheel–rail contact forces is dominant (Nielsen et al. 2005). To analyze the vehicle dynamic responses at mid- and high frequencies, the vehicle structural flexibility should be taken into account in the modeling. It is obvious that wheelset structural flexibility has an influence on wheel–rail contact behaviors at mid- and high frequencies. Different flexible wheelset models have been set up due to various motivations in the past (Chaar 2007).

The methods applied to modeling flexible wheelset can be summarized as three major categories (Chaar 2007). The first is a lumped model developed in a simple and convenient way, in which a wheelset is divided into several parts interconnected with springs and dampers. This model can describe the bending and torsional motions of the wheelset with only a few degrees of freedom, which could not be applied to studying wear phenomena on wheel treads or rails (Popp et al. 1999). The second is a continuous model developed by Szolc (1998a, b), in which the wheelset axle was modeled as a beam, and two wheels and brake disks were modeled as rigid rings attached to the axle through a massless, elastically isotropic membrane. The model can characterize the wheelset dynamic behavior in the frequency range of 30–300 Hz. In the model proposed by Popp et al. (2003), the wheelset axle was considered as a 1D continuum, having the properties of a bar, a torsional rod, and a Rayleigh beam. The wheel was considered as a 2D continuum, having the properties of a disk and a Kirchhoff plate. The third was developed based on finite element method (FEM) , which simulates wheelset flexibility more realistically than the first two categories of model. The wheelset modes and corresponding natural frequencies were obtained through the modal analysis of the finite element (FE) model by using the commercial software, and they were input into the simulation by means of the commercial codes (SIMPACK, NEWEUL) (Meinders and Meinker 2003) or some non-commercial multi-body dynamic system codes. The non-commercial code developed by Fayos et al. (2007) and Baeza et al. (2008, 2011) introduced the Eulerian coordinate system to replace the Lagrangian coordinate system in the flexible wheelset modeling. In this way, it is convenient to obtain the motion of fixed physical nodes and consider the inertial effect due to wheelset rotation. Relying on current computing power, it is feasible to use FEM to consider the effect of flexible wheelset in modeling a railway vehicle coupling with a track.

Regarding the wheel–rail contact treatment in considering flexible wheelset influence, wheel–rail rolling contact condition is simplified based on different prior assumptions, especially in the detection of wheel–rail contact points. This is the prerequisite for the calculation of wheel–rail creepages and contact forces. Baeza et al. (2011) neglected the effect of the high-frequency deformation and the deviation of a rotating flexible wheelset rolling over a flexible track model on the wheel–rail contact point in the investigation into the effect of the rotating flexible wheelset on rail corrugation. Through the detailed calculation, Kaiser and Popp (2006) found that the contact point was in the location where the wheel and the rail had positive penetration maxima, and the penetration direction was orthogonal to the common tangent plane of the wheel and the rail before their deformations. A linear wheel–rail contact model was proposed and used to carry out the detection of wheel–rail contact point and the contact zone's normal direction (Andersson and Abrahamsson 2002). In the detection, the functions were created using a first-order Taylor expansion around a reference state described by a group of parameters which represent a configuration, in which the train was in static equilibrium and the wheel and the track were free from geometric imperfections. The advantage of this approach is that the contact position and orientation in each time step can be calculated by interpolation replacing iterations, which results in a low computational cost. But the approach is only suitable for the case that the effect of all the parameters is very small on the contact point position and the contact patch orientation around the references is in static equilibrium. The wheel–rail contact point position and the contact patch orientation greatly depend on parameters, such as the curvatures of wheel and rail. In (Torstensson et al. 2012; Torstensson and Nielsen 2011), the contact point detection was done before the simulation and used in the subsequent time integration analysis in the form of lookup table. The commercial software GENSYS allows for such calculations using the preprocessor KPF (from Swedish contact point function). In the KPF, the location and orientation of the contact patch were assumed to be dependent only on the relative displacement in the lateral direction between the wheelset and the rails, and hence, the influence of the wheelset yaw angle was not taken into account. In some other papers, detailed discussions on the wheel–rail contact model were omitted. In this study, the wheel–rail contact model considering the effect of wheelset flexibility (Zhong et al. 2013, 2014) is further improved and the new contact model is suitable for the analysis on the effect of the local higher-frequency deformation of the wheels on the wheel–rail contact behavior.

## 2 Vehicle–Track Coupling Dynamic System

A flexible wheelset model (to be illustrated in Sect. 2.1) and a suitable wheel–rail contact model (to be discussed in Sect. 2.2) are integrated into the vehicle–track coupling dynamic system model. All parts of the vehicle system, except for its four wheelsets, are considered as rigid bodies. The primary and secondary suspension systems of the vehicle are modeled with spring–damper elements. A triple-layer model of discrete elastic support is adopted to simulate the ballasted track. The rails are modeled as Timoshenko beams. The sleepers are modeled as rigid bodies, and the ballast model consists of discrete equivalent masses. The equivalent spring–damper elements are used as the connections between the rails and the sleepers, the sleepers and the equivalent ballast bodies, and the ballast bodies and the roadbed. Figure 1 shows the vehicle–track coupling dynamic system model. The equations

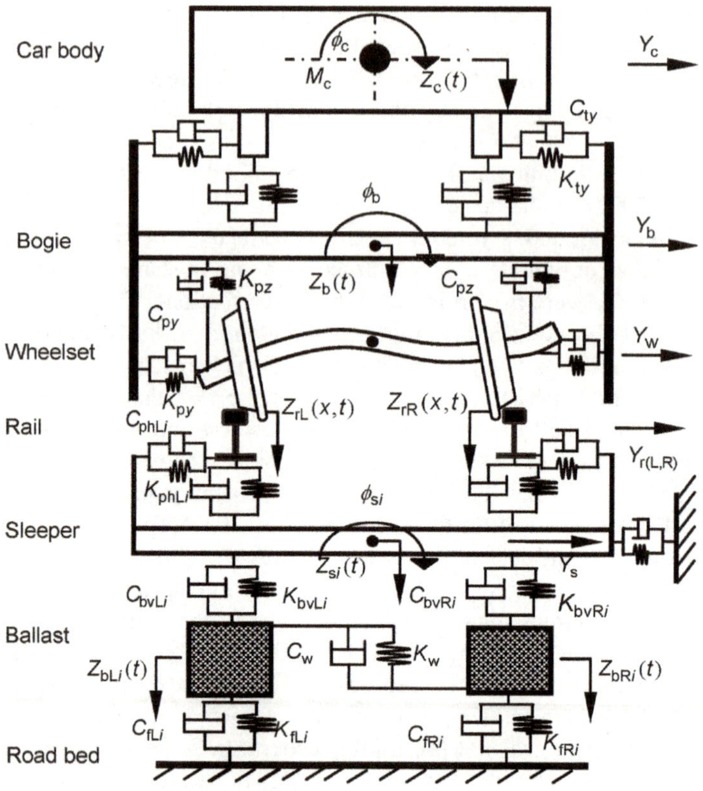


**Fig. 1** Vehicle–track coupling model (elevation)

of motion of each component of the vehicle excluding wheelsets and the track are illustrated in detail in (Xiao et al. 2007, 2008, 2010). The parameters and their values describing the dynamic models are given in Appendix A.

## *2.1 Flexible Wheelset Model*

The wheelset structural flexibility is considered by modeling the wheelset axle as an Euler–Bernoulli beam in two planes, one perpendicular to the track centerline and the other parallel to the track level. The crossing effect of the bending deformations in the two planes is ignored. In the first two bending modes obtained using the modal analysis of the FE model of a wheelset, two wheels have little deformation (Fig. 2), and their frequencies are in the available frequency range (0–500 Hz) of an Euler–Bernoulli beam model. Therefore, two wheels can be treated as rigid bodies in this study.

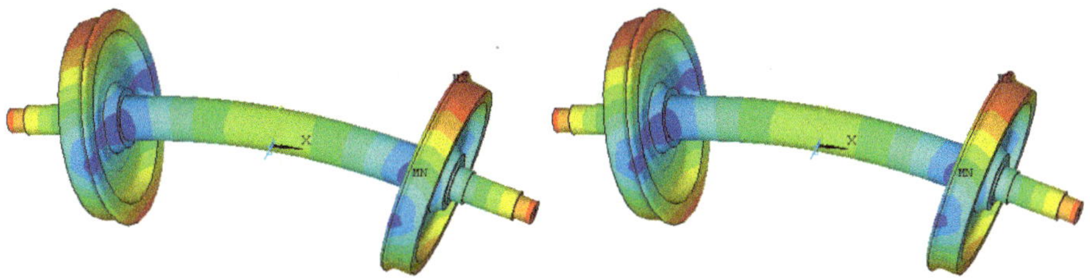

Fig. 2 First two bending modes obtained using FE model

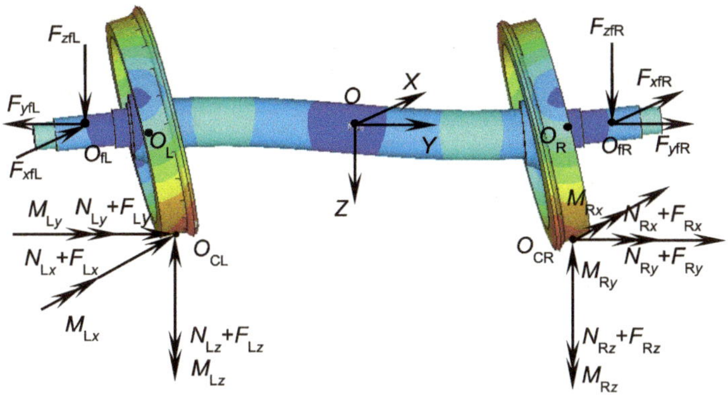


Fig. 3 Force analysis diagram of the flexible wheelset

There are two force systems acting on the wheelset, one is the wheel–rail contact forces and the other is the forces of the primary suspension system (Fig. 3).

In Fig. 3, $O_{\mathrm{fL}}$ and $O_{\mathrm{fR}}$ are the left and right points on the axle, respectively, where the primary suspension force systems are applied. $O_{\mathrm{CL}}$ and $O_{\mathrm{CR}}$ are the left and right contact points of wheel–rail, respectively. $O$ indicates the origin of the coordinate system $O$-$XYZ$ that is a coordinate system with a translational motion along the tangent track centerline at operational speed. If the speed is constant, this coordinate system is an inertial coordinate system and therefore regarded as an absolute coordinate system (geodetic coordinate system).

To analyze the axle's deformation, the force systems from wheel–rail interaction acting on the left and right wheel treads are translated to the nominal circle centers $O_{\mathrm{L}}$ and $O_{\mathrm{R}}$, respectively, and extra moments are produced in the procedure of translating contact forces. Thus, the force systems acting on the axle in the two planes are obtained in Fig. 4.

The notations of the variables and symbols are defined in Table 1. The subscript p denotes the primary suspension, the subscripts $x$, $y$, and $z$ denote $X$-, $Y$-, and $Z$-direction, respectively, and $A$ denotes the axle.

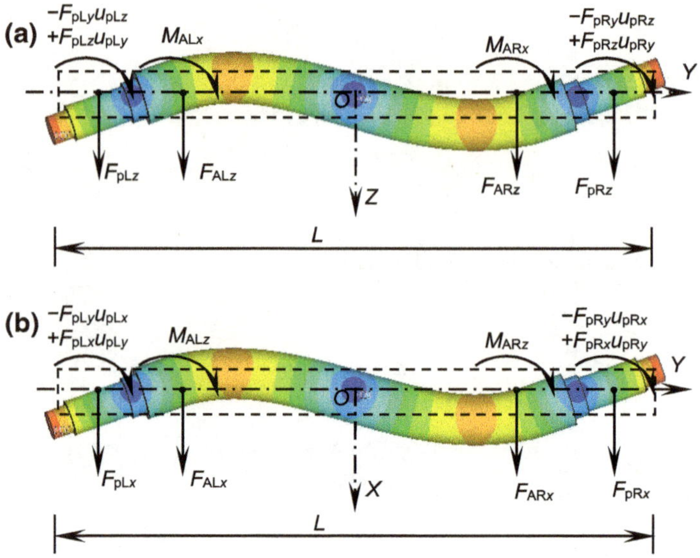


**Fig. 4** Force analysis diagram in the plane *O-YZ* (**a**) and in the *O-XY* plane (**b**)

The differential equation for the flexural vibration of an Euler–Bernoulli beam (the axle) in the plane *O-YZ* is written as

$$\mathrm{EI}_x \frac{\partial^4 u_z(y,t)}{\partial y^4} + \rho A \frac{\partial^2 u_z(y,t)}{\partial t^2} = Q_z(y,t) - \frac{\partial M_x(y,t)}{\partial y}, \tag{1}$$

where

$$Q_z(y,t) = F_{\mathrm{ALz}}\delta(y - y_{\mathrm{wL}}) + F_{\mathrm{pLz}}\delta(y - y_{\mathrm{pL}}) + F_{\mathrm{ARz}}\delta(y - y_{\mathrm{wR}}) + F_{\mathrm{pRz}}\delta(y - y_{\mathrm{pR}}), \tag{2}$$

$$\begin{aligned} M_x(y,t) = {} & M_{\mathrm{AL}x}\delta(y - y_{\mathrm{wL}}) \\ & + (-F_{\mathrm{pL}y}u_{\mathrm{pL}z} + F_{\mathrm{pL}z}u_{\mathrm{pL}y})\delta(y - y_{\mathrm{pL}}) \\ & + M_{\mathrm{AR}x}\delta(y - y_{\mathrm{wR}}) \\ & + (-F_{\mathrm{pR}y}u_{\mathrm{pR}z} + F_{\mathrm{pR}z}u_{\mathrm{pR}y})\delta(y - y_{\mathrm{pR}}). \end{aligned} \tag{3}$$

The force analysis diagram of the two wheels including the D'Alembert forces is shown in Fig. 5, based on which differential equations of motion of the two wheels are written as

$$m_{\mathrm{w}} \frac{\partial^2}{\partial t^2} u_z(y_{\mathrm{w(L,R)}}, t) = m_{\mathrm{w}} g - F_{\mathrm{A(L,R)}z} - F_{\mathrm{wr(L,R)}z}, \tag{4}$$

**Table 1** Notations of the variables

| Variable | Explanation |
|---|---|
| $u_{pLz}$, $u_{pRz}$ | *Z*-direction components of the displacements of the nodes where the left and right primary suspension forces are applied on the axle, respectively |
| $u_{pLy}$, $u_{pRy}$ | *Y*-direction components of the displacements of the nodes where the left and right primary suspension forces are applied on the axle, respectively |
| $u_{pLx}$, $u_{pRx}$ | *X*-direction components of the displacements of the nodes where the left and right primary suspension forces are applied on the axle, respectively |
| $L$ | Length of the wheelset axle |
| $F_{pLx}$, $F_{pLy}$, $F_{pLz}$ | *X*-, *Y*-, and *Z*-direction components of the primary suspension forces on the left sides of a wheelset |
| $F_{pRx}$, $F_{pRy}$, $F_{pRz}$ | *X*-, *Y*-, and *Z*-direction components of the primary suspension forces on the right sides of a wheelset |
| $F_{ALx}$, $F_{ALy}$, $F_{ALz}$ | *X*-, *Y*-, and *Z*-direction components of the forces between the left wheel and the axle of a wheelset |
| $F_{ARx}$, $F_{ARy}$, $F_{ARz}$ | *X*-, *Y*-, and *Z*-direction components of the forces between the right wheel and the axle of a wheelset |
| $M_{ALx}$, $M_{ALz}$ | *X*- and *Z*-direction components of the moments between the left wheel and the axle of a wheelset |
| $M_{ARx}$, $M_{ARz}$ | *X*- and *Z*-direction components of the moments between the right wheel and the axle of a wheelset |
| $E$ | Young's modulus |
| $I_x$ | Cross-sectional area moment of inertia about the X-axis |
| $I_z$ | Cross-sectional area moment of inertia about the *Z*-axis |
| $t$ | Time |
| $u_z(y,t)$, $u_x(y,t)$ | *X*- and *Z*-direction components of the displacements of the nodes on the axle at time *t*, respectively |
| $Q_z(y,t)$, $Q_x(y,t)$ | *X*- and *Z*-direction components of the forces on the axle at time *t*, respectively |
| $M_z(y,t)$, $M_x(y,t)$ | *X*- and *Z*-direction components of the moments on the axle at time *t*, respectively |
| $m_w$ | Mass of a wheel |
| $g$ | Gravity acceleration |
| $\boldsymbol{a}_{Lz}$, $\boldsymbol{a}_{Rz}$ | *Z*-direction components of the accelerations of the left and right wheels, respectively |
| $\boldsymbol{a}_{Lx}$, $\boldsymbol{a}_{Rx}$ | *X*-direction components of the accelerations of the left and right wheels, respectively |
| $J_w$ | Mass moment of inertia about the diameter of the wheel |
| $\alpha_{Lx}$, $\alpha_{Rx}$ | *X*-direction components of the angular acceleration of the left and right wheels, respectively |
| $\alpha_{Lz}$, $\alpha_{Rz}$ | *Z*-direction components of the angular acceleration of the left and right wheels, respectively |
| $u'_z(y,t)$, $u'_x(y,t)$ | The first derivative of $u_z(y, t)$, $u_x(y, t)$ with respect to *y*, respectively |

(continued)

**Table 1** (continued)

| Variable | Explanation |
|---|---|
| $y_{wL}$, $y_{wR}$ | *y* coordinates of the joints of the left and right wheels and the axle, respectively |
| $F_{wrLx}$, $F_{wrLy}$, $F_{wrLz}$ | *X*-, *Y*-, and *Z*-direction components of the left wheel–rail contact forces, respectively |
| $F_{wrRx}$, $F_{wrRy}$, $F_{wrRz}$ | *X*-, *Y*-, and *Z*-direction components of the right wheel–rail contact forces, respectively |
| $O_{cL}$, $O_{cR}$ | Left and right wheel–rail contact points, respectively |
| $O_{wL}$, $O_{wR}$ | Centers of the nominal circles of the left and right wheels, respectively |
| $O_{wL}$-$X_{wL}Y_{wL}Z_{wL}$, $O_{wR}$-$X_{wR}Y_{wR}Z_{wR}$ | Body coordinate systems attached to the left and right wheels, respectively |
| $q_{zk}$, $\ddot{q}_{zk}$ | The *k*th generalized coordinate and the *k*th generalized acceleration coordinate in the plane *O-YZ* |
| $q_{xk}$, $\ddot{q}_{xk}$ | The *k*th generalized coordinate and the *k*th generalized acceleration coordinate in the plane *O-XY* |
| $\omega_k$ | The *k*th circular frequency |
| $N$ | Considered number of the modes |
| $U_{zk}(y)$, $U'_{zk}(y)$ | The *k*th mode function of the axle in the *O-YZ* plane and its first derivative with respect to *y* |
| $U_{xk}(y)$, $U'_{xk}(y)$ | The *k*th mode function of the axle in the *O-XY* plane and its first derivative with respect to *y* |

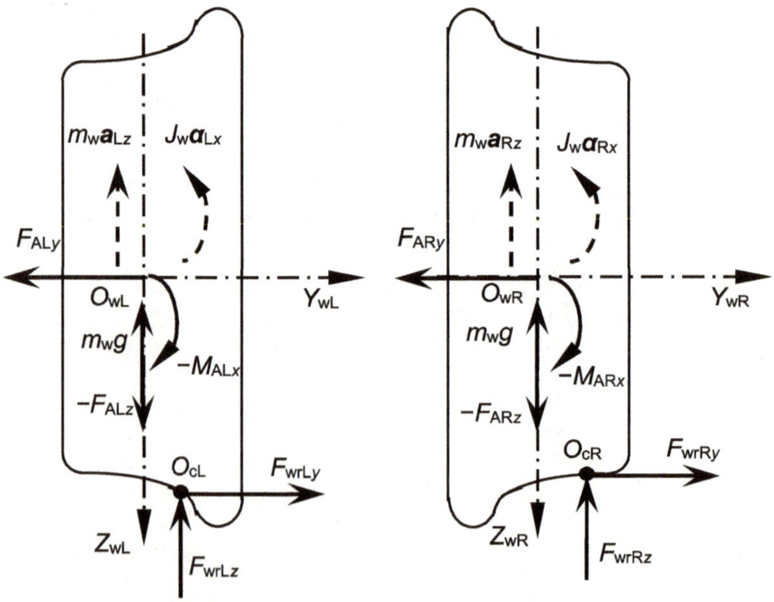


**Fig. 5** Force analysis diagram of the two wheels

$$J_{\mathrm{w}}\frac{\partial^2}{\partial t^2}u_z'(y_{\mathrm{w(L,R)}},t) = -F_{\mathrm{wr(L,R)}y}u_{\mathrm{c(L,R)}z} - F_{\mathrm{wr(L,R)}z}u_{\mathrm{c(L,R)}y} - M_{\mathrm{A(L,R)}x}. \tag{5}$$

Note that the lateral accelerations of the wheels are assumed to be the same as the wheelset axle so there is no relative motion between wheels and axle.

Substituting the expressions of $F_{\mathrm{A(L,R)}z}$ and $M_{\mathrm{A(L,R)}x}$ obtained through Eqs. (4) and (5) into Eqs. (2) and (3), respectively, we can obtain:

$$\begin{aligned} Q_z(y,t) = {} & F_{\mathrm{pL}z}\delta(y-y_{\mathrm{pL}}) \\ & + \left(m_{\mathrm{w}}g - F_{\mathrm{wrL}z} - m_{\mathrm{w}}\frac{\partial^2}{\partial t^2}u_z(y_{\mathrm{wL}},t)\right)\delta(y-y_{\mathrm{wL}}) \\ & + \left(m_{\mathrm{w}}g - F_{\mathrm{wrR}z} - m_{\mathrm{w}}\frac{\partial^2}{\partial t^2}u_z(y_{\mathrm{wR}},t)\right)\delta(y-y_{\mathrm{wR}}) \\ & + F_{\mathrm{pR}z}\delta(y-y_{\mathrm{pR}}), \end{aligned} \tag{6}$$

$$\begin{aligned} M_x(y,t) = {} & (-F_{\mathrm{pL}y}u_{\mathrm{pL}z} + F_{\mathrm{pL}z}u_{\mathrm{pL}y})\delta(y-y_{\mathrm{pL}}) \\ & + \left(-F_{\mathrm{wrL}y}u_{\mathrm{cL}z} - F_{\mathrm{wrL}z}u_{\mathrm{cL}y} - J_{\mathrm{w}}\frac{\partial^2}{\partial t^2}u_z'(y_{\mathrm{wL}},t)\right)\delta(y-y_{\mathrm{wL}}) \\ & + (-F_{\mathrm{pR}y}u_{\mathrm{pR}z} + F_{\mathrm{pR}z}u_{\mathrm{pR}y})\delta(y-y_{\mathrm{pR}}) \\ & + \left(-F_{\mathrm{wrR}y}u_{\mathrm{cR}z} - F_{\mathrm{wrR}z}u_{\mathrm{cR}y} - J_{\mathrm{w}}\frac{\partial^2}{\partial t^2}u_z'(y_{\mathrm{wR}},t)\right)\delta(y-y_{\mathrm{wR}}), \end{aligned} \tag{7}$$

$$\begin{aligned} & EI_x\frac{\partial^4 u_z(y,t)}{\partial y^4} + \rho A\frac{\partial^2 u_z(y,t)}{\partial t^2} + m_{\mathrm{w}}\frac{\partial^2}{\partial t^2}u_z(y,t)\delta(y-y_{\mathrm{wL}}) \\ & + m_{\mathrm{w}}\frac{\partial^2}{\partial t^2}u_z(y,t)\delta(y-y_{\mathrm{wR}}) - J_{\mathrm{w}}\frac{\partial^2}{\partial t^2}u_z'(y,t)\delta(y-y_{\mathrm{wL}}) \\ & - J_{\mathrm{w}}\frac{\partial^2}{\partial t^2}u_z'(y,t)\delta(y-y_{\mathrm{wR}}) = W_0, \end{aligned} \tag{8}$$

where

$$\begin{aligned} W_0 = {} & (m_{\mathrm{w}}g - F_{\mathrm{wrL}z})\delta(y-y_{\mathrm{wL}}) + F_{\mathrm{pL}z}\delta(y-y_{\mathrm{pL}}) \\ & + (m_{\mathrm{w}}g - F_{\mathrm{wrR}z})\delta(y-y_{\mathrm{wR}}) + F_{\mathrm{pR}z}\delta(y-y_{\mathrm{pR}}) \\ & - \frac{\partial}{\partial y}[(-F_{\mathrm{wrL}y}u_{\mathrm{cL}z} - F_{\mathrm{wrL}z}u_{\mathrm{cL}y})\delta(y-y_{\mathrm{wL}}) \\ & + (-F_{\mathrm{pL}y}u_{\mathrm{pL}z} + F_{\mathrm{pL}z}u_{\mathrm{pL}y})\delta(y-y_{\mathrm{pL}}) \\ & + (-F_{\mathrm{wrR}y}u_{\mathrm{cR}z} - F_{\mathrm{wrR}z}u_{\mathrm{cR}y})\delta(y-y_{\mathrm{wR}}) \\ & + (-F_{\mathrm{pR}y}u_{\mathrm{pR}z} + F_{\mathrm{pR}z}u_{\mathrm{pR}y})\delta(y-y_{\mathrm{pR}})]. \end{aligned} \tag{9}$$

Consider a solution of Eq. (8) in the form:

$$u_z(y,t) = U_z(y)\sin(\omega t + \sigma). \tag{10}$$

Using the calculus of variation (Qiu et al. 2009), the modal function satisfies:

$$\begin{aligned} m_{ij} = &\int_0^L \rho A(U_{zi}U_{zj})\mathrm{d}y \\ &+ m_{\mathrm{w}}\left(U_{zi}(y_{\mathrm{wL}})U_{zj}(y_{\mathrm{wL}}) + U_{zi}(y_{\mathrm{wR}})U_{zj}(y_{\mathrm{wR}})\right) \\ &+ J_{\mathrm{w}}\left(U'_{zi}(y_{\mathrm{wL}})U'_{zj}(y_{\mathrm{wL}}) + U'_{zi}(y_{\mathrm{wR}})U'_{zj}(y_{\mathrm{wR}})\right) = \delta_{ij}, \end{aligned} \tag{11}$$

$$k_{ij} = \int_0^L EI_x\left(U''_{zi}U''_{zj}\right)\mathrm{d}y = \omega_j^2\delta_{ij} \tag{12}$$

$$\begin{aligned} EI_x &\int_0^L U_{zj}U''''_{zi}\mathrm{d}y \\ &+ J_{\mathrm{w}}\left(U'_{zj}(y_{\mathrm{wL}})U'_{zi}(y_{\mathrm{wL}}) + U'_{zj}(y_{\mathrm{wR}})U'_{zi}(y_{\mathrm{wR}})\right) \\ &+ J_{\mathrm{w}}\left(U_{zj}(y_{\mathrm{wL}})U'_{zi}(y_{\mathrm{wL}}) + U_{zj}(y_{\mathrm{wR}})U'_{zi}(y_{\mathrm{wR}})\right) = \omega_i^2\delta_{ij}, \end{aligned} \tag{13}$$

where $\delta_{ij}$ is the Kronecker delta. For $i = j$, Eq. (11) can be written as

$$\begin{aligned} m_{jj} = &\int_0^L \rho A(U_{zj}^2)\mathrm{d}y + m_{\mathrm{W}}\left(U_{zj}^2(y_{\mathrm{wL}}) + U_{zj}^2(y_{\mathrm{wR}})\right) \\ &+ J_{\mathrm{W}}\left(U_{zj}'^2(y_{\mathrm{wL}}) + U_{zj}'^2(y_{\mathrm{wR}})\right) = 1. \end{aligned} \tag{14}$$

To obtain the mode shape functions with the wheelset axle modeled as a uniform Euler–Bernoulli beam carrying two particles (wheels), the segment of the beam from the left end to the first particle is referred to as the first portion, in between the two particles as the second portion, and from the second particle to the right end as the third portion. The beam mode shape will be the superposition of the mode shapes of the three portions. The derivation of the mode shape functions is presented in Appendix B. The first three modes have the frequencies of $f_1$ = 111 Hz,

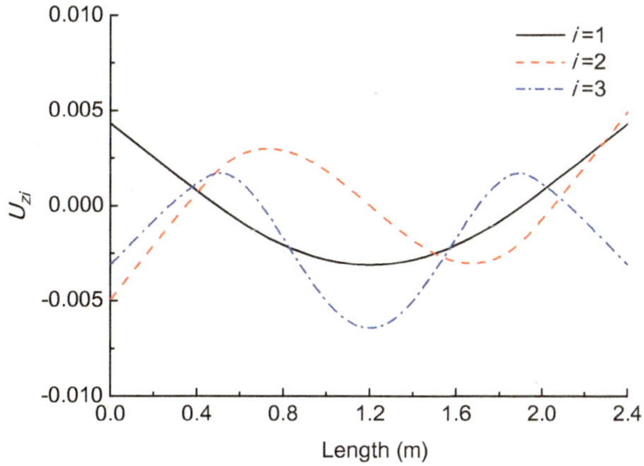


**Fig. 6** First three bending mode shapes of the wheelset

$f_2$ = 245 Hz, and $f_3$ = 547 Hz, respectively. These mode shape functions are normalized so as to satisfy Eq. (14), as shown in Fig. 6. The third mode is not in the frequency range of 0–500 Hz where the Euler–Bernoulli beam is available to analyze the system. Hence, the effect of the first two modes on dynamic responses is conducted in this study.

According to the modal analysis, we let the solution of Eq. (8) have the form:

$$u_z = \sum_{i=1}^{N} U_{zi} q_{zi}. \tag{15}$$

Substituting Eq. (15) into Eq. (8), the differential equation can be written as

$$\begin{aligned} & EI_x \sum_{i=1}^{N} U_{zi}'''' q_{zi} + \rho A \sum_{i=1}^{N} U_{zi} \ddot{q}_{zi} \\ & + m_{\mathrm{w}} \sum_{i=1}^{N} U_{zi}(y_{\mathrm{wL}}) \ddot{q}_{zi} + m_{\mathrm{w}} \sum_{i=1}^{N} U_{zi}(y_{\mathrm{wR}}) \ddot{q}_{zi} \\ & - J_{\mathrm{w}} \sum_{i=1}^{N} U_{zi}'(y_{\mathrm{wL}}) \ddot{q}_{zi} - J_{\mathrm{w}} \sum_{i=1}^{N} U_{zi}''(y_{\mathrm{wR}}) \ddot{q}_{zi} = W_0. \end{aligned} \tag{16}$$

Multiplying both sides of Eq. (16) by $U_{zj}$ and integrating over the domain $0 < y < L$, we can obtain:

$$EI_x \sum_{i=1}^{N} \left[ q_{zi} \int_0^L U_{zj} U_{zi}'''' \mathrm{d}y \right] + \rho A \sum_{i=1}^{N} \left[ \ddot{q}_{zi} \int_0^L U_{zj} U_{zi} \mathrm{d}y \right]$$
$$\cdot\, m_{\mathrm{w}} \sum_{i=1}^{N} \left[ \ddot{q}_{zi} \int_0^L U_{zj} U_{zi} \delta(y - y_{\mathrm{wL}}) \mathrm{d}y \right]$$
$$+ m_{\mathrm{w}} \sum_{i=1}^{N} \left[ \ddot{q}_{zi} \int_0^L U_{zj} U_{zi} \delta(y - y_{\mathrm{wR}}) \mathrm{d}y \right]$$
$$- J_{\mathrm{w}} \sum_{i=1}^{N} \left[ \ddot{q}_{zi} \int_0^L U_{zj} U_{zi}' \delta(y - y_{\mathrm{wL}}) \mathrm{d}y \right]$$
$$- J_{\mathrm{w}} \sum_{i=1}^{N} \left[ \ddot{q}_{zi} \int_0^L U_{zj} U_{zi}' \delta(y - y_{\mathrm{wR}}) \mathrm{d}y \right] = W_{1j}. \quad (17)$$

Using the orthogonality of the modal shape function as expressed in Eqs. (11) and (13), Eq. (17) can be written as

$$\sum_{i=1}^{N} \Big\{ q_{zi} \Big[ \omega_i^2 \delta_{ij} - J_{\mathrm{w}} \Big( U_{zj}'(y_{\mathrm{wL}}) U_{zi}'(y_{\mathrm{wL}})$$
$$+ U_{zj}'(y_{\mathrm{wR}}) U_{zi}'(y_{\mathrm{wR}}) \Big) - J_{\mathrm{w}} \left( U_{zj}(y_{\mathrm{wL}}) U_{zi}'(y_{\mathrm{wL}}) \right.$$
$$\left. + U_{zj}(y_{\mathrm{wR}}) U_{zi}'(y_{\mathrm{wR}}) \right) \Big]$$
$$+ \ddot{q}_{zi} \Big[ \delta_{ij} - J_{\mathrm{w}} \Big( U_{zj}'(y_{\mathrm{wL}}) U_{zi}'(y_{\mathrm{wL}})$$
$$+ U_{zj}'(y_{\mathrm{wR}}) U_{zi}'(y_{\mathrm{wR}}) \Big) - J_{\mathrm{w}} \left( U_{zj}(y_{\mathrm{wL}}) U_{zi}'(y_{\mathrm{wL}}) \right.$$
$$\left. + U_{zj}(y_{\mathrm{wR}}) U_{zi}'(y_{\mathrm{wR}}) \right) \Big] \Big\} = W_{1j}, \quad (18)$$

where

$$W_{1j} = \int_0^L U_{zj} W_0 \mathrm{d}y$$
$$= (m_{\mathrm{w}} g - F_{\mathrm{wrL}z}) U_{zj}(y_{\mathrm{wL}}) + F_{\mathrm{pL}z} U_{zj}(y_{\mathrm{pL}})$$
$$+ (m_{\mathrm{w}} g - F_{\mathrm{wrR}z}) U_{zj}(y_{\mathrm{wR}}) + F_{\mathrm{pR}z} U_{zj}(y_{\mathrm{pR}})$$
$$+ (-F_{\mathrm{wrL}y} u_{\mathrm{cL}z} - F_{\mathrm{wrL}z} u_{\mathrm{cL}y}) U_{zj}'(y_{\mathrm{wL}})$$
$$+ (-F_{\mathrm{pL}y} u_{\mathrm{pL}z} + F_{\mathrm{pL}z} u_{\mathrm{pL}y}) U_{zj}'(y_{\mathrm{pL}})$$
$$+ (-F_{\mathrm{wrR}y} u_{\mathrm{cR}z} - F_{\mathrm{wrR}z} u_{\mathrm{cR}y}) U_{zj}'(y_{\mathrm{wR}})$$
$$+ (-F_{\mathrm{pR}y} u_{\mathrm{pR}z} + F_{\mathrm{pR}z} u_{\mathrm{pR}y}) U_{zj}'(y_{\mathrm{pR}}). \quad (19)$$

Equation (18) can be expressed as:

$$
\begin{aligned}
&\ddot{q}_{zj} + \omega_j^2 q_{zj} - J_\mathrm{w} \sum_{i=1}^{N} \Big\{ q_{zi} \Big[ U'_{zj}(y_\mathrm{wL}) U'_{zi}(y_\mathrm{wL}) \\
&+ U'_{zj}(y_\mathrm{wR}) U'_{zi}(y_\mathrm{wR}) \\
&+ U_{zj}(y_\mathrm{wL}) U'_{zi}(y_\mathrm{wL}) + U_{zj}(y_\mathrm{wR}) U'_{zi}(y_\mathrm{wR}) \Big] \\
&+ \ddot{q}_{zi} \Big[ U'_{zj}(y_\mathrm{wL}) U'_{zi}(y_\mathrm{wL}) + U'_{zj}(y_\mathrm{wR}) U'_{zi}(y_\mathrm{wR}) \\
&+ U_{zj}(y_\mathrm{wL}) U'_{zi}(y_\mathrm{wL}) + U_{zj}(y_\mathrm{wR}) U'_{zi}(y_\mathrm{wR}) \Big] \Big\} = W_{1j}.
\end{aligned}
\tag{20}
$$

Equation (20) can be written in the matrix form:

$$
\boldsymbol{M}_1 \left[ \ddot{q}_{zj} \right] + \boldsymbol{M}_2 \left[ q_{zj} \right] = \left[ W_{1j} \right], \tag{21}
$$

where

$$
\begin{aligned}
(\boldsymbol{M}_{11})_{(i,j)} &= (\boldsymbol{M}_{21})_{(i,j)} \\
&= -J_\mathrm{w} \Big[ U'_{zj}(y_\mathrm{wL}) U'_{zi}(y_\mathrm{wL}) + U'_{zj}(y_\mathrm{wR}) U'_{zi}(y_\mathrm{wR}) \\
&\quad + U_{zj}(y_\mathrm{wL}) U'_{zi}(y_\mathrm{wL}) + U_{zj}(y_\mathrm{wR}) U'_{zi}(y_\mathrm{wR}) \Big], \\
\boldsymbol{M}_{12} &= \boldsymbol{I}, \quad \boldsymbol{M}_{22} = \left[ \omega_j^2 \right] \boldsymbol{I}, \\
\boldsymbol{M}_1 &= \boldsymbol{M}_{11} + \boldsymbol{M}_{12}, \quad \boldsymbol{M}_2 = \boldsymbol{M}_{21} + \boldsymbol{M}_{22}.
\end{aligned}
\tag{22}
$$

The explicit integral method illustrated in Zhai (2007) is used to obtain the vector $\left[ \ddot{q}_{zj} \right]$ of each acceleration coordinate.

For the vibration in the plane *YOX*, the differential equation expressed with respect to $\left[ \ddot{q}_{xy} \right]$ can be written as:

$$
\begin{aligned}
&\ddot{q}_{xj} + \omega_{xj}^2 q_{xj} - J_\mathrm{w} \sum_{i=1}^{N} \Big\{ q_i \Big[ U'_{xj}(y_\mathrm{wL}) U'_{xi}(y_\mathrm{wL}) \\
&+ U'_{xj}(y_\mathrm{wR}) U'_{xi}(y_\mathrm{wR}) \\
&+ U_{xj}(y_\mathrm{wL}) U'_{xi}(y_\mathrm{wL}) + U_{xj}(y_\mathrm{wR}) U'_{xi}(y_\mathrm{wR}) \Big] \\
&+ \ddot{q}_i \Big[ U'_{xj}(y_\mathrm{wL}) U'_{xi}(y_\mathrm{wL}) + U'_{xj}(y_\mathrm{wR}) U'_{xi}(y_\mathrm{wR}) \\
&+ U_{xj}(y_\mathrm{wL}) U'_{xi}(y_\mathrm{wL}) + U_{xj}(y_\mathrm{wR}) U'_{xi}(y_\mathrm{wR}) \Big] \Big\} = W_{1j}^{xoy}.
\end{aligned}
\tag{23}
$$

The derivation of Eq. (23) is similar to that of Eq. (20) and omitted here. Equation (23) can be expressed in matrix form:

$$\boldsymbol{M}_1^{xoy}\left[\ddot{q}_{xj}\right] + \boldsymbol{M}_2^{xoy}\left[q_{xj}\right] = \left[W_{1j}^{xoy}\right], \tag{24}$$

where

$$\begin{aligned}
(\boldsymbol{M}_{11})_{(i,j)}^{xoy} &= (\boldsymbol{M}_{21})_{(i,j)}^{xoy} \\
&= -J_{\mathrm{w}}\Big(U'_{xj}(y_{\mathrm{wL}})U'_{xi}(y_{\mathrm{wL}}) + U'_{xj}(y_{\mathrm{wR}})U'_{xi}(y_{\mathrm{wR}}) \\
&\quad + U_{xj}(y_{\mathrm{wL}})U'_{xi}(y_{\mathrm{wL}}) + U_{xj}(y_{\mathrm{wR}})U'_{xi}(y_{\mathrm{wR}})\Big), \\
\boldsymbol{M}_{12}^{xoy} &= \boldsymbol{I}, \quad \boldsymbol{M}_{22}^{xoy} = \left[\omega_j^2\right]\boldsymbol{I}, \\
\boldsymbol{M}_1^{xoy} &= \boldsymbol{M}_{11}^{xoy} + \boldsymbol{M}_{12}^{xoy}, \quad \boldsymbol{M}_2^{xoy} = \boldsymbol{M}_{21}^{xoy} + \boldsymbol{M}_{22}^{xoy}.
\end{aligned} \tag{25}$$

## 2.2 Wheel–Rail Contact Model

As mentioned in Sect. 2.1, the main concern in this work is the wheelset axle bending. The wheels are assumed to be rigid, and their nominal rolling circles are always perpendicular to the deformed wheelset axle at their interference fit surfaces. Figure 7 shows that the flexible wheelset moves from its initial reference state ($O_1(t_1)$) to its $t_2$ status ($O_2(t_2)$), which is described in the plane of *O-YZ*. $O_1$ is the center of the undeformed wheelset at $t_1$, and $O_2$ is the center of the deformed

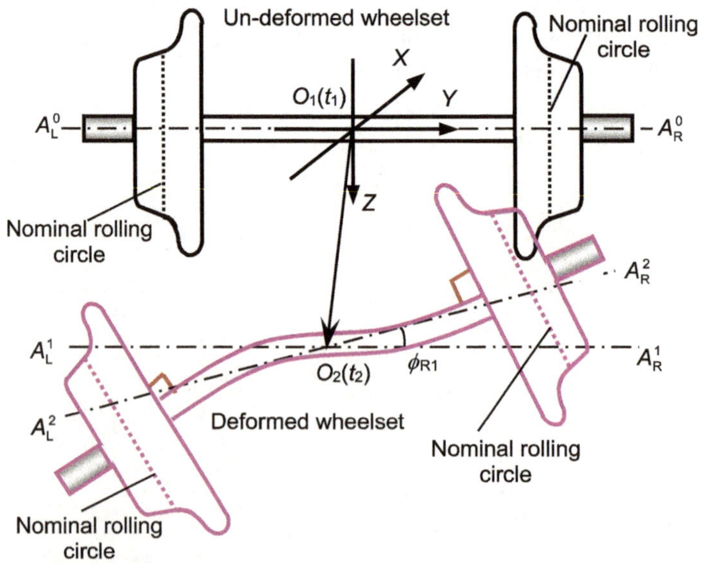


**Fig. 7** A flexible wheelset moving from its initial reference state ($O_1(t_1)$) to its any status ($O_2(t)$) in the plane of *O-YZ*

wheelset at any time $t$. $\boldsymbol{O}_1\boldsymbol{O}_2$ is the displacement vector of the wheelset center due to its rigid motion, and $\phi_{R1}$ is the roll angle due to the wheelset rigid motion. The auxiliary line, $A_L^0A_R^0$, is the central line of the undeformed wheelset axle, $A_L^1A_R^1$ is obtained by moving $A_L^0A_R^0$ from $O_1(t_1)$ to $O_2(t_2)$, and $A_L^2A_R^2$ is obtained through rotating $A_L^1A_R^1$ by $\phi_{R1}$. $A_L^2A_R^2$ is actually the central line of the rigid wheelset axle at $t_2$. Figure 7 shows that the wheels are assumed to be rigid and always perpendicular to the deformed axle line at their connections at any time $t_2$.

To clearly describe the new wheel–rail contact model, the dummies of the two rigid half wheelsets, as shown in Fig. 8, are employed to describe wheel–rail rolling contact behavior affected by the wheelset bending. The two dummies are indicated by DWL and DWR, respectively, and the wheels of the DWL and the DWR are assumed to overlap the left and right wheels of the flexible wheelset, respectively; all the time, namely, the motion of the assumed rigid wheels of the flexible wheelset can be described by the DWL and the DWR (Fig. 8). $\phi_{R2}$ is the roll angle of the right wheel due to the bending deformation of the flexible wheelset. It is exactly the included angle between the line $A_L^2A_R^2$ and the axle line of the right wheel or the wheel of the DWR.

It is not difficult to calculate the wheel–rail contact geometry considering the effect of the flexible deformation of the wheelset or the local high-frequency deformations of the wheels if the spatial positions of the DWL and the DWR are determined. Determining the spatial positions of the DWL and the DWR involves calculating their motion parameters, such as the lateral displacements of the centers of the wheels of the DWL and the DWR, indicated by $y_{DWL}$ and $y_{DWR}$, respectively, the vertical displacements, $z_{DWL}$ and $z_{DWR}$, the roll angles, $\phi_{DWL}$ and $\phi_{DWR}$, and

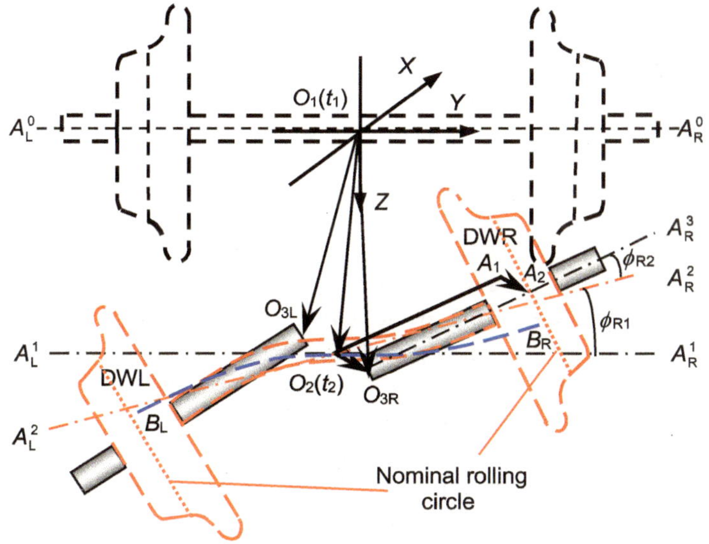


**Fig. 8** Relationship between the two rigid half wheelset dummies and the flexible wheelset

the yaw angles, $\psi_{\mathrm{DWL}}$ and $\psi_{\mathrm{DWR}}$. These parameters are key to calculating the contact geometry of the flexible wheelset in rolling contact with a pair of rails by using this new wheel–rail contact model. This will now be demonstrated in detail.

Figure 8 describes the motion of the DWL and the DWR influenced by the wheelset bending and its rigid motion in the *O-YZ* plane only. After the rigid wheelset moves with the center displacement of $\boldsymbol{O_1O_2}$ and the rolling angle of $\phi_{\mathrm{R1}}$ in the *O-YZ* plane of the global reference, *O-XYZ*, its center position $O_1(t_1)$ reaches the position $O_2(t_2)$ and $A_L^0A_R^0$ reaches (or becomes) $A_L^2A_R^2$. Note that the vector $\boldsymbol{O_1O_2}$ and the roll angle $\phi_{\mathrm{R1}}$ around axis *X* are described in the *O-YZ* plane. The dash-dot line $A_L^1A_R^1$ is through point $O_2(t_2)$ and parallel to $A_L^0A_R^0$. From Fig. 6, it is obvious that the rolling angle of the DWR caused by the wheelset rigid motion is just $\phi_{\mathrm{R1}}$ and that caused by the wheelset bending deformation is $\phi_{\mathrm{R2}}$, so the total rolling angle of the DWR is $\phi_{\mathrm{DWR}} = \phi_{\mathrm{R1}} + \phi_{\mathrm{R2}}$, as shown in Fig. 6.

In addition, the displacement of the DWR is the vector $\boldsymbol{O_1O_{3\mathrm{R}}}$, which could be written as:

$$\boldsymbol{O_1O_{3\mathrm{R}}} = \boldsymbol{O_1O_2} + \boldsymbol{O_2O_{3\mathrm{R}}}. \tag{26}$$

In Fig. 8, the vector $\boldsymbol{O_2A_1}$ is parallel to $\boldsymbol{O_{3\mathrm{R}}A_2}$ with the same length $l_0$. $l_0$ is actually the distance between the center of the wheel nominal circle and the center of the undeformed wheelset. The vector $\boldsymbol{O_2O_{3\mathrm{R}}}$ is parallel to $\boldsymbol{A_1A_2}$, with the same length. Thus, $\boldsymbol{O_1O_{3\mathrm{R}}}$ can be written as:

$$\boldsymbol{O_1O_{3\mathrm{R}}} = \boldsymbol{O_1O_2} + \boldsymbol{A_1A_2} = \boldsymbol{O_1O_2} + (\boldsymbol{O_2A_2} - \boldsymbol{O_2A_1}). \tag{27}$$

Moreover, the vector $\boldsymbol{O_2A_1}$ is described by $\{x_1\ y_1\ z_1\}[\boldsymbol{i\,j\,k}]^{\mathrm{T}}$ in *O-XYZ* and can be obtained by rotating the vector $\{0\ l_0\ 0\}[\boldsymbol{i\,j\,k}]^{\mathrm{T}}$ (coinciding with the line $A_L^1A_R^1$) about the *X*-axis by $\phi_{\mathrm{DWR}}$. $\boldsymbol{O_2A_1}$ is written as:

$$\begin{aligned}\boldsymbol{O_2A_1} &= \{x_1\quad y_1\quad z_1\}[\boldsymbol{i}\quad \boldsymbol{j}\quad \boldsymbol{k}]^{\mathrm{T}}\\ &= \begin{Bmatrix}0\\ l_0\\ 0\end{Bmatrix}^{\mathrm{T}}\begin{bmatrix}1 & 0 & 0\\ 0 & \cos(\phi_{\mathrm{R1}}+\phi_{\mathrm{R2}}) & \sin(\phi_{\mathrm{R1}}+\phi_{\mathrm{R2}})\\ 0 & -\sin(\phi_{\mathrm{R1}}+\phi_{\mathrm{R2}}) & \cos(\phi_{\mathrm{R1}}+\phi_{\mathrm{R2}})\end{bmatrix}\begin{bmatrix}\boldsymbol{i}\\ \boldsymbol{j}\\ \boldsymbol{k}\end{bmatrix}.\end{aligned} \tag{28}$$

The curve $\widehat{B_LB_R}$ (Fig. 6) is the deformed axle centerline of the wheelset, which does not consider the influence of the rotation caused by the wheelset rigid motion. The point $B_{\mathrm{R}}$ is the center of the right nominal circle. The axle centerline ($\widehat{O_2A_2}$) of the deformed wheelset can be obtained by rotating $\widehat{B_LB_R}$ about the *X*-axis by $\phi_{\mathrm{R1}}$. According to the definition of the curve $\widehat{B_LB_R}$, the vector $\boldsymbol{O_2B_{\mathrm{R}}}$ is defined as:

$$\boldsymbol{O_2B_{\mathrm{R}}} = \begin{Bmatrix}x_2\\ y_2\\ z_2\end{Bmatrix}^{\mathrm{T}}\begin{bmatrix}\boldsymbol{i}\\ \boldsymbol{j}\\ \boldsymbol{k}\end{bmatrix} = \begin{Bmatrix}\Delta x_2\\ \Delta y_2 + l_0\\ \Delta z_2\end{Bmatrix}^{\mathrm{T}}\begin{bmatrix}\boldsymbol{i}\\ \boldsymbol{j}\\ \boldsymbol{k}\end{bmatrix}, \tag{29}$$

where $\{\Delta x_2\ \Delta y_2\ \Delta z_2\}[\boldsymbol{i}\ \boldsymbol{j}\ \boldsymbol{k}]^{\mathrm{T}}$ is the displacement vector of the center of the right nominal circle due to the axle bending. Then, the vector $\boldsymbol{O}_2\boldsymbol{A}_2$ is defined as $\{x_3\ y_3\ z_3\}$ $[\boldsymbol{i}\ \boldsymbol{j}\ \boldsymbol{k}]^{\mathrm{T}}$ and can be written as:

$$\boldsymbol{O}_2\boldsymbol{A}_2 = \begin{Bmatrix} x_3 \\ y_3 \\ z_3 \end{Bmatrix}^{\mathrm{T}} \begin{bmatrix} \boldsymbol{i} \\ \boldsymbol{j} \\ \boldsymbol{k} \end{bmatrix} = \begin{Bmatrix} x_2 \\ y_2 \\ z_2 \end{Bmatrix}^{\mathrm{T}} \begin{bmatrix} 1 & 0 & 0 \\ 0 & \cos\phi_{\mathrm{R1}} & \sin\phi_{\mathrm{R1}} \\ 0 & -\sin\phi_{\mathrm{R1}} & \cos\phi_{\mathrm{R1}} \end{bmatrix} \begin{bmatrix} \boldsymbol{i} \\ \boldsymbol{j} \\ \boldsymbol{k} \end{bmatrix}, \tag{30}$$

which is obtained according to the relationship between $\widehat{O_2A_2}$ and $\widehat{O_2B_R}$ or $\widehat{O_2A_2}$ obtained by rotating $\widehat{O_2B_R}$ by $\phi_{\mathrm{R1}}$. The wheelset center displacement vector $\boldsymbol{O}_1\boldsymbol{O}_2$ is defined as $\{x_0\ y_0\ z_0\}[\boldsymbol{i}\ \boldsymbol{j}\ \boldsymbol{k}]^{\mathrm{T}}$.

Substituting Eqs. (28) and (30) and the expression of $\boldsymbol{O}_1\boldsymbol{O}_2$ into Eq. (27), the vector $\boldsymbol{O}_1\boldsymbol{O}_{3\mathrm{R}}$ can be written as:

$$\begin{aligned} \boldsymbol{O}_1\boldsymbol{O}_{3\mathrm{R}} &= \left( \begin{Bmatrix} x_0 \\ y_0 \\ z_0 \end{Bmatrix}^{\mathrm{T}} + \begin{Bmatrix} \Delta x_2 \\ \Delta y_2 + l_0 \\ \Delta z_2 \end{Bmatrix}^{\mathrm{T}} \boldsymbol{M}_1 - \begin{Bmatrix} 0 \\ l_0 \\ 0 \end{Bmatrix}^{\mathrm{T}} \boldsymbol{M}_2 \right) \begin{bmatrix} \boldsymbol{i} \\ \boldsymbol{j} \\ \boldsymbol{k} \end{bmatrix}, \\ \boldsymbol{M}_1 &= \begin{bmatrix} 1 & 0 & 0 \\ 0 & \cos\phi_{\mathrm{R1}} & \sin\phi_{\mathrm{R1}} \\ 0 & -\sin\phi_{\mathrm{R1}} & \cos\phi_{\mathrm{R1}} \end{bmatrix}, \\ \boldsymbol{M}_2 &= \begin{bmatrix} 1 & 0 & 0 \\ 0 & \cos(\phi_{\mathrm{R1}}+\phi_{\mathrm{R2}}) & \sin(\phi_{\mathrm{R1}}+\phi_{\mathrm{R2}}) \\ 0 & -\sin(\phi_{\mathrm{R1}}+\phi_{\mathrm{R2}}) & \cos(\phi_{\mathrm{R1}}+\phi_{\mathrm{R2}}) \end{bmatrix}. \end{aligned} \tag{31}$$

Similarly, when considering the wheelset bending deformation in the plane *O-XY*, the vector $\boldsymbol{O}_1\boldsymbol{O}_{3\mathrm{R}}$ should be given as

$$\begin{aligned} \boldsymbol{O}_1\boldsymbol{O}_{3\mathrm{R}} &= \left( \begin{Bmatrix} x_0 \\ y_0 \\ z_0 \end{Bmatrix}^{\mathrm{T}} + \begin{Bmatrix} \Delta x_2 \\ \Delta y_2 + l_0 \\ \Delta z_2 \end{Bmatrix}^{\mathrm{T}} \boldsymbol{M}_1\boldsymbol{M}_3 - \begin{Bmatrix} 0 \\ l_0 \\ 0 \end{Bmatrix}^{\mathrm{T}} \boldsymbol{M}_2\boldsymbol{M}_4 \right) \begin{bmatrix} \boldsymbol{i} \\ \boldsymbol{j} \\ \boldsymbol{k} \end{bmatrix}, \\ \boldsymbol{M}_3 &= \begin{bmatrix} \cos\psi_{\mathrm{R1}} & \sin\psi_{\mathrm{R1}} & 0 \\ -\sin\psi_{\mathrm{R1}} & \cos\psi_{\mathrm{R1}} & 0 \\ 0 & 0 & 1 \end{bmatrix}, \\ \boldsymbol{M}_4 &= \begin{bmatrix} \cos(\psi_{\mathrm{R1}}+\psi_{\mathrm{R2}}) & \sin(\psi_{\mathrm{R1}}+\psi_{\mathrm{R2}}) & 0 \\ -\sin(\psi_{\mathrm{R1}}+\psi_{\mathrm{R2}}) & \cos(\psi_{\mathrm{R1}}+\psi_{\mathrm{R2}}) & 0 \\ 0 & 0 & 1 \end{bmatrix}, \end{aligned} \tag{32}$$

where $\psi_{\mathrm{R1}}$ and $\psi_{\mathrm{R2}}$ are the yaw angles caused by the rigid motion and the bending deformation in the plane *O-XY*, respectively.

$\psi_{DWR} = \psi_{R1} + \psi_{R2}$ is the total yaw angle of the DWR. Similarly, the position of the DWL can be obtained. When the positions of the two dummies are known at $t_2$, the wheel–rail contact geometry can be calculated. Then, the positions of the wheel–rail contact points are easily found, and the wheel–rail contact forces can be calculated. The normal wheel–rail contact forces are calculated by the Hertzian nonlinear contact spring model, and the tangent contact forces and spin moments are calculated by means of the model by Shen et al. (1983). Compared with the conventional wheel–rail contact model (Wang 1984; Zhai 2007), this new wheel–rail contact model can characterize the independent high-frequency deformations of the two wheels of the flexible wheelset more conveniently.

# 3 Results and Discussion

When a vehicle is running on an ideal track, it is only excited by sleepers. Note that the "flexible" wheelset model used in this section denotes the model considering the first two bending modes. The dynamic system with flexible wheelset models is used in the simulation on an ideal track at the speed of 300 km/h. Figure 9a, b shows the vertical forces in the frequency domain in steady and unsteady stages, respectively. In the unsteady stage, the peaks appear not only at a set of harmonic frequencies $nf_s$ ($n = 1, 2, 3, \ldots$) produced by passing sleeper but also at $f_{b1}$, while the influence of the second bending mode is small since there is no peak at $f_{b2}$. In the steady stage, the contribution of the component at $f_{b1}$ is weakened and only the peaks at $nf_s$ ($n = 1, 2, 3, \ldots$) remain. These results are reasonable because when a system comes to a steady stage, its responses only contain the component at the excitation frequency.

Based on a large range of site measurements, the components of roughness on rails mostly appear in the range of 1–20 m. The natural frequencies of the first two bending modes are below 250 Hz, meaning the available frequency of this model is limited. Therefore, the components of the random irregularity on the rails are

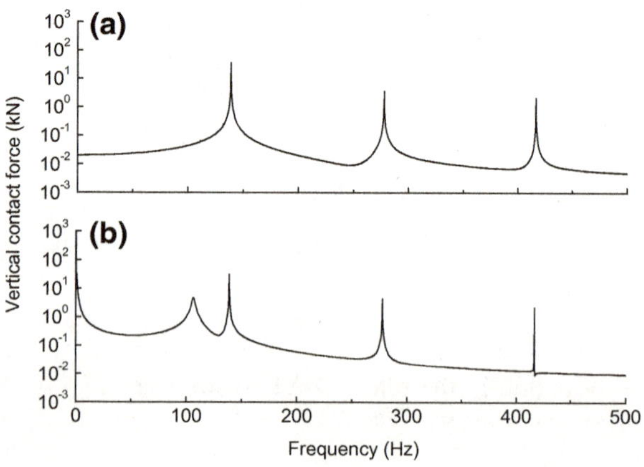


**Fig. 9** Vertical contact force in steady (**a**) and unsteady (**b**) stages

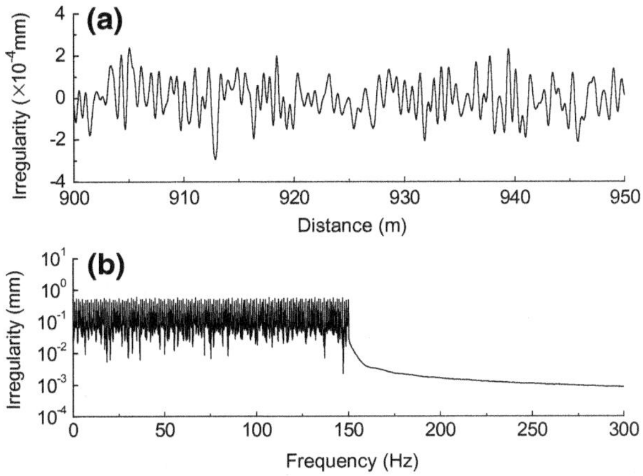


**Fig. 10** Random irregularity in the time domain (**a**) and frequency domain (**b**)

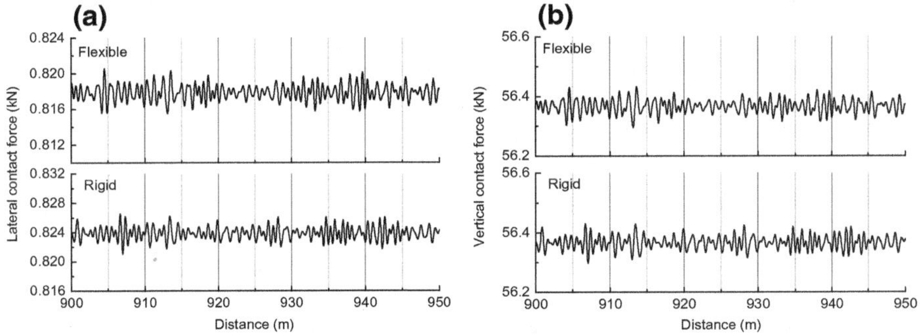


**Fig. 11** Lateral contact forces (**a**) and vertical contact forces (**b**) in the time domain

mainly in the frequency range of 0–150 Hz at the speed of 300 km/h. Figure 10a presents the local section of 900–950 m in the time domain, and Fig. 10b shows the irregularity in the frequency domain. Note that the results below are from the steady stage.

Figures 11 and 12 show the wheel–rail contact forces acting on the rigid and flexible wheelsets in the time and frequency domains, respectively. As shown in Fig. 11a, the average of the oscillation of the lateral contact force acting on the flexible model is little smaller than that on the rigid wheelset model, and the shapes of the oscillation are different. As shown in Fig. 11b, the vertical contact forces acting on the two models oscillate around a similar average, while their shapes are different. These differences are caused by the wheelset flexibility.

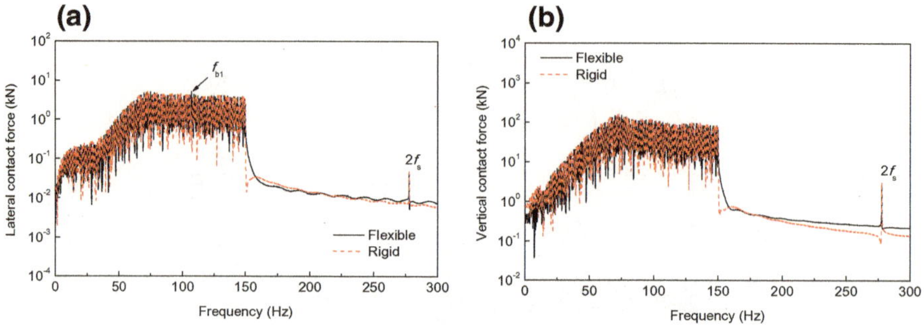


**Fig. 12** Lateral contact forces (**a**) and vertical contact forces (**b**) in the frequency domain

In the frequency domain, the distributions of the components contained in both the lateral and vertical contact forces are in the excitation frequency range of the random irregularity. A peak at frequency $2f_s$ appears in Fig. 11a, b. The contribution of the component at frequency $f_s$ is overwhelmed by the effect of the irregularity. In addition, the uniform distribution in 0–150 Hz of the irregularity results in the non-uniform distribution of contact forces. As shown in Fig. 12a, b, the components in 80–150 Hz are higher than those in 0–80 Hz. This shows that under this present irregularity, this dynamic system is more sensitive to the excitation in 80–150 Hz than to that in 0–80 Hz.

In the frequency domain, the component at $f_{b1}$ of the lateral contact force acting on the flexible model is a little larger than that on the rigid model, as marked using the arrow in Fig. 12a. This shows that the first bending mode is excited, and the availability of the model to characterize the wheelset bending is proved. However, there is no evident difference at $f_{b1}$ for vertical contact forces acting on the two models. This shows that the wheelset bending deformation has a stronger effect on the lateral contact force than on the vertical contact force.

The wheel–rail contact force is affected by the position of the lateral contact points. Figure 13a, b shows the oscillations of the contact points in lateral direction described in the body coordinate system attached to the rail cross section in the time and frequency domains, respectively. The average of the magnitudes of oscillation of the contact points on the flexible model in the time domain is larger than that on the rigid model. This is caused by the wheelset bending. Moreover, it can weaken the relative movement between rail and wheel caused by the irregularity. Therefore, it is one cause of the smaller average of the lateral contact force acting on the flexible model (Fig. 11a). As shown in Fig. 13b, the difference of the components between the two models at $f_{b1}$ is evident. This explains the difference in the time domain (Fig. 13a) and again shows the effectiveness of the proposed model.

**Fig. 13** Oscillation of contact points in lateral direction in the time domain (**a**) and frequency domain (**b**)

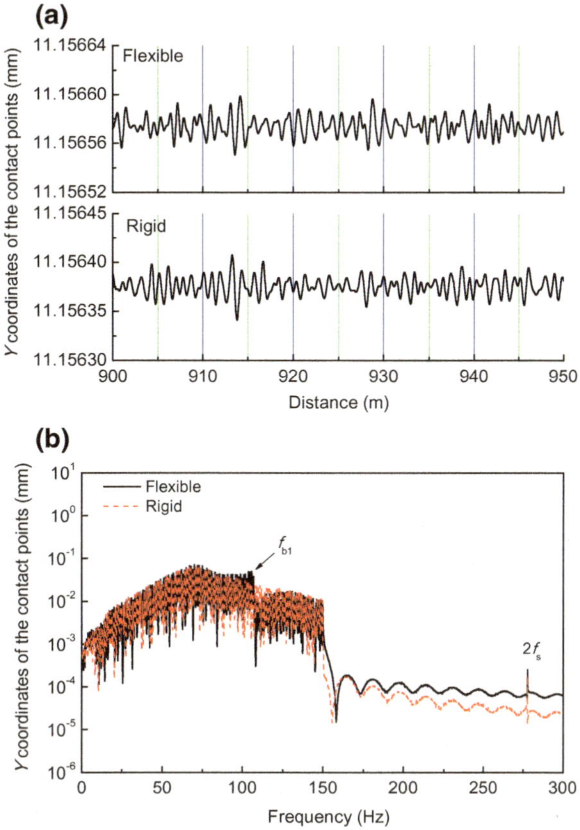


## 4 Conclusions

In this study, a new wheel–rail contact model is integrated into the high-speed vehicle–track coupling dynamics system model, which takes into account the effect of wheelset structural flexibility. Based on the new vehicle–track model, the effect of the first two bending modes of the wheelset on wheel–rail contact behavior is analyzed under the random irregularity in a frequency range of 0–150 Hz. The numerical results of the rigid wheelset model and the flexible wheelset model are compared in detail. The following conclusions can be drawn from the results:

1. The present vehicle–track model considering flexible wheelsets can very well characterize the effect of the flexible wheelset on wheel–rail dynamic behavior.
2. Under the excitation, the shapes of the oscillations of the wheel–rail contact forces and contact points for the new and conventional vehicle–track models are different. The difference is caused by the excited first bending mode of the wheelset.

For future work, the first improvement to be considered is to model a wheelset using the FEM or the Timoshenko beam theory to broaden the model's available frequency range. This could allow it to help investigate the mechanisms behind the generation and development of wheel–rail wear and noise.

## Appendix A

The vehicle notations and track parameters are given in Table 2.

## Appendix B

The axle is modeled as a uniform Euler-Bernoulli beam carrying two particles (wheels). The segment of the beam from the left end to the first particle is referred to as the first portion, in between the two particles as the second portion and from the second particle to the right end as the third portion. The beam mode shape will be the superposition of the mode shapes of the three portions. The mode shape of each portion has four constants of integration, i.e., a total of 12 for the three portions. It is necessary to satisfy: the boundary conditions; continuity of deflection and continuity of slope at the two 'locations'; and compatibility of bending moments and compatibility of forces acting on the two particles.

Here we take the calculation of the mode shape functions in the plane $O\text{-}YZ$ as an example. Fig. 14 shows a uniform Euler-Bernoulli beam $O_1O_3$ of flexural rigidity $EI_x$, and length $(R_1 + R_2 + R_3)L$ carrying the first particle of mass $m_w$ at axial coordinate $R_1L$ from $O_1$ and the second particle of mass $m_w$ at axial coordinate $R_3L$ from $O_3$.

To write the equations of transverse vibrations of the system, three coordinate systems are chosen with origin at $O_1$, $O_2$, and $O_3$. The choice of these coordinate systems has some algebraic advantages. In the text, the subscripts $k = 1, 2,$ and $3$ refer to the first portion, the second portion, and the third portion of the beam, respectively. For free vibration of the beam at frequency, if the amplitude of vibration of the beam is $U_{zk}(y_k)$ at axial coordinate $y_k$ (in the range $0 < y_k < R_kL$), then based on the Euler-Bernoulli bending theory, the bending moment $M_{xk}(y_k)$, the shearing force $Q_{zk}(y_k)$, and the mode shape differential equation for the three portions are

$$\begin{cases} M_{xk}(y_k) = EI_x \dfrac{\mathrm{d}^2 U_{zk}(y_k)}{\mathrm{d}y_k^2}, \\ Q_{zk}(y_k) = -EI_x \dfrac{\mathrm{d}^3 U_{zk}(y_k)}{\mathrm{d}y_k^3}, \\ EI_x \dfrac{\mathrm{d}^4 U_{zk}(y_k)}{\mathrm{d}y_k^4} - \rho A \omega^2 U_{zk}(y_k) = 0. \end{cases} \tag{33}$$

**Table 2** The vehicle notations and track parameters

| Physical parameter | Value | Notation |
|---|---|---|
| $M_c$ (kg) | $3.38 \times 10^4$ | Car body mass |
| $M_{bi}$ (kg) | $2.4 \times 10^3$ | The *i*th bogie mass |
| $M_{wi}$ (kg) | $1.85 \times 10^3$ | The *i*th wheelset mass |
| $C_{ty}$ (N·s/m) | $2.0 \times 10^4$ | Equivalent lateral damping of the secondary suspension (considering damping of lateral shock absorber joint) |
| $K_{ty}$ (N/m) | $1.813 \times 10^7$ | Equivalent lateral stiffness of the secondary suspension (considering stiffness of lateral shock absorber joint and lateral stiffness of air spring) |
| $C_{tz}$ (N·s/m) | $4.0 \times 10^4$ | Equivalent vertical damping of the secondary suspension (considering vertical damping of air spring) |
| $K_{tz}$ (N/m) | $2.99 \times 10^5$ | Equivalent vertical stiffness of the secondary suspension (considering vertical stiffness of air spring) |
| $C_{fy}$ (N·s/m) | 0 | Equivalent lateral damping of the primary suspension |
| $K_{fy}$ (N/m) | $6.47 \times 10^6$ | Equivalent lateral stiffness of the primary suspension (considering the lateral stiffness locating node of the axle-box rotary arm) |
| $C_{fz}$ (N·s/m) | $1.5 \times 10^4$ | Equivalent vertical damping of the primary suspension (considering damping of vertical shock absorber joint) |
| $K_{fz}$ (N/m) | $6.076 \times 10^6$ | Equivalent vertical stiffness of the primary suspension (considering stiffness of vertical shock absorber joint and steel spring) |
| $M_r$ (kg/m) | 60.64 | Rail mass per unit length |
| $M_s$ (kg) | 349 | Mass of sleeper |
| $M_b$ (kg) | 466 | Mass of ballast element |
| $L_s$ (m) | 0.6 | Sleeper bay |
| $E$ (N/m²) | $2.06 \times 10^{11}$ | Young's modulus |
| $K_{pLi}$ (N/m) | $2.0 \times 10^7$ | Lateral stiffness of the *i*th pad |
| $C_{pLi}$ (N/m) | $5 \times 10^4$ | Lateral damping of the *i*th pad |
| $K_{pVi}$ (N/m) | $4.0 \times 10^7$ | Vertical stiffness of the *i*th pad |
| $C_{pVi}$ (N/m) | $5 \times 10^4$ | Vertical damping of the *i*th pad |
| $K_{bv(L,R)i}$ (N/m) | $8.0 \times 10^7$ | Vertical stiffness between sleeper and the *i*th ballast element |
| $C_{bv(L,R)i}$ (N·s/m) | $1 \times 10^5$ | Vertical damping between sleeper and the *i*th ballast element |
| $K_w$ (N/m) | $7.8 \times 10^7$ | Vertical stiffness between the *i*th ballast elements on the left and right |
| $C_w$ (N·s/m) | $8 \times 10^4$ | Vertical damping between the *i*th ballast elements on the left and right |
| $K_{fv(L,R)i}$ (N/m) | $6.5 \times 10^7$ | Vertical stiffness between road bed and the *i*th ballast element |
| $C_{fv(L,R)i}$ (N/m) | $3.1 \times 10^4$ | Vertical damping between road bed and the *i*th ballast element |

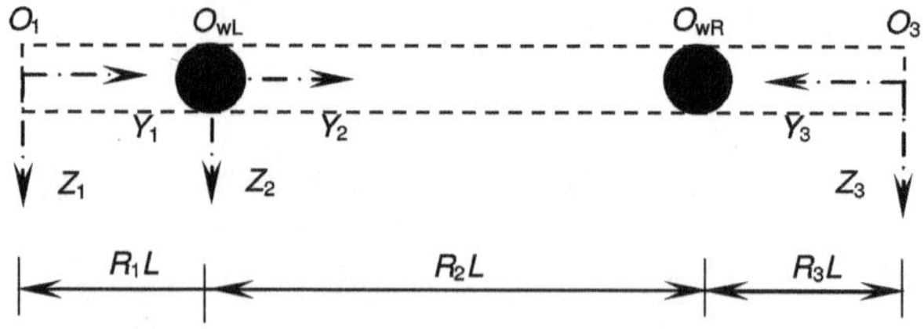


**Fig. 14** Coordinate systems attached to the three sections of the wheelset axle

To express these equations in dimensionless form, one defines the dimensionless axial coordinate $Y_k$, amplitude $Z_k(Y_k)$, operator $D^n$, dimensionless bending moment $M_{xk}(Y_k)$, shearing force $Q_{zk}(Y_k)$, and a dimensionless natural frequency $\Omega$ as follows:

$$\begin{cases} Y_k = \frac{y_k}{L}, Z_k(Y_k) = \frac{U_{zk}(y_k)}{L}, \\ D^n = \frac{\mathrm{d}^n}{\mathrm{d}Y_k^n}, M_{xk}(Y_k) = \frac{M_{xk}(y_k)L}{EI_x}, \\ Q_{zk}(Y_k) = \frac{Q_{zk}(y_k)L^2}{EI_x}, \Omega^2 = \frac{\rho A\omega^2 L^4}{EI_x} = \alpha^4. \end{cases} \tag{34}$$

Therefore, Eq. (33) can be expressed in the dimensionless form:

$$\begin{cases} M_{xk}(Y_k) = D^2 Z_k(Y_k), \\ Q_{zk}(y_k) = -D^3 Z_k(Y_k), \\ D^4 Z_k(Y_k) - \Omega^2 Z_k(Y_k) = 0. \end{cases} \tag{35}$$

Consider the solution of the previous equation as

$$\begin{aligned} Z_k(Y_k) = {} & C_{k1}\sin(\alpha Y_k) + C_{k2}\cos(\alpha Y_k) \\ & + C_{k3}\sinh(\alpha Y_k) + C_{k4}\cosh(\alpha Y_k). \end{aligned} \tag{36}$$

There are 12 unknown constants $C_{ki}$ ($i$ = 1, 2, 3, 4) for the three segments.

For free vibration the D'Alembert force and moment acting on the left wheel is $m_{\mathrm{w}}\omega^2 U_{z1}(R_1L)$ and $J_{\mathrm{w}}\omega^2 U'_{z1}(R_1L)$, respectively (Fig. 15). Continuity of deflection and continuity of slope at $O_{\mathrm{wL}}$ together with compatibility of bending moments and compatibility of forces acting on the left wheel results in

$$\begin{cases} U_{z1}(R_1L) = U_{z2}(0), \\ \frac{\mathrm{d}U_{z1}(R_1L)}{\mathrm{d}y_1} = \frac{\mathrm{d}U_{z2}(0)}{\mathrm{d}y_2}, \\ M_{x1}(R_1L) = M_{x2}(0) + J_{\mathrm{w}}\omega^2 U'_{z1}(R_1L), \\ Q_{z1}(R_1L) = Q_{z2}(0) + m_{\mathrm{w}}\omega^2 U_{z1}(R_1L). \end{cases} \tag{37}$$

The D'Alembert force and moment acting on the left wheel is $m_{\mathrm{w}}\omega^2 U_{z2}(R_2L)$ and $J_{\mathrm{w}}\omega^2 U'_{z2}(R_2L)$, respectively. Continuity of deflection and continuity of slope at

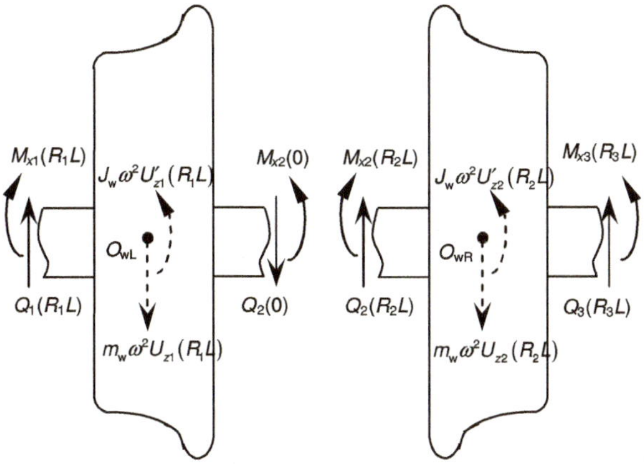

**Fig. 15** Wheel diagrams including D'Alembert forces

$O_{\mathrm{wR}}$ together with compatibility of bending moments and compatibility of forces acting on the right wheel results in

$$\begin{cases} U_{z2}(R_2L) = U_{z3}(R_3L), \\ \frac{\mathrm{d}U_{z2}(R_2L)}{\mathrm{d}y_2} = -\frac{\mathrm{d}U_{z3}(R_3L)}{\mathrm{d}y_3}, \\ M_{x2}(R_2L) = M_{x3}(R_3L) + J_{\mathrm{w}}\omega^2 U'_{z2}(R_2L), \\ Q_{z2}(R_2L) = -Q_{z3}(R_3L) + m_{\mathrm{w}}\omega^2 U_{z2}(R_2L). \end{cases} \tag{38}$$

Note that Eq. (38) takes into account the contra directions of the axial coordinates $y_2$ and $y_3$. Eqs. (37) and (38) in dimensionless form are

$$\begin{cases} Z_1(R_1) = Z_2(0), \\ DZ_1(R_1) = DZ_2(0), \\ D^2Z_1(R_1) = D^2Z_2(0) + \frac{J_{\mathrm{w}}\Omega^2}{m_{\mathrm{a}}L^2} DZ_1(R_1), \\ D^3Z_1(R_1) = D^3Z_2(0) - \frac{m_{\mathrm{w}}}{m_{\mathrm{a}}}\Omega^2 Z_1(R_1), \end{cases} \tag{39}$$

$$\begin{cases} Z_2(R_2) = Z_3(R_3), \\ DZ_2(R_2) = -DZ_3(R_3), \\ D^2Z_2(R_2) = D^2Z_3(R_3) + \frac{J_{\mathrm{w}}\Omega^2}{m_{\mathrm{a}}L^2} DZ_2(R_2), \\ D^3Z_2(R_2) = -D^3Z_3(R_3) - \frac{m_{\mathrm{w}}}{m_{\mathrm{a}}}\Omega^2 Z_2(R_2). \end{cases} \tag{40}$$

For the free boundary condition at the left end of the first portion and the right end of the third portion, the coefficients of the dimensionless mode shape functions satisfy

$$C_{11} = C_{13},\ C_{12} = C_{14},\ C_{31} = C_{33},\ C_{32} = C_{34}. \tag{41}$$

Then one can write the dimensionless mode shape functions of the first and third portions as

$$
\begin{aligned}
Z_k(Y_k) &= C_{k1}(\sin(\alpha Y_k) + \sinh(\alpha Y_k)) \\
&\quad + C_{k2}(\cos(\alpha Y_k) + \cosh(\alpha Y_k)) \\
&= B_{k1}P_k(Y_k) + B_{k2}V_k(Y_k), \\
k &= 1,\ 3.
\end{aligned} \tag{42}
$$

Substituting Eqs. (42) and (36) into Eqs. (39), we can obtain:

$$
\begin{cases}
B_{11}P_1(R_1) + B_{12}V_1(R_1) = C_{22} + C_{24}, \\
B_{11}DP_1(R_1) + B_{12}DV_1(R_1) = \alpha(C_{21} + C_{23}), \\
B_{11}(D^2P_1(R_1) - \frac{J_w\Omega^2}{m_aL^2}DP_1(R_1)) \\
+ B_{12}(\mathrm{D}^2V_1(R_1) - \frac{J_w\Omega^2}{m_aL^2}DV_1(R_1)) = \alpha^2(-C_{22} + C_{24}), \\
B_{11}(D^3P_1(R_1) + \frac{m_w}{m_a}\Omega^2P_1(R_1)) \\
+ B_{12}(D^3V_1(R_1) + \frac{m_w}{m_a}\Omega^2V_1(R_1)) = \alpha^3(-C_{21} + C_{23}).
\end{cases} \tag{43}
$$

Then one can write the dimensionless mode shape function of the second portions as

$$
Z_2(Y_2) = B_{11}P_2(Y_2) + B_{12}V_2(Y_2), \tag{44}
$$

where

$$
\begin{cases}
P_2(Y_2) = P_{21}\sin(\alpha Y_2) + P_{22}\cos(\alpha Y_2) \\
\quad + P_{23}\sinh(\alpha Y_2) + P_{24}\cosh(\alpha Y_2), \\
V_2(Y_2) = V_{21}\sin(\alpha Y_2) + V_{22}\cos(\alpha Y_2) \\
\quad + V_{23}\sinh(\alpha Y_2) + V_{24}\cosh(\alpha Y_2).
\end{cases} \tag{45}
$$

The coefficients of $\sin(\alpha Y_2)$, $\cos(\alpha Y_2)$, $\sinh(\alpha Y_2)$, and $\cosh(\alpha Y_2)$ in Eq. (36) (when $k = 2$) correspond to those in the expression obtained by substituting Eq. (45) into Eq. (44), so we can obtain

$$
C_{2i} = B_{11}P_{2i} + B_{12}V_{2i}, \quad i = 1,\ 2,\ 3,\ 4. \tag{46}
$$

The coefficients of $B_{11}$ and $B_{12}$ in Eq. (46) correspond to those in the expression by simplifying Eq. (43), so we can obtain:

$$
\begin{cases}
P_{21} = \frac{DP_1(R_1)}{2\alpha} - \frac{D^3P_1(R_1) + \frac{m_w}{m_a}\Omega^2DP_1(R_1)}{2\alpha^3}, \\
P_{23} = \frac{DP_1(R_1)}{2\alpha} + \frac{D^3P_1(R_1) + \frac{m_w}{m_a}\Omega^2DP_1(R_1)}{2\alpha^3}, \\
P_{22} = \frac{P_1(R_1)}{2} - \frac{D^2P_1(R_1) - \frac{J_w\Omega^2}{m_aL^2}DP_1(R_1)}{2\alpha^2}, \\
P_{24} = \frac{P_1(R_1)}{2} + \frac{D^2P_1(R_1) - \frac{J_w\Omega^2}{m_aL^2}DP_1(R_1)}{2\alpha^2},
\end{cases} \tag{47}
$$

$$\begin{cases} V_{21} = \frac{DV_1(R_1)}{2\alpha} - \frac{D^3V_1(R_1) + \frac{m_w}{m_a}\Omega^2 DV_1(R_1)}{2\alpha^3}, \\ V_{23} = \frac{DV_1(R_1)}{2\alpha} + \frac{D^3V_1(R_1) + \frac{m_w}{m_a}\Omega^2 DV_1(R_1)}{2\alpha^3}, \\ V_{22} = \frac{V_1(R_1)}{2} - \frac{D^2V_1(R_1) - \frac{J_w\Omega^2}{m_a L^2} DV_1(R_1)}{2\alpha^2}, \\ V_{24} = \frac{V_1(R_1)}{2} + \frac{D^2V_1(R_1) - \frac{J_w\Omega^2}{m_a L^2} DV_1(R_1)}{2\alpha^2}. \end{cases} \tag{48}$$

So far the mode shape functions of the second and third portions have four unknown constants in total. These four unknown constants can be calculated using Eq. (40). The first two equations of Eq. (40) can be written as

$$\begin{cases} B_{11}P_2(R_2) + B_{12}V_2(R_2) \\ \quad = B_{31}P_3(R_3) + B_{32}V_3(R_3), \\ B_{11}DP_2(R_2) + B_{12}DV_2(R_2) \\ \quad = -B_{31}DP_3(R_3) - B_{32}DV_3(R_3). \end{cases} \tag{49}$$

One considers

$$\begin{cases} B_{31} = B_{11}P_{31} + B_{12}P_{32}, \\ B_{32} = B_{11}V_{31} + B_{12}V_{32}, \end{cases} \tag{50}$$

where

$$\begin{aligned} P_{31} &= \frac{DV_3(R_3)P_2(R_2) + V_3(R_3)DP_2(R_2)}{DV_3(R_3)P_3(R_3) - V_3(R_3)DP_3(R_3)}, \\ P_{32} &= \frac{DV_3(R_3)V_2(R_2) + V_3(R_3)DV_2(R_2)}{DV_3(R_3)P_3(R_3) - V_3(R_3)DP_3(R_3)}, \\ V_{31} &= -\frac{DP_3(R_3)P_2(R_2) + P_3(R_3)DP_2(R_2)}{DV_3(R_3)P_3(R_3) - V_3(R_3)DP_3(R_3)}, \\ V_{32} &= -\frac{DP_3(R_3)V_2(R_2) + P_3(R_3)DV_2(R_2)}{DV_3(R_3)P_3(R_3) - V_3(R_3)DP_3(R_3)}. \end{aligned} \tag{51}$$

Using the last two equations of Eq. (40), we can obtain:

$$\begin{cases} B_{11}E_{11} + B_{12}E_{12} = 0, \\ B_{11}E_{21} + B_{12}E_{22} = 0, \end{cases} \Leftrightarrow \begin{bmatrix} E_{11} & E_{12} \\ E_{21} & E_{22} \end{bmatrix} \begin{Bmatrix} B_{11} \\ B_{12} \end{Bmatrix} = 0, \tag{52}$$

where

$$\begin{cases} E_{11} = D^2P_2(R_2) - \frac{J_w\Omega^2}{m_aL^2}DP_2(R_2) \\ \quad -P_{31}D^2P_3(R_3) - V_{31}D^2V_3(R_3), \\ E_{12} = D^2V_2(R_2) - \frac{J_w\Omega^2}{m_aL^2}DV_2(R_2) \\ \quad -P_{32}D^2P_3(R_3) - V_{32}D^2V_3(R_3), \\ E_{21} = D^3P_2(R_2) + \frac{m_w}{m_a}\Omega^2P_2(R_2) \\ \quad +P_{31}D^3P_3(R_3) + V_{31}D^3V_3(R_3), \\ E_{22} = D^3V_2(R_2) + \frac{m_w}{m_a}\Omega^2V_2(R_2) \\ \quad +P_{32}D^3P_3(R_3) + V_{32}D^3V_3(R_3). \end{cases} \tag{53}$$

Using the matrix form of Eq. (52), one can obtain

$$\left| \begin{bmatrix} E_{11} & E_{12} \\ E_{21} & E_{22} \end{bmatrix} \right| = E_{11}E_{22} - E_{12}E_{21} = 0. \tag{54}$$

Eq. (54) is the frequency equation, which is a transcendental equation. By using an iterative procedure based on linear interpolation, the first three natural frequencies are $f_1$ = 111 Hz, $f_2$ = 245 Hz, and $f_3$ = 547 Hz, respectively.

The calculation of the coefficients of the three mode shape functions are demonstrated in detail in the following.

The dimensionless mode shape functions can be written as

$$Z_k(Y_k) = B_{11}P_k(Y_k) + B_{12}V_k(Y_k), \quad k = 1, 2, 3. \tag{55}$$

One may set the deflection of the first particle to be $A$ and without loss of generality one may choose $A$ = 1, hence

$$A = 1 = Z_1(R_1) = B_{11}P_1(R_1) + B_{12}V_1(R_1). \tag{56}$$

From the above equation and Eq. (52), one can obtain the following equations:

$$\begin{cases} B_{11} = \frac{AE_{12}}{P_1(R_1)E_{12} - V_1(R_1)E_{11}}, \\ B_{12} = \frac{-AE_{11}}{P_1(R_1)E_{12} - V_1(R_1)E_{11}}. \end{cases} \tag{57}$$

Subsequently, substituting the last equations into Eq. (55) (assuming $k$ = 1) one can obtain the dimensionless mode shape function of the first portion $Z_1(Y_1)$ $(0 \le Y_1 \le R_1)$. Substituting Eq. (55) into Eq. (55) (assuming $k$ = 3) one can obtain the dimensionless mode shape function of the third portion.

$Z_3(Y_3)$ $(0 \le Y_3 \le R_3)$. By inserting Eqs. (47) and (48) into Eq. (36) (assuming $k$ = 2) one can obtain the dimensionless mode shape function of the second portion $Z_2(Y_2)$ $(0 \le Y_2 \le R_2)$.

Hence, the coefficient of the three mode shape functions can be calculated in Table 3.

**Table 3** Coefficients of the three modes

| Mode | $C_{11}$ | $C_{12}$ | $C_{13}$ | $C_{14}$ | $C_{21}$ | $C_{22}$ | $C_{23}$ | $C_{24}$ | $C_{31}$ | $C_{32}$ | $C_{33}$ | $C_{34}$ |
|---|---|---|---|---|---|---|---|---|---|---|---|---|
| 1st | −8.25 | 6.67 | −8.25 | 6.67 | −11.61 | −3.91 | −4.16 | 4.91 | −8.25 | 6.67 | −8.25 | 6.67 |
| 2nd | 2.13 | −1.81 | 2.13 | −1.81 | 1.03 | 3.53 | 2.66 | −2.53 | −2.13 | 1.81 | −2.13 | 1.81 |
| 3rd | 1.02 | −0.99 | 1.02 | −0.99 | −1.40 | 3.55 | 2.53 | −2.55 | 1.02 | −0.99 | 1.02 | −0.99 |

1st mode $f = 111$ Hz, $\alpha = 4.01$; 2nd mode $f = 245$ Hz, $\alpha = 5.96$; 3rd mode $f = 547$ Hz, $\alpha = 8.89$

## References


Andersson, C., & Abrahamsson, T. (2002). Simulation of interaction between a train in general motion and a track. *Vehicle System Dynamics, 38*(6), 433–455. doi:10.1076/vesd.38.6.433.8345.

Baeza, L., Vila, P., Rodaa, A., et al. (2008). Prediction of corrugation in rails using a non-stationary wheel-rail contact model. *Wear, 265*(9–10), 1156–1162. doi:10.1016/j.wear.2008.01.024.

Baeza, L., Vila, P., Xie, G., et al. (2011). Prediction of rail corrugation using a rotating flexible wheelset coupled with a flexible track model and a non-Hertzian/non-steady contact model. *Journal of Sound and Vibration, 330*(18–19), 4493–4507. doi:10.1016/j.jsv.2011.03.032.

Chaar, N. (2007). *Wheelset structural flexibility and track flexibility in vehicle/track dynamic interaction*. PhD Thesis, Royal Institute of Technology, Stockholm, Sweden.

Fayos, J., Baeza, L., Denia, F. D., et al. (2007). An Eulerian coordinate-based method for analysing the structural vibrations of a solid of revolution rotating about its main axis. *Journal of Sound and Vibration, 306*(3–5), 618–635. doi:10.1016/j.jsv.2007.05.051.

Jin, X. S., Xiao, X. B., Ling, L., et al. (2013). Study on safety boundary for high-speed train running in severe environments. *International Journal of Rail Transportation, 1*(1–2), 87–108. doi:10.1080/23248378.2013.790138.

Kaiser, I., & Popp, K. (2006). Interaction of elastic wheelsets and elastic rails: Modeling and simulation. *Vehicle System Dynamics, 44*(S1), 932–939. doi:10.1080/00423110600907675.

Meinders, T., & Meinker, P. (2003). Rotor dynamics and irregular wear of elastic wheelsets, system dynamics and long-term behavior of railway vehicles, track and subgrade. *System Dynamics and Long-term Behaviour of Railway Vehicles, Track and Subgrade, 6,* 133–152. doi:10.1007/978-3-540-45476-2_9.

Nielsen, J. C. O., Lundén, R., Johansson, A., et al. (2003). Train-track interaction and mechanisms of irregular wear on wheel and rail surfaces. *Vehicle System Dynamics, 40*(1–3), 3–54. doi:10.1076/vesd.40.1.3.15874.

Nielsen, J. C. O., Ekberg, A., & Lundén, R. (2005). Influence of short-pitch wheel/rail corrugation on rolling contact fatigue of railway wheels. *Journal of Rail and Rapid Transit, 219*(3), 177–188. doi:10.1243/095440905X8871.

Popp, K., Kruse, H., & Kaiser, I. (1999). Vehicle/track dynamics in the mid-frequency range. *Vehicle System Dynamics, 31*(5–6), 423–463. doi:10.1076/vesd.31.5.423.8363.

Popp, K., Kaiser, I., & Kruse, H. (2003). System dynamics of railway vehicles and track. *Archive of Applied Mechanics, 72*(11–12), 949–961. doi:10.1007/s00419-002-0261-6.

Qiu, J. B., Xiang, S. H., & Zhang, Z. P. (2009). *Computational structural dynamics*. Hefei, China: Press of University of Science and Technology of China. (in Chinese).

Shen, Z. Y., Hedrick, J. K., & Elkins, J. A. (1983). A comparison of alternative creep force models for rail vehicle dynamic analysis. *Vehicle System Dynamics, 12*(1–3), 79–83. doi:10.1080/00423118308968725.

Szolc, T. (1998a). Medium frequency dynamic investigation of the railway wheelset-track system using a discrete-continuous model. *Archive of Applied Mechanics, 68*(1), 30–45. doi:10.1007/s004190050144.

Szolc, T. (1998b). Simulation of bending-torsional-lateral vibrations of the railway wheelset-track system in the medium frequency range. *Vehicle System Dynamics, 30*(6), 473–508. doi:10.1080/00423119808969462.

Torstensson, P. T., & Nielsen, J. C. O. (2011). Simulation of dynamic vehicle-track interaction on small radius curves. *Vehicle System Dynamics, 49*(11), 1711–1732. doi:10.1080/00423114.2010.499468.

Torstensson, P. T., Pieringer, A., & Nielsen, J. C. O. (2012). *Simulation of rail roughness growth on small radius curves using a non-Hertzian and non-steady wheel-rail contact model* (pp. 223–230). In 9th International Conference on Contact Mechanics and Wear of Rail/Wheel Systems, Chengdu, China. doi:10.1016/j.wear.2013.11.032.

Wang, K. W. (1984). Wheel contact point trace line and wheel/rail contact geometry parameters computation. *Journal of Southwest Jiaotong University, 1,* 89–99. (in Chinese).

Xiao, X. B., Jin, X. S., & Wen, Z. F. (2007). Effect of disabled fastening systems and ballast on vehicle derailment. *Journal of Vibration and Acoustics, 129*(2), 217–229. doi:10.1115/1.2424978.

Xiao, X. B., Jin, X. S., Deng, Y. Q., et al. (2008). Effect of curved track support failure on vehicle derailment. *Vehicle System Dynamics, 46*(11),1029–1059. doi:10.1080/00423110701689602.

Xiao, X. B., Jin, X. S., Wen, Z. F., et al. (2010). Effect of tangent track buckle on vehicle derailment. *Multibody System Dynamics, 25*(1), 1–41. doi:10.1007/s11044-010-9210-2.

Zhai, W. M. (2007). *Vehicle/track coupling dynamics* (3rd ed.). Beijing, China (in Chinese): Chinese Science Press.

Zhong, S. Q., Xiao, X. B., Wen, Z. F., et al. (2013). The effect of first-order bending resonance of wheelset at high speed on wheel-rail contact behavior. *Advances in Mechanical Engineering, 2013,* 296106. doi:10.1155/2013/296106.

Zhong, S. Q., Xiao, X. B., Wen, Z. F., et al. (2014). *A new wheel-rail contact model integrated into a coupled vehicle-track system model considering wheelset bending* (pp. 1–10). In 2nd International Conference on Railway Technology Research, Development and Maintenance, Ajacocia, France. doi:10.4203/ccp.104.10.

## Author Biography

**Xue-song Jin** Ph.D., is a professor at Southwest Jiaotong University, Chengdu, China. He is a leading scholar of wheel–rail interaction in China and is a State Council Expert for Special Allowance. He is the author of three academic books and over 200 articles. He is an editorial board member of several journals and has been a member of the Committee of the International Conference on Contact Mechanics and Wear of Rail/Wheel Systems for more than 10 years. He has been a visiting scholar at the University of Missouri-Rolla for 2.5 years. Now, his research focuses on wheel–rail interaction, rolling contact mechanics, vehicle system dynamics, vibration, and noise.

# Influence of Wheel Polygonal Wear on Interior Noise of High-Speed Trains

**Jie Zhang, Guang-xu Han, Xin-biao Xiao, Rui-qian Wang, Yue Zhao and Xue-song Jin**

## Contents



J. Zhang · G. Han · X. Xiao · Y. Zhao · X. Jin (✉)
State Key Laboratory of Traction Power,
Southwest Jiaotong University, Chengdu 610031, China
e-mail: xsjin@home.swjtu.edu.cn

R. Wang
School of Urban Rail Transit, Changzhou University, Changzhou 213164, China

 
Y. Fang and Y.H. Zhang (eds.), *China's High-Speed Rail Technology*,
Advances in High-speed Rail Technology, https://doi.org/10.1007/978-981-10-5610-9_20

## 1 Introduction

Wheel polygonalization is one type of irregular wear of railway wheels. Until recently, it seemed that wheel polygonalization leads to a major problem of not only an increase in track and possibly vehicle maintenance, but also vehicle interior noise and passenger comfort reduction. This problem has not been completely solved. The polygonal phenomenon occurring on the rolling circles of railway wheels is often called wheel corrugation or wheel harmonic wear or wheel periodic out-of-roundness (OOR). Nielsen and Johansson (2000) discussed why out-of-round railway wheels develop and the damage they cause to track and vehicle components, and Nielsen et al. (2003) surveyed high-frequency train-track interaction and mechanisms of wheel/rail wear that is nonuniform in magnitude around/along the running surface. Johansson and Andersson (2005) and Johansson (2006) extended an existing multi-body system model for simulation of general 3D train-track interaction, which considered wheel/rail rolling contact mechanics and measured the transverse profile and surface hardness of 99 wheels on passenger trains, freight trains, commuter trains, and underground trains, and investigated wheel tread polygonalization. Furthermore, a series of site tests and numerical simulations about polygonal wheels were carried out by Morys (1999), Meinke and Meinke (1999), and Jin et al. (2012). While there is a lot of research on the effect of wheel polygonal wear on the dynamic behavior of the vehicle/track, there are few studies on its noise problems, especially of high-speed trains. The few studies on the noise problem related to wheel polygonal wear are mainly divided into two categories: (1) for the wheel polygonal wear problem, because it is very complex and has not been completely solved, researchers focus on the vehicle/track system dynamics to study its mechanism. They have findings on such things as the effects of tread braking, stiffness of axle, wheel material, etc.; (2) for the wheel/rail noise problem, others have focused on their acoustic characteristics and actively have designed a low-noise wheel/rail (Bouvet et al. 2000; Jones and Thompson 2000; Thompson and Gautier 2006; Behr and Cervello 2007). However, wheel polygonal wear leads to ever more vehicle noise problems on the high-speed railways of China (Zhang et al. 2013). The focus here is on the characteristics of high-order polygonal wear and its influence on wheel/rail noise, and the interior noise of high-speed trains. We also discuss whether the current criteria used for wheel re-profiling of China's high-speed trains are suitable from the point of view of noise control. This work presents a detailed investigation through extensive experiments and numerical simulations.

## 2 Measurement of Wheel Polygon and Vehicle Noise and Vibration

### 2.1 Test Overview

From long-term field experiments on high-speed trains, it was found that wheel polygonal wear caused a series of interior noise problems. These occurred suddenly and seriously, and the railway operation departments of China called these "abnormal interior noise", but they did not know the mechanism of their productions (Zhang et al. 2013). The present work conducted a typical "abnormal interior noise" analysis. The high-speed train under investigation is made up of 16 coaches and its business operational speed is 300 km/h. The sketch of the coach generating "abnormal interior noise" is shown in Fig. 1.

There is a microphone at a vertical height of 1.5 m above the interior floor used for testing the interior noise and a surface microphone installed on the exterior floor for testing the exterior noise. An accelerometer was fixed on the interior floor to measure the vertical vibration of the floor, and three accelerometers were fixed on the axle box, the bogie frame, and the car body, respectively, to measure the vertical vibration of the bogie.

Figure 2 shows the test photos. Both the noise and the vibration before and after re-profiling were tested, as well as the wheel roughness.

The vibration and acoustic measurements were conducted using a B&K PULSE platform, including a B&K 4190 microphone, a B&K 4948 surface microphone, four B&K 4508 accelerometers, and a B&K Type 3560D data acquisition hardware. The wheel roughness measurement was carried out using a Müller-BBM's m| wheel.

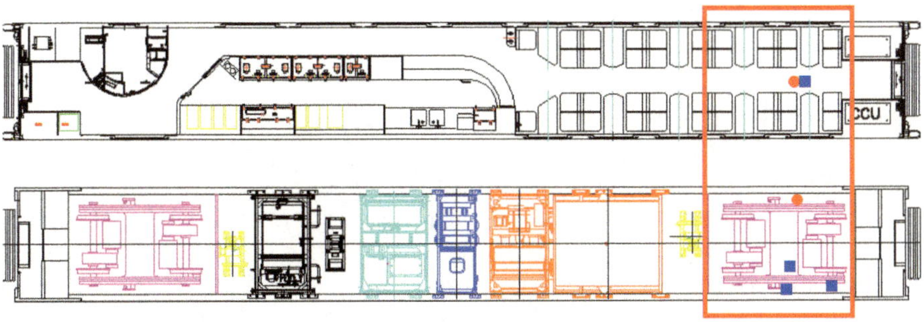


**Fig. 1** Measuring points on the high-speed coach generating "abnormal interior noise". "*Red circle*" refers to the acoustic measuring points, and "*blue square*" denotes the vibration measuring points

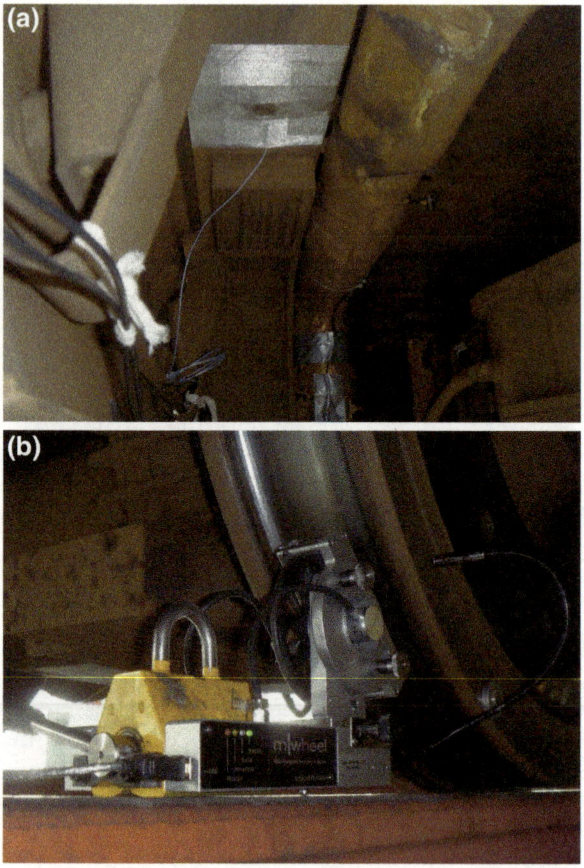


**Fig. 2** Test photos: **a** surface microphone installed on the exterior floor; **b** wheel roughness measurement

## 2.2 *Characteristics of Wheel Diameter Difference and Polygon*

Figure 3 shows the roughness and the polygon order of the wheel circumference of the left 4th axle of the coach bogie before and after re-profiling. The roughness results (Fig. 3a) are based on the wheel diameter before and after re-profiling.

From Fig. 3a, it can be seen that the roughness of the wheel before re-profiling is very large. For example, at 240°, the roughness is about −0.573 mm of the wheel before re-profiling, and it is about 0.001 mm after re-profiling. Here, −0.573 mm means its roughness is much bigger than 0.001 mm. The diameter difference of the wheel before re-profiling is up to 0.795 mm. The wheel diameter difference here is defined by

$$D = R_{\max} - R_{\min}, \tag{1}$$

where $R_{\max}$ and $R_{\min}$ are the maximum and minimum wheel roughness in numerical value, respectively.

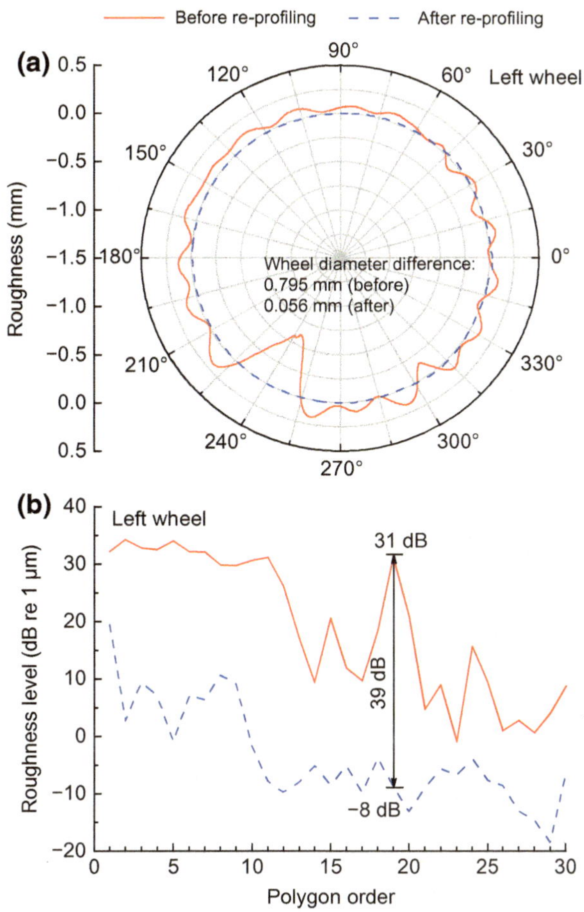


**Fig. 3** Wheel roughness and polygon order: **a** wheel roughness; **b** polygon order

The statistical results of wheel roughness tests show that the diameter differences of all other wheels are less than 0.1 mm except the wheels of the 4th axle before re-profiling. After re-profiling, the diameter difference of all the wheels are less than 0.1 mm. Figure 3b indicates the polygon order distributions corresponding to the measured results as shown in Fig. 3a. In Fig. 3b, the horizontal axis illustrates the polygon order and the vertical axis denotes the amplitudes of the wheel polygons. The peaks mean that the corresponding polygons have a large contribution to the uneven wear of the wheels. Figure 3b shows the high peak at the ordinate of 19, which indicates that the 19th order polygonal wear of the wheel is very serious. After re-profiling, the roughness level of the 19th order polygon reduced nearly 39 dB.

## 2.3 *Effect of Re-profiling on Vehicle Noise and Vibration*

Here, the train is running at 293 km/h (the actual tested speed). Figure 4 shows the measured results of noise and vibration before and after re-profiling. Before and after re-profiling with polygonal wear, the acceleration level differences of the axle box, the bogie frame, and the car body reach almost 16–19 dB (Fig. 4a). This phenomenon is particularly obvious on the axle box whose acceleration level decreases from 46 to 27 dB. The wheel polygonal wear can cause a very high acceleration level of the axle box. In addition, such a fierce vibration due to the wheel polygonal wear excitation is further transmitted into the coach and the track infrastructure. The acceleration level of the interior floor increases by 7 dB.

The wheel polygonal wear increases the extra exterior noise in bogie area by 9 dB(A). The big exterior noise and the strong vehicle vibration eventually result in the 11 dB(A) increase of the interior noise (Fig. 4a). The interior noise level directly affects the ride comfort of a high-speed train. In this test, the 11 dB(A) overall level increase of the interior noise is mainly caused by the significant noise

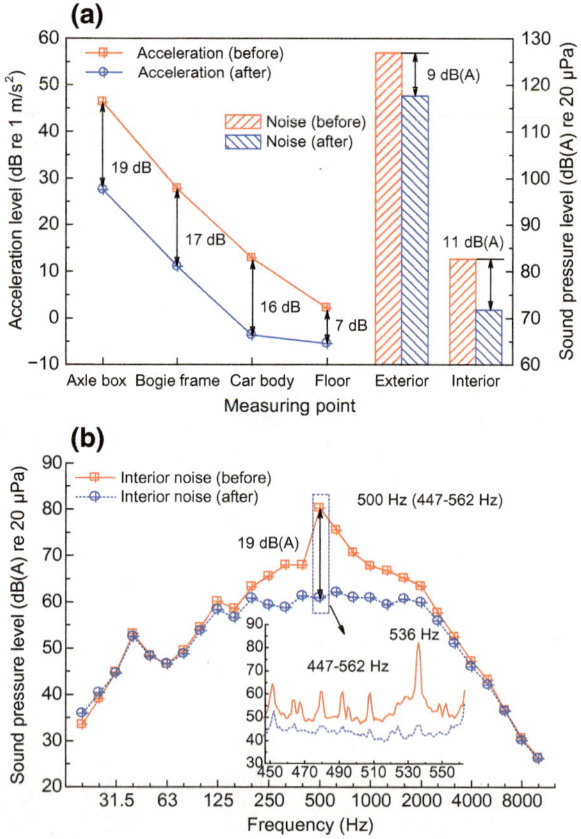


**Fig. 4** Noise and vibration before and after re-profiling: **a** overall levels; **b** 1/3 octave band and FFT

components in the 1/3 octave band centred at 500 Hz (Fig. 4b). The sound pressure level in this 1/3 octave band before re-profiling is nearly 19 dB(A) bigger than after re-profiling, and the 1/3 octave band centred at 500 Hz includes those at 447–562 Hz frequencies. In this range, it is found, by fast Fourier transformation (FFT) analysis, that there is the highest peak at 536 Hz as shown in Fig. 4b. Here 536 Hz is just the passing frequency of the 19th order polygon when the train is running at 293 km/h.

For a given train speed $v$, the passing frequency of the wheel polygon is calculated by

$$f = \frac{v/3.6}{\pi d} \times i, \tag{2}$$

where $d$ ($d$ = 0.92 m) is the wheel diameter, and $i$ is the polygon order. Hence, the 19th order polygon of the wheel wear can excite the strong wheel/rail vibration and the big noise at about 540 Hz at about 300 km/h.

## 3 Model of Polygonal Wheel for Wheel/Rail Rolling Noise Calculation

From the above, it is clear that the wheel polygonal wear has a major impact on the interior noise and exterior noise of the high-speed train. However, up to now there has been little understanding or research on the effect of the wheel polygonal wear on the interior and exterior noise of high-speed trains, and the wheel/rail rolling noise. The maintenance regulation of China's high-speed trains now has a criterion of high-speed wheel re-profiling based on wheel diameter difference due to uneven wear, regardless of the wheel polygonal wear characteristics and its mechanisms. The current wheel re-profiling criterion involves only the effect of the wheel uneven wear on the vehicle dynamic performance, but neglects that of the interior noise problem related to the uneven wear characteristics of the wheel. There are two problems with this criterion: (1) the serious uneven wheel wear generally occurs due to the wheel polygonal wear of low order (the low passing frequencies) and high amplitudes, which can cause the annoying interior noise; (2) in spite of mild uneven wheel wear (i.e., not a large wear amplitude), the wheel polygonal wear of high order (high passing frequencies) still creates a large interior noise. If the wheel uneven wear situation is classified as (2) in maintenance, it can avoid re-profiling according to the present criterion. So the re-profiling criteria for high-speed wheels need to involve the effect of the polygonal wear on not only the dynamic behavior but also on the noise of the vehicle.

## 3.1 *Characteristics of Two Wheels with the Same Diameter Difference*

Figure 5 indicates the measured results of nearly the same diameter difference but the different wheel polygon distributions of the two high-speed wheels, A and B. Figure 5a shows that the diameter differences of the two wheels are about 0.054 mm. However, their wheel polygonal wear is different (Fig. 5b). In general, the roughness amplitudes of most order polygons of the wheel B are larger than those of the wheel A. However, wheel A shows the 20th order polygonal wear with a high peak, as indicated in Fig. 5b.

To investigate the noise difference caused by a very similar diameter difference of the high-speed wheels with different polygon distributions, the high-speed wheel-track noise software (HWTNS) is used to analyze wheel/rail noise of the two wheels running at 300 km/h.

**Fig. 5** Wheel roughness and polygon order: **a** wheel roughness; **b** polygon order

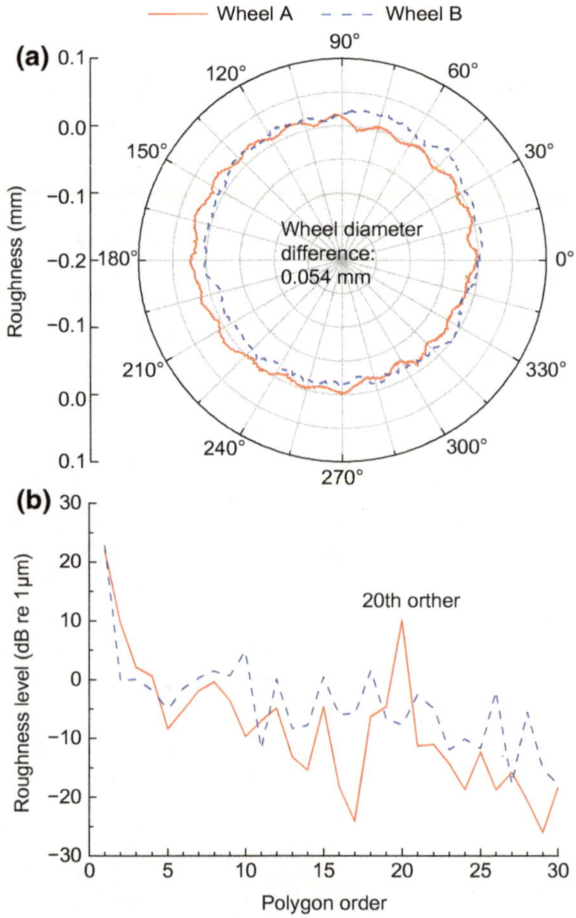

## 3.2 *Theory of Wheel/Rail Rolling Noise Prediction*

The HWTNS was developed by Wu and Thompson (1999, 2000, 2001). It uses a similar conformation to the track-wheel interaction noise software (TWINS) (Thompson et al. 1996a, b), but adds the ballastless track model of high-speed railway (Wu 2012). Figure 6 shows the flowchart of the HWTNS.

For the ballastless track model in the HWTNS, a wheel/rail interaction model is used to calculate wheel/rail dynamic force based on wheel/rail combined roughness. This dynamic force has effects on the rolling contact of the wheel and the rail and causes the vibration and the noise radiation. The dynamic force can also transmit to the track infrastructure through the rails and fasteners, which causes vibration and noise radiation of the sleepers or slabs. More details are given in (Wu 2012).

The vibration and sound radiation of the wheel is calculated using a semi-analytical method. They are divided into three parts, namely vibration and sound radiation caused by (i) wheel axial modes, (ii) wheel radial modes, and (iii) wheel rim rotation modes. The overall sound power level of the wheel is obtained by the summation of the sound radiation of each mode (Thompson 1997):

$$W_{\mathrm{w}} = \rho c \sum_{n} \left\{ \sigma_{\mathrm{a}}(n) \sum_{j} S_{\mathrm{a}j} \overline{\tilde{v}_{\mathrm{a}jn}^{2}} + \sigma_{\mathrm{r}}(n) S_{\mathrm{r}} \overline{\tilde{v}_{\mathrm{r}n}^{2}} + \sigma_{\mathrm{t}}(n) S_{\mathrm{t}} \overline{\tilde{v}_{\mathrm{t}n}^{2}} \right\}, \quad (3)$$

where $\rho$ is the density of the air and $c$ is the sound velocity in the air. $\sigma_{\mathrm{a}}(n)$, $\sigma_{\mathrm{r}}(n)$, and $\sigma_{\mathrm{t}}(n)$ are the sound radiation ratios of the wheel axial vibration, the wheel radial

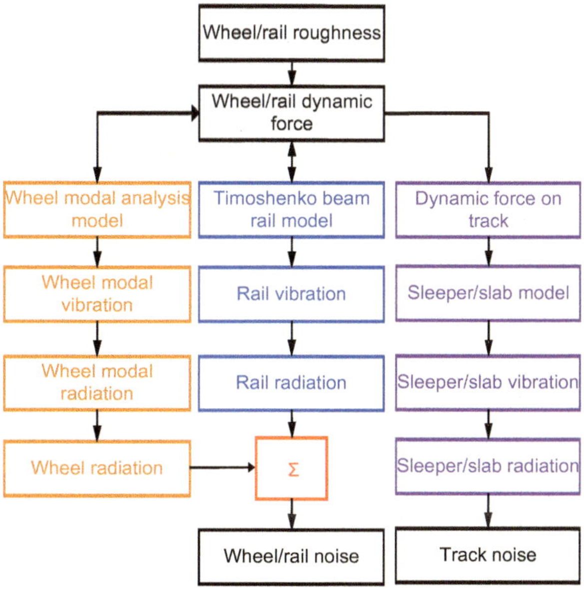


**Fig. 6** Wheel/rail noise prediction using the HWTNS (Wu 2012), with permission from the author

vibration, and the wheel rim rotation of $n$ nodal diameter, respectively. $\overline{v^2_{ajn}}$, $\overline{v^2_{rn}}$, and $\overline{v^2_{tn}}$ are the mean square velocities of the wheel axial vibration, the wheel radial vibration, and the wheel rim rotation of $n$ nodal diameter, respectively. $S_{aj}$, $S_r$, and $S_t$ are the sound radiation areas of these mean square velocities, respectively. For the wheel axial vibration, the area of the wheel web is large and its velocity generates great variation, so in the calculation of the vibration and sound radiation of the wheel it needs to be divided into circles with different radii and then the vibration and sound radiation of the circles are summed.

Figure 7 shows the divided circles of the wheel in the HWTNS.

The sound radiation areas are given by Thompson (1997):

$$S_{aj} = 2\pi(r_j^2 - r_{j+1}^2), \tag{4}$$

$$S_r = 2\pi rw + 2\pi r_2(w - w_{web}), \tag{5}$$

$$\begin{aligned} S_t = {} & 2\pi r\frac{w^3}{12} + 2\pi r_2\left(\frac{w^3}{12} - \frac{w_{web}^3}{12}\right) \\ & + 4\pi\left(\frac{r+r_2}{2}\right)\frac{(r-r_2)^3}{12}, \end{aligned} \tag{6}$$

where $r$ is the radius of the wheel, $r_j$ ($j$ = 1, 2, …, 7) is the radius of each divided circle, $w$ is the width of the wheel tire in its cross-section, and $w_{web}$ is the width of the wheel web in its cross-section (Fig. 7).

The sound radiation ratio of each nodal diameter mode is given by Thompson (1997):

(i) Wheel axial vibration

$$\begin{aligned} \sigma_a(n) &= \frac{1}{1 + (f_{ca}(n)/f)^{2n+4}}, \\ f_{ca}(n) &= \frac{c}{2\pi r}\sqrt{(n+1)(n+4)}; \end{aligned} \tag{7}$$

(ii) Wheel radial vibration

$$\begin{aligned} \sigma_r(n) &= \frac{1}{1 + (f_{cr}(n)/f)^{2n+2}}, \\ f_{cr}(n) &= \frac{c}{2\pi w}\sqrt{(n+1)(n+4)}; \end{aligned} \tag{8}$$

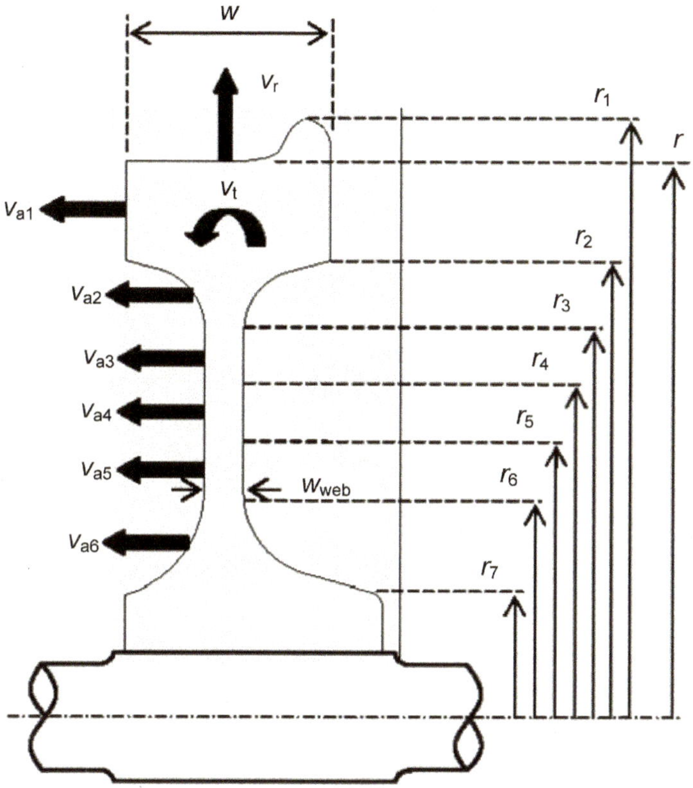


**Fig. 7** The divided circles of the wheel in the HWTNS (Wu 2012), with permission from the author

(iii) Wheel rim rotation

$$\sigma_t(n) = \frac{1}{1 + (f_{ct}(n)/f)^{2n+4}},$$
$$f_{ct}(n) = \frac{c}{\pi w}\sqrt{(n+1)(n+4)}, \tag{9}$$

where $f$ is the frequency. and $f_{ca}$, $f_{cr}$, and $f_{ct}$ are the critical frequencies when the bending wavelength of the wheel vibration and the acoustic wavelength are equal.

The rail is defined as the infinite line sound source and its radiation is given by Wu (2012):

$$W_r = \rho c L h \left\langle \overline{v_r^2} \right\rangle \sigma_r, \tag{10}$$

where $L$ is the length of the rail, $h$ is the length of the rail cross-section profile in the vertical projection, and $\sigma_r$ is the rail sound radiation ratio. $\left\langle \overline{v_r^2} \right\rangle$ is the mean square velocity of vertical vibration of the rail and is given by

$$\left\langle \overline{v_r^2} \right\rangle = \frac{1}{L} \sum v_r^2(z) \Delta z, \tag{11}$$

where $v_r$ is the vertical velocity of the rail in the length of $\Delta z$.

The slab sound radiation is given by Wu (2012):

$$\begin{aligned} W_p &= \rho c B L \left\langle \overline{v_p^2} \right\rangle \frac{1}{1 + (f_{cp}/f)^2}, \\ f_{cp} &= \frac{c}{\sqrt{2\pi B L_p}}, \end{aligned} \tag{12}$$

where $\left\langle \overline{v_p^2} \right\rangle$ is the mean square velocity of vertical vibration of the slab in the track length of $L$, $L_p$ is the length of the slab, and $B$ is the width of the half slab. When the slab is considered as a rigid body, the mean square velocity of vertical vibration of the slab in Eq. (12) is given by

$$\begin{aligned} \left\langle \overline{v_p^2} \right\rangle &= \frac{1}{N_p} \sum_{i=1}^{N_p} \overline{v_{pi}^2}, \\ \overline{v_{pi}^2} &= \frac{1}{L_p} \int_{-L_p/2}^{L_p/2} (v_{ci} + \dot{\theta}_{ci} x)^2 \mathrm{d}x = v_{ci}^2 + \frac{1}{12} \dot{\theta}_{ci}^2 L_p^2, \end{aligned} \tag{13}$$

where $\overline{v_{pi}^2}$ is the mean square velocity of vertical vibration of single slab, $N_p$ is the quantity of the slabs in the track length of $L$, and $v_{ci}$ and $\dot{\theta}_{ci}$ are the vertical velocity of the slab mass center and the rotational velocity around its mass center, respectively.

Predicting the wheel/rail noise using the HWTNS needs three input files including wheel data, wheel/rail combined roughness, and rail sound radiation ratio. The wheel data is analyzed using the wheel finite element (FE) model based on the commercial ANSYS. The wheel/rail combined roughness is calculated using the method described in (Thompson et al. 1996a, 1996). The rail roughness is the

**Table 1** Parameters in the HWTNS

| Parameter | Value |
|---|---|
| *Rail* | |
| Density (kg/m$^3$) | 7850 |
| Tensile modulus (MPa) | $2.1 \times 10^5$ |
| Shear modulus (MPa) | $7.7 \times 10^4$ |
| Shear factor | 0.4 |
| Cross-sectional area (m$^2$) | $7.69 \times 10^{-3}$ |
| Area moment of inertia (m$^4$) | $3.055 \times 10^{-5}$ |
| Projected length of radiation (m) | 0.413 |
| *Slab* | |
| Density (kg/m$^3$) | 2500 |
| Stiffness of cement mortar (MPa/m) | 4000 |
| Damping loss factor | 0.2 |
| Length (m) | 6.0 |
| Width (m) | 2.5 |
| Thickness (m) | 0.2 |
| *Fastener* | |
| Stiffness (MN/m) | 60 |
| Damping loss factor | 0.2 |
| Spacing (m) | 0.6 |

Class A selected from the HARMONOISE project (van Beek and Verheijen 2003). The rail sound radiation ratio calculation uses the data of UIC 60 rail.

The other parameters used are shown in Table 1.

## 3.3 Wheel/Rail Rolling Noise Prediction Results

Figure 8 shows the wheel/rail noise prediction results of wheel A and wheel B at 300 km/h.

From Fig. 8, it can be seen that the sound power levels of wheel B and the rail in rolling contact are higher than that of the wheel A and the rail in all 1/3 octave bands except for one band centred at 500 Hz. It is because the roughness amplitudes of most order polygons of the wheel B are higher than those of wheel A, but the peak of the 20th order polygon of wheel A is very prominent (Fig. 5b, also see the small picture on the left in Fig. 8). When the train operates at 300 km/h, the 20th order polygon of the wheel would excite vibration with frequency at about 577 Hz, which is in the 1/3 octave band centred at 630 Hz. However, the simulation result (Fig. 8) shows that the sound power level of wheel A and the rail is higher than that of wheel B and the rail in the 1/3 octave band centred at 500 Hz, not 630 Hz. The reason for this 1/3 octave frequency band difference is attributed to

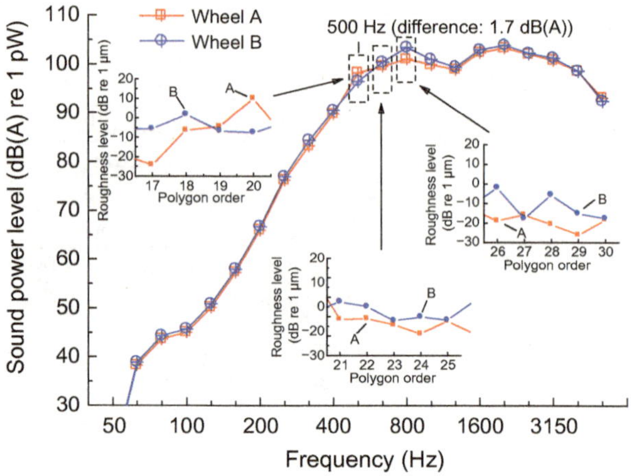


**Fig. 8** Sound power level of wheel/rail noise

**Table 2** The difference of 1/3 octave frequency bands between the simulated and actual cases

| Polygon order | Actual case | | Simulated case | | |
|---|---|---|---|---|---|
| | Frequency (Hz) | 1/3 octave frequency band (Hz) | Wavelength (m) | 1/3 octave wavelength band (m) | 1/3 octave frequency band (Hz) |
| 13 | 375.01 | 400 | 0.2222 | 0.2 | 400 |
| 14 | 403.86 | 400 | 0.2063 | 0.2 | 400 |
| 15 | 432.71 | 400 | 0.1926 | 0.2 | 400 |
| 16 | 461.55 | 500 | 0.1806 | 0.2 | 400 |
| 17 | 490.40 | 500 | 0.1699 | 0.16 | 500 |
| 18 | 519.25 | 500 | 0.1605 | 0.16 | 500 |
| 19 | 548.09 | 500 | 0.1520 | 0.16 | 500 |
| 20 | 576.94 | 630 | 0.1444 | 0.16 | 500 |
| 21 | 605.79 | 630 | 0.1376 | 0.125 | 630 |
| 22 | 634.63 | 630 | 0.1313 | 0.125 | 630 |
| 23 | 663.48 | 630 | 0.1256 | 0.125 | 630 |
| 24 | 692.33 | 630 | 0.1204 | 0.125 | 630 |
| 25 | 721.18 | 800 | 0.1156 | 0.125 | 630 |

the correspondence between the wavelengths and the frequencies in the wheel/rail noise calculation. Table 2 shows the difference of 1/3 octave frequency bands between the simulated and actual cases.

As shown in Table 2, when the train operates at 300 km/h, the actual excitation (passing) frequencies of the 13th–25th order polygons of the wheel and their 1/3 octave frequency bands are shown, and their correspondences used in the numerical simulation are also shown. In the wheel/rail noise calculation (in the HWTNS), part of the input data is the wheel/rail combined roughness in 1/3 octave frequency band (some of the bands are as indicated in Table 2). So the FFT analysis on the measured wheel circle irregularity samples is first carried out to translate them to the samples expressed with different wavelengths. Then the different wavelengths are allocated into several 1/3 octave wavelength bands [Table 2, and Eqs. (20) and (21)], and further transformed into 1/3 octave frequency bands at the specific speed (300 km/h). Obviously, the wavelength of the 20th order polygon is 0.1444 m which belongs to the 1/3 octave wavelength band centered at 0.16 m, and this 1/3 octave wavelength band is transformed into 1/3 octave frequency band centered at 500 Hz (not 630 Hz) at 300 km/h. That is the root cause of the 1/3 octave frequency band difference between the numerical simulation and the actual result of the 20th order polygon. In addition, the wavelength 0.1444 m of the 20th order polygon is just the lower boundary value of the 1/3 octave wavelength band centered at 0.16 m, and neighbors the 1/3 octave wavelength band centered at 0.125 m. During the transformation from the measured wheel circle irregularity samples to the wavelength samples through the FFT analysis, the partial power of the 20th order polygon leaks into the 1/3 octave wavelength band centred at 0.125 m (corresponding to 630 Hz) though it belongs to the 1/3 octave wavelength band centred at 0.16 m (corresponding to 500 Hz). Therefore, in spite of the visible 20th order polygon occurring on wheel A, the wheel/rail rolling noise calculation results in the 1/3 octave band centred at 500 Hz and is only a little higher than that of the wheel B.

The sound power levels of wheel B in the 1/3 octave bands centred between 630 Hz and 800 Hz are much higher than those of wheel A owing to the larger roughness of the 21st to the 25th and the 26th to the 30th order polygons (see the small pictures at the bottom and in the right of Fig. 8, respectively).

In summary, although there are some differences between the actual data and the calculated sound power levels of the wheel polygons in 1/3 octave frequency bands, the calculation results can still show well the characteristics of the wheel polygon excitation at high-speed. The overall sound power levels of wheel A and wheel B are 110.9 dB(A) and 111.6 dB(A), respectively. It can be seen that a very similar wheel diameter difference can cause different wheel/rail rolling noises because of different distribution of the wheel polygon order.

## 4 Prediction Model of Interior Noise of Coach

To further study the influence of wheel polygonal wear on the interior noise of a high-speed train, the simulation model of the coach end is built up by means of the vibro-acoustic analysis software VA One 2012. The model is developed based on a hybrid of the finite element method and the statistic energy analysis (Langley and Bremner 1999; Shorter and Langley 2005; Cotoni et al. 2007), which is called the hybrid FE-SEA for short.

## 4.1 Theory of the Hybrid FE-SEA

The FE method and the SEA are at present the two main methods for solving the acoustic problem of complex structural systems. The FE method usually works well in solving the structural acoustic problem in the low-frequency range, where the modal density is low and the system exhibits a global modal behavior. The SEA is usually used for solving the acoustic problem of complex structural systems at high frequencies, where the modal densities of all the subsystems are very high. For the high-speed train, because its large size leads to the generation of numerous elements in modeling the coach for solving its noise problem, it is difficult to solve the vibro-acoustic issue only using the FE method. Furthermore, the coach interior noise caused by wheel polygonal wear is always prominent in the middle frequency range, which is defined as from about 200 to 1000 Hz in the present analysis (Fig. 4b). So the hybrid FE-SEA is used in the noise calculation.

However, the coupling of FE and SEA in a single model is difficult because the two methods differ in two ways: (i) the FE method is based on dynamic equilibrium while the SEA is based on the conservation of energy flow, and (ii) the FE method is a deterministic method while the SEA is an inherently statistical method. Here the two main equations of the hybrid method are given as follows (Shorter and Langley 2005; Cotoni et al. 2007):

$$\omega(\eta_j + \eta_{\mathrm{d},j})E_j + \sum_k \omega\eta_{jk}n_j(E_j/n_j - E_k/n_k) = P^{\mathrm{ext}}_{\mathrm{in},j}, \tag{14}$$

$$S_{qq} = D_{\mathrm{tot}}^{-1}\left[S_{ff} + \sum_k \left(\frac{4E_k}{\omega\pi n_k}\right)\mathrm{Im}\left\{D_{\mathrm{dir}}^{(k)}\right\}\right]D_{\mathrm{tot}}^{-1*T}, \tag{15}$$

where

$$P^{\mathrm{ext}}_{\mathrm{in},j} = \frac{\omega}{2}\sum_{rs}\mathrm{Im}\left\{D^{(j)}_{\mathrm{dir},rs}\right\}\left(D_{\mathrm{tot}}^{-1}S_{ff}D_{\mathrm{tot}}^{-1*T}\right)_{rs}, \tag{16}$$

$$\omega\eta_{jk}n_j = \frac{2}{\pi}\sum_{rs}\mathrm{Im}\left\{D^{(j)}_{\mathrm{dir},rs}\right\}\left(D_{\mathrm{tot}}^{-1}\mathrm{Im}\left\{D_{\mathrm{dir}}^{(k)}\right\}D_{\mathrm{tot}}^{-1*T}\right)_{rs}, \tag{17}$$

$$\omega\eta_{\mathrm{d},j} = \frac{2}{\pi n_j}\sum_{rs}\mathrm{Im}\left\{D_{\mathrm{d},rs}\right\}\left(D_{\mathrm{tot}}^{-1}\mathrm{Im}\left\{D_{\mathrm{dir}}^{(j)}\right\}D_{\mathrm{tot}}^{-1*T}\right)_{rs}. \tag{18}$$

The definitions of the variables in Eqs. (14)–(18) can be found in (Langley and Bremner 1999; Shorter and Langley 2005; Cotoni et al. 2007). Equation (14) represents the subsystem energy balance in the SEA, in which $P^{\mathrm{ext}}_{\mathrm{in},j}$ is the external power input, and Eq. (15) describes the system response by the FE method, while $S_{qq}$ is the ensemble average cross-spectral response. Equations (14)–(18) couple the FE with the SEA. Equation (14) has a precise form of the SEA, but the coupling loss factor $\eta_{jk}$ and the loss factor $\eta_{d,j}$ are calculated using the FE model. Furthermore, Eq. (15) has the form of a standard random FE analysis, but additional forces arise from the reverberant energies in the SEA subsystems. If the SEA subsystems are not included, it is obvious that the present hybrid method is just the FE method. On the other hand, if only the junctions between the SEA subsystems are modeled by the FE method, the hybrid method becomes the SEA method.

## *4.2 Interior Noise Simulation Model of the Coach End*

The hybrid FE-SEA simulation model of the high-speed coach end includes the beam, the plate, and the acoustic cavity subsystems. Figure 9 presents the hybrid model in VA One 2012.

Because the modal densities of the beam subsystems are very low, the beam subsystems are meshed into the subsystems characterized by the FE models. The noise sources, including the wheel/rail noise and the aerodynamic noise, are extracted from the field pass-by test data of the high-speed trains operating at 300 km/h. In the test, B&K 4948 surface microphones are fixed on the external floor above the bogie and on the external side wall of the car body. So the wheel/rail

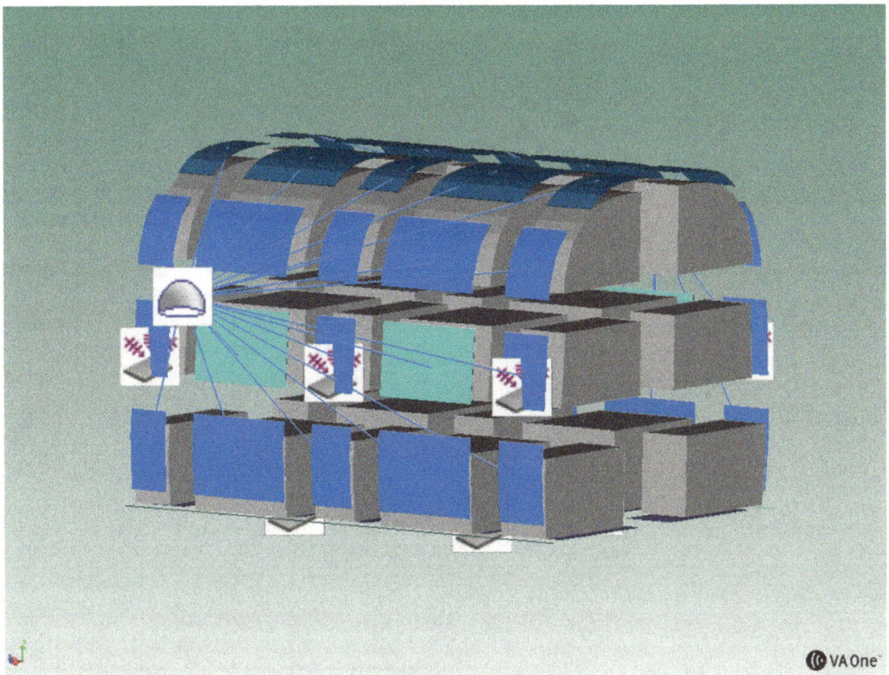


**Fig. 9** The hybrid FE-SEA simulation model of coach end

noise and the aerodynamic noise are treated approximately as the diffuse acoustic field (DAF) loaded on the plate. Transmission loss (TL) data, including from the floor, the side wall, and the roof, is taken from the test results in the acoustics laboratory, and defined through the area junction. A semi-infinite fluid (SIF) is created and connected to the car body to simulate the outside acoustic environment.

Figure 10 shows the interior noise results of the numerical simulation and the field experiment, including before and after re-profiling. In the simulation, the different cases change only the wheel/rail noise inputs.

From Fig. 10, it can be seen that both before and after re-profiling, the interior acoustic energy is dominant in the frequency range from 200 to 2000 Hz, which belongs to the middle frequency range. The predicted responses and the measured sound pressure level are generally in good agreement. The difference of the overall sound pressure level is less than 2 dB(A). Thus, the interior noise hybrid FE-SEA model established in this study is reliable and effective. However, in the low-frequency range (frequencies lower than 200 Hz), simulation results do not

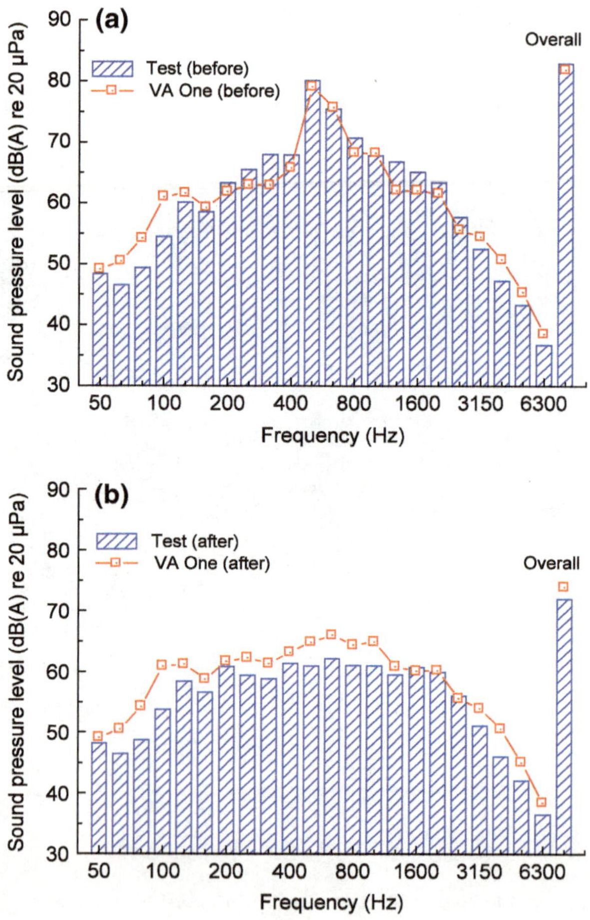


**Fig. 10** Results of the numerical simulation and the experiment before (**a**) and after (**b**) re-profiling

agree well with experimental results. This is because the hybrid FE-SEA model is more suitable for the mid-frequency. For low-frequency, simulation results could be improved by the use of a boundary element method (BEM) fluid and a set of plane waves. Further investigation on this topic is being conducted.

The wheel polygons which are going to be discussed have the characteristics of high-order polygons (13–30). They can excite strong vibration with frequencies from 375 to 865 Hz (at 300 km/h), which are typical mid-frequency. So in the following sections, the hybrid model is used to calculate the effect of different wheel polygon characteristics on the coach interior noise. We examine different wheel/rail noise sources due to different wheel polygon characteristics.

# 5 Influence of Different Wheel Polygonal Wear on Noise

## 5.1 Characteristics of Wheel Polygon

Usually, the wheel polygonal wear has three characteristics, which should be discussed before carrying out the analysis. They are (i) the different polygon order with very similar roughness levels, (ii) very similar polygon order with different roughness levels, and (iii) very similar polygon order and very similar roughness levels with different distribution phases of the polygons. These characteristics can be found from the detailed analysis of the measured wheel circle irregularity data.

A MATLAB program is developed to construct a series of different wheel polygons. The code technique routine is indicated with Fig. 11.

The created wheel circle irregularity sample consists of a series of sine curves, $A_i\sin(\omega_i x + \varphi_i)$ ($i = 1, 2, \ldots, n$), where $A_i$ is the roughness amplitude of wave $i$, $\omega_i$ is the wavenumber and $\varphi_j$ is the phase angle. First, use a bandpass filter [here we use the Chebyshev filter (Williams and Taylors 2006)], which filters out each polygon and obtains its roughness level. Then all filtered wheel circle irregularity samples are summed and these samples correspond to their polygon order. It is checked whether the summed sample is correct for the original sample. If not, the bandpass filter should be redesigned. In addition, the FFT analysis can translate wheel circle irregularity samples into wavenumber samples using

$$\mathbf{X}(k) = \sum_{j=1}^{N} \mathbf{x}(j)\omega_N^{(j-1)(k-1)}. \tag{19}$$

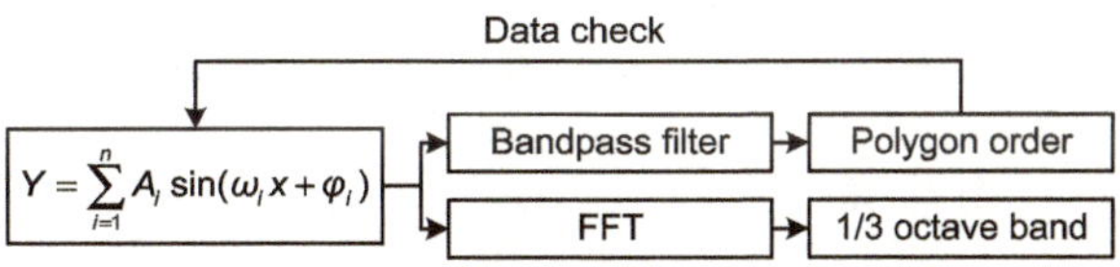


**Fig. 11** MATLAB code technique routine for creating wheel polygon data

Equation (19) implements the transform given for vectors of length $N$, where $\omega_N = e^{(2\pi i)/N}$, which is the $N$th root of the unit.

Through FFT analysis, the 1/3 octave wavelength band can be calculated. The relationship between the centre wavelength and the cut-off wavelength of 1/3 octave is given by

$$\lambda_{\text{upper}} = \sqrt{2^n}\lambda_c, \tag{20}$$

$$\lambda_{\text{lower}} = \lambda_c/\sqrt{2^n}, \tag{21}$$

where $\lambda_{\text{upper}}$, $\lambda_{\text{lower}}$, and $\lambda_c$ are the upper cut-off wavelength, the lower cut-off wavelength, and the centre wavelength, respectively, where $n$ = 1/3 represents a 1/3 octave.

## 5.2 *Effect of Different Order of Wheel Polygon on Noise*

According to the extensive investigation into high-speed wheel roughness and vehicle noise, mainly the 19th and the 20th order wear polygons are those which quite often cause serious noise problems. To study the influence of different polygon order of the worn wheels, a MATLAB program is used to generate polygon wheel shapes with order from 13 to 30. Their roughness levels are 30 dB.

Figure 12a shows the different polygons versus their order, and Fig. 12b shows their roughness levels versus their wavelengths.

Although there are 18 polygon order (Fig. 12a), there are only four main peaks with different wavelengths in 1/3 octave (Fig. 12b). The root cause of this phenomenon has already been discussed in Sect. 3.3 (Table 2). Because noise peaks are mainly due to wheel polygon peaks (there are only four main peaks as indicated in Fig. 12b), and to avoid some critical polygon order (Table 2), the 15th, the 19th, the 24th, and the 30th order polygons are selected in calculating the effect of them on wheel/rail noise and interior noise using the models described in the Sects. 3 and 4.

Figure 13a, b indicate the effect of different polygon order on the wheel/rail noise and the interior noise at 300 km/h, respectively.

By using the obvious wheel polygons as inputs in the calculation, there are four peak frequencies of 400, 500, 630, and 800 Hz in the calculated noise results. The noise peak frequencies correspond to the wheel polygon order. Notably, even though the roughness levels are 30 dB (Fig. 12a) and the peak amplitudes with different wavelengths are almost the same (Fig. 12b), the wheel/rail noise levels caused by the four different order polygons are different. As the wheel polygon order increases, in the different 1/3 octave bands, the wheel/rail noises increase by about 5 dB(A) per 1/3 octave. This is mainly because the higher order polygon has

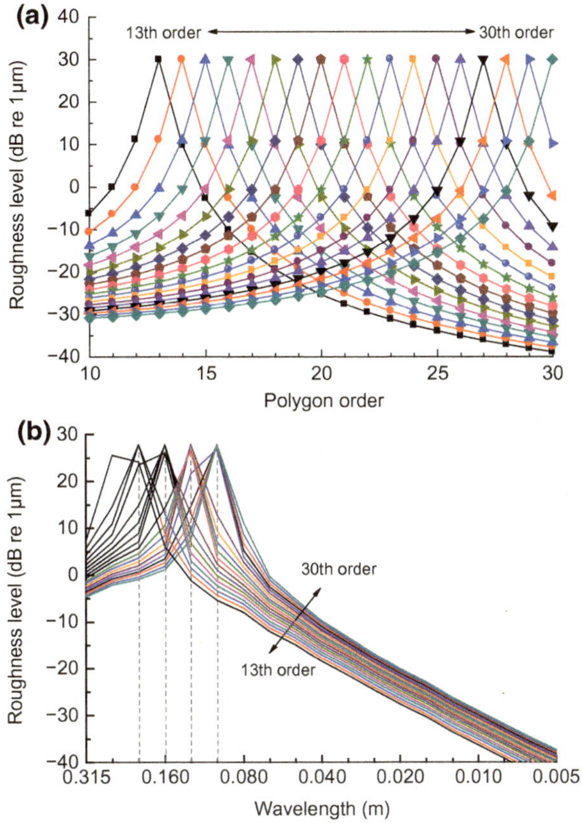


**Fig. 12** Different polygon orders: **a** polygon roughness levels characterized by order; **b** polygon roughness levels characterized with the wavelength

the higher passing frequency (corresponding to the shorter wavelength) and a larger excitation energy when the wheel rolls over the rail at the same speed, compared to the lower order polygon with the same roughness level. The other reason for this phenomenon is that as the excitation frequencies increase, the sound radiation ratios of both the wheel and the rail increase. Generally speaking, the higher sound radiation ratio causes the higher sound radiation power. Thus, although there are the same roughness levels, different polygon order can cause different wheel/rail noises. Therefore, the higher order polygon can cause the higher wheel/rail noise level.

From Fig. 13b, there are four peak frequencies in the interior noise, which is the same as in the wheel/rail noise. So the wheel/rail noise makes a great contribution to the interior noise. However, the increase in the ratio of the interior noise at the peak frequencies is different from that of the wheel/rail noise (as indicated in Fig. 13a). One reason for this phenomenon is that the interior noise is influenced by not only the wheel/rail noise, but also other exterior sources and the structure TL.

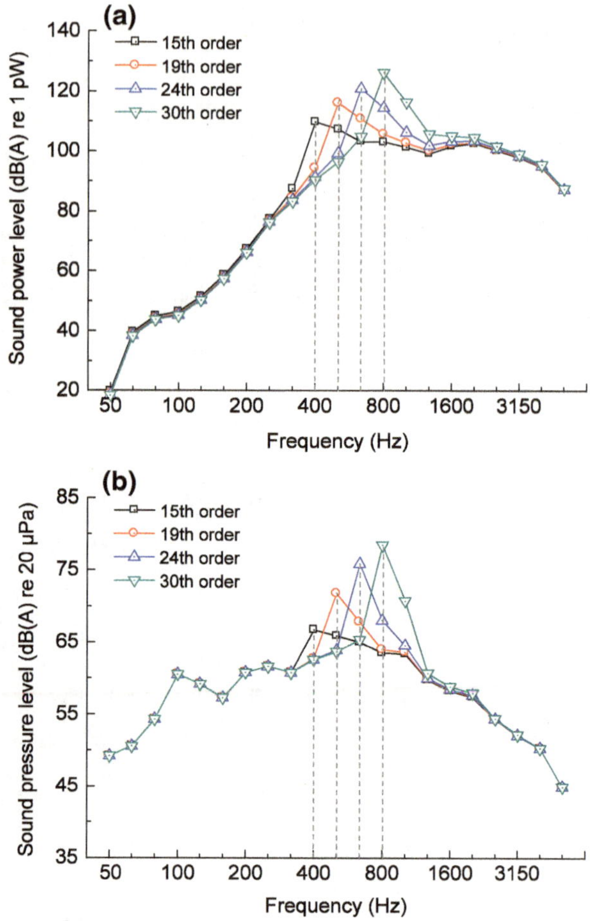


**Fig. 13** Different polygon orders: **a** wheel/rail noise; **b** interior noise

## 5.3 *Effect of Different Roughness Levels of Wheel Polygon on Noise*

With increase in operational mileage, the roughness levels and the distribution of the wear polygons will change. Here, this section discusses the influence of different roughness levels of the wear polygon on the noise level. Taking the frequently occurring wheel polygon order (the 19th order polygon) as a numerical example, in which the MATLAB code discussed above is used to make it with different roughness levels. Figure 14a shows the 19th order polygon with different roughness levels, characterized by order, and Fig. 14b shows the roughness level described with the wavelength.

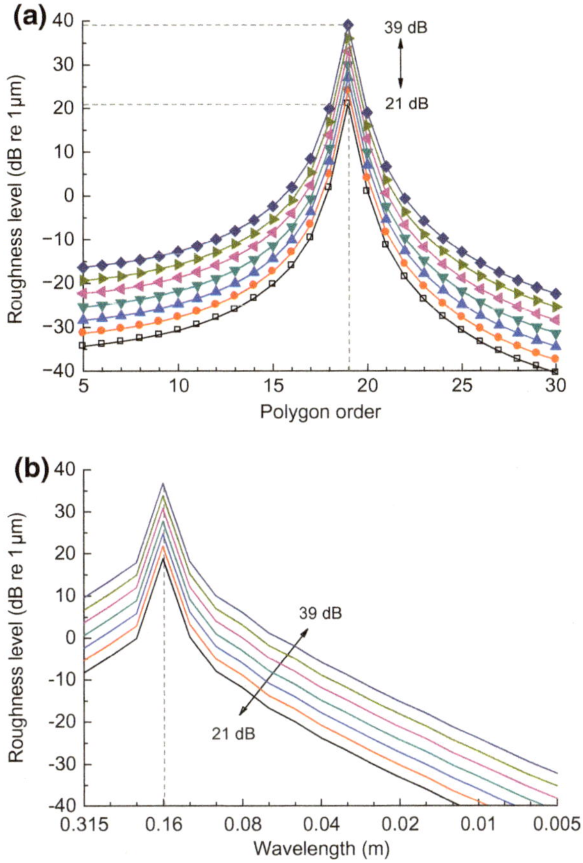


**Fig. 14** Different roughness levels: polygon roughness levels characterized by order (**a**) and with the wavelength (**b**)

The roughness levels of the 19th order polygon increase from 21 to 39 dB with a 3 dB interval, as indicated in Fig. 14a. The roughness levels in the 1/3 octave wavelength centred at 0.16 m also increase with about 3 dB interval, as indicated in Fig. 14b.

Figure 15 shows the results of the wheel/rail noise and the interior noise at 300 km/h.

Obviously, there is a peak frequency of 500 Hz in the wheel/rail noise that is the passing frequency of the 19th order polygon at 300 km/h. As the roughness level of the polygon increases from 21 to 39 dB with a 3 dB interval, the wheel/rail noise in the 1/3 octave band centred at 500 Hz increases from 107 to 125 dB(A) also with about a 3 dB interval (Fig. 15a).

As shown in Fig. 15b, there is a 500 Hz peak frequency in the interior noise which is the same as in the wheel/rail noise. With the increase of the roughness level of the 19th order polygon, the interior noise in the 1/3 octave band centred at 500 Hz increases from 66 to 80 dB(A), but with a nonlinear increase.

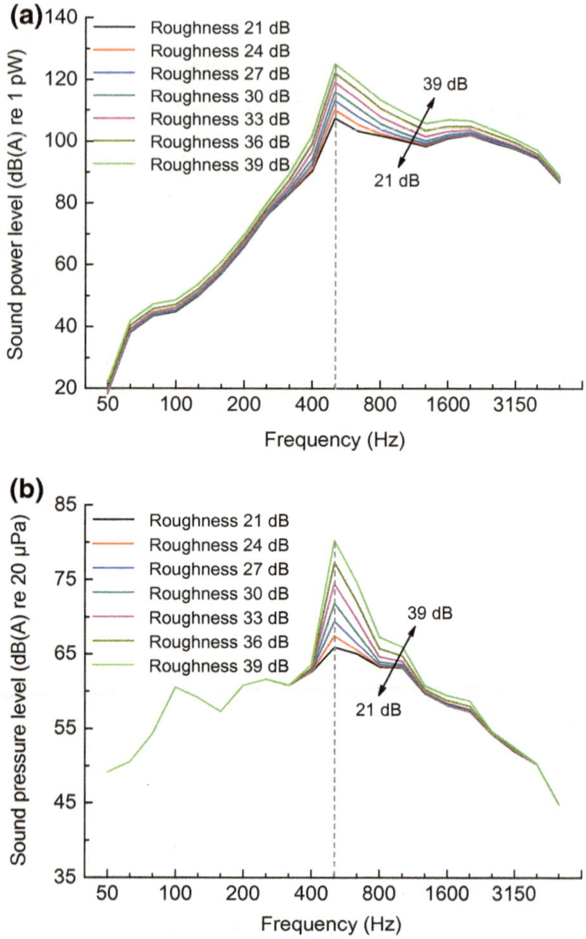


**Fig. 15** Different roughness levels: **a** wheel/rail noise; **b** interior noise

## 5.4 *Effect of Different Phases of Wheel Polygon on Noise*

As shown in Sects. 5.2 and 5.3, both the different order polygons and the different roughness levels have a great effect on the interior noise. Actually, the different polygon phases are also important, especially for the re-profiling. This is because the different polygon phases can be combined to make different wheel diameter differences even though the order distribution and roughness of the wheel polygons are the same. In other words, high-order wheel polygons, which can make the annoying interior noise, may occur with normal or low wheel diameter difference. To study the influence of the different polygon phases on the interior noise, this section discusses whether the present standard of wheel re-profiling is suitable for the noise reduction problem, being only based on the diameter difference. We consider the 17th to 19th order polygons and the different diameter differences are obtained by creating different combinations of the polygon phases.

Using three sine curves creates the different combinations of the different polygon phases. The equation including the three sine waves is given by

$$Y = A\sin(\omega_1 x + \varphi_1) + B\sin(\omega_2 x + \varphi_2) + C\sin(\omega_3 x + \varphi_3), \tag{22}$$

where $A$, $B$, and $C$ are used to define the roughness levels of the 17th, the 18th, and the 19th order polygons, respectively. $\omega_j$ ($j$ = 1, 2, 3) is used to indicate the $j$th order polygon and $\varphi_j$ ($j$ = 1, 2, 3) is used to denote the phase angle of the $j$th order polygon. $Y_{max} - Y_{min}$ is the wheel diameter difference. To make the diameter difference higher than 0.3 mm which is caused by the combination of the three polygons described with Eq. (22), the calculation uses the 30 dB roughness level of the 17th order polygon and 33 dB roughness levels of both the 18th and the 19th order polygons. Here 0.3 mm is the upper limit of the wheel diameter difference according to the maintenance standard for the high-speed trains of China (Zeng 2011). The numerical analysis sets $\varphi_1 = 0$ rad while $\varphi_2$ and $\varphi_3$ change from 0 rad to $2\pi$ rad. Figure 16 shows the effect of the variations of $\varphi_2$ and $\varphi_3$ on the wheel diameter differences with $\varphi_1 = 0$ rad.

From Fig. 16, it can be seen that even though the roughness levels of the wheel polygons do not change, their different phases can be combined to make the different wheel diameter differences. When $\varphi_1 = \varphi_2 = \varphi_3 = 0$ rad, there is the biggest diameter difference at 0.33 mm. When $\varphi_1 = 0$ rad, $\varphi_2 = 4.54$ rad, and $\varphi_3 = 5.93$ rad, there is the smallest diameter difference, 0.24 mm. The difference between the biggest and the smallest is about 0.1 mm.

Figure 17 shows the wheel/rail noise and the interior noise of the two cases at 300 km/h which indicate the biggest and the smallest diameter differences (indicated by Max and Min in the figure) discussed above.

As shown in Fig. 17a, there is a peak frequency at 500 Hz in the wheel/rail noise as the 17th to 19th order polygons have an excitation frequency in the 1/3 octave band centred at 500 Hz. In spite of the different polygon phases (different diameter difference), their wheel/rail noise sound power levels in the 1/3 octave frequency centred at 500 Hz are nearly the same. Thus, the polygon phase's change cannot change its wheel/rail noise level in the excitation frequency band. However, there

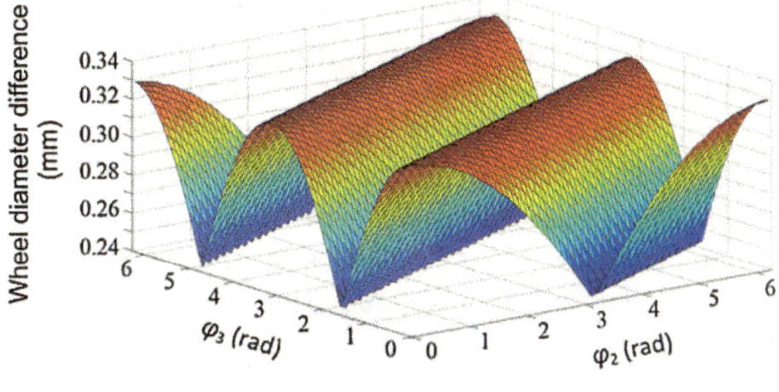


**Fig. 16** Wheel diameter differences caused by combination of different phase angles

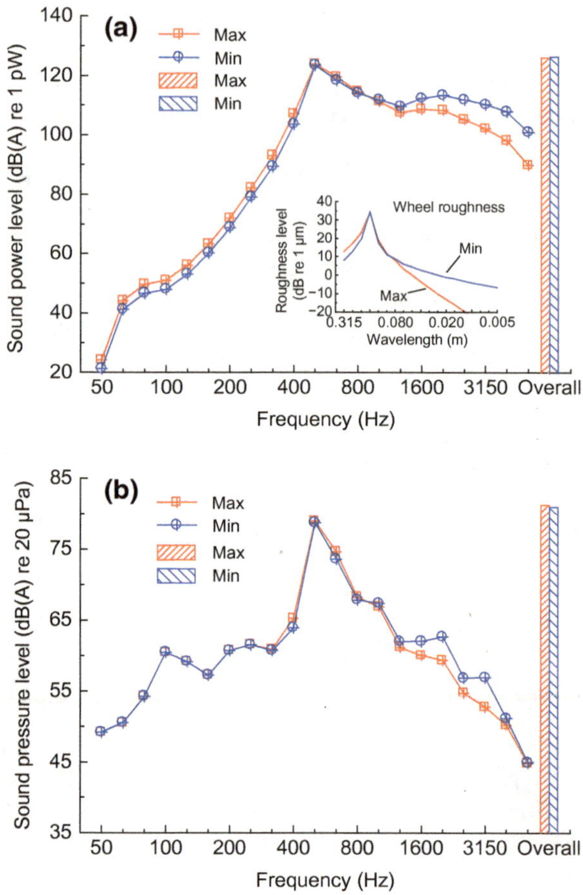


**Fig. 17** Different polygon phases: **a** wheel/rail noise; **b** interior noise

are some differences between the wheel/rail noises of the two cases in the 1/3 octave bands below 500 Hz and above 1000 Hz. These are due to the differences between their wavelength spectra which are caused by the characteristic of the filter in the program. Because the wheel/rail noises in the 1/3 octave band centred at 500 Hz are almost the same, there is little difference between the overall levels of these two cases.

From Fig. 17b, it can be seen that there is also little difference between the interior noises, similar to the wheel/rail noises. Nevertheless, the interior noise in the 1/3 octave bands below 500 Hz are nearly the same which is a little different from the wheel/rail noise. That is because wheel/rail noise in the 1/3 octave bands below 500 Hz is not as important as other exterior sources. The overall sound pressure levels of the interior noises are nearly the same in the two cases. So the influence of the polygon phases on interior noise is small.

However, the maintenance regulation of China's high-speed train now uses a criterion for re-profiling based only on diameter difference, regardless of the wheel polygonal wear status. It is generally considered that when the wheel circle diameter difference is smaller than 0.1 mm, the polygon status is good. When the diameter difference due to the polygonal wear is between 0.1 and 0.2 mm, it is normal. Furthermore, when the diameter difference is bigger than 0.3 mm, it is bad. If the diameter difference is bigger than 0.3 mm, the wheel needs to be re-profiled (Zeng 2011). Although the interior noise is almost the same for the above two cases, the wheel with 0.33 mm diameter difference needs to be re-profiled and the other wheel with 0.24 mm diameter difference could continue working.

From the test and simulation results shown in Sect. 3, we find that although the wheel circle diameter differences are nearly the same, different wheel polygons can cause different wheel/rail noise levels. Furthermore, we can find that even if the wheel polygon order are the same, their different polygon phases can be combined to make a different wheel circle diameter difference. However, the wheel/rail noise, especially the interior noise, is more determined by the order of the wheel polygons, not the phase of wheel polygons. In summary, a criterion based on only the wheel diameter difference is not suitable for carrying out re-profiling for noise reduction. The wheel polygon characteristics should also be considered.

## 6 Conclusions

The present work conducts a detailed investigation into the relationships between high-speed wheel polygonal wear and wheel/rail noise, and the interior noise of high-speed trains through extensive field experiments and numerical simulations. The conclusions can be drawn:

1. The field experiments show that when the 19th order polygonal wear is prominent and common, the vibration and noise of the high-speed coach are dominant at about 540 Hz. That is just the passing frequencies of the 19th order polygon at about 300 km/h. The root cause of the 19th order polygonal wear has not so far been understood. Further investigation is underway on this topic and is not in the scope of this paper.
2. Through test and simulation, in cases where the wheel circle diameter differences due to the wheel polygonal wear are nearly the same, the different wheel polygonal wear patterns can cause different wheel/rail noise levels.
3. The numerical simulation shows that different polygon order with nearly the same roughness levels can cause different wheel/rail noise levels and interior noise levels. Namely, the wheel polygons with higher order can make more serious wheel/rail noise and interior noise. This is because the higher order has the higher passing frequency at a certain operational speed, and generates higher wheel/rail vibration energy.
4. Changing the phases or the distribution of the wheel polygons can change the wheel diameter difference caused by the wheel polygonal wear. However, the effect of the change of the polygon phases is not great on wheel/rail noise and interior noise.

5. The criterion for re-profiling of high-speed trains needs to involve not only the wheel diameter difference due to the wear but also the characteristics of the wheel polygons.

**Acknowledgements** The authors are grateful for Da-bin CUI, Wei LI, Shuo-qiao Zhong, Heng-yu Wang, and Yong-guo Deng (Southwest Jiaotong University, China) for their assistance in this study.

## References

Behr, W., & Cervello, S. (2007). Optimization of a wheel damper for freight wagons using FEM simulation. In: Proceedings of the 9th international workshop on railway noise, Munich, pp 334–340. doi:10.1007/978-3-540-74893-9_47.

Bouvet, P., Vincent, N., Coblentz, A., et al. (2000). Optimization of resilient wheels for rolling noise control. *Journal of Sound and Vibration, 231*(3), 765–777. doi:10.1006/jsvi.1999.2561.

Cotoni, V., Shorter, P., & Langley, R. (2007). Numerical and experimental validation of a hybrid finite element-statistical energy analysis method. *The Journal of the Acoustical Society of America, 122*(1), 259–270. doi:10.1121/1.2739420.

Jin, X. S., Wu, L., Fang, J. Y., et al. (2012). An investigation into the mechanism of the polygonal wear of metro train wheels and its effect on the dynamic behaviour of a wheel/rail system. *Vehicle System Dynamics, 50*(12), 1817–1834. doi:10.1080/00423114.2012.695022.

Johansson, A. (2006). Out-of-round railway wheels-assessment of wheel tread irregularities in train traffic. *Journal of Sound and Vibration, 293*(3–5), 795–806. doi:10.1016/j.jsv.2005.08.048.

Johansson, A., & Andersson, C. (2005). Out-of-round railway wheels—A study of wheel polygonalization through simulation of 3D wheel/rail interaction and wear. *Vehicle System Dynamics, 43*(8), 539–559. doi:10.1080/00423110500184649.

Jones, C. J. C., & Thompson, D. J. (2000). Rolling noise generated by railway wheels with visco-elastic layers. *Journal of Sound and Vibration, 231*(3), 779–790. doi:10.1006/jsvi.1999.2562.

Langley, R. S., & Bremner, P. (1999). A hybrid method for the vibration analysis of complex structural-acoustic systems. *The Journal of the Acoustical Society of America, 105*(3), 1657–1671. doi:10.1121/1.426705.

Meinke, P., & Meinke, S. (1999). Polygonalization of wheel treads caused by static and dynamic imbalances. *Journal of Sound and Vibration, 227*(5), 979–986. doi:10.1006/jsvi.1999.2590.

Morys, B. (1999). Enlargement of out-of-round wheel profiles on high speed trains. *Journal of Sound and Vibration, 227*(5), 965–978. doi:10.1006/jsvi.1999.2055.

Nielsen, J. C. O., & Johansson, A. (2000). Out-of-round railway wheels—A literature survey. *Proceedings of the Institution of Mechanical Engineers, Part F: Journal of Rail and Rapid Transit, 214*(2), 79–91. doi:10.1243/0954409001531351.

Nielsen, J. C. O., Lunden, R., Johansson, A., et al. (2003). Train-track interaction and mechanisms of irregular wear on wheel and rail surface. *Vehicle System Dynamics, 40*(1–3), 3–54. doi:10.1076/vesd.40.1.3.15874.

Shorter, P. J., & Langley, R. S. (2005). Vibro-acoustic analysis of complex systems. *Journal of Sound and Vibration, 288*(3), 669–699. doi:10.1016/j.jsv.2005.07.010.

Thompson, D. J. (1997). *TWINS Theoretical Manual (v2.4).* TNO Report, TPN-HAG-RPT-93-0214, Delft.

Thompson, D. J., & Gautier, P. E. (2006). Review of research into wheel/rail rolling noise reduction. *Proceedings of the Institution of Mechanical Engineers, Part F: Journal of Rail and Rapid Transit, 220*(4), 385–408. doi:10.1243/0954409JRRT79.

Thompson, D. J., Hemsworth, B., & Vincent, N. (1996a). Experimental validation of the TWINS prediction program for rolling noise, Part I: Description of the model and method. *Journal of Sound and Vibration, 193*(1), 123–135. doi:10.1006/jsvi.1996.0252.

Thompson, D. J., Fodiman, P., & Mahe, H. (1996b). Experimental validation of the TWINS prediction program for rolling noise, Part II: Results. *Journal of Sound and Vibration, 193*(1), 137–147. doi:10.1006/jsvi.1996.0253.

van Beek, A., & Verheijen, E. (2003). *Definition of track influence: Roughness in rolling noise*. Harmonoise Report, HAR12TR-020813-AEA10, the Netherlands.

Williams, A. B., & Taylors, F. J. (2006). *Electronic filter design handbook*. New York: McGraw-Hill.

Wu, T. X. (2012). *HWTNS theoretical manual (v1.0)*. Shanghai (in Chinese): Shanghai Jiaotong University.

Wu, T. X., & Thompson, D. J. (1999). A double Timoshenko beam model for vertical vibration analysis of railway track at high frequencies. *Journal of Sound and Vibration, 224*(2), 329–348. doi:10.1006/jsvi.1999.2171.

Wu, T. X., & Thompson, D. J. (2000). Theoretical investigation of wheel/rail non-linear interaction due to roughness excitation. *Vehicle System Dynamics, 34*(4), 261–282. doi:10.1076/vesd.34.4.261.2060.

Wu, T. X., & Thompson, D. J. (2001). Vibration analysis of railway track with multiple wheels on the rail. *Journal of Sound and Vibration, 239*(1), 69–91. doi:10.1006/jsvi.2000.3157.

Zeng, J. (2011). *The failure mechanism and optimization of high-speed wheel/rail in rolling contact*. Technical Report of National Basic Research Program (973) of China, No. 2007CB714702, Chengdu (in Chinese).

Zhang, J., Xiao, X. B., Han, G. X., et al. (2013). Study on abnormal interior noise of high-speed trains. In *Proceedings of the 11th international workshop on railway noise*, Uddevalla, Sweden, pp. 691–698. doi:10.1007/978-3-662-44832-8_82.

## Author Biography

**Xue-song Jin** Ph.D., is a professor at Southwest Jiaotong University, Chengdu, China. He is a leading scholar of wheel-rail interaction in China, and is a State Council Expert for Special Allowance. He is the author of three academic books and over 200 articles. He is an editorial board member of several journals, and has been a member of the Committee of the International Conference on Contact Mechanics and Wear of Rail/Wheel Systems for more than 10 years. He has been a visiting scholar at the University of Missouri-Rolla for 2.5 years. Now his research focuses on wheel-rail interaction, rolling contact mechanics, vehicle system dynamics, vibration, and noise.

# Investigation into External Noise of a High-Speed Train at Different Speeds

**Bin He, Xin-biao Xiao, Qiang Zhou, Zhi-hui Li and Xue-song Jin**

## Contents



## 1 Introduction

The noise generated by high-speed trains is a sensitive issue, as high-speed train tracks are built in densely populated areas, where prior noise levels were very low. There is a need for an accurate description of the main characteristics (position, strength, spectrum, energy contribution, etc.) of different noise sources of moving high-speed trains, since this information is typically used as input for noise prediction and to guide studies on noise mitigation measures.

Over the past decades, many studies have been performed to determine the noise sources of high-speed trains and their precise characteristics. It has been recognized that the main noise sources are rolling noise, aerodynamic noise, and traction noise (Thompson 2008). However, so far it has been difficult to clearly distinguish the


B. He · X. Xiao · Q. Zhou · Z. Li · X. Jin (✉)
State Key Laboratory of Traction Power, Southwest Jiaotong University,
Chengdu 610031, China
e-mail: xsjin@home.swjtu.edu.cn;

B. He
e-mail: bin_he1986@163.com





Y. Fang and Y.H. Zhang (eds.), *China's High-Speed Rail Technology*,
Advances in High-speed Rail Technology, https://doi.org/10.1007/978-981-10-5610-9_21

location of these noise sources and their independent contribution to the total noise of a traveling high-speed train. In this respect, the beam-forming method is able to effectively identify individual noise sources. As a signal processing technique, this method has been used in microphone arrays for signal transmission and reception (Christensen and Hald 2004). Recently, Dittrich and Janssens (2000) developed a T-like microphone array to identify the noise sources of trains. Based on near-field acoustical holography, Schulte-Werning et al. (2003) developed a spiral-like microphone array with a diameter of 4.0 m. Within the Deutsche Bahn 'low-noise railway' project, noise sources from ICE high-speed trains were recognized in a frequency range of 200–3150 Hz. Furthermore, noise source identification experiments on TGV trains moving at speeds from 250 to 320 km/h were carried out, and the distribution of aerodynamic and wheel/rail rolling noise was analyzed (Mellet et al. 2006). Silence (2005) carried out source identification for TFS tramcars and a CITADIS 302 train for traveling speeds of 20–100 km/h, and reported that wheel/rail noise was the major noise source, and 5.0–10.0 dB(A) larger than other noise sources. The spiral-like microphone array was also used for the Japanese Shinkansen train and for noise source identification tests of Fastech 360S trains (Wakabayashi et al. 2008). When the speed of the trains reached 340 km/h, the maximum noise came from their wheels. Poisson et al. (2008) carried out noise source identification for TGV trains and arrived at the conclusion that the first bogie and the pantograph gradually became the main sound sources as the train speed increased. Thron (2010) studied the sound source identification of the IC2000 for a train velocity of 190 km/h. Below 2000 Hz, the main sound source is the bogies. Above 4000 Hz, the sound originates along the entire train height, especially at the first coach. Koh et al. (2007) also carried out a noise source identification of Korean high-speed trains. In the frequency range of 2500–4500 Hz, the noise was distributed along the train height. Another noise source identification of Korean high-speed trains was carried out by Noh et al. (2011). Also in a report about the TR08 maglev system, sound sources were identified and their vertical distribution was given (Bernd et al. 2002). The intensity difference in this direction lessened with increasing train velocity. All the source identification tests used the classic delay and sum beam-forming technique, which is recognized for the identification of moving sound sources.

In this study, the delay and sum beam-forming method is used for noise measurements of Chinese high-speed trains, with test speeds ranging from 270 to 390 km/h. In the analysis of the test results, the main characteristics of the different noise sources are discussed. The sound exposure level (SEL) in every 1/3 octave frequency is analyzed for speeds of 271, 341, and 386 km/h. In addition, the frequency characteristics of the pass-by noises and assessment of the wayside noises are provided.

## 2 Noise Source Identification of High-Speed Train

### 2.1 *Facility and Its Principle*

As a signal processing technique, the beam-forming method is always used in sensor arrays for localization and separation of noise sources (Johnson and Dudgeon 1993; Christensen and Hald 2004). When sound propagates from an arbitrary direction, its signal can be measured using a sensor array. The sensor signals are associated with time delays. The distances between the sources and sensor positions determine the length of the time delays. The delays are adjusted according to the source locations, enabling the correct reinforcement of the sum of the signals.

As illustrated in Fig. 1, when a planar array consisting of $M$ microphones at locations $\boldsymbol{r}_{\mathrm{m}}$ is applied for both the localization and separation of the noise sources, the measured sensor signals $y_{\mathrm{m}}$ are individually delayed and subsequently summed (Johnson and Dudgeon 1993):

$$z(\boldsymbol{\kappa}, t) = \sum_{m=1}^{M} w_m y_m(t - \Delta_m(\boldsymbol{\kappa})), \tag{1}$$

where $\boldsymbol{\kappa}$ is a unit vector, $t$ is transient time, $w_{\mathrm{m}}$ is the weighting or shading coefficient applied to the microphone signal, and $\Delta_{\mathrm{m}}$ is the time delay, the value of which is obtained by adjusting the time delays such that the sound signals are associated with a plane wave incident from that specific direction. The signals are time-aligned before being summed. As shown in Fig. 1, geometrical considerations indicate that the time delays can be obtained by

$$\Delta_m = \frac{\boldsymbol{\kappa} \cdot \boldsymbol{r}_m}{c}, \tag{2}$$

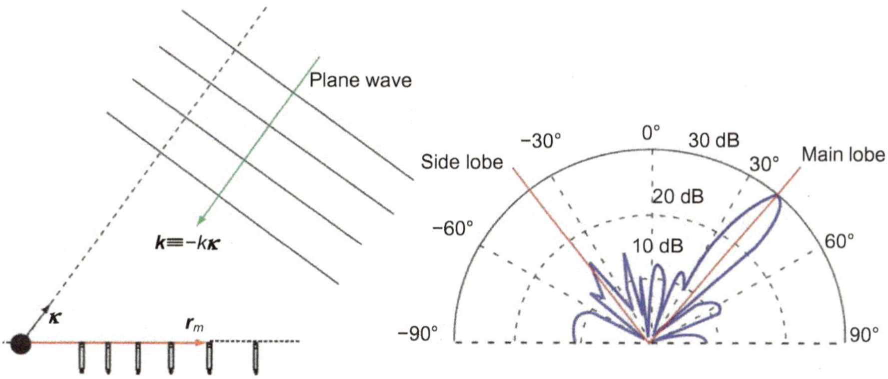


**Fig. 1** Basic principle of external sound source localization (Christensen and Hald 2004). $k$ is the amplitude of $\boldsymbol{k}$. Reprinted from Christensen and Hald (2004) by permission of Brüel & Kjær

where $c$ is the speed of sound. The sound signals arriving from other far-field directions are not aligned before their summation, and are therefore not added up coherently. The frequency-domain version of Eq. (1) for the delay and sum beam-forming method is expressed as

$$Z(\boldsymbol{\kappa}, \omega) = \sum_{m=1}^{M} w_m Y_m(\omega) \mathrm{e}^{-j\omega\Delta_m(\boldsymbol{\kappa})} = \sum_{m=1}^{M} w_m Y_m(\omega) \mathrm{e}^{j\boldsymbol{k}\cdot\boldsymbol{r}_m}, \tag{3}$$

where $Y_{\mathrm{m}}(\omega)$ is the frequency-domain sound pressure expression of the $m$th microphone, $\omega$ is the angular frequency, and $\boldsymbol{k}$ is the wavenumber vector of a plane wave incident from the focus direction $\boldsymbol{\kappa}$.

The presence of side lobes can cause waves from unfocused directions to leak into the main lobe direction $\boldsymbol{\kappa}$, resulting in false noise sources. A good planar array design can overcome this weakness. The performance of the beam-forming array is determined by the array pattern, the spatial resolution, and the maximum side lobe levels (MSLs). When a plane wave with a wave vector $\boldsymbol{k}_0$ arrives from a direction different from the focus direction, the pressure measured by the array sensors is given by

$$Y_m(\omega) = Y_0 \mathrm{e}^{-j\boldsymbol{k}_0\cdot\boldsymbol{r}_m}, \tag{4}$$

where $Y_0$ is the pressure amplitude. According to Eq. (3), the output in the frequency domain is

$$Z(\boldsymbol{\kappa}, \omega) = Y_0 \sum_{m=1}^{M} w_m \mathrm{e}^{j(\boldsymbol{k}-\boldsymbol{k}_0)\cdot\boldsymbol{r}_m} = Y_0 W(\boldsymbol{k}-\boldsymbol{k}_0). \tag{5}$$

Assuming that $\boldsymbol{K} = \boldsymbol{k} - \boldsymbol{k}_0$, the function $W$ can be expressed by

$$W(\boldsymbol{K}) = \sum_{m=1}^{M} w_m \mathrm{e}^{j(\boldsymbol{K})\cdot\boldsymbol{r}_m}. \tag{6}$$

This expression is the so-called array pattern and represents the amount of "leakage" obtained from plane waves incident from other directions, being something that strongly depends on the array geometry.

The ability of a beam-forming array to distinguish incident waves from various directions is indicated by its resolution. Under the assumption that two waves with wave vectors $\boldsymbol{k}_1$ and $\boldsymbol{k}_2$ have a unitary amplitude, the output can be expressed as

$$Z(\boldsymbol{\kappa}, \omega) = W(\boldsymbol{k}-\boldsymbol{k}_1) + W(\boldsymbol{k}-\boldsymbol{k}_2). \tag{7}$$

Then, considering a small angle difference $\mathrm{d}\theta$ between $\boldsymbol{k}_1$ and $\boldsymbol{k}_2$, the resolution at a finite distance $z$ is given by

$$R(\theta) = \frac{z\mathrm{d}K}{k}\frac{1}{\cos^3\theta}, \tag{8}$$

where $\mathrm{d}K$ is the width of the main lobe in the array pattern, whose value is $|\boldsymbol{k}_1 - \boldsymbol{k}_2|$, $\theta$ is the incident angle of the wave, and $k$ is the amplitude of $\boldsymbol{k}_1$ and $\boldsymbol{k}_2$. It can be seen that the resolution deteriorates when the incidence angle increases. In practice, the useful open angle is typically restricted to 30°.

Another important parameter is the side lobe magnitude. For existing side lobes, waves from unfocused directions leak into the measurement direction (the main lobe direction), resulting in the possible production of false noise sources. A good planar array design can therefore be characterized by a low MSL. The radial profile of the array pattern is defined by

$$W_{\mathrm{P}}(K) = 10\,\mathrm{lg}\left(\underset{|\boldsymbol{K}|=K}{max}|W(\boldsymbol{K})|^2/M^2\right). \tag{9}$$

Furthermore, based on this expression, the MSL function can be written as

$$\begin{aligned}\mathrm{MSL}(K) &= \max_{K^0_{\min} < K' \le K} W_{\mathrm{P}}(K') \\ &= 10\,\mathrm{lg}\left(\max_{K^0_{\min} < K' \le K}|W(\boldsymbol{K})|^2/M^2\right),\end{aligned} \tag{10}$$

where $K^0_{\min}$ is the width of the main lobe. For low MSL values, the array will exhibit a better performance. In this test, an optimized wheel microphone array with 78 channels is used (Fig. 2). Its geometrical design has improved the spatial resolution and reduced the MSL. The microphones are arranged in a series of tilted linear spokes. In practice, such a wheel array has an excellent performance and is easily operated.

## 2.2 *Measurement of High-Speed Train Noise Sources*

The beam-forming method can effectively locate noise sources in the far-field (Kook et al. 2000). Hence, to measure the noise emitted from a train passing at high speed, the microphone array with 78 channels is placed at a horizontal distance of 7.5 m from the center line of the track (Fig. 2). The diameter of the array is 4.0 m and the height of the array center is located 2.0 m above the rail head. A photoelectric sensor is installed to measure the speed of the train. The speed is

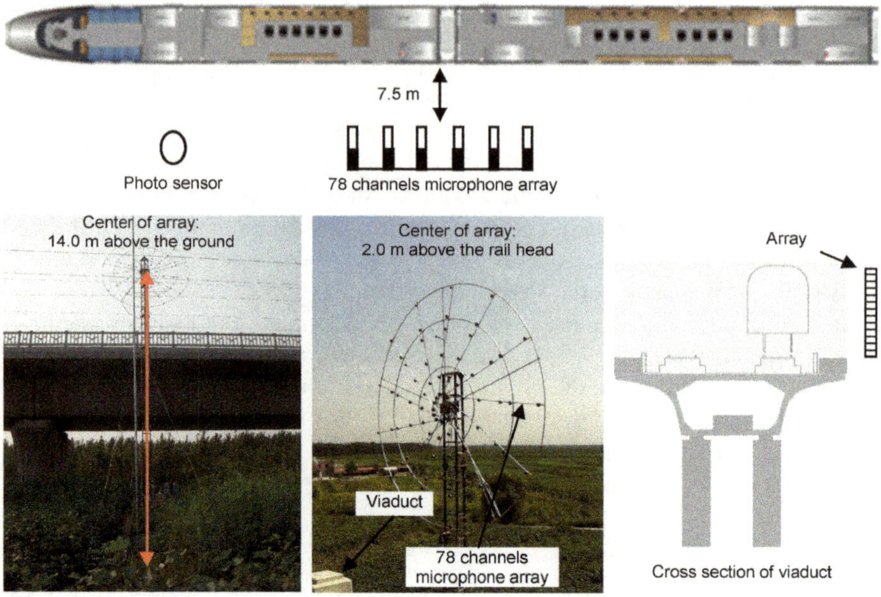


**Fig. 2** Microphone array measurement setup

determined by calculating the ratio of the measured train length to its pass-by time. Processing of the measurement data is conducted using B&K devices.

The test speeds range from 270 to 390 km/h. Figure 3 indicates the distribution of the external noise sources of the train at 386 km/h, which is identified by the microphone array discussed above. The frequency range of the noise shown in Fig. 3 is from 500 to 5000 Hz, being the frequency range in which main external noise components are present. From the noise map, it can be deduced that the main noise sources over the entire frequency range are located at the bogies, the elevated pantograph, and the inter-coach gaps (Fig. 4).

The maximum noise comes from the bogie areas, followed by the pantograph and inter-coach gap areas. The noise generated from the bogie areas includes the wheel/rail rolling noise, gear noise, and aerodynamic noise generated at the bogies under the carriages. The wheel/rail rolling noise is caused by the surface roughness of the wheels and the rails, and their mutual friction at the running surface. The gear noise is caused by the meshing impact and friction between the gears, and structural vibrations. The aerodynamic noise originates from the airflow around the entire bogie, including the influence of severe turbulence due to the complex geometry of the bogie. The noise coming from the bogie areas also includes contributions from structural vibrations. However, in these bogie areas, the wheel/rail rolling noise is always the dominant noise source when the train is moving although, so far, it has been very difficult to accurately identify to which extent (by percentage) each noise source contributes to the overall noise coming from the bogie area.

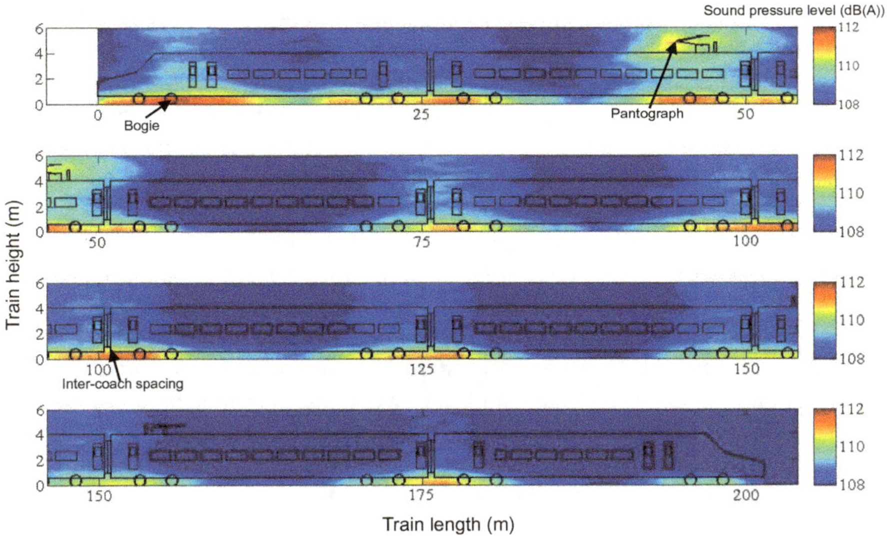


**Fig. 3** Noise source distribution of a high-speed train traveling at 386 km/h on viaduct (frequency range: 500–5000 Hz)

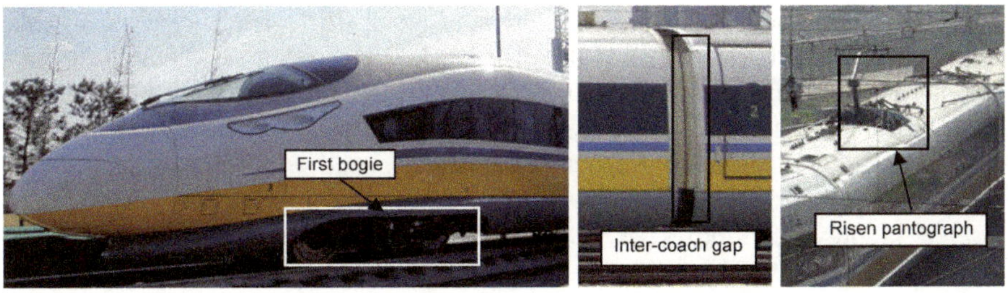


**Fig. 4** Noise sources: the first bogie, inter-coach gap, and pantograph

The noise in the pantograph area mainly includes aerodynamic noise, friction noise, and sparking noise. The aerodynamic noise is generated from interactions of the pantograph frame and shield with the airflow around them. The frictional noise is caused by the pantograph collector sliding on the catenary, while the sparking noise is produced by interaction between the collector and the catenary. In the pantograph area, the aerodynamic noise contribution is the greatest, followed by the sparking noise. As is the case for the bogie areas, it is also difficult to accurately identify the contribution percentages of each source in the overall noise generated in this area.

The noise of the inter-coach gaps includes aerodynamic noise and noise coming from structural vibrations. In this regard, the aerodynamic noise is the dominant source, due to the sags and crests of the outer windshield surfaces in the inter-coach gaps.

Hence, the external noise emitted by high-speed trains can be divided into two main source families: rolling noise and aerodynamic noise. The rolling noise is one of the most important noise sources of high-speed trains and is caused by the excitation at the wheel/rail contact patch. The aero-acoustic sources are generated by vortex shedding and flow disturbances around the train structure; these are mainly located at the bogies, the pantograph and its recess, the inter-coach gaps, and the doors.

Figure 3 shows that the noise produced at the first bogie of the train is evidently greater than that at the other bogies. Obviously, when the train is running at high speed, the vortex shedding from the first bogie generates a much larger aerodynamic noise than the other bogies. From the color map shown in Fig. 3, it can be seen that the noise from the inter-coach gaps is mainly distributed underneath the coach roof and on the apron board. In these two areas, no windshields are used to decrease flow disturbances at the coach edges and the gaps (Fig. 4). Note that the color map in Fig. 3 is a summation of every 1/3 octave frequency-band results and represents the overall noise distribution. However, the predominant noise components of the noise sources vary for different frequency ranges and train speeds.

Other results involving external noise measurements are shown in Fig. 5, in which the noise distribution of the first two coaches is highlighted at 630, 1600, and 5000 Hz.

At 630 Hz, the noise sources are mainly located at the bogies and consist mainly of rolling and gear noises, which are produced by wheel/rail contact and friction/meshing mechanisms, respectively. At 1600 Hz, the noise is mainly located at the bogies and is formed by wheel/rail contact. Lastly, at 5000 Hz, noise is mainly located at the train body and inter-coach gaps, and is caused by flow disturbance around the train structure and vortex shedding. Hence, the dominant noise sources differ in all three analyzed frequencies. Similar results have been reported in previous studies (Koh et al. 2007; Thron 2010; Noh et al. 2011).

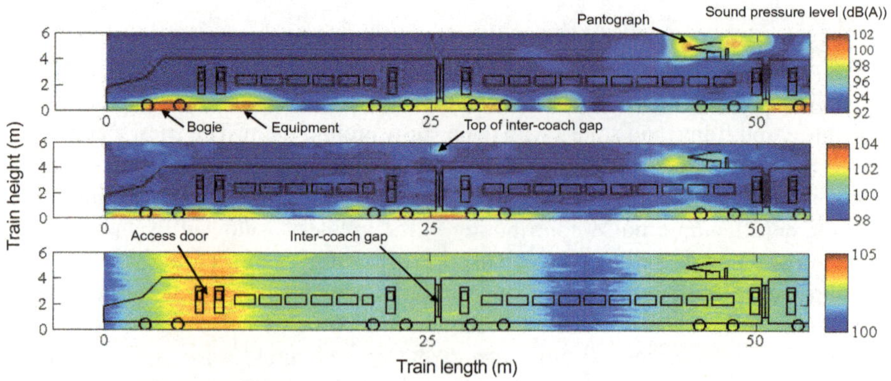


**Fig. 5** Noise distribution of the first two cars at 386 km/h with the frequency of 630 Hz (**a**), 1600 Hz (**b**), and 5000 Hz (**c**)

Distinction between the test results discussed in this study and those in previous studies are attributed to the train type and its operational speed. The train type affects the intensity, distribution, and frequency characteristics of the noise sources. Furthermore, as the train velocity increases, the aerodynamic noise becomes more significant, especially at high frequencies. Thus, for a speed of 386 km/h, the noise source intensity at 5000 Hz changes negligibly and does not depend on the height.

## 2.3 Frequency Characteristics of Main Noise Sources

The previous section provided a qualitative analysis of the noise sources, their locations, and their relative intensity. In this section, a quantitative analysis of the frequency characteristics of the main noise sources is provided, in which the noises at the bogies, the elevated pantograph, and the inter-coach gaps are being considered. The formula for averaging the noise pressure can be expressed as

$$\begin{aligned} L_{\mathrm{p}} &= 20\lg\left(\frac{1}{s}\cdot\frac{\int_s p\mathrm{d}s}{p_0}\right) \\ &= 20\lg\left[\left(\sum_{i=1}^{N_i}\sum_{j=1}^{N_j} p(y_i,z_j)\mathrm{d}y\mathrm{d}z\right)\Big/(sp_0)\right], \end{aligned} \tag{11}$$

where $s$ is the area of the analyzed region, $p$ is the A-weighted sound pressure in the considered area, $p_0$ is the reference sound pressure, $N_i$ and $N_j$ represent discrete points, and d$y$ and d$z$ represent the increment of diversity along the length and the height directions, respectively. The analyzed regions and their areas for the three noise sources have been documented elsewhere (He 2010). When combined with the source regions, the intensity of the three noise sources can be evaluated. As the sound pressure at a given field point is an average value, the average sound pressure $L_{\mathrm{p}}$ deserves more attention, since this data is also used as the sound source in noise prediction models.

The frequency characteristics of the three sources are shown in Figs. 6, 7 and 8, corresponding to speeds given by 271, 341, and 386 km/h, respectively. When the speed is 271 km/h, the average sound pressures in the wheel/rail areas, the pantograph area, and at the inter-coach gaps are 106.2, 104.2, and 104.3 dB(A), respectively. The sound pressure increases as the frequencies increase, peaking at 1600 Hz with values of 99.5, 95.9, and 95.0 dB(A) for the wheel/rail areas, the inter-coach gaps, and the pantograph area, respectively. Below 1600 Hz, the noise originating at the wheel/rail area is the largest, followed by the inter-coach gap noise and the pantograph noise. The maximum pressure difference between the noises originating at the wheel/rail area and the inter-coach gaps is approximately 6.8 dB(A). The pressures at frequencies above 3150 Hz are very similar, with a maximum value of approximately 97.7 dB(A).

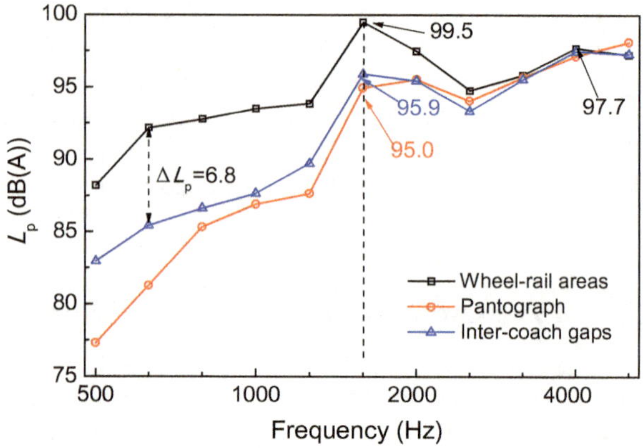


**Fig. 6** Frequency characteristics of the main noises at 271 km/h (frequency range: 500–5000 Hz)

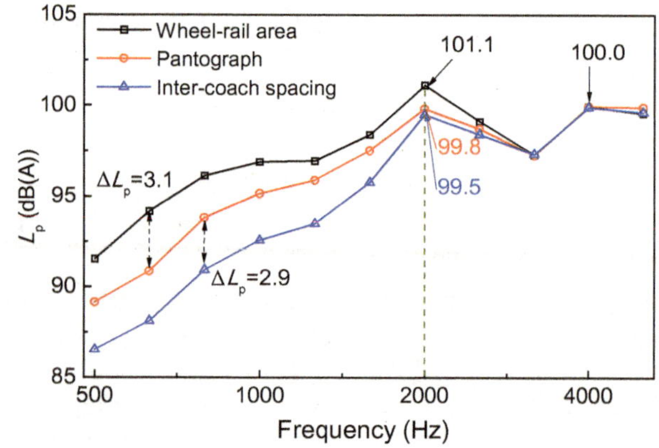


**Fig. 7** Frequency characteristics of the main noises at 341 km/h (frequency range: 500–5000 Hz)

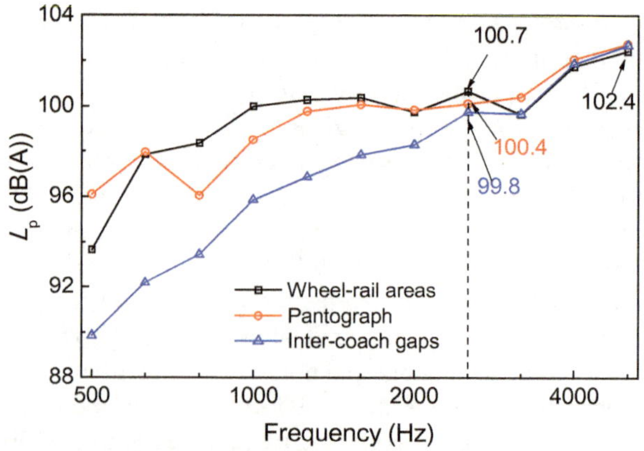


**Fig. 8** Frequency characteristics of the main noises at 386 km/h (frequency range: 500–5000 Hz)

For a speed of 341 km/h, the average sound pressures in the wheel/rail areas, the pantograph area, and the inter-coach gap areas are 108.5, 107.7, and 107.0 dB(A), respectively. The sound pressures reach their maxima at 2000 Hz, with values of 101.1, 99.8, and 99.5 dB(A) for the wheel/rail areas, the pantograph area, and the inter-coach gap areas, respectively. Below 2000 Hz, the noise coming from the wheel/rail areas is the highest, followed by the noise originating at the pantograph area. The noise at the inter-coach gaps has the smallest contribution in this frequency range. The maximum pressure differences between them are approximately 3.1 and 2.9 dB(A). The resulting pressures above 3150 Hz are quite similar, with a maximum value of 100.0 dB(A).

Finally, for a train velocity of 386 km/h, the average sound pressures in the wheel/rail areas, the pantograph area, and the inter-coach gap areas are 110.4, 110.3, and 109.0 dB(A), respectively. The largest sound pressure, at approximately 102.4 dB(A), is reached at 5000 Hz. At 2500 Hz, there are local sound pressure peaks with values of 100.7, 100.4, and 99.8 dB(A) for the wheel/rail areas, the pantograph area, and the inter-coach gap area, respectively. Below 2500 Hz, the wheel/rail area is the largest noise contributor, followed by the pantograph areas and the inter-coach gap area. Above 3150 Hz, the measured pressures are very similar, with a maximum measured value of approximately 102.4 dB(A).

From the above analysis, it can be seen that for low train velocities, the average noise pressure coming from the wheel/rail areas is the largest, while the noise pressure produced by the inter-coach gaps takes the second place. As the speed increases, the average pantograph noise exceeds that of the inter-coach gaps. When the speed is 386 km/h, the noise produced in the wheel/rail areas is still the largest, although the noise produced in the pantograph area is almost consistent with that from the wheel/rail areas. Furthermore, as the velocity increases, the frequency of the maximum noise moves to higher values. In this regard, for speeds of 271, 341, and 386 km/h, the maximum noise is measured at 1600, 2000, and 5000 Hz, respectively.

Figure 9 shows the average pressures of the three noise sources as a function of the train speed. The aerodynamic noise at the pantograph area increases faster than the other two noise sources. So the power law relationship for this area between noise and train speed is larger than that for other areas. The wheel/rail noise contains rolling noise and aerodynamic noise, so the produced average sound pressure in this area is higher than that of the inter-coach gap areas. From the sound levels and the spectrum of the three noise sources, it can be seen that for a speed of 271 km/h, the major noise type generated in the wheel/rail areas is rolling noise. However, as the train speed increases, the share of the aerodynamic noise in the wheel/rail noise is greater. For a velocity of 386 km/h, the aerodynamic noise in wheel/rail areas approaches the rolling noise.

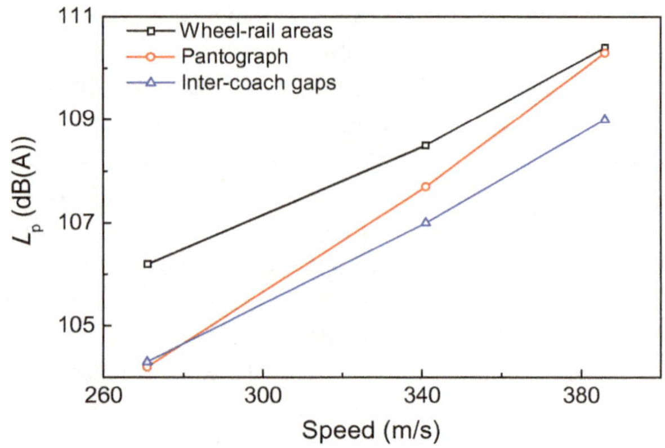


**Fig. 9** Sound pressures versus speed

## 2.4 Characteristics of SEL at Different Speeds

The vertical noise distribution of high-speed trains reflects the intensity of noise sources in the vertical direction along the train and is expressed by the SEL. This quantity measures the energy contained in transient noise. According to the ISO 3095:2013 standard (ISO 2013), the formula to calculate the SEL is given by

$$\mathrm{SEL} = 10\lg\left(\frac{1}{T_0}\int_0^T \frac{p_A^2(t)}{p_0^2}\mathrm{d}t\right) = 10\lg\left(\int_0^T \frac{p_A^2(t)}{p_0^2}\mathrm{d}t\right), \tag{12}$$

where $T_0$ is the reference time interval ($T_0 = 1$ s), $p_A(t)$ is the A-weighted instantaneous sound pressure, $p_0$ is the reference sound pressure, and $T$ is the measurement time interval.

Figures 10, 11 and 12 show the SEL distribution along the vertical height direction for train velocities of 271, 341, and 386 km/h, respectively. The parameter $T$ is the pass-by time of the high-speed train. The abscissa represents the value of the SEL, the ordinate represents the vertical height, and the height coordinate corresponding to 0 indicates the height of the rail head.

The vertical distribution of the SEL in the frequency range of 500–5000 Hz is illustrated in Fig. 10a for a train speed of 271 km/h. It can be clearly observed that in every 1/3 octave frequency band, the SEL corresponding to the wheel/rail area within a height of 1.0 m is the highest. Below 1600 Hz, SEL values of all heights increase as the frequencies increase. Furthermore, the SELs at 4000 and 5000 Hz are approximately similar, showing negligible variation along the vertical direction. At a height of 1.0 m, the SEL at 1600 Hz is maximized, and can be attributed to wheel/rail noise. Above 1.0 m, the SELs corresponding to 4000 and 5000 Hz exhibit the largest values. As revealed by the noise source identification, it is known that the main noise contributor corresponding to these frequencies is aerodynamic noise produced by the pantograph, the bogies, the inter-coach gaps, and the first

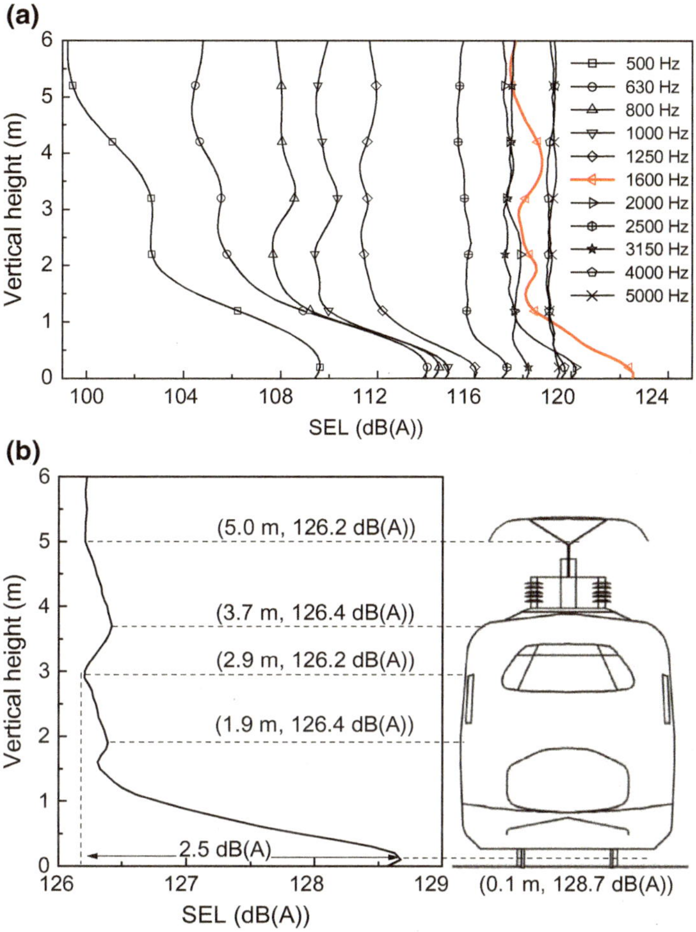


**Fig. 10** SEL distribution at 271 km/h. **a** SEL in every 1/3 octave frequency band; **b** SEL in the full frequency range

access door. Figure 10b shows the total SEL values for a speed of 271 km/h. Here the maximum SEL is 128.7 dB(A), corresponding to a height of 0.1 m. At the base of the pantograph, a local maximum SEL is observed with a value of 126.4 dB(A). The minimum SEL value is 126.2 dB(A) and is measured at heights given by 2.9 and 5.0 m, corresponding to the center of the train body and the pantograph, respectively. Hence, during the pass-by time, the total SEL is maximized in the wheel/rail region, followed by the inter-coach gaps and the pantograph. Furthermore, the SEL difference between the wheel/rail area and the pantograph (from the bottom up) is 2.5 dB(A), being the maximum measured SEL difference.

As shown in Fig. 11a, for a speed of 341 km/h, SEL values below 1.0 m exhibit the largest values for every 1/3 octave frequency band. This is attributed to the

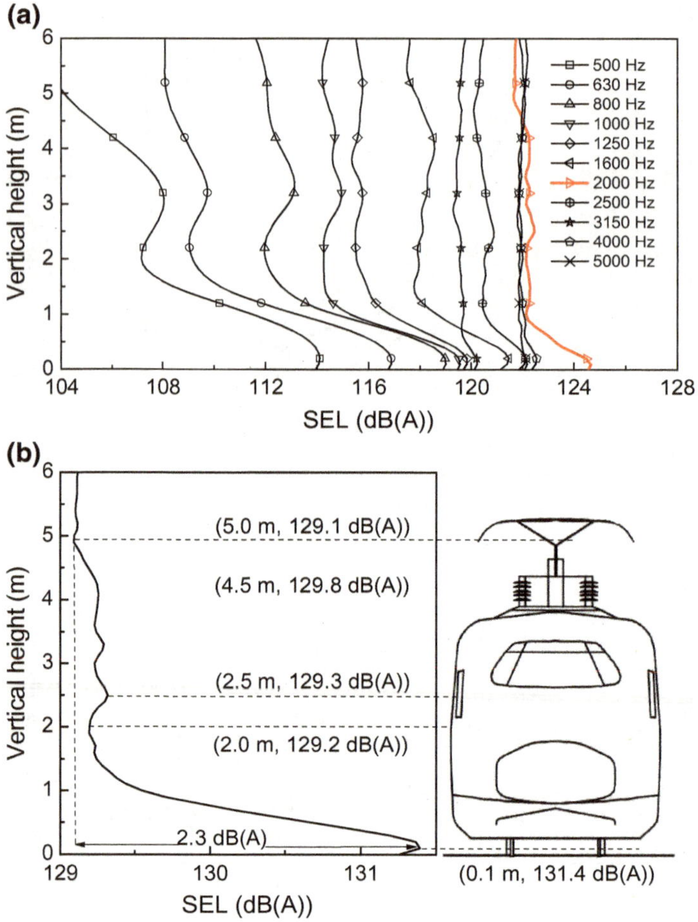


**Fig. 11** SEL distribution at 341 km/h. **a** SEL in every 1/3 octave frequency band; **b** SEL in the full frequency range

contribution of intensive wheel/rail noise. Below 2000 Hz, the SEL increases along all heights as the frequencies increase. The SEL at 4000 and 5000 Hz are approximately equal and only change little along the vertical direction. Below 4.5 m, the SEL at 2000 Hz is larger than the values corresponding to 4000 and 5000 Hz. This is due to a large noise contribution at this frequency from the wheel/rail area and the carriage structure. Above 4.5 m, the contrary is the case, something that can be attributed to the large aerodynamic noise generated by the pantograph at 4000 and 5000 Hz. Figure 11b shows the total SEL for a speed of 341 km/h, showing a maximum SEL of 131.4 dB(A) located 0.1 m above the rail head. Two local SEL minima are found at a height of 2.0 m and at the pantograph with values given by 129.2 and 129.1 dB(A), respectively. Hence, the largest SEL

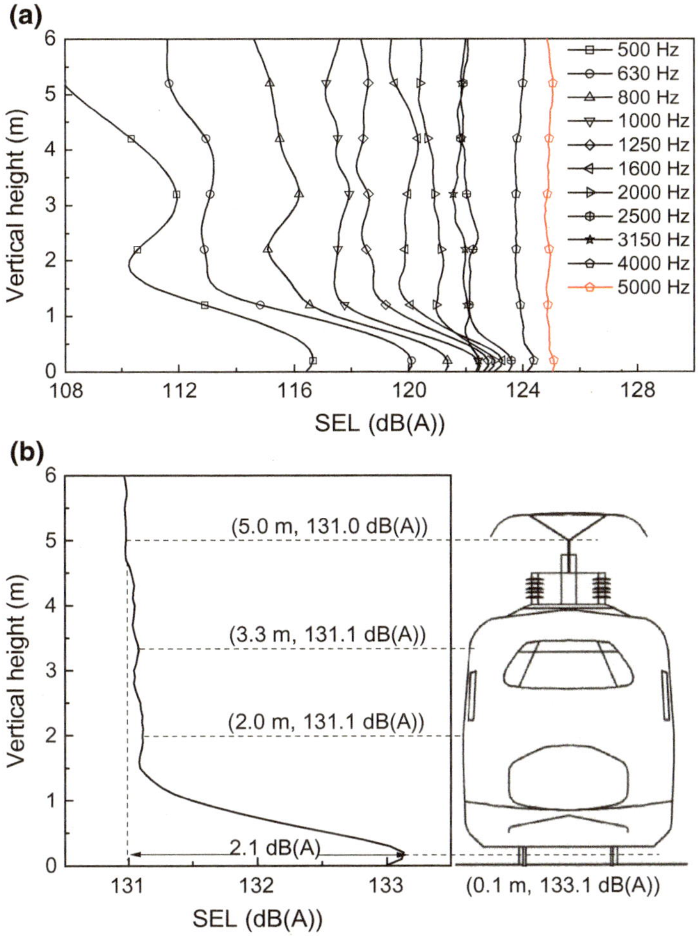


**Fig. 12** SEL distribution at 386 km/h. **a** SEL in every 1/3 octave frequency band; **b** SEL in the full frequency range

is obtained in the wheel/rail area, while the SELs corresponding to the regions of the pantograph and the inter-coach gaps are similar to each other. For a speed of 341 km/h, the maximum SEL difference is obtained between the wheel/rail area and the pantograph (from the bottom up) and equals 2.3 dB(A).

Similarly to Figs. 10a, 11a, and 12a shows that the SELs corresponding to a train velocity of 386 km/h are maximized below a height of 1.0 m for every 1/3 octave frequency band. However, in this case the SELs at all heights increase for larger 1/3 octave frequency bands. The SEL at 5000 Hz is the highest compared to all other frequencies. From the results obtained for the source identification, it became apparent that the main noises at this frequency are of an aerodynamic nature, originating at the pantograph, the inter-coach gaps, and the first access door.

Figure 12b shows the total SEL for a train moving at 386 km/h, with a maximum SEL value of 133.1 dB(A) that again occurs at a height of 0.1 m relative to the rail head. The minimum SEL is 131.0 dB(A) and is obtained at the pantograph. The total SEL difference above 1.0 m lies within 0.1 dB. Hence, the total SEL at the wheel/rail is the largest, while the SEL values measured at the pantograph and the inter-coach gaps are very similar. Lastly, the maximum SEL difference is 2.1 dB(A) and is between the wheel/rail area and the pantograph.

The SEL discussed above reflects the characteristics of the sound energy distribution along the vertical height of the high-speed train during the pass-by time. For low frequencies, the observed SEL difference is larger in this direction, since the noise mainly comes from the wheel/rail regions. As the frequency increases, the aerodynamic noise in the high frequency range that is caused by the train body, the inter-coach gaps, and the pantograph starts to emerge. Table 1 depicts the maximum SEL differences for each 1/3 octave frequency band. From the results shown in Figs. 10, 11 and 12 and Table 1, it can be deduced that the frequency corresponding to the maximum sound energy increases with increasing train velocity. In every 1/3 octave frequency band, the SEL exhibits its maximum in the wheel/rail region, especially at a height of 0.1 m above the rail head. As the frequency and the speed increase, the aerodynamic noise caused by disturbed flow and turbulence emerges along the entire height, resulting in a decrease of the maximum SEL differences by 9.3–9.9 dB among the analyzed frequency components that range from 500 Hz to 5000 Hz. For the full frequency range, the maximum SEL differences in the vertical direction equal 2.5, 2.3, and 2.1 dB for speeds of 271, 341, and 386 km/h, respectively. Hence, the SEL difference along the vertical direction decreases as the train speed increases. A similar result has been reported in another study (Bernd et al. 2002). Furthermore, faster train speeds also push down the ratio between the wheel/rail noise and the aerodynamic noise.

**Table 1** Maximum SEL differences at different train speeds and different frequencies

| Frequency (Hz) | Maximum SEL difference [dB(A)] | | |
|---|---|---|---|
| | $v$ = 271 km/h[a] | $v$ = 341 km/h | $v$ = 386 km/h |
| 500 | 10.4 | 10.6 | 9.6 |
| 630 | 9.7 | 8.8 | 8.5 |
| 800 | 6.9 | 7.4 | 6.8 |
| 1000 | 5.5 | 5.4 | 5.4 |
| 1250 | 4.8 | 4.5 | 4.6 |
| 1600 | 5.1 | 4.0 | 4.0 |
| 2000 | 3.0 | 3.0 | 2.6 |
| 2500 | 2.1 | 2.0 | 1.9 |
| 3150 | 1.0 | 0.9 | 0.9 |
| 4000 | 0.8 | 0.7 | 0.6 |
| 5000 | 0.5 | 0.4 | 0.3 |
| Overall SEL | 2.5 | 2.3 | 2.1 |

[a]$v$ is the operating speed of the train

## 3 Pass-by Noise Magnitude and Its Characteristics

When noises generated by different sources of a high-speed train radiate outward based on their precise characteristics (especially directivity) and reach different field points, the corresponding sound pressure responses can be measured. The sound pressure levels at standard measuring points are important for evaluating the noise of high-speed trains. In the ISO 3095:2013 standard, M1 (7.5, 1.2 m), M2 (7.5, 3.5 m), and M3 (25, 3.5 m) are three specified field points. The numbers in brackets represent the distances with respect to the center line of the track and the height of the rail head.

Figure 13 illustrates the time histories of the sound pressure levels of the high-speed train measured at M1, M2, and M3, obtained during a single pass-by at 271 km/h. These results can be used for further identification of the noise sources. The time histories of field points M1 and M2 have a strong correlation with the characteristics of the train structure, showing nine sound pressure peaks (Fig. 13). The first peak corresponds to the entrance of the train head, while the last peak is attributed to the exit of the train tail. The seven peaks between them represent the seven gaps between the eight coaches, making the number of observed valleys equal to the number of coaches. It is confirmed that the pressure peaks are predominantly generated at the train head, the train tail, the bogie areas, and the inter-coach gaps. Note that the pantograph is located close to one of the inter-coach gaps (Fig. 13). During the pass-by time, the highest noise pressure level is measured at M2, subsequently followed by M1 and M3. This situation is closely related to the characteristics of the near-field sound directivity, except for the aerodynamic noise generated in the inter-coach gaps, the pantograph, and the bogies.

Figure 14 depicts the time histories of the sound pressure levels measured at M2 for different speeds. The noise peak distribution and its qualitative explanation are similar to the results shown in Fig. 13. The noise peak caused by the pantograph is the third peak located near the second inter-coach gap. For a train speed of 271 km/h the maximum noise during a single pass-by is caused by the fourth inter-coach gap, close to the wheel/rail region. When the speed is 341 km/h, the

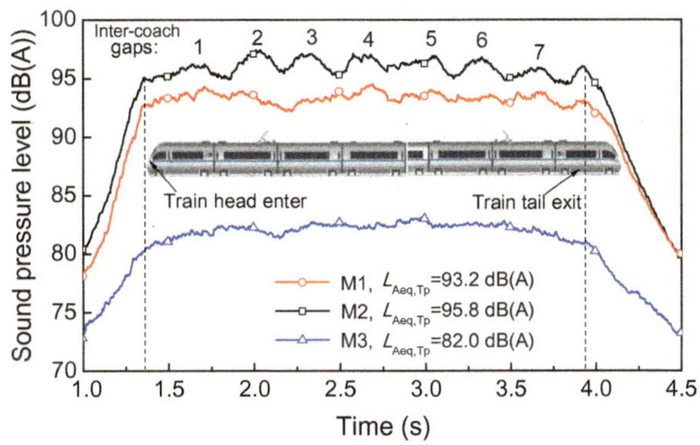


**Fig. 13** Time histories of the sound pressure levels during a single pass-by of a high-speed train at 271 km/h

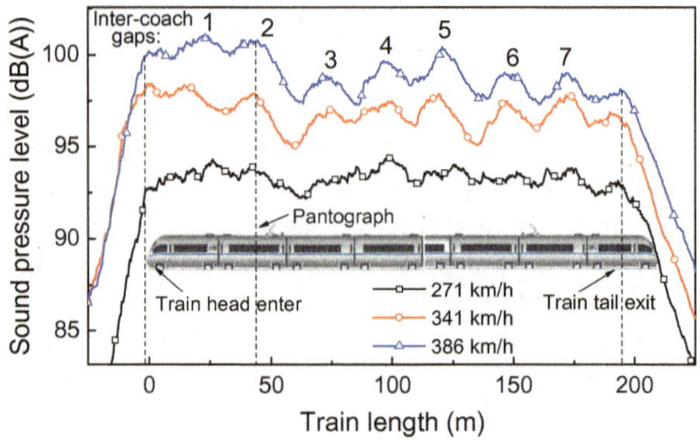


**Fig. 14** Time histories of the sound pressure level at M2 at different speeds

**Table 2** Sound pressure levels of M1, M2, and M3 at different speeds

| Speed (km/h) | $L_{Aeq,Tp}$ [dB(A)] | | |
|---|---|---|---|
| | M1 | M2 | M3 |
| 271 | 93.2 | 95.8 | 82.0 |
| 341 | 96.5 | 98.0 | 85.5 |
| 386 | 98.5 | 100.1 | 88.1 |

maximum noise comes from the train head and the first bogie region. The noise peak caused by the second inter-coach gap and the pantograph is larger than all other peaks, except for the first and second peaks. When the speed is 386 km/h, the second and third peaks are dominant, corresponding to the first inter-coach gap, and the combination of the second inter-coach gap and the pantograph, respectively.

As the speed increases, it can be seen that the noise caused by the first two coaches increases, becoming significantly larger than the noises caused by all other coaches. The main reason for this is the fierce interaction of airflows with the first bogie, the pantograph, and the first two inter-coach gaps, thus generating a strong aerodynamic noise around these structures when the train travels at higher speeds.

The A-weighted equivalent continuous sound pressure level ($L_{Aeq,Tp}$) reflects the average noise energy during the pass-by time. Table 2 shows these values at the three field points M1, M2, and M3 for different train speeds, while Fig. 15 depicts the frequency characteristics of the noise pressures at these three field points for different speeds. The frequency range shown in Fig. 15 goes from 160 to 10,000 Hz, covering the main external noise frequencies of the train. The sound sources spread through the air, and subsequently cause a response at the field points. According to sound propagation theory, the sound pressures measured at the field points are closely related to the frequency, the directivity of the sound source, and the environment of the sound field (airflow, temperature, humidity, terrain, etc.). Thus, the frequency characteristics of the field points are not solely determined by the frequency characteristics of the sound sources. The pressure levels detected at the field points increase with increasing speeds. For equal velocities, the noise level at M3 is approximately 10.0–13.0 dB lower than that at M1 and M2.

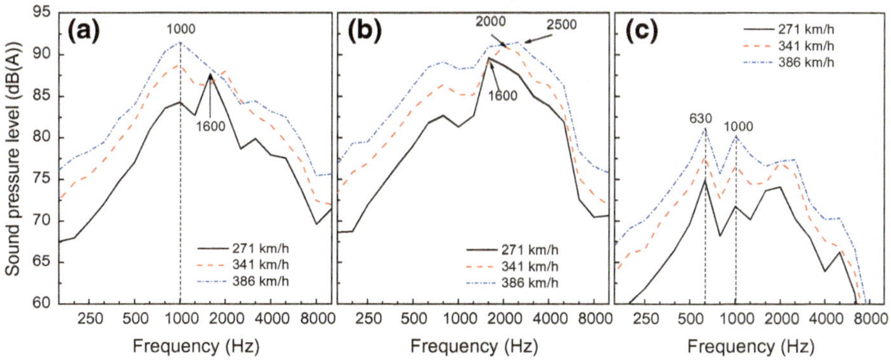


**Fig. 15** Frequency characteristic of three field points at different speeds: **a** M1; **b** M2; **c** M3

As shown in Fig. 15a, at M1, for a speed of 271 km/h the noise is maximized at 1600 Hz, while for velocities of 341 and 386 km/h this happens at 1000 Hz. However, when analyzed as a function of the speed, the noise level at 1600 Hz does not change significantly and the peak disappears for larger velocities. For M2 (Fig. 15b), the noise component characteristics at 1600 Hz are similar to those measured at M1, while the noise levels corresponding to speeds of 341 and 386 km/h reach their maxima at 2000 and 2500 Hz, respectively. At both M1 and M2, the noise components measured at 1600 and 2000 Hz do not change much as a function of the train velocity. At M3 (Fig. 15c), the levels of all noise components increase with increasing speeds along the entire analyzed frequency range.

Since the measuring points M1 and M2 have the same distances with respect to the center line of the track, but different distances with respect to the rail head, their frequency characteristics are different, which is exemplified in Figs. 15a, b. The main difference is that the dominant components of the noise or the peak frequencies move forward. At M1 the noise components from 600 to 2000 Hz are predominant, while at M2 the dominant noise components are distributed between 1600 and 4000 Hz. This is caused by the fact that M1 is located close to the wheel/rail area, hence demonstrating that noise generated in this area has a greater contribution to the measured results, when compared to other noise sources. However, M2 is located near the pantograph, the carriage body, and the inter-coach gaps. Since at these regions much aerodynamic noise is generated, this demonstrates that this noise type greatly contributes to the measured results at M2, when compared to other noise sources.

When comparing the results at M1, M2, and M3, it can be concluded that the results measured at M1 and M2 are quite similar. Although the heights of M2 and M3 are approximately the same, at M3, which is placed at a much larger distance from the central line of the track, the wheel/rail noise has a much larger contribution to the total noise level, when compared to the aerodynamic noise. At this measurement point, the dominant noise components range from 600 to 2500 Hz and their characteristics clearly change with increasing train speeds. Usually, these noise components mainly consist of wheel/rail noise. Hence, for moving high-speed trains, the control and reduction of wheel/rail rolling noise should be considered first.

Figure 15 clearly shows that the differences in level of the noise components at points M1, M2, and M3 have the tendency to decrease for larger frequencies and increasing speeds. This explains that the aerodynamic noise of the train increases in an extensive frequency range as the train velocity becomes larger. In addition, Fig. 15 shows that the noise components at 630, 1000, 1600, 2000, and 2500 Hz are always activated. Their precise generation mechanisms, which up to this point are not known, will be further investigated.

To effectively reduce the external noise, the maximum noise in the full frequency range, the location of its source, and the mechanism generating it should first be clearly identified. Taking M3 as an example, maximum noise peaks occur at 630 and 1000 Hz. From Fig. 12, it can be deduced that the SEL below a vertical height of 1.0 m is about 9.0 and 5.0 dB larger than those at other heights for 630 and 1000 Hz, respectively. The vertical height of 1.0 m belongs to the wheel/rail area. Therefore, the noise generated in this area needs to be controlled and reduced preferentially. Major measures to suppress wheel/rail noise include the employment of a bogie skirt to reduce the aerodynamic noise, the use of damping measures for both wheels and rails to reduce rolling noise, and the installation of sound barriers to change the noise propagation path.

## 4 External Noise Behaviors as a Function of Speed

The characteristics of the noise radiated by high-speed trains can be established with an approximate function that depends on the train speed. The function can be approximated by a second-order polynomial of the variable lg($v$) (Mellet et al. 2006). The regression equation can be expressed as

$$L_{\mathrm{Aeq,T_p}}(v) = A[\lg(v)]^2 + B\lg(v) + C, \tag{13}$$

where $A$, $B$, and $C$ are regression coefficients. For simplicity, this piecewise linear function can be used to replace the nonlinear function displayed in Eq. (13). According to the characteristics of the wheel/rail noise and the aerodynamic noise with respect to the train speed, the piecewise linear regression function is considered in two speed ranges, one being defined below 300 km/h, and the other above 300 km/m. The speed of 300 km/h is considered the break speed (Mauclaire 1990; Krylov 2001). Below 300 km/h, the slope of the linear function is approximately 30, while above the break speed a slope of 60 is found. For both speed ranges, 30 and 60 denote the regression coefficient $B$ in Eq. (13). If the value of $L_{\mathrm{Aeq,Tp}}$ at the reference speed $v_0$ is known, the linear regression law can be expressed as

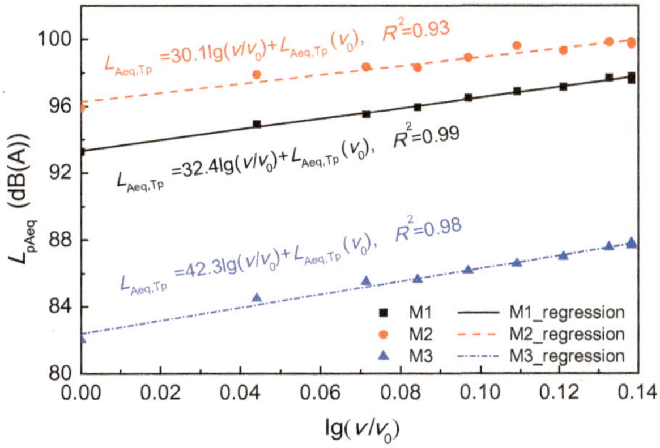


**Fig. 16** Linear regressions of the measured data from 280 to 390 km/h ($v_0$ = 280 km/h)

$$L_{\mathrm{Aeq,T_p}}(v) = B\lg(v/v_0) + L_{\mathrm{Aeq,T_p}}(v_0). \tag{14}$$

Figure 16 shows the linear regressions of $L_{\mathrm{Aeq,Tp}}$ for the three different field points. Note that from 280 to 390 km/h, the regression coefficients $B$ are 32.4, 30.1, and 42.3, while the corresponding correlation coefficients $R^2$ are 0.99, 0.93, and 0.98 for M1, M2, and M3, respectively.

Using these measured results, it can be assumed that this linear expression is valid up to velocities of 390 km/h. The transition speed for which aerodynamic noise becomes as important as the rolling noise, which generally is considered to lie around 300 km/h (Mauclaire 1990; Krylov 2001), is not clearly observed. The regression coefficients for M1 and M2 are close to 30, a value which is commonly used in the prediction formula for wheel/rail rolling noise. Hence, the rolling noise is dominant when considering the total noise measured at M1 and M2 for a speed close to 390 km/h. For M3, the regression coefficient is approximately 42. To reduce the pass-by noise at this field point, one needs to take measures to suppress both rolling noise and aerodynamic noise, although the rolling noise still makes a greater contribution to the total noise detected at M3 compared with the aerodynamic noise.

## 5 Conclusions

This paper presents the test results and an analysis of external noise characteristics produced by a Chinese high-speed train traveling at different speeds. From this study, the following conclusions can be drawn:

1. External noise identification of the high-speed train shows that main noise originates at three areas: the wheel/rail systems (or bogies), the pantograph, and the inter-coach gaps. The wheel/rail area produces the dominant rolling noise

and the aerodynamic noise caused by airflow around the bogie. The pantograph and the inter-coach gaps of the train mainly generate aerodynamic noise. For speeds below 386 km/h, the SEL of the wheel/rail area is the greatest in the frequency range below 3150 Hz, while the SELs of the three noise sources are quite similar for larger frequencies.

2. For both the total external noise and the predominant noise components analyzed at different frequencies, the wheel/rail noise has the greatest contribution compared with the aerodynamic noise sources located at the pantograph and the inter-coach gaps. The measured noise from the wheel/rail area mainly includes the wheel/rail rolling noise and the aerodynamic noise coming from the bogie area. However, it has so far been difficult to experimentally determine their exact relative proportions in the total noise.
3. Along the vertical train height, maximum noise levels are found in the wheel/rail area. At distances far from the central track line, the wheel/rail rolling noise still makes a greater contribution than the aerodynamic noise for the entire train velocity range analyzed.
4. The measured results at all field points show that the noise components from 630 to 2500 Hz, which are typically attributed to wheel/rail rolling noise, always dominate. Therefore, it is suggested that the design of low-noise high-speed trains and external noise control should be focused on the control and reduction of this type of noise.
5. The measured results at all different field points clearly indicate that the noise level differences have a tendency to decrease with increasing frequencies and train speeds. This explains the growth of aerodynamic noise over an extensive frequency range as the train runs faster. In addition, the test results show that noise components at 630, 1000, 1600, 2000, and 2500 Hz are always activated. Nevertheless, their exact generation mechanisms are currently unknown and will be further investigated.

## References


Bernd, B., Daniel, R. D., Hanson, C. E., et al. (2002). *Noise characteristics of the transrapid TR08 Maglev system*. Dot-vntsc-fra-02-13, US Department of Transportation.

Christensen, J. J., & Hald, J. (2004). *Technical review beamforming* (pp. 3–12). Danmark: Bruel & Kjear.

Dittrich, M. G., & Janssens, M. H. A. (2000). Improved measurement methods for railway rolling noise. *Journal of Sound and Vibration, 231*(3), 595–609. doi:10.1006/jsvi.1999.2547.

He, B. (2010). *Primary investigation into external noise distribution characteristics of high-speed train and damping control measures of wheel*. MS Thesis, Southwest Jiaotong University, Chengdu, China (in Chinese).

ISO (International Organization for Standardization). (2013). *Acoustics-railway applications—Measurement of noise emitted by railbound vehicles, ISO 3095:2013*. Switzerland: ISO.

Johnson, D. H., & Dudgeon, D. E. (1993). *Array Signal Processing: Concepts and Techniques* (pp. 94–99). New Jersey: Prentice Hall.

Koh, H., You, W., Kwon, H., et al. (2007). Noise source identification of Korean high speed train. In *The 14th International Congress on Sound and Vibration*, Cairns, pp. 1498–1503.

Kook, H., Moebs, G. B., Davies, P., et al. (2000). An efficient procedure for visualizing the sound field radiated by vehicle during standardized pass by tests. *Journal of Sound and Vibration, 233*(1), 137–156. doi:10.1006/jsvi.1999.2794.

Krylov, V. (2001). *Noise and vibration from High-speed trains* (pp. 78–85). London: Thomas Telford.

Mauclaire, B. (1990). Noise generated by high speed trains. In *Proceedings of Internoise*, Gothenburg, Sweden, p. 371–374.

Mellet, C., Létourneaux, F., Poisson, F., et al. (2006). High speed train noise emission: latest investigation of the aerodynamic/rolling noise contribution. *Journal of Sound and Vibration, 293*(3–5), 535–546. doi:10.1016/j.jsv.2005.08.069.

Noh, H. M., Choi, S., Hong, S. Y., et al. (2011). Designing a microphone array system for noise measurements on high-speed trains. *Journal of the Korean Society for Railway, 14*(6), 477–483 (in Korean). doi:10.7782/JKSR.2011.14.6.477.

Poisson, F., Gautier, P. E., & Letourneaux, F. (2008). *Noise sources for high speed trains: A review of results in the TGV case. Noise and vibration mitigation for rail transportation systems* (pp. 71–77). Berlin, Heidelberg: Springer.

Schulte-Werning, B., Jäger, K., Strube, R., et al. (2003). Recent developments in noise research at Deutsche Bahn (noise assessment, noise source localization and specially monitored track). *Journal of Sound and Vibration, 267*(3), 689–699. doi:10.1016/S0022-460X(03)00733-8.

Silence. (2005). *Report on source ranking on state of the art validation platforms and final priorities for research effort*. Austria: European Commission.

Thompson, D. (2008). *Railway noise and vibration: Mechanisms. Modelling and means of control* (pp. 3–9). New York: Elsevier.

Thron, T. (2010). *A contribution to the noise prediction based on recognized metrological model parameters*. PhD Thesis, Technical University of Berlin, Berlin, Germany (in German).

Wakabayashi, Y., Kurita, T., Yamada, H., et al. (2008). *Noise measurement results of Shinkansen high-speed test train (FASTECH360S, Z). Noise and vibration mitigation for rail transportation systems* (pp. 63–70). Berlin, Heidelberg: Springer.

## Author Biography

**Xue-song Jin** Ph.D., is a professor at Southwest Jiaotong University, Chengdu, China. He is a leading scholar of wheel-rail interaction in China, and is a State Council Expert for Special Allowance. He is the author of three academic books and over 200 articles. He is an editorial board member of several journals, and has been a member of the Committee of the International Conference on Contact Mechanics and Wear of Rail/Wheel Systems for more than 10 years. He has been a visiting scholar at the University of Missouri-Rolla for 2.5 years. Now his research focuses on wheel-rail interaction, rolling contact mechanics, vehicle system dynamics, vibration, and noise.

# Effect of Softening of Cement Asphalt Mortar on Vehicle Operation Safety and Track Dynamics

**Jian Han, Guo-tang Zhao, Xin-biao Xiao, Ze-feng Wen, Qing-hua Guan and Xue-song Jin**

## Contents



---

J. Han · G. Zhao · X. Xiao · Z. Wen · Q. Guan · X. Jin (✉)
State Key Laboratory of Traction Power, Southwest Jiaotong University,
Chengdu 610031, China
e-mail: xsjin@home.swjtu.edu.cn

G. Zhao
China Railway Corporation, Beijing 100844, China



Y. Fang and Y.H. Zhang (eds.), *China's High-Speed Rail Technology*,
Advances in High-speed Rail Technology, https://doi.org/10.1007/978-981-10-5610-9_22

## 1 Introduction

In the operation of high-speed trains, different degrees of damage are suffered by the cement asphalt mortar (CAM) that forms the filling layer between the slab and the concrete base (Lin 2009; Liu 2013). The damage includes cracks, shelling, aging, and rain soaking (Fig. 1). Much research work on vehicle-track coupling dynamics and track-subgrade dynamics was carried out (Chen et al. 2014; Ling et al. 2014; Zhong et al. 2014). However, there have been few studies of vehicle-track coupling systems that consider CAM damage. Xiang et al. (2009) studied the effect of a voided slab induced by the deterioration of the CAM layer on vibration responses of a slab track at variable vehicle speeds. Based on the perspective of the system energy, he used the Wilson-$\vartheta$ numerical integral method to solve the track vibration equations. Wang et al. (2014) analyzed the effect of CAM debonding on the dynamic properties of a CRTS-II slab track, using LS-DYNA to solve the dynamic equations. Both Xiang et al. (2009) and Wang et al. (2014) treated the rail as a continuous Euler beam, and their models considered only vertical vibration. Zhu and Cai (2014) investigated interface damage and its effect on vibrations of a slab track under different temperature and vehicle dynamic loads. The loads were obtained using the developed vehicle-track coupling dynamic model, and the track model was developed using ABAQUS software. The model assumed that the influence of temperature is important to CAM damage after a period of time. However, when CAM damage had already occurred, its effect on train running safety was not discussed.

The CAM softening not only leads to track structural failure but also becomes a potential factor responsible for increasing the probability of vehicle derailment.

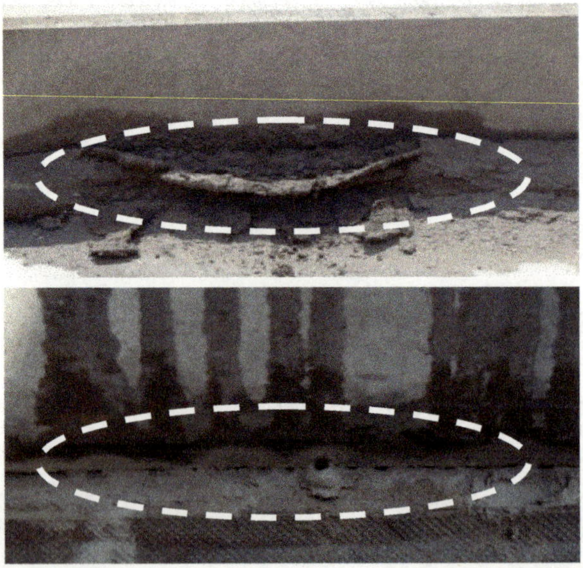

**Fig. 1** CAM damage (*softening*). Reprinted from (Zhu et al. 2014), with kind permission from Springer Science + Business Media

Zhou and Shen (2013) and Xiao et al. (2007) studied the effect of disabled fastening systems or unsupported sleepers on ballast tracks on vehicle derailment using a vehicle-track coupling dynamic model. CAM damage is particularly common in slab tracks. It is important to study the influences of CAM softening on the dynamic characteristics of a 3D vehicle-track system because CAM damage endangers the safety of train operation, especially for curved lines.

CAM softening seriously affects vehicle operation safety and track interface shear failure. In this paper, a 3D coupling dynamic model of a vehicle and a CRTS-I slab track is developed. The vehicle runs on a curved track at 300 km/h. Using the proposed model, the wheel-rail contact forces, derailment coefficient, wheelset loading reduction ratio, and the track displacements are calculated to study the influences of CAM softening on the dynamic characteristics of the vehicle-track system. A track-subgrade finite difference model is developed to investigate the effect of CAM softening on slab stress and track interface failure.

# 2 Coupling Dynamic Model of Vehicle and CRTS-I Slab Track

Figure 2 illustrates the coupling dynamic model of a high-speed vehicle and the CRTS-I slab track to study the effect of CAM softening on the dynamic behavior of a vehicle-track system. In the numerical simulation, different degrees of CAM softening are considered under one slab. A moving rail-support is adopted as a new vehicle-track coupling interface excitation model (called the “Tracking Window”) (Jin and Wen 2008; Xiao et al. 2011; Jin 2014). This excitation model is closer to a real moving vehicle under the excitation of discrete sleepers and saves a lot of computation time. The vehicle-track coupling system equations are solved by means of a new explicit integration method (Zhai 1996).

## *2.1 Dynamic Model of Vehicle Subsystem*

The high-speed railway vehicle is considered as a rigid multi-body model, in which the car body is supported by two double-axle bogies with the primary and the secondary suspension systems. For the connecting parts (the primary vertical damper, the secondary lateral damper, the secondary yaw damper, and the lateral stopping block) with nonlinear characteristics, a piecewise linear simulation is used. Each component of the vehicle has six degrees of freedom (DOFs): longitudinal motion, lateral motion, vertical motion, roll angle, yaw angle, and pitch angle (Fig. 2). The vehicle has a total of 42 DOFs. Based on the coordinate system, moving along the track at the constant speed of the vehicle, the equation of the

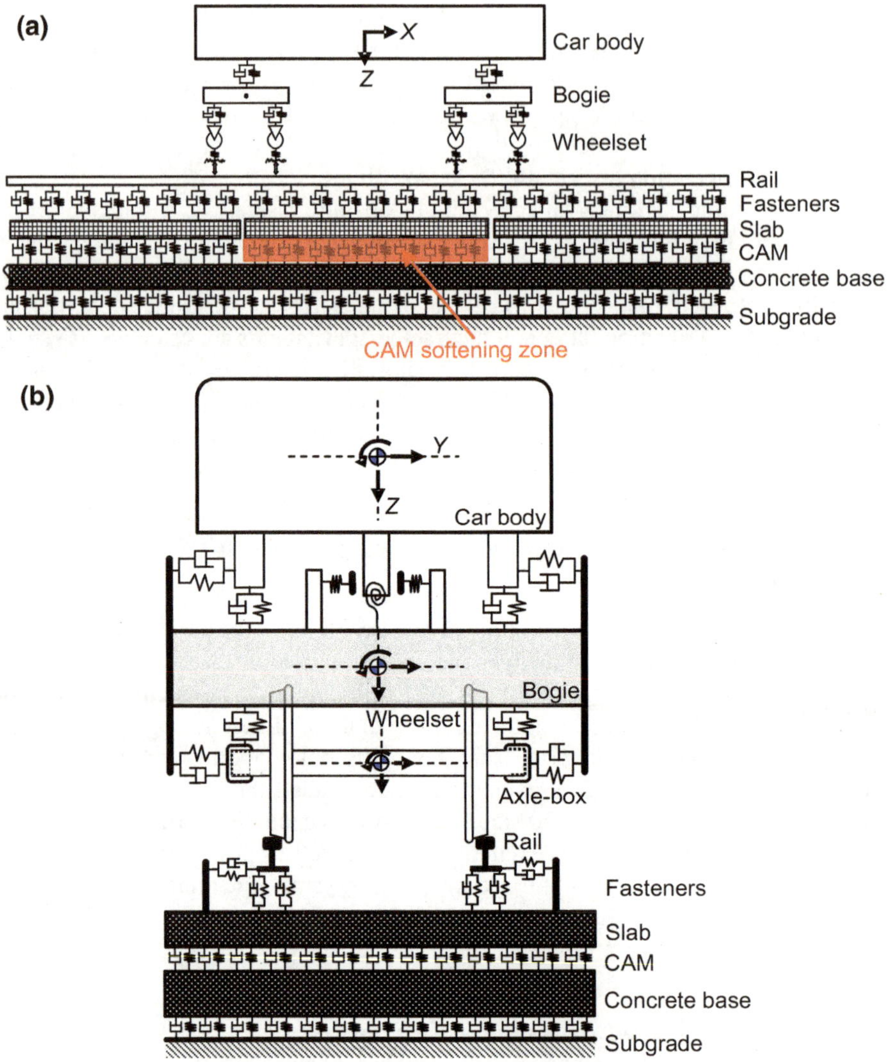


**Fig. 2** Vehicle-track coupling dynamic model. **a** Elevation; **b** end view

vehicle subsystem can be described in the second-order differential equation in the time domain as follows:

$$\boldsymbol{M}_{\mathrm{v}}\ddot{\boldsymbol{u}}_{\mathrm{v}} + \boldsymbol{C}_{\mathrm{v}}\dot{\boldsymbol{u}}_{\mathrm{v}} + \boldsymbol{K}_{\mathrm{v}}\boldsymbol{u}_{\mathrm{v}} = \boldsymbol{F}_{\mathrm{v}}, \tag{1}$$

where $\boldsymbol{M}_{\mathrm{v}}$ is the mass matrix of the vehicle, and $\boldsymbol{C}_{\mathrm{v}}$ and $\boldsymbol{K}_{\mathrm{v}}$ are the damping and the stiffness matrices. $\boldsymbol{u}_{\mathrm{v}}$, $\dot{\boldsymbol{u}}_{\mathrm{v}}$, and $\ddot{\boldsymbol{u}}_{\mathrm{v}}$ are the vectors of displacement, velocity, and acceleration, respectively, of the vehicle subsystem, and $\boldsymbol{F}_{\mathrm{v}}$ is the vector of generalized loads acting on the vehicle subsystem.

## 2.2 Dynamic Model of Slab Track Subsystem

The dynamic model of the slab track subsystem includes rails, fastener systems, slabs, CAM layers, and concrete base (Fig. 2). The rail is treated as a continuous Timoshenko beam resting on rail pads, and the lateral, vertical, and torsion motions of rails are simultaneously taken into account (Xiao et al. 2008). The slabs and the concrete base are modeled using the 3D finite element method. The rail fastener systems and the CAM layer are modeled using periodic discrete viscoelastic units. The finite element model of the slab has 20,600 solid elements and 26,520 DOFs. The length of the slab is 4.962 m. The geometric dimensions of its cross section are 2.4 m × 0.19 m. The vibration of the slab can be easily described in the second-order differential equation in terms of generalized coordinates, as expressed by Eq. (2). Modal analysis of the slab is carried out by means of ANSYS to obtain 20 order modes, by which Eq. (2) is decoupled and solved according to the modal superposition principle, as follows:

$$\boldsymbol{M}_{\mathrm{s}i}\ddot{\boldsymbol{u}}_{\mathrm{s}i}+\boldsymbol{C}_{\mathrm{s}i}\dot{\boldsymbol{u}}_{\mathrm{s}i}+\boldsymbol{K}_{\mathrm{s}i}\boldsymbol{u}_{\mathrm{s}i}=\boldsymbol{F}_{\mathrm{rs}i}+\boldsymbol{F}_{\mathrm{sc}i}, \tag{2}$$

where $\boldsymbol{M}_{\mathrm{s}i}$, $\boldsymbol{C}_{\mathrm{s}i}$, and $\boldsymbol{K}_{\mathrm{s}i}$ are the mass, damping, and stiffness matrices, respectively, of the $i$th slab. $\ddot{\boldsymbol{u}}_{\mathrm{s}i}$, $\dot{\boldsymbol{u}}_{\mathrm{s}i}$, and $\boldsymbol{u}_{\mathrm{s}i}$ are the acceleration vector, velocity vector, and displacement vector, respectively. $\boldsymbol{F}_{\mathrm{rs}i}$ is the load vector between the rail and the $i$th slab, and $\boldsymbol{F}_{\mathrm{sc}i}$ is the load vector between the slab and the concrete base.

The model of the concrete base is similar to that of the slab. The concrete base model has 433,956 solid elements and 515,424 DOFs. The length of the concrete base is 60 m. The geometric dimensions of its cross section are 0.3 m × 3 m.

## 2.3 Model of Wheel-Rail Interaction in Rolling Contact

Wheel/Rail dynamic interaction modeling is the key to the vehicle-track coupling dynamic model. The calculation of wheel/rail contact forces includes a normal model and a tangent model. The normal model, which characterizes the relationship law of a normal load and deformation between the wheel and rail, is described by a nonlinear Hertz contact spring with a unilateral restraint:

$$N(t)=\begin{cases}\left[\frac{1}{G_{\mathrm{Hertz}}}Z_{\mathrm{wrnc}}(t)\right]^{3/2}, & Z_{\mathrm{wrnc}}(t)>0,\\ 0, & Z_{\mathrm{wrnc}}(t)\leq 0,\end{cases} \tag{3}$$

where $G_{Hertz}$ is the wheel/rail contact constant (m/N$^{2/3}$), which can be obtained using the Hertz contact theory. $Z_{wrnc}(t)$ is the normal amount of compression at the wheel/rail contact point. $Z_{wrnc}(t)$ is strictly defined as an approach between two distant points, one belonging to the wheel, and the other belonging to the rail. The wheel and the rail are assumed to be an elastic half-space. This approach is confined to the normal direction at the contact point of the wheel and the rail. $Z_{wrnc}(t) > 0$ indicates the wheel/rail in contact, and $Z_{wrnc}(t) \leq 0$ indicates their separation.

The tangential wheel-rail creep forces are calculated using the Shen-Hedrick-Elkins non-linear theory (Shen et al. 1983). In this paper, when calculating the dynamic response of the vehicle-track, the tracing-curve-method (Chen and Zhai 2004) is adopted to locate the wheel-rail spatial contact geometry. This can greatly reduce the computational time.

## 2.4 CAM Softening in the Vehicle-Track Coupling Dynamic Model

CAM softening, including CAM aging or rain soaking, is considered. CAM softening leads to changes in the vertical and lateral supporting stiffnesses of the slab and becomes a potential factor responsible for increasing the probability of vehicle derailment.

CAM softening is simulated by changing the stiffness coefficient of the CAM layer, i.e., the parameters considered are multiplied by "softening coefficients" in the coupled vehicle-track model. The damping used in this paper is assumed to be structural damping. So the same softening coefficient is applied to damping, as shown in Eqs. (4) and (8).

$$\begin{cases} K'_{scl} = K_{scl}/\lambda_{scl}, \ C'_{scl} = C_{scl}/\lambda_{scl}, & 1 < \lambda_{scl} < +\infty, \\ K'_{scv} = K_{scv}/\lambda_{scv}, \ C'_{scv} = C_{scv}/\lambda_{scv}, & 1 < \lambda_{scv} < +\infty, \end{cases} \tag{4}$$

where $K'_{scl}$ is the softening lateral stiffness, $K_{scl}$ is the original lateral stiffness, $C'_{scl}$ is the softening lateral damping, $C_{scl}$ is the original lateral damping, $K'_{scv}$ is the softening vertical stiffness, $K_{scv}$ is the original vertical stiffness, $C'_{scv}$ is the softening vertical damping, $C_{scv}$ is the original vertical damping, $\lambda_{scl}$ is the lateral softening coefficient, and $\lambda_{scv}$ is the vertical softening coefficient.

## 2.5 Evaluation Criteria of Railway Vehicle Derailment

At present, two important criteria are widely used to evaluate the dynamic behavior and safety operation of high-speed trains (Xiao et al. 2007, 2014; Zhou and Shen

2013). One is Nadal's criterion (derailment coefficient), denoted by Eq. (5), and the other is the wheelset loading reduction, indicated by Eq. (6):

$$\left(\frac{L}{V}\right)_{\text{Critical}} = \frac{\tan\delta_{\max} - \mu}{1 + \mu\tan\delta_{\max}}, \tag{5}$$

$$\frac{\Delta V}{V} = \frac{\frac{1}{2}(V_L - V_R)}{\frac{1}{2}(V_L + V_R)} = \frac{V_L - V_R}{V_L + V_R}, \tag{6}$$

where $\delta_{\max}$ is the maximum flange angle of the wheel, and $\mu$ indicates the friction coefficient between the wheel and the rail. $L$ and $V$ denote the lateral and vertical forces, respectively, of the wheel and the rail, and $\Delta V$ indicates the normal loading difference between the left and right wheels of the same wheelset.

## 3 Track/Subgrade Coupling Model

### *3.1 Finite Difference Model of Slab Track and Subgrade*

Figure 3 shows the 3D finite difference model of the CRTS-I slab track system and its subgrade built in this study. The slab track includes three layers: the slab, the CAM, and the concrete base. The subgrade includes three layers: the upper, middle, and bottom layers. The layers have different properties (Table 1). The constitutive relation of the CAM is the Mohr-Coulomb elastic-plastic model. Those of the other parts are linear elastic models. Table 2 shows the material parameters of the CRTS-I slab track components. A viscoelastic artificial boundary (Liu et al. 2006) was applied to the bottom and sides along the longitudinal direction of the subgrade. This can characterize the real behaviour of the subgrade bottom support, and avoid wave reflection on the boundary and model infinite track length along the longitudinal direction. The boundary is simulated using a normal and tangential spring and damping. One end of the spring-damping is connected to the subgrade boundary, and the other end is fixed. Equation (7) describes their stiffness and

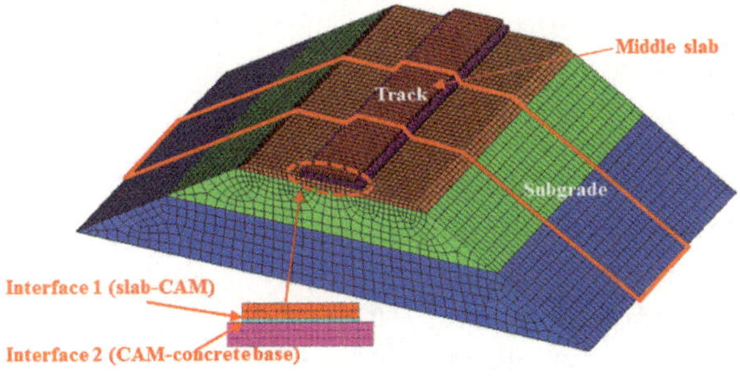


**Fig. 3** 3D model of a CRTS-I slab track and its subgrade

damping. The subgrade slope is free. There are three slabs in Fig. 3. The middle slab was chosen for the analysis.

$$\begin{cases} K_t = 0.5G\Delta s/R, \\ K_n = G\Delta s/R, \\ C_t = \rho C_s \Delta s, \\ C_n = \rho C_p \Delta s, \end{cases} \tag{7}$$

where $K_t$ and $K_n$ are, respectively, the tangent and normal stiffness, $C_t$ and $C_n$ are, respectively, the tangent and normal damping, and $G$ is the shear stiffness. $R$ is the equivalent length between the source and the bottom (5.7 m), and $\Delta s$ is the smallest mesh size. $C_s$ is the shear wave velocity, and $C_p$ is the press wave velocity.

### 3.2 CAM Softening in the Track Finite Difference Model

In the track finite difference model, the CAM is modeled by a solid layer. CAM softening is characterized by changing Young's modulus and the cohesion of the CAM, i.e., the parameters considered are multiplied by "softening coefficients" in the model, as follows:

$$\begin{cases} E'_{CAM} = E_{CAM}/\lambda_{CAM}, & 1 \le \lambda_{CAM} < +\infty, \\ c'_{CAM} = c_{CAM}/\lambda_{CAM}, & 1 \le \lambda_{CAM} < +\infty, \end{cases} \tag{8}$$

where $E'_{CAM}$ is the softening Young's modulus, $E_{CAM}$ is the original Young's modulus, $c'_{CAM}$ is the softening cohesion, $c_{CAM}$ is the original cohesion, and $\lambda_{CAM}$ is the softening coefficient.

### 3.3 Contact Model of Track

There are two interfaces (i.e., slab-CAM and CAM-concrete base) in the track system. The interfaces of the slab-CAM and the CAM-concrete base are simulated by zero thickness elements. The constitutive relation is the Coulomb shear model. In this model, the interfaces have the properties of friction, cohesion, normal stiffness, and shear stiffness. The interface is represented as a collection of

**Table 1** Material parameters of subgrade components

| Component | Poisson's ratio | Young's modulus (MPa) | Density (kg/m$^3$) |
|---|---|---|---|
| Upper layer of subgrade | 0.25 | 150 | 1900 |
| Middle layer of subgrade | 0.25 | 110 | 1950 |
| Bottom layer of subgrade | 0.30 | 70 | 1950 |

**Table 2** Material parameters of CRTS-I slab track components

| Component | Poisson's ratio | Young's modulus (MPa) | Density ($kg/m^3$) | Internal friction angle | Cohesion (kPa) |
|---|---|---|---|---|---|
| Slab | 0.2 | 36,000 | 2400 | – | – |
| CAM | 0.3 | 150 | 2100 | 35° | 1000 |
| Concrete base | 0.2 | 32,500 | 2400 | – | – |

triangular elements (interface elements), each of which consists of three nodes (interface nodes). Two triangular interface elements form a quadrilateral zone face. Interface nodes are then created automatically at every interface element vertex. When another grid surface comes into contact with an interface element, the contact is detected at the interface node and is characterized by normal and shear stiffnesses, and sliding properties. Each interface element distributes its area to its nodes in a weighted way. Each interface node has an associated representative area. The entire interface is thus divided into active interface nodes representing the total area of the interface as shown in Fig. 4 (Han et al. 2015). Han et al. (2015) used this contact model to study the relationship between track and subgrade surface considering water. The model in this paper is mainly used to study the CAM layer without considering water. The normal and shear forces that describe the elastic interface response are determined at the calculation time ($t + \Delta t$) using the following relations (Han et al. 2015):

$$\begin{cases} F_{\rm n}^{t+\Delta t} = k_{\rm n} u_{\rm n} A + \sigma_{\rm n} A, \\ F_{\rm si}^{t+\Delta t} = F_{\rm si}^{t} + k_{\rm s} \Delta u_{\rm si}^{t+0.5\Delta t} A + \sigma_{\rm si} A, \end{cases} \tag{9}$$

where $F_{\rm n}^{t+\Delta t}$ is the normal force at time $t + \Delta t$, $F_{\rm si}^{t+\Delta t}$ is the shear force at time $t + \Delta t$, $u_{\rm n}$ is the absolute normal penetration of the interface node into the target face, $\Delta u_{\rm si}$ is the incremental relative shear displacement, $\sigma_{\rm n}$ is the additional normal stress added due to interface stress initialization, $k_{\rm n}$ is the normal stiffness, $k_{\rm s}$ is the shear stiffness, $\sigma_{\rm si}$ is the additional shear stress due to interface stress, and $A$ is the representative area associated with the interface node initialization.

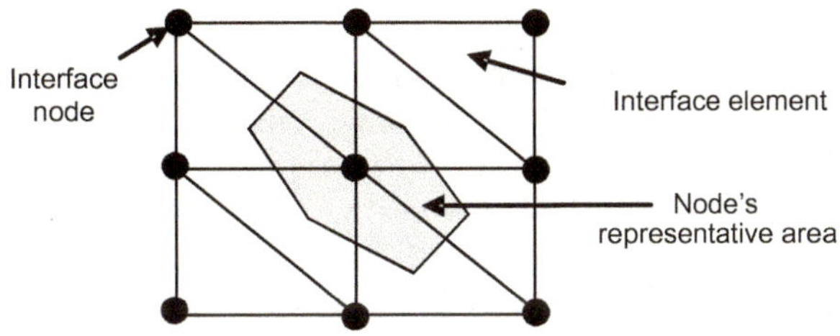


**Fig. 4** Distribution of representative areas in relation to interface nodes (Han et al. 2015)

The Coulomb shear-strength criterion limits the shear force by the following relation without considering water pressure (Han et al. 2015):

$$F_{\mathrm{smax}} = cA + F_{\mathrm{n}} \tan \phi, \tag{10}$$

where $c$ is the cohesion along the interface, and $\phi$ is the friction angle of the interface surface. If the criterion is satisfied (if $|F_s| \geq F_{smax}$), then sliding is assumed to occur.

### 3.4 *Loading on the Track-Subgrade Finite Difference Model*

The rail-supporting forces at fastener $i$ can be calculated using Eq. (5) by the coupling dynamic model of the vehicle and CRTS-I slab track (Sect. 2). The rail-supporting forces are applied to the fasteners on the slab of the 3D track-subgrade coupling model in Fig. 3.

$$F_{\mathrm{sup},i}(t) = k_{\mathrm{sup}} \Delta Z_{\mathrm{sup},i} + c_{\mathrm{sup}} \Delta \dot{Z}_{\mathrm{sup}}(t), \tag{11}$$

where $F_{\mathrm{sup}}$ is the discrete rail-supporting force, $k_{\mathrm{sup}}$ and $c_{\mathrm{sup}}$ are, respectively, the supporting stiffness and damping, and $\Delta Z_{\mathrm{sup}}$ and $\Delta \dot{Z}_{\mathrm{sup}}$ are, respectively, the relative displacement and the relative velocity between the rail and slab.

## 4 Results and Discussion

In the analysis, the considered curved track has a radius of 7000 m and a super-elevation of 150 mm. The left rail is the high rail, and the right rail is the low rail. It is assumed that different degrees of CAM softening occur when the vehicle is running on the curved track. The usual track geometry irregularity is not considered. The train speed is 300 km/h. In Sects. 4.1–4.3, the CAM softening coefficients were chosen as 1 (Good: without CAM softening), 10, 100, 1000, and 10,000 (Empty: the CAM has almost completely failed). Although the CAM damage condition corresponding to each CAM softening coefficient was not tested and discussed in this paper, it is very important to study the effect of the percentage of CAM softening on the dynamic behavior of the vehicle-track system. Once the influencing factors, such as track age, loading cycles, and weather cycles, are determined and shown by testing correspond to the softening coefficient, the limit value discussed below will provide a helpful reference for the safe running of high-speed trains and track maintenance.

## 4.1 Effect of CAM Softening on High-Speed Vehicle Operation Safety

Figure 5 shows the lateral and vertical forces between the rails and the first wheelset of the vehicle when the high-speed train passes over the curved track with different degrees of CAM softening. The section with CAM softening is shaded in Fig. 5. When the high-speed vehicle passes through the CAM softening area, considerable impact vibrations occur between the wheels and the rails, which then gradually decay and reach a steady-state similar to that of the track without CAM softening. Figure 5 shows clearly that the forces of the wheel/rail fluctuate dramatically in the case where CAM softening occurs on the curved track. When the vehicle passes through the track area at a CAM softening coefficient of 10,000 (Empty) (when the CAM has almost completely failed), the maximum lateral and vertical forces are generated on the left wheel of the wheelset. Wheel/Rail separation is generated on both the left and right wheels when the CAM softening coefficient is larger than 1000. Due to the impact of CAM softening and the external centrifugal inertial force of the vehicle body when the train is running through the curved track, the lateral and vertical forces on the left wheel (on the high rail) are much larger than those on the right wheel of the same wheelset. Thus, the right wheel easily jumps and loses contact with the low rail, and the high-speed train risks a jumping derailment.

Figure 6 shows the derailment coefficients (*L*/*V*) with different degrees of CAM softening. The section with CAM softening is shaded in Fig. 6. Compared with the case without CAM softening, the absolute values of the derailment coefficient increase by about 0.02, 0.19, 0.36, and 0.84 on the high rail and by about 0.01, 0.10, 0.96, and 0.96 on the low rail with increasing degrees of CAM softening (Fig. 6a and b). When the softening coefficient is larger than 1000, the derailment coefficients exceed their limit value. The limit value of $L/V$ is $\pm 0.8$ (Zhang 2011) according to the standard of Chinese high-speed railways.

Figure 7 shows the wheelset loading reduction ratio ($\Delta V/V$) with different degrees of CAM softening. Compared with the case without CAM softening, the absolute values of wheelset loading reduction increase by about 0.01, 0.31, 0.88, and 0.88 with increasing degrees of CAM softening. When the softening coefficient is larger than 1000, the wheelset loading reduction ratio exceeds its limit value. The limit value of $\Delta V/V$ is 0.6 (Zhang 2011) according to the standard of Chinese high-speed railways.

## 4.2 Effect of CAM Softening on Track Displacement

Figure 8 shows the rail displacement at the first wheelset in different cases of CAM softening. Compared with the case without CAM softening, the displacements of the high rail and the low rail increase as the degree of CAM softening increases.

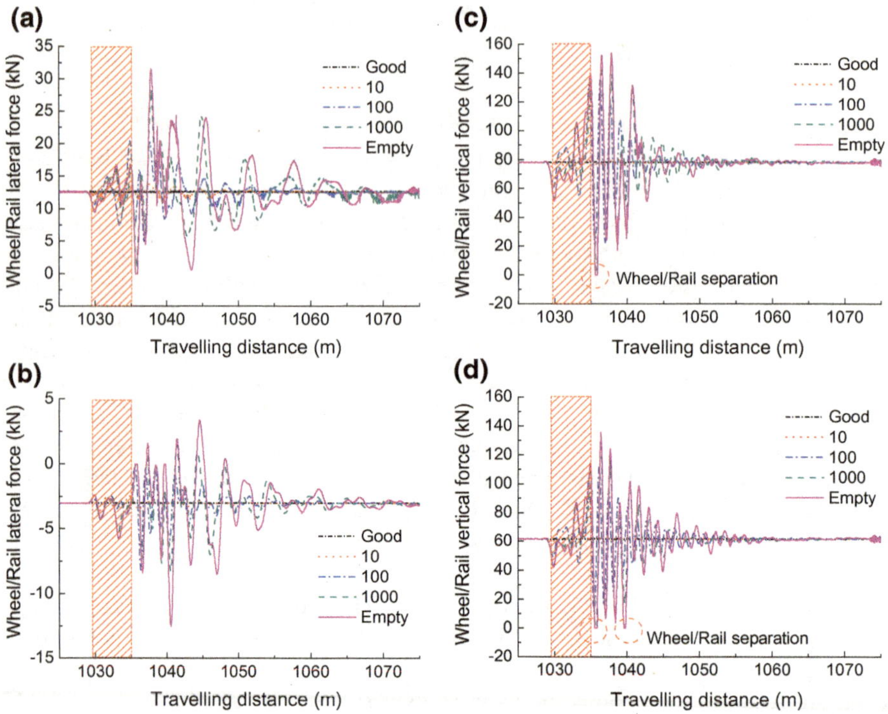


**Fig. 5** Wheel/Rail forces with different degrees of CAM softening. **a** Lateral forces on high rail; **b** lateral forces on low rail; **c** vertical forces on high rail; **d** vertical forces on low rail

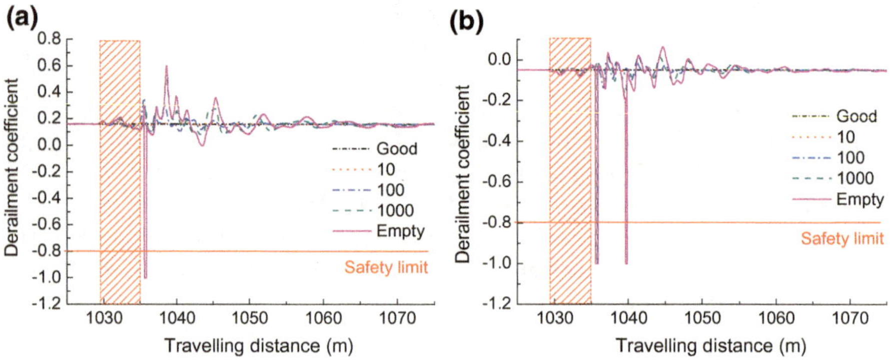


**Fig. 6** Derailment coefficient (*L*/*V*) with different degrees of CAM softening. **a** High rail; **b** low rail

The vertical displacement of the rail is usually less than the benchmark 1.5 mm and should not be greater than the maximum limit of 2 mm (MRPRC 2013). None of the lateral rail displacements exceed the benchmark and the maximum limit. When

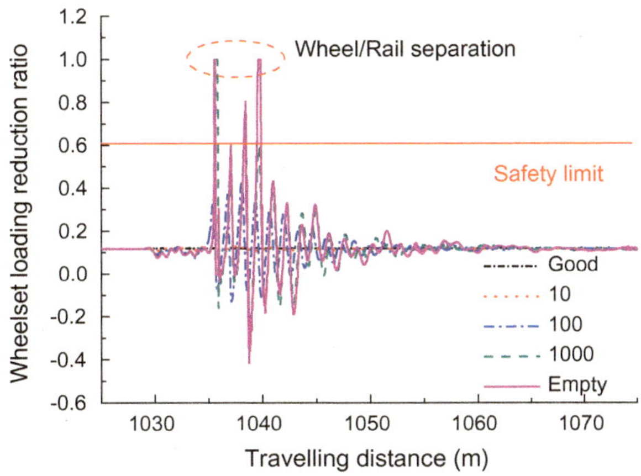


**Fig. 7** Wheelset loading reduction ratio ($\Delta V/V$) with different degrees of CAM softening

the softening coefficient is larger than 100, the vertical displacement of the rail exceeds the maximum limit.

Figure 9 shows the slab displacement in different cases of CAM softening. Compared with the case without CAM softening, the displacements of the slab increase at the end and in the middle as the degree of CAM softening increases. For the slab lateral displacement, the benchmark is 0.5 mm and the maximum limit is 1 mm. For the vertical slab displacement at the end, the benchmark is 0.4 mm and the maximum limit is 0.5 mm. For the vertical slab displacement in the middle, the benchmark is 0.2 mm and the maximum limit is 0.3 mm (MOHURD 2010). None of the lateral slab displacements exceed the benchmark and the maximum limit. When the softening coefficient is larger than 10, the vertical slab displacement in the middle exceeds the corresponding benchmark. When the softening coefficient is larger than 100, the vertical slab displacements in the middle and in the end both exceed the corresponding maximum limit.

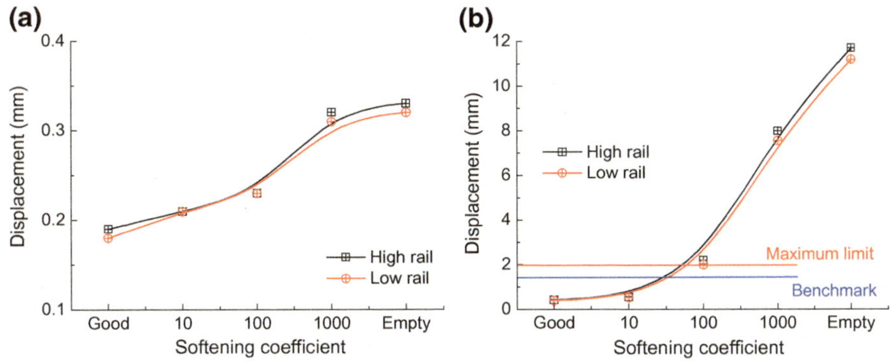


**Fig. 8** Rail displacement with different degrees of CAM softening. **a** Lateral displacement; **b** vertical displacement

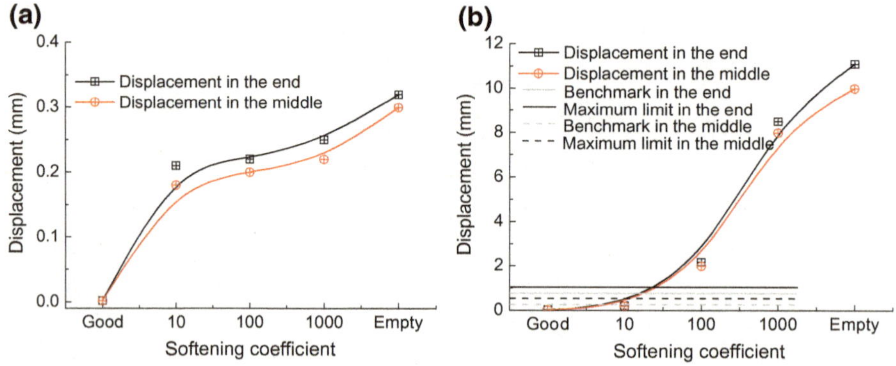


**Fig. 9** Slab displacement with different degrees of CAM softening. **a** Lateral displacement; **b** vertical displacement

## 4.3 *Effect of CAM Softening on Slab Stress and Track Interface Failure*

Figure 10 shows the maximum tensile stress and shear stress of the slab in different cases of CAM softening. X, Y, and Z represent tensile stress in the lateral direction, longitudinal direction (traveling direction), and vertical direction, respectively. In Fig. 10a, SXX, SYY, and SZZ are tensile stresses in three directions. In Fig. 10b, SXY, SXZ, and SYZ are shear stresses in three directions. Compared with the case without CAM softening, the maximum tensile stresses and shear stresses increase as the degree of CAM softening increases. The slab material is C60 (concrete 60). The tensile strength is 2.85 MPa and the shear strength is 4.1 MPa (MOHURD, 2010). Thus, in the four cases of CAM softening, the maximum tensile stress and shear stress do not exceed their allowable strength. The compressive strength of C60 is

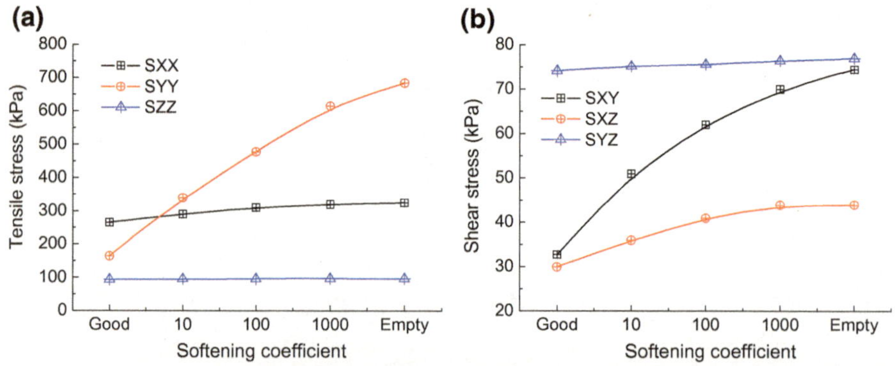


**Fig. 10** Slab stress with different degrees of CAM softening: **a** tensile stress; **b** shear stress

much larger than the maximum compressive stress of the slab in the four cases of CAM softening. The compressive stresses are not given in this study.

Figure 11a shows the interface shear failure percentage caused by CAM softening. According to Eqs. (3) and (4), when the shear force exceeds the shear-strength criterion limit, interface shear failure occurs. As the degree of CAM softening increases, the interface shear failure percentage increases slowly when the softening coefficient is larger than 1000 or smaller than 10 and increases quickly when the softening coefficient is between 10 and 1000. The relationship between interface shear failure percentage and the softening coefficient can be fitted using the GaussAmp Formula (Amplitude version of Gaussian peak function). The whole fitting curve is similar to an S-shaped curve. Compared with the interface between the slab and CAM, the interface between CAM and the concrete base is more vulnerable to shear failure (Fig. 11a). Figure 11b and c show the interface shear failure distributions corresponding to softening coefficients 10 and 100. When the

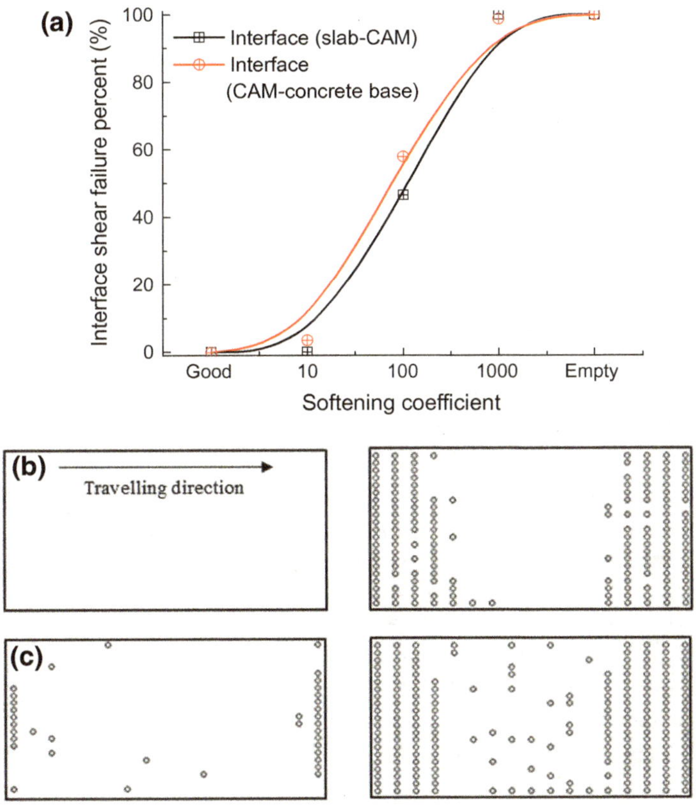


**Fig. 11** Interface shear failure. **a** Interface shear failure percent vs CAM softening coefficient; **b** interface shear failure distribution (slab-CAM) (*left* 10; *right* 100); **c** interface shear failure distribution (CAM-concrete base) (*left* 10; *right* 100)

softening coefficient is 10, the interface between the slab and the CAM does not show shear failure, and the interface between the CAM and the concrete base shows only a small partial shear failure. However, when the softening coefficient is 100, the shear failure percentages of both the interfaces between the slab and the CAM and between the CAM and the concrete base reach about 45–60%. The failure percentage of the CAM-concrete base interface is higher than that of the slab-CAM interface because the cohesion of the interface between the CAM and the concrete base is smaller. When the softening coefficients change from 10 to empty, the gap between the CAM-concrete base and the slab-CAM lines decreases. As this progresses, as discussed above, the interface between the CAM and the concrete base fails first. With the failure increasing significantly, even reaching complete failure, lateral movement of the slab may easily occur. This lateral movement of the slab will then speed up the relative motion between the slab and the CAM, increasing the risk of interface failure. Finally, the interface between the CAM-concrete base and the interface between the slab-CAM fail completely. Figure 11b and c also show that the interface shear failure develops from the end to the middle. This is because the relative shear displacement at the end is larger than that in the middle, which leads to a larger shear force at the end. The shear force at the end then more easily exceeds the shear-strength limit.

## 5 Conclusions

In this paper, a 3D coupling dynamic model of a vehicle and a CRTS-I slab track is developed. Using the proposed model, the wheel-rail contact forces, derailment coefficient, wheelset loading reduction ratio, and the track displacements are calculated to study the influence of CAM softening on the dynamic characteristics of the vehicle-track system. A track-subgrade finite difference model is developed to study the effect of CAM softening on track damage. The following conclusions can be drawn:

1. Wheel-rail contact forces fluctuate dramatically when a high-speed train runs over the curved track with CAM softening. When the CAM softening coefficient is larger than 1000, wheel/rail separation occurs, and the derailment coefficient and wheelset loading reduction ratio both exceed their safety limits.
2. As CAM softening increases, slab displacement more easily exceeds its geometric limit than rail displacement. When the CAM softening coefficient is larger than 10, slab vertical displacement in the middle exceeds the corresponding benchmark. When the softening coefficient is larger than 100, the vertical displacements of both the rail and slab exceed their corresponding maximum limits.
3. CAM softening cannot lead to slab damage based on a simple strength analysis. When the CAM softening coefficient reaches 10, a small partial slip occurs between the CAM and the concrete base. When the CAM softening coefficient

is larger than 100, at both the slab-CAM interface and the CAM-concrete base interface, serious damage occurs due to slippage.

According to these conclusions, when the CAM softening coefficients reach 10–100, track interface shear failure develops. The CAM softening coefficient should not be less than 1000, otherwise a high-speed running vehicle may risk derailment.

In future work, we propose to conduct a series of tests to obtain the relationship between the softening coefficients of CAM and loading cycles, temperature cycles, and weather conditions.

## References


Chen, R. P., Chen, J. M., & Wang, H. L. (2014). Recent research on the track-subgrade of high-speed railways. *Journal of Zhejiang University-SCIENCE A (Applied Physics & Engineering), 15*(12), 1034–1038. doi:10.1631/jzus.A1400342.

Chen, G., & Zhai, W. M. (2004). A new wheel/rail spatially dynamic coupling model and its verification. *Vehicle System Dynamics, 41*(4), 301–322. doi:10.1080/00423110412331315178.

Han, J., Zhao, G. T., Xiao, X. B., et al. (2015). Contact behaviour between slab track and its subgrade under high-speed train loading and water-soil interaction. *Electronic Journal of Geotechnical Engineering, 20*(2), 709–722.

Jin, X. S. (2014). Key problems faced in high-speed train operation. *Journal of Zhejiang University-SCIENCE A (Applied Physics & Engineering), 15*(12), 936–945. doi:10.1631/jzus.A1400338.

Jin, X. S., & Wen, Z. F. (2008). Effect of discrete track support by sleepers on rail corrugation at a curved track. *Journal of Sound and Vibration, 315*(1–2), 279–300. doi:10.1016/j.jsv.2008.01.057.

Lin, H. S. (2009). Research on the static and dynamic property of Ballastless Track based on fracture and damage mechanics. PhD Thesis, Southwest Jiaotong University, Chengdu, China (in Chinese).

Ling, L., Xiao, X. B., Xiong, J. Y., et al. (2014). A 3D model for coupling dynamics analysis of high-speed train/track system. *Journal of Zhejiang University-SCIENCE A (Applied Physics & Engineering), 15*(12), 964–983. doi:10.1631/jzus.A1400192.

Liu, Y. (2013). Study on characteristics and influences of CRTS II slab track early temperature field. PhD Thesis, Southwest Jiaotong University, Chengdu, China (in Chinese).

Liu, J. B., Gu, Y., & Du, Y. X. (2006). Consistent viscous-spring artificial boundaries and viscous-spring boundary elements. *Chinese Journal of Geotechnical Engineering, 28*(9), 1070–1075. (in Chinese).

MOHURD (Ministry of Housing and Urban-Rural Development of the People's Republic of China). (2010). *Code for Design of Concrete Structures, GB 50010-2010*. Beijing, China (in Chinese): China Architecture and Building Press.

MRPRC (Ministry of Railways of the People's Republic of China). (2013). *Technical Regulations for Dynamic Acceptance for High-speed Railways Construction, TB 10716-2013*. Beijing, China (in Chinese): Railway Publishing House.

Shen, Z. Y., Hedrick, J. K., & Elkins, J. A. (1983). A comparison of alternative creep-force models for rail vehicle dynamic analysis. *Vehicle System Dynamics, 12*(1–3), 79–83. doi:10.1080/00423118308968725.

Wang, P., Xu, H., & Chen, R. (2014). Effect of cement asphalt mortar debonding on dynamic properties of CRTS II slab Ballastless Track. *Advances in Materials Science and Engineering, 2014,* 193128. doi:10.1155/2014/193128.

Xiang, J., He, D., & Zeng, Q. Y. (2009). Effect of cement asphalt mortar disease on dynamic performance of slab track. *Journal of Central South University, 40*(3), 791–796. (in Chinese).

Xiao, X. B., Jin, X. S., Deng, Y. Q., et al. (2008). Effect of curved track support failure on vehicle derailment. *Vehicle System Dynamics, 46*(11), 1029–1059. doi:10.1080/00423110701689602.

Xiao, X. B., Jin, X. S., & Wen, Z. F. (2007). Effect of disabled fastening systems and ballast on vehicle derailment. *Journal of Vibration and Acoustics, 129*(2), 217–229. doi:10.1115/1.2424978.

Xiao, X. B., Jin, X. S., Wen, Z. F., et al. (2011). Effect of tangent track buckle on vehicle derailment. *Multibody System Dynamics, 25*(1), 1–41. doi:10.1007/s11044-010-9210-2.

Xiao, X.B., Ling, L., Xiong, J.Y., *et al.*, 2014. Study on the safety of operating high-speed railway vehicles subjected to crosswinds. *Journal of Zhejiang University-SCIENCE A (Applied Physics & Engineering)*, **15**(9):694–710. [doi:10.1631/jzus.A1400062].

Zhai, W. M. (1996). Two simple fast integration methods for large-scale dynamic problems in engineering. *International Journal for Numerical Methods in Engineering, 39*(24), 4199–4214.

Zhang, W. H. (2011). *Overall Technique of EMU and Bogies*. Beijing, China (in Chinese): Chinese Railway Press.

Zhong, S. Q., Xiong, J. Y., Xiao, X. B., et al. (2014). Effect of the first two wheelset bending modes on wheel-rail contact behavior. *Journal of Zhejiang University-SCIENCE A (Applied Physics & Engineering), 15*(12), 984–1001. doi:10.1631/jzus.A1400199.

Zhou, L., & Shen, Z. Y. (2013). Dynamic analysis of a high-speed train operating on a curved track with failed fasteners. *Journal of Zhejiang University-SCIENCE A (Applied Physics & Engineering), 14*(6), 447–458. doi:10.1631/jzus.A1200321.

Zhu, S. Y., & Cai, C. B. (2014). Interface damage and its effect on vibrations of slab track under temperature and vehicle dynamic loads. *International Journal of Non-Linear Mechanics, 58*, 222–232. doi:10.1016/j.ijnonlinmec.2013.10.004.

Zhu, S. Y., Fu, Q., Cai, C. B., et al. (2014). Damage evolution and dynamic response of cement asphalt mortar layer of slab track under vehicle dynamic load. *Science China Technological Sciences, 57*(10), 1883–1894. doi:10.1007/s11431-014-5636-8.

## Author Biography

**Xue-song Jin** PhD, is a professor at Southwest Jiaotong University, Chengdu, China. He is a leading scholar of wheel-rail interaction in China and is a State Council Expert for Special Allowance. He is the author of three academic books and over 200 articles. He is an editorial board member of several journals and has been a member of the Committee of the International Conference on Contact Mechanics and Wear of Rail/Wheel Systems for more than 10 years. He has been a visiting scholar at the University of Missouri-Rolla for 2.5 years. Now his research focuses on wheel-rail interaction, rolling contact mechanics, vehicle system dynamics, vibration, and noise.

# Part V
# Advances in Traction Power Supply and Transportation Organization Technologies

# A Two-Layer Optimization Model for High-Speed Railway Line Planning

**Li Wang, Li-min Jia, Yong Qin, Jie Xu and Wen-ting Mo**

## Contents



---

L. Wang
School of Traffic and Transportation, Beijing Jiaotong University,
Beijing 100044, China
e-mail: wangli@bjtu.edu.cn

L. Jia · Y. Qin (✉) · J. Xu
State Key Laboratory of Rail Traffic Control and Safety, Beijing Jiaotong University,
Beijing 100044, China
e-mail: qinyong2146@126.com

L. Jia
e-mail: jialm@vip.sina.com

J. Xu
e-mail: jxu1@bjtu.edu.cn

W. Mo
IBM China Research Lab, 2F Diamond A, Zhong Guan Cun Software Park,
Beijing 100193, China
e-mail: mowent@cn.ibm.com



Y. Fang and Y.H. Zhang (eds.), *China's High-Speed Rail Technology*,
Advances in High-speed Rail Technology, https://doi.org/10.1007/978-981-10-5610-9_23

## 1 Introduction

Generally, line planning is a procedure of allocating trains with specific travel demands of many origins and destinations to appropriate lines or line sections. As the basis of successive decisions, such as rolling stock planning and timetable planning, line planning is a classical optimization problem in order to obtain stop-schedules and service frequencies.

Today, China is extensively developing the infrastructure of a high-speed railway. The target is to cover its major economic areas with a high-speed railway (HSR) network, with four horizontal and four vertical lines in the next several years. The network scale is much larger than any existing ones in the world. Considering the high train speed and high train frequency of this railway, the impact of capacity loss would be more serious than on existing lines (in this paper, the existing non-high-speed railway line is called existing line for short). On the other hand, high-speed lines are more passenger-oriented than existing lines (Mo et al. 2011), where serving different types of passengers, such as regional, interregional, and intercity, is extraordinarily important. Thus, line planning, as the basis of more detailed planning problems, such as the construction of timetables, rolling stock planning, and crew scheduling, is of great theoretical and practical significance for safe and efficient operation of China's HSR network.

The literature describes the line planning problem in two ways. One is to find train routes and service frequencies in the railway network. Most papers focus on balancing the train route and passenger line assignment to reduce the passenger transfer times (Baaj and Mahmassani 1991; Bussieck 1998; Nielsen 2000; Chakroborty and Wivedi 2002; Poon et al. 2004; Pfetsch and Borndorfer 2005; Cepeda et al. 2006; Goossens et al. 2006; Guan et al. 2006; Borndŏrfer et al. 2007; Hamdouch and Lawphongpanich 2008; Schmŏcker et al. 2008, 2011; Laporte et al. 2010). The objective is to find a set of routes that maximizes the number of direct travelers from a service perspective or minimizes the operational costs from the railway company perspective. The other is to optimize halting stations with a given route between an origin and a destination station (Goossens et al. 2004b; Deng et al. 2009), and the objective is to reduce the total travel time. Chang et al. (2000) developed a multi-objective programming model for the optimal allocation of passenger train services on an intercity high-speed rail line without branches. For a given travel demand and a specified operating capacity, the model is solved by a fuzzy mathematical programming approach to determine the best train service plan, including the train stop-schedule plan, service frequency, and fleet size. However, this may be feasible for short rail lines with few stations, but is not adaptable to long rail lines with dozens of stations. Complex computation is also a significant problem in this model.

Many different algorithms for line planning have been put forward. The integer programming (Goossens et al. 2004a; Guan et al. 2006; Borndŏrfer et al. 2007), such as branch and bound, and other traditional optimization methods are usually used for train route optimization and passenger assignment. In recent years, some computational intelligence methods, like genetic algorithm (GA) and particle swarm optimization algorithm, have also been used to solve the large-scale combinatorial optimization problems. Chakroborty and Wivedi (2002) proposed a GA-based evolutionary optimization technique to develop the optimal transit route networks. They developed an initial route set generation procedure with a pre-specified number of routes and gave a measure of goodness of a route set according to the travel time and the passenger demand satisfaction. However, the technique neither involves the service frequency of different routes, nor refers to the influence of different stop-schedules on the passenger assignment. Game theory is also used in line planning. Laporte et al. (2010) proposed a game theoretic framework for the problem of designing an incapacitated railway transit network in the presence of link failures and a competing mode. It is assumed that when a link fails, another path or another transportation mode will be provided to transport passengers between the endpoints of the affected link. The goal is to build a network that optimizes a certain utility function when failures occur. The problem is posed as a non-cooperative, two-player zero-sum game with perfect information. Schŏebel and Schwarze (2006) presented a game theoretic model for line planning with line players. Each player aims to minimize its own delay, which depends on the traffic load along its edges. Equilibrium exists to minimize the sum of delays of the transportation system. Deng (2007) and Shi et al. (2007) proposed a bi-level model through balancing the profit of the railway corporation and demand of passengers, combining passenger train operation plan with a passenger transfer plan and considering the flow assignment on the railway passenger transfer network.

In this paper, we investigate a two-layer model with a decision support mechanism (DSM) for line planning. Note the three prominent features of this mechanism: realizing interaction with dispatchers, emphasizing passengers' satisfaction, and reducing computation complexity with a two-layer modeling approach.

## 2 Decision Support Mechanism (DSM) for Line Planning

In order to handle operation tasks, there is usually a train line planning system for dispatchers. When a new rail line is constructed, dispatchers usually issue a line plan through this system before timetable planning. The proposed DSM is shown in Fig. 1.

The line planning consists of five components executed in a loop. From the optimization perspective, the line planning problem is decomposed into two layers. The first layer optimizes the stop-schedule and service frequency, and the second layer assigns the passengers to the trains. The detailed descriptions of the components are given below.

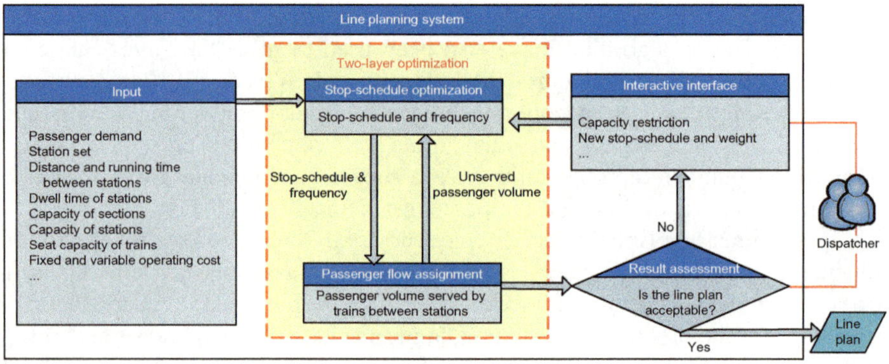


**Fig. 1** Architecture of the decision support mechanism

## 2.1 *Stop-Schedule Optimization*

This layer aims at finding the optimal stop-schedule and service frequency. Thereafter, the optimization objective is to minimize the operation cost, the weighted unserved passengers, and used capacities, where the weights correspond to the passengers' priorities. In practice, HSR lines in China usually span several provinces, with dozens of stations. There are two kinds of stations. One is the start or end station of a train trip (such a station is called major station for short); the other is the station that is not used as a start or end of a train trip (such a station is called minor station for short). Four kinds of stop-schedule patterns are usually adopted for the long rail line according to the passengers' traveling habits.

Non-stopping-schedule (NSS): The train taking this schedule runs between two major stations and does not stop at any other stations. Thus, this schedule uses the least travel time to serve the passengers with the highest priority.

Stop at major stations (SMS): The train taking this schedule can stop at intermediate major stations besides the two endpoint stations. This schedule also has a high priority.

Stop at staggered stations (SSS): The train taking this schedule staggeringly stops at minor stations besides intermediate major stations. This schedule can satisfy most of the passenger demands with a middle priority.

Stop at all stations (SAS): The train taking this schedule must stop at all stations between the two endpoint stations. This schedule can satisfy the short-path passenger demands with a lower priority.

## 2.2 *Passenger Assignment Optimization*

Given the stop-schedule and service frequency, the passenger assignment decides the number of passengers taking different trains between different origins and destinations. This layer builds two optimization objectives: one is to maximize the product of the number of passengers and their travel length (in this paper, the product is called served passenger volume); the other is to minimize the total travel time.

The advantages of two-layer decomposition can be interpreted from different angles. From the system perspective, it facilitates the interaction with dispatchers. They are enabled to explicitly control the capacity restrictions and add some new stops for the stop-schedule. From the problem-solving perspective, it effectively reduces the complexity of optimization. Long computation time is usually a curse for line planning problems because of the large number of decision variables, which prevents the application of optimization-based automatic line planning algorithms.

## 3 Modeling of Line Planning

The line planning optimization is decomposed into two layers as mentioned above. Data transference between the layers is shown in Fig. 1. The top layer aims at finding the optimal stop-schedule and its service frequency subject to weighted unserved passenger volume calculated from the bottom layer. GA is adopted to solve the nonlinear stop-schedule optimization model in this layer. Finally, mixed integer linear programming (MILP) is used to model the passenger assignment problem in the bottom layer. We will describe the key parts of the problem as follows. First, the numerical inputs of the line planning model are described.

### *3.1 Input Data*

$T$ The planned operating period, i.e., one day.
$S$ The station set in the railway network.
$E$ The section set in the railway network.
$l$ The train type. There are two kinds of trains running on the HSR in China, high-speed trains and quasi-high-speed trains.
$L_{o,d}$ The distance between stations $o$ and $d$.
$P_{o,d,l}$ The travel demand of multiple origins and destinations with different train types for the planned operating period $T$.
$T_{o,d,l}$ The running time of train $l$ between stations $o$ and $d$.
$T_i^s$ The dwell time of station $i$.
$C_i^s$ The carrying capacity of station $i$ for the planned operating period $T$.
$C_k^e$ The carrying capacity of section $k$ for the planned operating period $T$.
$C_l$ The seat capacity of train $l$.
$K$ Attendance ratio for the trains. If all the passengers always have seats, then $K < 1$.

$C_i^d$ If the station $i$ can be used as a start or an end station from which a train trip starts or ends, then $C_i^d = 1$; otherwise, $C_i^d = 0$.
$D$ The fixed overhead cost for one train.
$F$ The variable operating cost for one train running one kilometer.
$M$ The total service frequency for all stop-schedules of the plan. It is a large integer for the passenger demand.

## 3.2 Model of the Stop-Schedule Optimization

Note that not all the stations in China can be used as a start or end station from which a train trip starts or ends, and this layer adds many terminal restrictions, which is different from other studies. The objective of this layer is to minimize the total operation cost and unserved passenger volume. Some decision variables are described as follows.

### 3.2.1 Decision Variables

$y_j$ If train $j$ exists in the line plan, then $y_j = 1$; otherwise, $y_j = 0$.
$x_{j,i}$ If train $j$ stops at station $i$ in the line plan, then $x_{j,i} = 1$; otherwise, $x_{j,i} = 0$.
$o_{j,i}$ If the start station of train $j$ is station $i$, then $o_{j,i} = 1$; otherwise, $o_{j,i} = 0$.
$d_{j,i}$ If the end station of train $j$ is station $i$, then $d_{j,i} = 1$; otherwise, $d_{j,i} = 0$.
$u_{j,o,d}$ The passengers served by train $j$ between stations $o$ and $d$.
$z_{j,l}$ If the type of train $j$ is $l$, then $z_{j,l} = 1$; otherwise, $z_{j,l} = 0$.
$s_{j,i}$ If station $i$ exists in the stop-schedule of train $j$, then $s_{j,i} = 1$; otherwise, $s_{j,i} = 0$.
$e_{j,k}$ If section $k$ exists in the stop-schedule of train $j$, then $e_{j,k} = 1$; otherwise, $e_{j,k} = 0$.

### 3.2.2 Objective Functions

1. To minimize the total operation cost:

$$C_{\text{cost}} = \min \sum_{j=1}^{M} D \cdot y_j + \sum_{j=1}^{M} \sum_{o=1}^{S-1} \sum_{d=o+1}^{S} y_j \cdot F \cdot L_{o,d} \cdot o_{j,o} \cdot d_{j,d}. \quad (1)$$

2. To minimize the unserved passengers:

$$C_{\text{passenger}} = \min \sum_{l} \sum_{o=1}^{S-1} \sum_{d=o+1}^{S} \left( P_{o,d,l} - \sum_{j=1}^{M} u_{j,o,d} \cdot z_{j,l} \cdot y_j \right). \tag{2}$$

### 3.2.3 Constraints

1. Trains cannot start from a station without original capacity:

$$C_i^d \geq o_{j,i}, \quad i = 1, 2, \ldots, S-1, \; j = 1, 2, \ldots, M. \tag{3}$$

2. Trains cannot end a trip at a station without destination capacity:

$$C_i^d \geq d_{j,i}, \quad i = 2, 3, \ldots, S, \; j = 1, 2, \ldots, M. \tag{4}$$

3. Trains cannot pass by any station before the start station:

$$\sum_{j=1}^{M} \sum_{i=2}^{S-1} \sum_{i1=0}^{i-1} y_j \cdot o_{j,i} \cdot x_{j,i1} = 0. \tag{5}$$

$$\sum_{j=1}^{M} \sum_{i=2}^{S-1} \sum_{i1=0}^{i-1} y_j \cdot o_{j,i} \cdot s_{j,i1} = 0. \tag{6}$$

4. Trains cannot pass by any station after the end station:

$$\sum_{j=1}^{M} \sum_{i=1}^{S-1} \sum_{i1=i+1}^{S} y_j \cdot d_{j,i} \cdot x_{j,i1} = 0. \tag{7}$$

$$\sum_{j=1}^{M} \sum_{i=1}^{S-1} \sum_{i1=i+1}^{S} y_j \cdot d_{j,i} \cdot s_{j,i1} = 0. \tag{8}$$

5. There is only one start station for each train:

$$\sum_{i=1}^{S} y_i \cdot o_{j,i} = 1, \quad j = 1, 2, \ldots, M. \tag{9}$$

6. There is only one end station for each train:

$$\sum_{i=1}^{S} y_i \cdot d_{j,i} = 1, \quad j = 1, 2, \ldots, M. \tag{10}$$

7. The train must dwell at its start station:

$$o_{j,i} \leq x_{j,i}. \quad j = 1, 2, \ldots, M, \; i = 1, 2, \ldots, S. \tag{11}$$

8. The train must dwell at its end station:

$$d_{j,i} \leq x_{j,i}, \quad j = 1, 2, \ldots, M, \; i = 1, 2, \ldots, S. \tag{12}$$

9. The train must pass by the stations that it has dwelt at:

$$s_{j,i} \geq x_{j,i}, \quad j = 1, 2, \ldots, M, \; i = 1, 2, \ldots, S. \tag{13}$$

10. Station capacity restrictions:

$$C_i^s \geq \sum_{j=0}^{M} y_j \cdot s_{j,i} \quad i = 1, 2, \ldots, S. \tag{14}$$

11. Section capacity restrictions:

$$C_k^e \geq \sum_{j=0}^{M} y_j \cdot e_{j,k}, \quad k = 1, 2, \ldots, E. \tag{15}$$

The stop-schedule optimization is a multi-objective, discrete, nonlinear program. It is very difficult to obtain a solution through traditional optimization techniques. Thus, GA is used to solve the difficult problem.

### 3.2.4 Coding and Initialization

The solution in this layer means a description of all stop-schedules and their service frequencies. The stop-schedules are represented as a two-dimensional matrix $\boldsymbol{X}$; each row of the matrix means a schedule and the columns represent the stations. The element value of the matrix is 1 or 0. If a train, based on the stop-schedule $j$, stops at station $i$, then $X_{j,i} = 1$; otherwise, $X_{j,i} = 0$. The service frequencies are represented as a string, each element of which corresponds to the appropriate stop-schedule. The element value of the service frequency string is an integer, and 0 means that no train will take the stop-schedule in the line plan. Figure 2 shows a typical solution. There are $N$ stations and $M$ stop-schedules with $U$ service frequency strings. Thus, the chromosome size is $M$ and the population size is $U$.

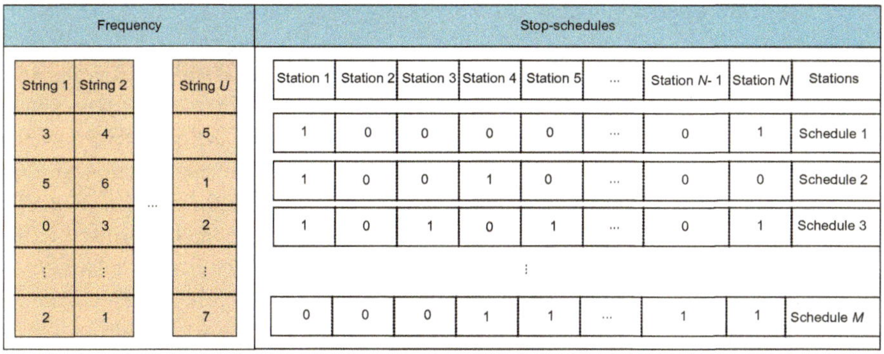


**Fig. 2** Representation of stop-schedules

As proposed in DSM for line planning section, four kinds of stop-schedule patterns are usually used in the railway line. Thus, the solution matrix is initialized as all the possible stop-schedules. The service frequency is created randomly subject to an upper bound. Note that if a service frequency with the stop-schedule goes beyond the capacity of some sections or stations, the initialization of this string should repeat at once.

Since the service frequency, designed as a string, participates in the crossover and mutation process with the fixed stop-schedule matrix, and the solution coding covering all the possible stop-schedules can meet lots of the terminal restrictions, the complexity of the algorithm is effectively reduced.

### 3.2.5 Crossover Operator

The purpose of crossover is to exchange different features of good strings with the hope of obtaining better strings (Eiben and Smith 2003). In the present scenario, features of strings are service frequencies. First, we select two different parent strings and determine the starting position $s$ and end position $e$ for crossover randomly; then exchange the string fragments between $s$ and $e$ to form two new service frequency strings. Repeat the procedure until $U$ new service frequency strings are created. Note that this procedure also needs to verify the capacity constraints.

### 3.2.6 Mutation Operator

The purpose of the mutation operator is to slightly modify the frequencies. $U$ new chromosomes are created by crossover process, thus there are $2U$ strings to participate in mutation. First, select a gene of a chromosome randomly and then change it to any of its probable service frequency. In the model, the mutation probability is set to 20%, which means 20% of the $2U$ strings will be modified.

### 3.2.7 Reproduction Operator

The purpose of the reproduction operator is to select $U$ good strings for the next generation from the $2U$ strings. First, calculate the total objective value, $f = \lambda_1 C_{\text{cost}} + \lambda_2 C_{\text{passenger}}$, where $\lambda_i$ ($i$ = 1, 2) is the weight of the objective $i$, so the fitness function is $1/f$; then, select the string with the highest fitness function to form the next generation; finally, select the residuary $U - 1$ strings using the Roulette Wheel selection (Goldberg 1989). The fitness function is performed considering the operation cost, weighted unserved passengers, and used capacity, and the weight $\lambda_i$ shows the impact of different objectives in the final decision. Although it is a simple way to deal with the multi-objectives by taking the sum of the weighted objectives as the final objective, it works in the model, which is validated by the case in Sect. 4. The principle for selection of $\lambda_i$ is reflecting the optimization purpose, and the trial and error method or the Delphi method can be used in the selection of $\lambda_i$.

### 3.2.8 Termination

As proposed in the paper, DSM is designed to facilitate the interaction with dispatchers and effectively utilize the precious experience. Hence, there are some conditions for the dispatchers to terminate the generational process.

1. A solution is found with lower operation cost, less unserved passengers, and used capacity.
2. Computation time or fixed number of generations is reached.
3. The highest ranking solution's fitness has reached a plateau such that successive iterations no longer produce better results (Eiben and Smith 2003).
4. Combination of the above conditions occurs during the iterative process.

## 3.3 *Model of the Passenger Assignment*

Given the stop-schedule set and service frequencies, the passenger assignment optimization is abstracted as an MILP.

### 3.3.1 Objective Functions

1. To maximize the served passenger volume:

$$\max \sum_{j=1}^{M} \sum_{o=1}^{S-1} \sum_{d=o+1}^{S} y_j \cdot u_{j,o,d} \cdot L_{o,d}. \tag{16}$$

2. To minimize the total travel time for all passengers:

$$\min \sum_{j=1}^{M} \sum_{o=1}^{S-1} \sum_{d=o+1}^{S} y_j \cdot u_{j,o,d} \cdot \left( \sum_{l} T_{o,d,l} \cdot z_{j,l} + \sum_{o1=o+1}^{d-1} x_{j,o1} \cdot T_{o1}^{s} \right). \tag{17}$$

### 3.3.2 Constraints

1. Passenger demand restrictions:

$$\sum_{o=1}^{S-1} \sum_{d=o+1}^{S} \left( \sum_{j=1}^{M} y_j \cdot u_{j,o,d} \cdot z_{j,l} \leq P_{o,d,l} \right), \quad l = 1, 2. \tag{18}$$

2. Train seat capacity restrictions:

$$\sum_{o=1}^{k} \sum_{d=k+1}^{S} y_j \cdot u_{j,o,d} \leq \sum_{l} C_l \cdot K \cdot z_{j,l} \cdot y_j, \quad j = 1, 2, \ldots, M, \ k = 1, 2, \ldots, S-1. \tag{19}$$

The whole composite algorithm procedure of the two-layer model is shown in Fig. 3. Note that weighted passenger assignment, performed in the fitness function calculation, is integrated in the stop-schedule optimization.

## 4 Case Studies

In this section, we complete two case studies. The first one is simulated on the Taiwan HSR, aiming to compare the two-layer model proposed in this paper and the model in (Chang et al. 2000). The other case is illustrated with the Beijing-Shanghai HSR line with 20 stations, which proves the two-layer model in practical situations with a long rail line.

### 4.1 Taiwan HSR

The Taiwan HSR system is a 340-km intercity passenger service line without branches along the western corridor of the island. It connects two major cities, Taipei and Kaohsiung, with seven intermediate stations and three terminal stations. For more details of the Taiwan HSR system, we can refer to (Chang et al. 2000).

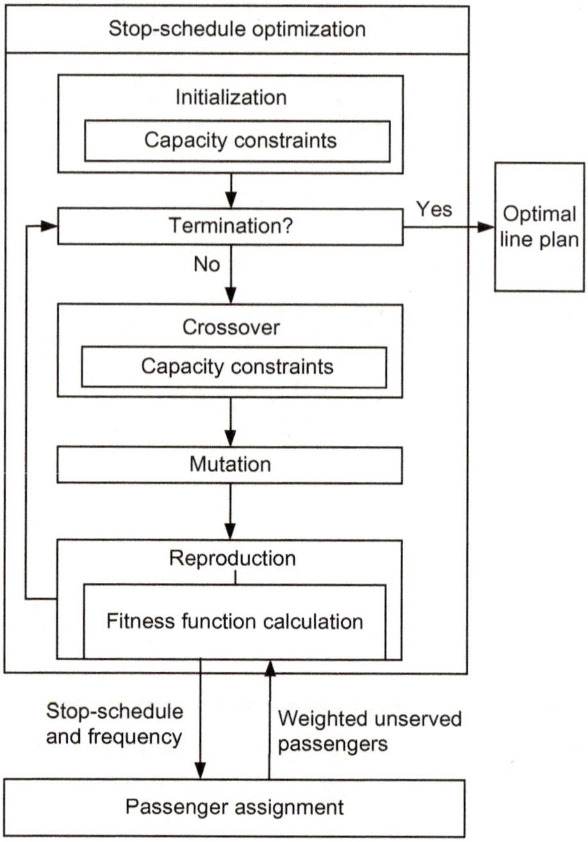


**Fig. 3** Composite algorithm procedure of the two-layer model

First, we obtain all the four kinds of stop-schedules according to the top-layer optimization rules. There are three stop-schedules of NSS, one stop-schedule of SMS, 14 stop-schedules of SSS, and three stop-schedules of SAS. Thus, the chromosome size is 21 and the population size is set to 50. Figure 4 shows that the solution converges at 18,277.2 after 40 iterations. The detail line plan is shown in Fig. 5. All the passengers have been served in the plan, and the total service frequency is 11. There is one train departing from station 4, which is different from the result of (Chang et al. 2000). That is because all trains must take the first station as the start in (Chang et al. 2000). Also, the computation time is 67 s, which is less than the model in (Chang et al. 2000). All the models are achieved on a Java platform, and the passenger assignment programming is solved by IBM ILOG CPLEX 12.2.

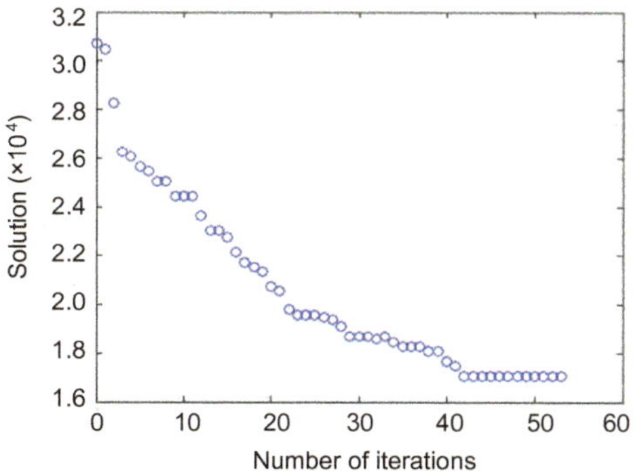


**Fig. 4** Iteration procedure of the solution of the Taiwan High-Speed Railway (HSR)

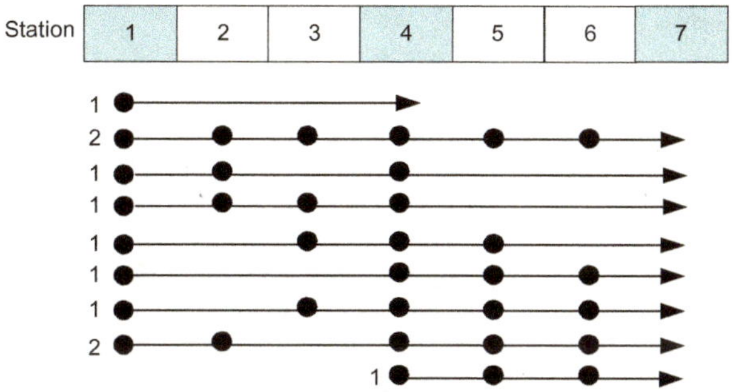


**Fig. 5** Optimal line plan of the Taiwan High-Speed Railway (HSR). Stations 1, 4, and 7 are the major stations which can be used as a start or an end station. Lines with *points* (stops) and *arrows* denote different stop-schedules with the figures on the *left* of the lines meaning the service frequencies of the stop-schedules

## *4.2 Beijing-Shanghai HSR*

This model is also simulated on the Beijing-Shanghai HSR. As the north-south aorta of China, the Beijing-Shanghai HSR connects Beijing (the capital of China) and Shanghai (the biggest economic center of China) and goes through the Yangtze River Delta region (the most developed area in China). Figure 6 shows three kinds of rail lines in the network: HSR, intercity line, and existing line. There are three close lines (the Shanghai-Nanjing intercity HSR, the Beijing-Tianjin intercity HSR, and the existing Beijing-Shanghai line), almost parallel to the Beijing-Shanghai HSR. As Fig. 6 shows, there are 20 stations on the Beijing-Shanghai HSR with five terminal stations: Beijing, Tianjin, Jinan, Nanjing, and Shanghai. The lengths of the 19 sections starting from Beijing to Shanghai are 59.3, 62.8, 87.9, 103.8, 92.3, 58.6, 70.3, 92.2, 65.3, 67.0, 88.2, 116.0, 32.0, 59.0, 65.3, 61.0, 57.4, 26.8, 31.4, and 43.6 km, respectively. The average running speed is 300 km/h, average dwell time

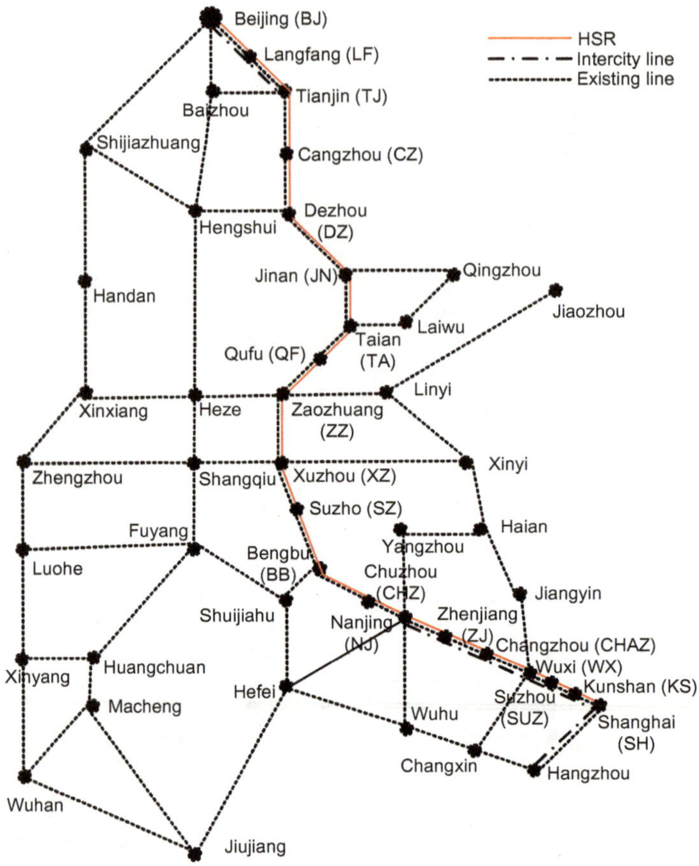


**Fig. 6** Railway network around the Beijing-Shanghai High-Speed Railway (HSR)

of a station is 2 min, train seat capacity is 1200, and the attendance ratio for the trains is set to 1. Daily passenger travel demand of multiple origins and destinations is shown in Table 1 (Wang 2006). The number of passengers with high priority is 52% with the reference of NSS and SMS, the number of passengers with middle priority is 28% with the reference of SMS and SSS, and the number of passengers with low priority is 20% with the reference of SSS and SAS. We only consider the direction from Beijing to Shanghai without loss of generality, since all of the lines are double-track.

First, we obtain all the four kinds of stop-schedules set according to the top-layer optimization rules. There are six stop-schedules of NSS, 19 stop-schedules of SMS, 38 stop-schedules of SSS, and four stop-schedules of SAS. Thus, the chromosome size is 67 and the population size is set to 150. Figure 7 shows that the solution converges at 42,714.398 after 103 iterations. The detail line plan is shown in Fig. 8. All the passengers have been served in the plan. The sum of service frequencies is 133, and the computation time is 316 s.

**Table 1** Passenger travel demand for the Beijing-Shanghai High-Speed Railway (HSR)

| Station | Number of passengers | | | | | | | | | | | | | | | | | | |
|---|---|---|---|---|---|---|---|---|---|---|---|---|---|---|---|---|---|---|---|
| | LF | TJ | CZ | DZ | JN | TA | QF | ZZ | XZ | SZ | BB | CHZ | NJ | ZJ | CHAZ | WX | SUZ | KS | SH |
| BJ | 135 | 792 | 297 | 213 | 741 | 306 | 429 | 99 | 438 | 81 | 132 | 63 | 783 | 84 | 120 | 114 | 105 | 99 | 1263 |
| LF | | 141 | 93 | 60 | 66 | 0 | 0 | 0 | 30 | 0 | 0 | 0 | 84 | 0 | 0 | 0 | 0 | 0 | 87 |
| TJ | | | 225 | 246 | 366 | 105 | 222 | 57 | 360 | 45 | 168 | 39 | 219 | 48 | 51 | 54 | 42 | 45 | 297 |
| CZ | | | | 42 | 75 | 18 | 18 | 12 | 36 | 9 | 21 | 6 | 72 | 12 | 21 | 21 | 18 | 15 | 87 |
| DZ | | | | | 147 | 54 | 66 | 45 | 42 | 24 | 33 | 0 | 78 | 21 | 27 | 27 | 24 | 24 | 96 |
| JN | | | | | | 207 | 210 | 192 | 171 | 30 | 51 | 21 | 339 | 54 | 72 | 75 | 69 | 75 | 390 |
| TA | | | | | | | 51 | 48 | 123 | 15 | 21 | 0 | 90 | 18 | 27 | 27 | 27 | 24 | 96 |
| QF | | | | | | | | 36 | 42 | 12 | 18 | 6 | 111 | 30 | 39 | 39 | 36 | 42 | 138 |
| ZZ | | | | | | | | | 72 | 15 | 57 | 6 | 48 | 51 | 36 | 36 | 45 | 81 | 102 |
| XZ | | | | | | | | | | 75 | 147 | 105 | 297 | 87 | 108 | 147 | 147 | 258 | 357 |
| SZ | | | | | | | | | | | 99 | 132 | 141 | 45 | 51 | 51 | 30 | 48 | 183 |
| BB | | | | | | | | | | | | 108 | 162 | 114 | 120 | 123 | 126 | 120 | 396 |
| CHZ | | | | | | | | | | | | | 72 | 30 | 39 | 84 | 96 | 60 | 240 |
| NJ | | | | | | | | | | | | | | 102 | 111 | 114 | 171 | 108 | 1281 |
| ZJ | | | | | | | | | | | | | | | 120 | 114 | 168 | 150 | 183 |
| CHAZ | | | | | | | | | | | | | | | | 81 | 300 | 120 | 255 |
| WX | | | | | | | | | | | | | | | | | 450 | 261 | 429 |
| SUZ | | | | | | | | | | | | | | | | | | 288 | 435 |
| KS | | | | | | | | | | | | | | | | | | | 315 |

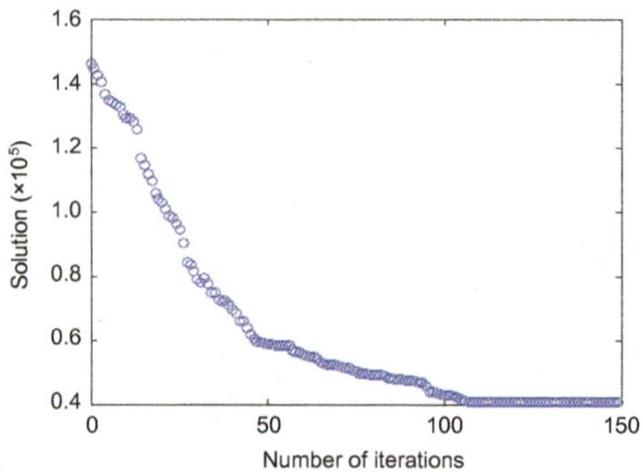


**Fig. 7** Iteration procedure of the solution of the Beijing-Shanghai High-Speed Railway (HSR)

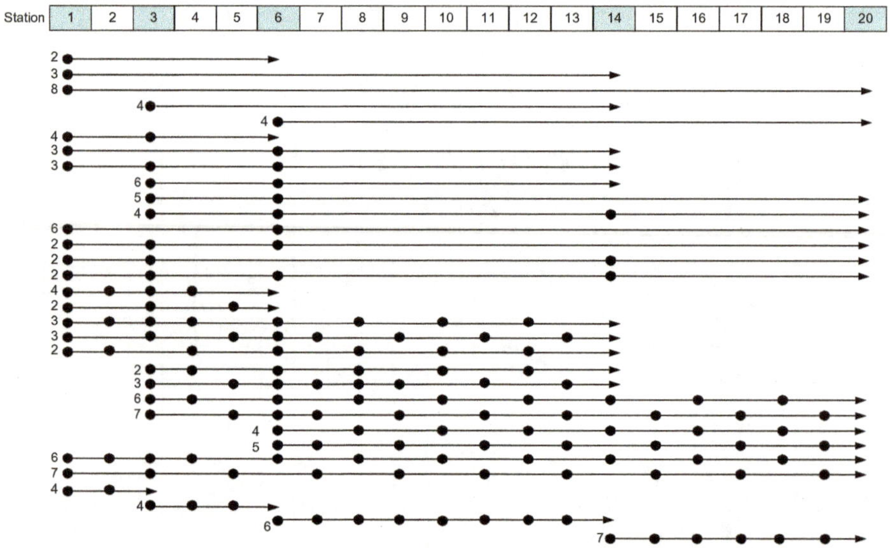


**Fig. 8** Optimal line plan of the Beijing-Shanghai High-Speed Railway (HSR). Stations 1, 3, 6, 14, and 20 are the major stations which can be used as a start or an end station. Lines with *points* (stops) and *arrows* denote different stop-schedules with the figures on the *left* of the lines meaning the service frequencies of the stop-schedules

## 5 Conclusions and Future Work

This paper deals with a DSM for dispatchers to use line planning in China. The main contributions include: (1) Passenger assignment is incorporated into the stop-schedule optimization with a two-layer optimization model; (2) GA with innovative solution coding is used to solve the nonlinear stop-schedule optimization model to reduce the computation complexity; (3) Weighted passengers are incorporated into the passenger assignment to optimize stop-schedules for different failure modes of the railway network; (4) Dispatchers are effectively involved into the line planning so that their valuable experience can be leveraged, while most research work on line planning focuses on the optimization model itself. The proposed mechanism, methods, and models are simulated on the Taiwan HSR and Beijing-Shanghai HSR. The proposed DSM shows good performance in the sense that different stop-schedules are created to match the railway station set.

There remain some interesting topics to explore concerning the proposed model. First, travel demand tends to change in spatial and temporal distribution in different situations. How to forecast the passenger demand is an issue. Second, the rolling stock rebalancing is yet to be considered together with the line planning. Finally, integrated optimization of line planning and timetable scheduling is our ultimate goal for safe and efficient operation of China's HSR.

## References


Baaj, M. H., & Mahmassani, H. (1991). An AI-based approach for transit route system planning and design. *Journal of Advance Transportation, 25*(2), 187–209. doi:10.1002/atr.5670250205.

Borndŏrfer, R., Grŏtschel, M., Pfetsch, M. E. (2007). *A column-generation approach to line planning in public transport*. Technical report no. ZIB-Report 05–18, Konrad-Zuse-Zentrum für Informationstechnik, Berlin.

Bussieck, M. R. (1998). *Optimal lines in public rail transport*. PhD Thesis, Technical University Braunschweig, Germany.

Cepeda, M., Cominetti, R., & Florian, M. (2006). A frequency-based assignment model for congested transit networks with strict capacity constraints: characterization and computation of equilibria. *Transportation Research Part B, 40*(6), 437–459. doi:10.1016/j.trb.2005.05.006.

Chakroborty, P., & Wivedi, T. (2002). Optimal route network design for transit systems using genetic algorithms. *Engineering Optimization, 34*(1), 83–100. doi:10.1080/03052150210909.

Chang, Y. H., Yeh, C. H., & Shen, C. C. (2000). A multiobjective model for passenger train services planning: application to Taiwan's high-speed rail line. *Transportation Research Part B, 34*(2), 91–106. doi:10.1016/S0191-2615(99)00013-2.

Deng, L. B. (2007). *Study on the optimal problems of passenger train plan for dedicated passenger traffic line*. PhD Thesis, Central South University, China.

Deng, L. B., Shi, F., & Zhou, W. L. (2009). Stop schedule plan optimization for passenger train. *China Railway Science, 30*(4), 102–106. (in Chinese).

Eiben, A. E., & Smith, J. E. (2003). *Introduction to Evolutionary Computing* (pp. 57–104). Berlin: Springer.

Goldberg, D. E. (1989). *Genetic Algorithms in Search, Optimization and Machine Learning* (pp. 23–47). Inc, Boston, MA, USA: Addison-Wesley Longman Publishing Co.

Goossens, J. W., van Hoesel, S., & Kroon, L. (2004a). A branch-and-cut approach for solving railway line-planning problems. *Transportation Science, 38*(3), 379–393. doi:10.1287/trsc.1030.0051.

Goossens, J. W., van Hoesel, S., Kroon, L. (2004b). *Optimising halting station of passenger railway lines*. Available from http://arno.unimaas.nl/show.cgi?fid=803. Accessed on September 1, 2011.

Goossens, J. W., van Hoesel, S., & Kroon, L. (2006). On solving multi-type railway line planning problems. *European Journal of Operational Research, 168*(2), 403–424. doi:10.1016/j.ejor.2004.04.036.

Guan, J. F., Yang, H., & Wirasinghe, S. C. (2006). Simultaneous optimization of transit line configuration and passenger line assignment. *Transportation Research Part B, 40*(10), 885–902. doi:10.1016/j.trb.2005.12.003.

Hamdouch, Y., & Lawphongpanich, S. (2008). Schedule-based transit assignment model with travel strategies and capacity constraints. *Transportation Research Part B, 42*(7–8), 663–684. doi:10.1016/j.trb.2007.11.005.

Laporte, G., Mesa, J. A., & Perea, F. (2010). A game theoretic framework for the robust railway transit network design problem. *Transportation Research Part B, 44*(4), 447–459. doi:10.1016/j.trb.2009.08.004.

Mo, W. T., Wang, L., Wang, B. H., Sun, W., Fei, X., Wang, F. J., Xu, J., & Qin, Y. (2011). *Two-layer optimization based timetable rescheduling in speed restriction for high speed railway*. In: Transportation Research Board 90th Annual Meeting, Report No. 11–2367, Washington DC, USA.

Nielsen, O. A. (2000). A stochastic transit assignment model considering differences in passengers utility functions. *Transportation Research Part B, 34*(5), 377–402. doi:10.1016/S0191-2615(99)00029-6.

Pfetsch, M. E., & Borndorfer, R. (2005). *Routing in line planning for public transport*. Technical report no. ZIB-Report 05–36, Konrad-Zuse-Zentrum für Informationstechnik, Berlin.

Poon, M. H., Wong, S. C., & Tong, C. O. (2004). A dynamic schedule-based model for congested transit networks. *Transportation Research Part B, 38*(4), 343–368. doi:10.1016/S0191-2615(03)00026-2.

Schmŏcker, J. D., Bell, M. G. H., & Kurauchi, F. (2008). A quasi-dynamic capacity constrained frequency-based transit assignment model. *Transportation Research Part B, 42*(10), 925–945. doi:10.1016/j.trb.2008.02.001.

Schmŏcker, J. D., Fonzone, A., Shimamoto, H., Kurauchi, F., & Bell, M. G. H. (2011). Frequency-based transit assignment considering seat capacities. *Transportation Research Part B, 45*(2), 392–408. doi:10.1016/j.trb.2010.07.002.

Schŏebel, A., & Schwarze, S. (2006). *A game-theoretic approach to line planning*. Available from http://drops.dagstuhl.de/opus/volltexte/2006/688/pdf/06002.SchoebelAnita.Paper.688.pdf. Accessed on September 1, 2011.

Shi, F., Deng, L. B., & Huo, L. (2007). Bi-level programming model and algorithm of passenger train operation plan. *China Railway Science, 28*(3), 110–116. (in Chinese).

Wang, H. Z. (2006). *Study on the train scheme of passenger transport special line*. MS Thesis, China Academy of Railway Sciences, Beijing, China.

## Author Biographies

**Li Wang** Ph.D., is a lecturer in the School of Traffic and Transportation from the Beijing Jiaotong University in China. Her research interests include railway transport operation and safety guarantee, such as passenger flow distribution, trend prediction, train operation, adjustment optimization in an emergency, and transportation safety comprehensive information integration.

**Li-min Jia** Ph.D., is professor and chief scientist in the State Key Lab of Rail Traffic Control and Safety at the Beijing Jiaotong University in China. He has chaired the Standing Expert Committee on National Cooperative Innovation Program for Chinese High Speed Train, and is a Member of the Standing Expert Committee on the National Innovation Program for Transportation Safety and State Consulting Committee on Intelligent Transportation Systems in China. He is the recipient of numerous grants and the author of over 200 articles and five books.

# Dynamic Performance of a Pantograph–Catenary System with the Consideration of the Appearance Characteristics of Contact Surfaces

**Ning Zhou, Wei-hua Zhang and Rui-ping Li**

**Contents**



## 1 Introduction

In the high-speed electric railway, the dynamic performance of a pantograph–catenary system plays an important role in maintaining good contact between the pantograph and catenary and improving the quality of current collection. Standard mathematical models and solution methods have been proposed for the dynamic performance of the pantograph–catenary system (Vinayagalingam 1983; Cai and Zhai 1997; Arnold and Simeon 2000; Collina and Bruni 2002; Mei and Zhang 2002; Liu et al. 2003; Park et al. 2003; Metrikine and Bosch 2006; Lee 2007; Lopez-Garcia et al. 2007). Furthermore, in recent years, more attention has been paid to the influence of contact wire unevenness, the wear between contact line and

N. Zhou (✉) · W. Zhang · R. Li
State Key Laboratory of Traction Power, Southwest Jiaotong University, Chengdu 610031, China
e-mail: zhou_ningbb@sina.com



Y. Fang and Y.H. Zhang (eds.), *China's High-Speed Rail Technology*,
Advances in High-speed Rail Technology, https://doi.org/10.1007/978-981-10-5610-9_24

collector, the flexible deformation at higher frequencies, and the influence of aerodynamics on the dynamic performance, etc. Zhang et al. (2000) investigated the influence of the irregularity of the contact wire on the contact state and discussed how to reduce the influence by modifying the design parameters of the pantograph. Nagasaka and Aboshi (2004) analyzed the influence of the contact wire unevenness on the contact forces between the catenary and pantograph and devised an instrument to measure the unevenness of contact wires, both accurately and continuously. They proposed a method to evaluate the conditions of contact wires. He et al. (1998) investigated the wear and electrical properties of contact wires and collectors used in lightweight systems, based on laboratory tests with a wear equipment. Bucca and Collina (2009) established a wear model for the contact between collectors and contact wires and designed a procedure to simulate the dynamic interaction between the pantograph and catenary. They predicted the wear of collectors and contact wires. The values of contact forces and current were the inputs of the wear model, and the amount of the wear of the collectors and contact wires was determined, generating an irregular profile of the contact wires. Collina et al. (2009) identified the modal parameters of the collectors by experiment and then investigated the dynamic contact behavior of the pantograph–catenary system, considering the deformation modes of the collectors. It is proved that there is an obvious influence of the deformable modes of the collectors on the dynamic behavior of the pantograph–catenary system. Based on the quasi-steady theory formulation of the drag and lift forces on the collectors, Bocciolone et al. (2006) analyzed the turbulence of the incoming flow and the dynamic variation of the contact force between the pantograph and catenary and investigated the influence of the aerodynamic action on the current collection.

However, these studies mainly focus on the dynamic contact behavior in the vertical direction. There have been few published papers on how to comprehensively evaluate the dynamic performance in space, considering the appearance characteristics of contact surfaces of the pantograph and catenary. Therefore, the objective of this research was to put forward a pantograph–catenary system model to investigate the dynamic contact behavior in space and, more importantly, to find a reasonable method to analyze the influence of the contact wire irregularity on the contact behavior and the vibration caused by the front pantograph on the dynamic performance of the rear pantograph for a pantograph–catenary system with double pantographs.

## 2 Model of the Pantograph–Catenary System

### 2.1 *Catenary Model*

As shown in Fig. 1, the finite element model of the stitched catenary is composed of a support wire, assistant wire, contact wire, and dropper. A beam element is defined

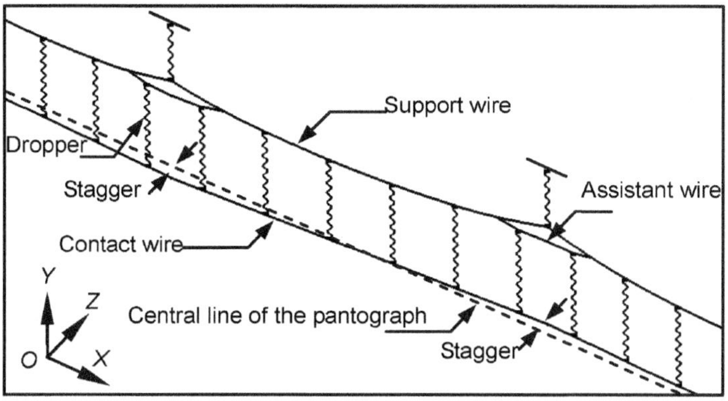


**Fig. 1** Catenary model with one span

**Table 1** Material parameters of the catenary

| Wire | Material | Density (kg/m) | Tension (kN) | Section area (mm$^2$) |
|---|---|---|---|---|
| Contact | CTMH-150 | 1.35 | 30 | 150 |
| Support | JTMH-120 | 1.07 | 21 | 120 |
| Assistant | JTMH-35 | 0.31 | 3.5 | 35 |

to simulate the support wire and contact wire, and a spring element is used to build the model of the dropper. The length of the 3D finite element model of the catenary is 500 m (ten spans), and the stagger is 300 mm. The material parameters of the catenary are shown in Table 1.

## 2.2 *Pantograph Model*

First, the pantograph is modeled with a rigid–flexible hybrid body, regarding two collectors of the pan-head as a flexible body and the other parts of the pantograph as a rigid body (Fig. 2). With the finite element method, the flexible body model of two collectors is established. A total of 3980 solid elements are used to depict the appearance characteristics of two collectors. Then, for comparison, the pantograph is completely considered as a multiple rigid body system. The two collectors of the pan-head are no longer a flexible body but a rigid body (Fig. 3). Thus, the appearance characteristics of contact surfaces of the pantograph and catenary are not involved.

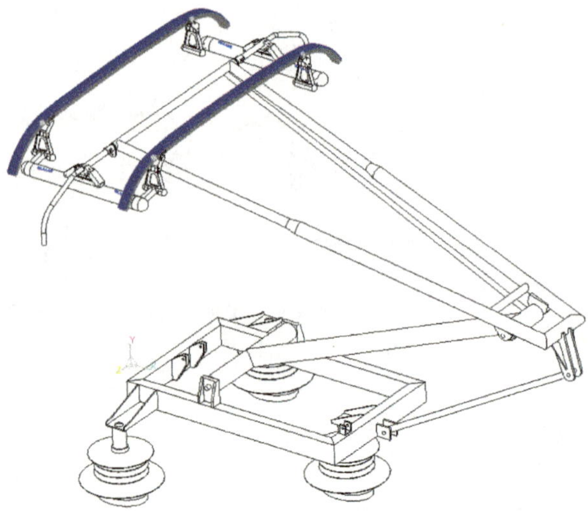

**Fig. 2** Rigid–flexible hybrid pantograph model

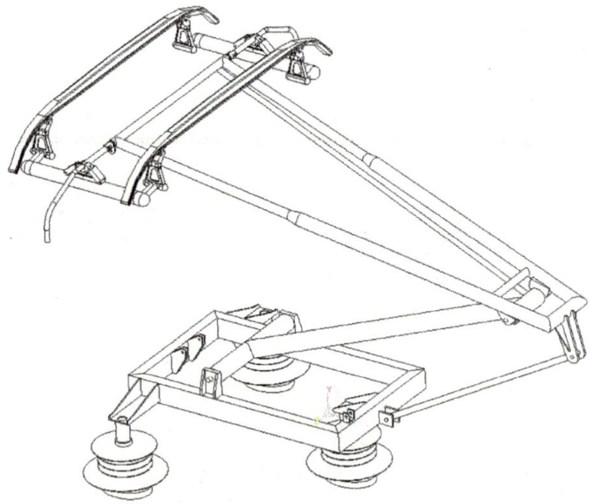

**Fig. 3** Rigid pantograph model

## *2.3 Pantograph–Catenary System Model*

Based on the catenary model and the pantograph model, the coupled model of the pantograph–catenary system is built. If the rigid–flexible hybrid pantograph model is employed, the action contact surfaces are approximated as multi-rectangular patches according to the solid elements of the collectors, and the base contact surfaces are approximated as multi-cylinders in the line of the beam element (Fig. 4). The lines of the base cylinders are examined to determine whether they are in contact with the surfaces of the action patches. And then, the contact force is generated with the compliance characteristics allowing penetration. Thus, the

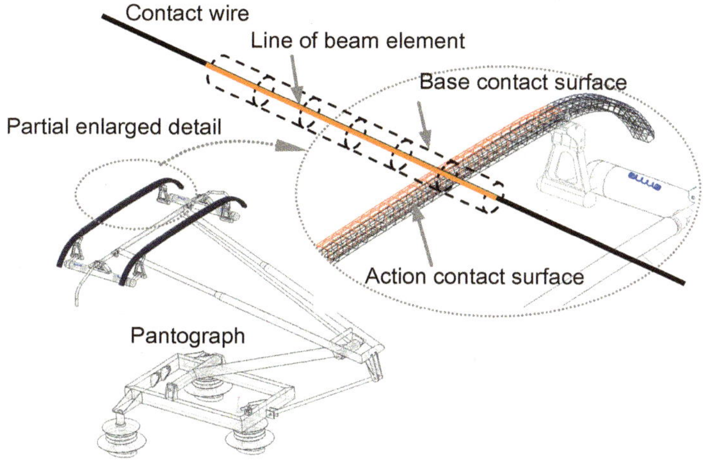


**Fig. 4** Contact description of the pantograph–catenary system

dynamic contact behavior, such as vertical vibration, longitudinal impact, and lateral oscillation of the pantograph and the catenary, may be exactly described by the line-to-surface contact. Compared to the rigid–flexible hybrid model, it can be seen that if the pantograph is simplified as a multiple rigid model, all the appearance characteristics of the collectors will be lost, the contact behavior is only a line-to-line contact, and the contact description will be not more accurate than that defined by the line-to-surface contact.

## 3 Results of the Dynamic Performance

For the pantograph–catenary system, the solution of the dynamic contact behavior has been carried out by means of two kinds of pantograph models. Furthermore, the results of the dynamic performance are obtained, including contact forces and accelerations in space.

Figure 5 shows the contact forces in the vertical (*Y*) direction at the speed of 350 km/h. It can be found that there is an obvious difference in the contact forces for two kinds of pantograph models, and the fluctuation of contact forces based on the rigid–flexible hybrid model is more volatile than that based on the rigid model. Meanwhile, it can be found that, for the rigid model, although there are contact losses on the rear collector, the total contact forces do not appear to be a zero value. It shows that the contact loss on the front and rear collectors does not occur at the same time. However, for the rigid–flexible hybrid model, the total contact forces already present the zero value, and there is a simultaneous contact loss on the front and rear collectors.

Figure 6 shows the total contact forces in the longitude (*X*) and lateral (*Z*) directions at the speed of 350 km/h. Similarly, it can be found that the obtained

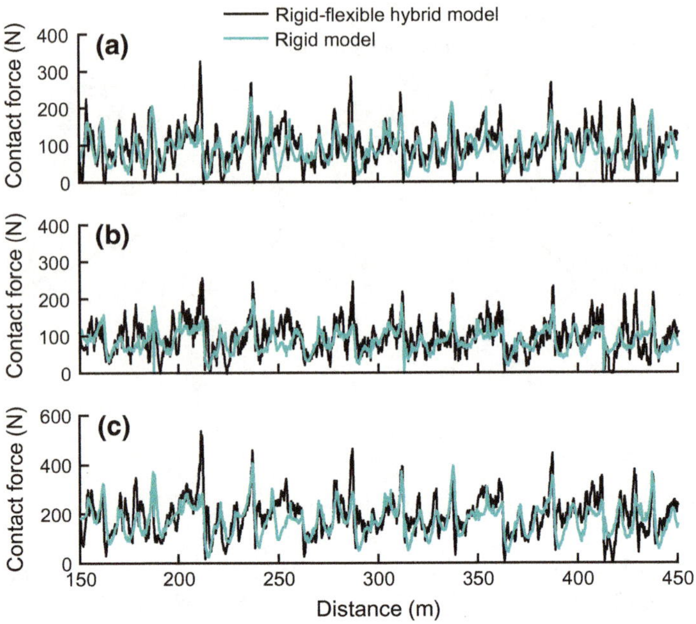


**Fig. 5** Contact forces of the front (**a**) and rear (**b**) collectors and the total contact force (**c**) in the *Y* direction

contact forces in the *X* and *Z* directions, based on two pantograph models, have a basically same rule, and the contact forces by means of the flexible–rigid hybrid model are slightly larger than those by means of the rigid model.

Figure 7 shows the contact loss between the pantograph and catenary at different speeds. It can be seen that the contact loss is not detected until the speed is higher than 325 km/h for the rigid–flexible hybrid model. However, for the rigid model, there is no contact loss when the speed is lower than 400 km/h. Thus, it indicates that the calculated maximum operating speed based on the rigid–flexible hybrid model is less than that based on the rigid model.

Figure 8 shows the accelerations and the corresponding spectra at the speed of 350 km/h. It can be found that the obtained maximum accelerations by means of the flexible–rigid hybrid model in the *X*, *Y*, and *Z* directions are much larger than those by means of the rigid model. In particular, in the *Y* direction, the former is about six times larger than the latter. For the spectrum of the former, the contribution from the lower and higher frequency bands can be observed. Furthermore, at a higher frequency, there is an obvious contribution from the frequency component of about 110–120 Hz.

For the rigid–flexible hybrid pantograph model, the modal analysis of the flexible body is performed to obtain natural frequencies and mode shapes. Table 2 shows the natural frequencies and corresponding mode shapes of the flexible collector.

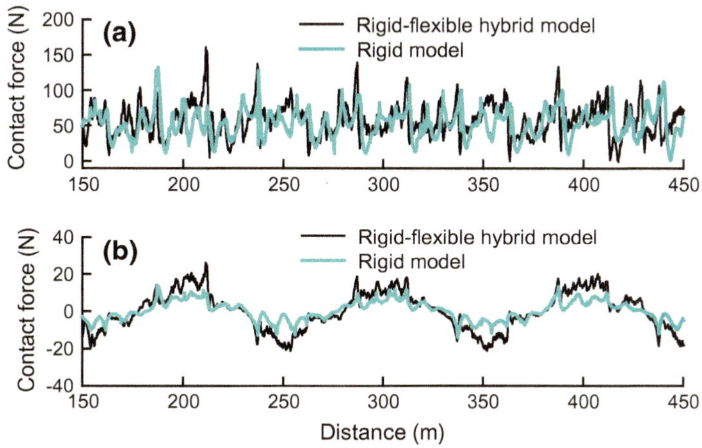


**Fig. 6** Total contact forces in the $X$ (**a**) and $Z$ (**b**) directions

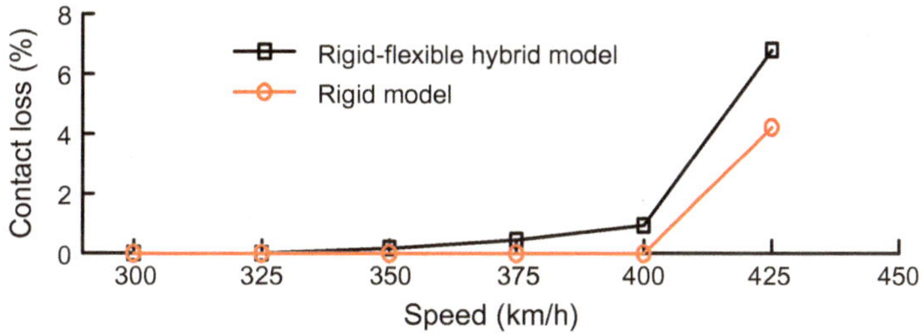


**Fig. 7** Contact loss at different speeds

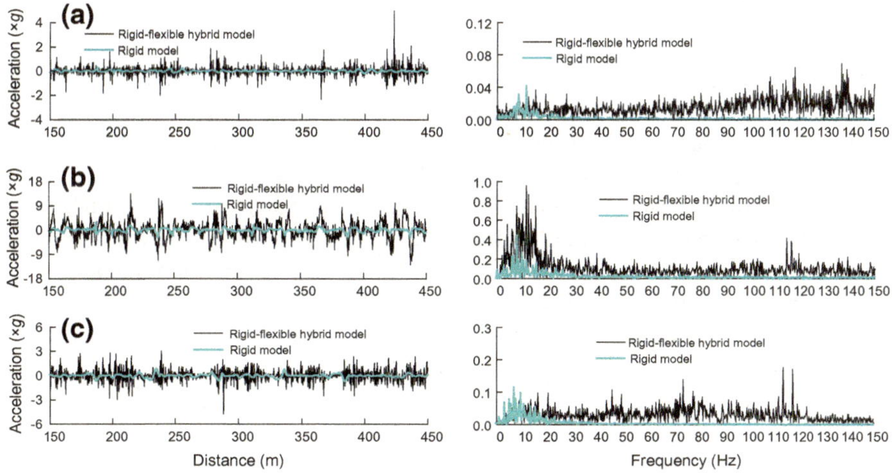


**Fig. 8** Accelerations and the corresponding spectra at the speed of 350 km/h in the $X$ (**a**), $Y$ (**b**), and $Z$ (**c**) directions

**Table 2** Natural frequencies and mode shapes of the flexible body

| Mode number | 1 | 2 | 3 | 4 | 5 |
|---|---|---|---|---|---|
| Mode shape |  |  |  | 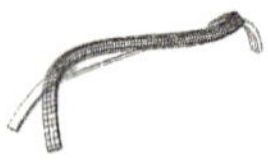 |  |
| Frequency (Hz) | 60.57 | 111.68 | 141.90 | 150.07 | 240.59 |

Combining the results of the modal analysis for the flexible body, it is obvious that the frequency component of about 110–120 Hz mainly comes from the contribution of the second mode of the collector. The flexible deformation of the collector in the $X$ and $Y$ directions has an important influence on the dynamic performance. Furthermore, for the rigid–flexible hybrid model, there are multiple degrees of freedom for the pan-head, and it can exactly describe the motion and the contact behavior of the pan-head and excite the flexible deformation at higher frequencies. However, if the pantograph model is considered as a rigid body system, all shape features of the pan-head are lost and its flexible deformation cannot be considered. Thus, it can be seen that it is the consideration of the appearance characteristics that may inevitably lead to the difference of the calculation results.

## 4 Validation by a Field Test

A field test of dynamic performances, aimed at identifying contact forces and accelerations of the pantograph, has been performed on a 350 km/h railway line. The contact forces between pantograph and catenary are measured by means of force sensors. Four force sensors are divided into two groups to, respectively, determine the contact forces of the front and rear collectors. The No. 3 force sensor between the collector and the triangular frame of the pantograph is shown in Fig. 9.

Two accelerometers are fixed to measure the acceleration of the front and rear collectors (Fig. 10). As mentioned above, the obtained force by the force sensors is actually the interaction force ($\boldsymbol{F}_t$) between the collector and the triangular frame and is not the contact force ($\boldsymbol{F}_c$) between the collector and the contact wire. For the collectors, the applied force may be written as

$$\boldsymbol{F}_c = \boldsymbol{F}_t + M_t \boldsymbol{a},$$

where $M_t$ is the mass of the collector, $\boldsymbol{a}$ is the vertical acceleration, and $M_t\boldsymbol{a}$ is the inertial force. Thus, the actual contact force is equal to the test interaction force plus

**Fig. 9** Force sensor of the pantograph

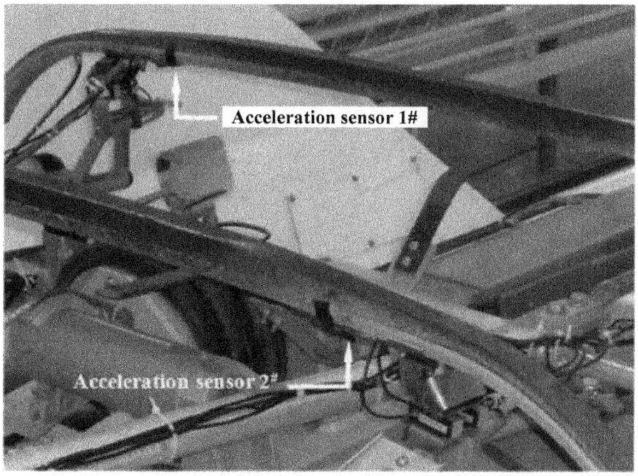


**Fig. 10** Accelerometers of the collector

the inertial force. Moreover, the inertial force is determined by the test acceleration of collectors. Figure 10 shows the contact force for the pantograph at the speeds of 300 and 350 km/h.

The results in Fig. 11 show that there is no contact loss at the speed of 300 km/h, and the pantograph may keep a steady contact with the catenary. However, when the operating speed is increased to 350 km/h, the contact forces between the pantograph and catenary vary more strongly than those at 300 km/h. The steady contact is lost, and the quality of current collection is worsened. Furthermore, the comparison with the calculated contact forces by means of the two pantograph models at the speed of 350 km/h is shown in Table 3. It can be seen that the contact forces obtained by means of the rigid–flexible hybrid model is basically consistent with the test results; however, for the rigid model, there is an obvious difference in the statistical results of the contact forces between tests and simulation. Thus, through the field test, it is proved that the rigid–flexible hybrid model, with consideration of the appearance characteristics, is more reasonable.

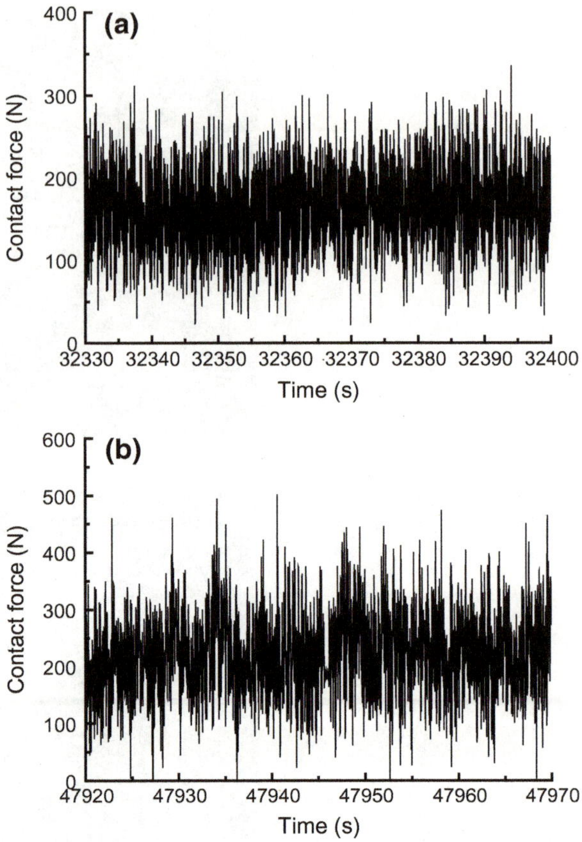


**Fig. 11** Contact forces in the *Y* direction at the speeds of 300 km/h (**a**) and 350 km/h (**b**)

**Table 3** Statistical results of the contact forces at the speed of 350 km/h

| Method | Contact force (N) | | |
|---|---|---|---|
| | Mean | Min | Max |
| Field test | 223.87 | 0 | 502.11 |
| Rigid–flexible hybrid model | 192.40 | 0 | 534.95 |
| Rigid model | 174.80 | 23.81 | 405.32 |

## 5 Analysis of the Influence of Contact Wire Irregularity

Based on the modeling and simulation method mentioned above for the pantograph–catenary system, the influence of contact wire irregularity on the dynamic performance has been analyzed. The catenary model composed of ten spans and the rigid–flexible hybrid pantograph model are established. The contact wire irregularity is artificially considered as the height error at the location of the third dropper of the eighth span (about 370 m from the initial location), as shown in Fig. 12.

Figure 13 shows the acceleration of the front and rear collectors at the speed of 350 km/h, considering the height error of 20 mm. Compared to the results in Fig. 8

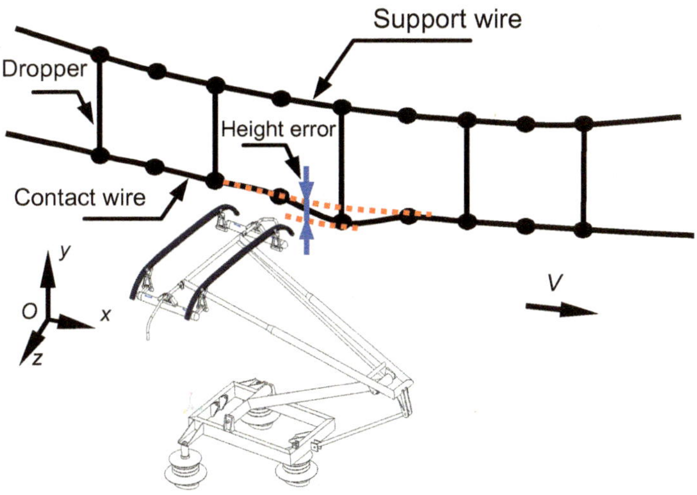


**Fig. 12** Model considering the contact wire irregularity

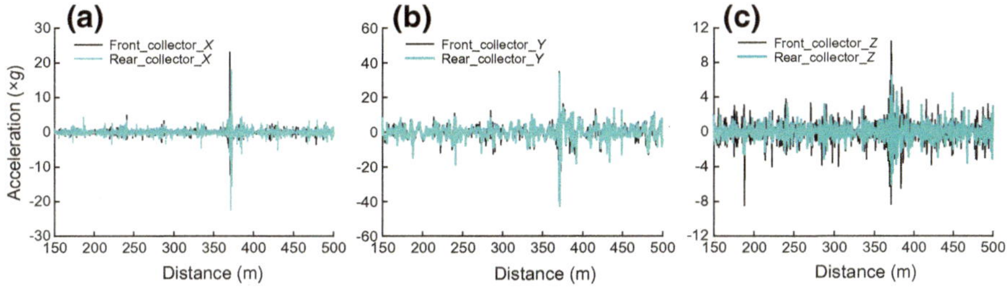


**Fig. 13** Acceleration of the front and rear collectors based on the rigid–flexible hybrid model in *X* (a), *Y* (b), and *Z* (c) directions

without consideration of height error, it can be found that when the pantograph passes through the eighth span of the catenary, there is an obvious difference in the acceleration. In particular, at the location of about 370 m, the acceleration in the *Y* direction is much larger than 40*g*, the acceleration in the *X* direction is up to 20*g*, and the acceleration in the *Z* direction is greatly increased.

By means of the rigid pantograph model, the influence of contact wire irregularity on the dynamic performance is similarly analyzed. Figure 14 shows the acceleration of the front and rear collectors based on the rigid model at the speed of 350 km/h. However, it can be seen that, when the pantograph runs through the location of about 370 m, a significantly evident difference in the acceleration is not observed. The influence of contact wire irregularity on the dynamic performance cannot be truly represented, due to the lack of appearance characteristics. Thus, the appearance characteristics should be taken into consideration to reasonably evaluate the influence of contact wire irregularity on the dynamic performance, using the rigid–flexible hybrid pantograph model.

# 6 Analysis of the Influence of Double Pantographs

By means of the similar modeling and simulation methods mentioned above, the influence of the vibration caused by the front pantograph on the rear pantograph has been analyzed. A pantograph–catenary system model with double pantographs was built, and the space between two pantographs is 200 m, as shown in Fig. 15.

Figure 16 shows the contact forces in the *Y* direction at the speed of 350 km/h by means of two pantograph models. It can be observed that, when the rigid pantograph model is employed, the contact forces of the rear pantograph fluctuate slightly and are basically consistent with those of the front pantograph. However, for the rigid–flexible hybrid pantograph model, there is an obvious difference in the contact forces between the rear pantograph and the front pantograph. The contact forces between the rear pantograph and the catenary vary more greatly, and the quality of current collection deteriorates. Thus, taking into consideration of the appearance characteristics plays an important role in the analysis of the influence of the vibration caused by the front pantograph on the rear pantograph for a pantograph–catenary system with double pantographs.

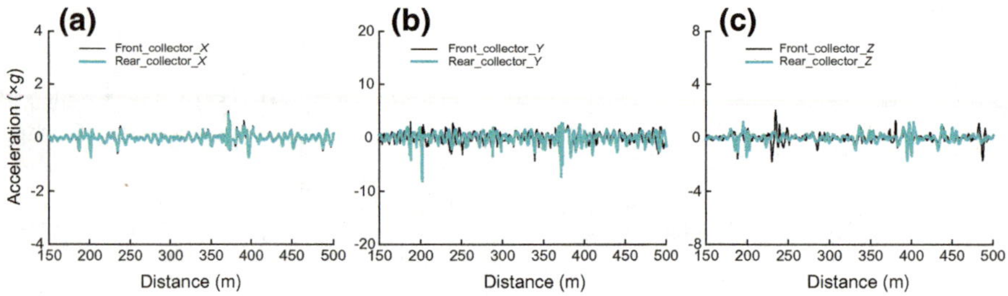


**Fig. 14** Acceleration of the front and rear collectors based on the rigid model in *X* (a), *Y* (b), and *Z* (c) directions

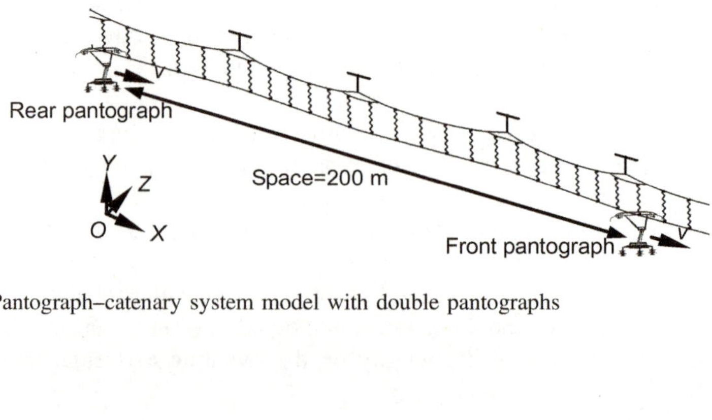


**Fig. 15** Pantograph–catenary system model with double pantographs

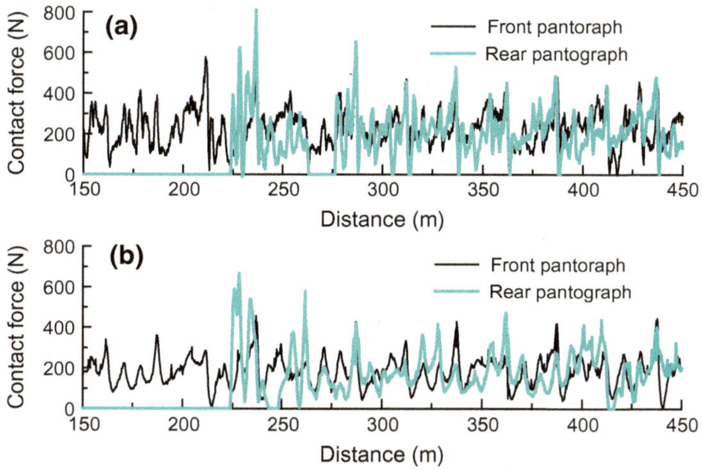


**Fig. 16** Contact forces in the *Y* direction at the speed of 350 km/h based on the rigid–flexible hybrid model (**a**) and rigid model (**b**)

## 7 Conclusions

Based on the conventional pantograph–catenary system model, a rigid–flexible hybrid pantograph model has been put forward to include a consideration of the appearance characteristics of contact surfaces of the pantograph and the catenary. The dynamic behavior of the pantograph–catenary system in space has been investigated by means of two levels of modeling, with and without consideration of the appearance characteristics. Furthermore, the influence of the contact wire irregularity and the vibration caused by the front pantograph on the rear pantograph for a pantograph–catenary system with double pantographs has been analyzed. The results show that the appearance characteristics of contact surfaces play an important role in the analysis of dynamic performance. The obvious difference of the contact force, the maximum operating speed, the acceleration, and the corresponding spectrum is observed. A consideration of the appearance characteristics is thus essential to reasonably evaluate the dynamic performance, using the rigid–flexible hybrid pantograph model.

## References

Arnold, M., & Simeon, B. (2000). Pantograph and catenary dynamics: A benchmark problem and its numerical solution. *Applied Numerical Mathematics, 34*(4), 345–362. doi:10.1016/S0168-9274(99)00038-0.

Bocciolone, M., Resta, F., Rocchi, D., Tosi, A., & Collina, A. (2006). Pantograph aerodynamic effects on the pantograph-catenary interaction. *Vehicle System Dynamics, 44*(S1), 560–570. doi:10.1080/00423110600875484.

Bucca, G., & Collina, A. (2009). A procedure for the wear prediction of collector strip and contact wire in pantograph-catenary system. *Wear, 266*(1–2), 46–59. doi:10.1016/j.wear.2008.05.006.

Cai, C. B., & Zhai, W. M. (1997). Study on simulation of dynamic performance of pantograph-catenary system at high speed railway. *Journal of the China Railway Society, 19* (5), 38–43. (in Chinese).

Collina, A., & Bruni, S. (2002). Numerical simulation of pantograph-overhead equipment interaction. *Vehicle System Dynamics, 38*(4), 261–291. doi:10.1076/vesd.38.4.261.8286.

Collina, A., Conte, A. L., & Carnevale, M. (2009). Effect of collector deformable modes in pantograph-catenary dynamic interaction. *Proceedings of the Institution of Mechanical Engineers, Part F: Journal of Rail and Rapid Transit, 223*(1), 1–14. doi:10.1243/09544097JRRT212.

He, D. H., Manory, R. R., & Grady, N. (1998). Wear of railway contact wires against current collector materials. *Wear, 215*(1–2), 146–155. doi:10.1016/S0043-1648(97)00262-7.

Lee, K. (2007). Analysis of dynamic contact between overhead wire and pantograph of a high-speed electric train. *Proceedings of the Institution of Mechanical Engineers, Part F: Journal of Rail and Rapid Transit, 221*(2), 157–166. doi:10.1243/0954409JRRT93.

Liu, Y., Zhang, W. H., & Mei, G. M. (2003). Study of dynamic stress of the catenary in the pantograph/catenary vertical coupling movement. *Journal of the China Railway Society, 25*(4), 23–26. (in Chinese).

Lopez-Garcia, O., Carnicero, A., & Marono, J. L. (2007). Influence of stiffness and contact modelling on catenary-pantograph system dynamics. *Journal of Sound and Vibration, 299*(4–5), 806–821. doi:10.1016/j.jsv.2006.07.018.

Mei, G. M., & Zhang, W. H. (2002). Dynamics model and behavior of pantograph/catenary system. *Journal of Traffic and Transportation Engineering, 2*(1), 20–25. (in Chinese).

Metrikine, A. V., & Bosch, A. L. (2006). Dynamic response of a two-level catenary to a moving load. *Journal of Sound and Vibration, 292*(3–5), 676–693. doi:10.1016/j.jsv.2005.08.026.

Nagasaka, S., & Aboshi, M. (2004). Measurement and estimation of contact wire unevenness. *Quarterly Report of RTRI, 45*(2), 86–91. doi:10.2219/rtriqr.45.86.

Park, T. J., Han, C. S., & Jang, J. H. (2003). Dynamic sensitivity analysis for the pantograph of a high-speed rail vehicle. *Journal of Sound and Vibration, 266*(2), 235–260. doi:10.1016/S0022-460X(02)01280-4.

Vinayagalingam, T. (1983). Computer evaluation of controlled pantographs for current collection from simple catenary overhead equipment at high speed. *Journal of Dynamics Systems, Measurement and Control, 105*(4), 287–294. doi:10.1115/1.3140673.

Zhang, W. H., Mei, G. M., & Chen, L. Q. (2000). Analysis of the influence of catenary's sag and irregularity upon the quality of current-feeding. *Journal of the China Railway Society, 22*(6), 50–54. (in Chinese).

## Author Biography

**Ning Zhou** is the assistant professor of the State Key Laboratory of Traction Power at Southwest Jiaotong University, China. In 2013, he graduated from this University and received a Ph.D. He has been engaged in the investigation on the dynamics theory, simulation and experiment of pantograph and catenary systems since 2005. He has participated in a number of national basic and applied research programs of China and published over 30 articles about high-speed railways.

# Design and Reliability, Availability, Maintainability, and Safety Analysis of a High Availability Quadruple Vital Computer System

**Ping Tan, Wei-ting He, Jia Lin, Hong-ming Zhao and Jian Chu**

## Contents



## 1 Introduction

A high-speed railway is an energy-saving, environmentally friendly, and sustainable transport mode. It has the advantages of being safe, punctual, fast, and comfortable. High-speed railway trunk lines in China will be completed in 2012. With the construction of intercity railways, a high-speed railway network, covering large

P. Tan (✉) · W. He · J. Lin (✉) · J. Chu
State Key Laboratory of Industrial Control Technology & Institute of Cyber-Systems and Control, Zhejiang University, Hangzhou 310027, China
e-mail: ptan@iipc.zju.edu.cn

J. Lin
e-mail: zeroplus_zju@zju.edu.cn

H. Zhao
Zhejiang Insigma-Supcon Co., Ltd., Hangzhou 310013, China

 
Y. Fang and Y.H. Zhang (eds.), *China's High-Speed Rail Technology*,
Advances in High-speed Rail Technology, https://doi.org/10.1007/978-981-10-5610-9_25

cities with a population of more than 500,000, will be gradually formed. There will then be more than 2000 high-speed trains put into operation. Safety and efficiency of railway transport is increasingly important. The train operation control system is the key signal system equipment to guarantee the safety of train operation and improve the transport efficiency. The system is composed of an on board automatic train protection (ATP) system and a ground control system. The onboard train control system is the so-called ATP, including onboard vital computer (VC), track circuit reader (TCR), balise transmission module (BTM), data recording unit (DRU), driver machine interface (DMI), train interface unit (TIU), and train and wayside communication unit (TWC). ATP is the final safety executant to ensure safe operation of a high-speed train, satisfying the requirements of safety integrity level 4 (SIL4), with fault-oriented safe attributes.

Commonly, the existing domestic ATP system uses a double 2-out-of-2 VC platform. This computer platform is built on a hot-standby redundant subsystem, which uses the 2-out-of-2 VCs (Qin et al. 2010). When a failure is found in any module of the subsystem, it will be in the fail-safe state, and the double 2-out-of-2 system is transformed into 2-out-of-2 redundancy system (Dou et al. 2007), whose hardware fault tolerance is 1. In the 2-out-of-3 VC platform, if a failure is found in a certain module, the system will be changed into the 2-out-of-2 redundancy system. If more than two modules failed, the system will be in the fail-safe state. The hardware fault tolerance of the 2-out-of-3 VC platform is also 1. There is no disparity between the double 2-out-of-2 VC platform and the 2-out-of-3 VC platform in the aspect of hardware fault tolerance (IEC 61508-6:2000).

There are two approaches to improve the reliability and safety of the system to block the failure of a system. The first is fault avoidance, and the second is fault tolerance (Kim et al. 2005). Because components may develop faults with time, a fault avoidance technique is very difficult to apply (Kim et al. 2002). However, with the fault tolerance technique, the system has a redundancy, and a fault is allowed without termination of its normal operation.

There are several types of fault tolerance techniques, such as hardware redundancy, software redundancy, time redundancy, and information redundancy techniques (Kim et al. 2005). In the high availability quadruple vital computer (HAQVC) system, we use a hardware redundancy technique, embedded software redundancy, and safe-bus redundancy. Technologies used in safety systems mainly include voting structure, or parallel structure, or both structures. The typical voting system is comprised of $n$ units. The $k/n$ system is that if the numbers of active units are no less than $k$ ($k$ is between 1 and $n$), the system will not be inactive. We assume that the reliability of $n$ units is $R$, and the reliability mathematical model is shown as

$$R_s = \sum_{i=k}^{n} \binom{n}{i} R^i (1 - R)^{n-i}, \tag{1}$$

where $\binom{n}{i} = \frac{n!}{i!(n-i)!}$ and $R_s$ is the system reliability. The parallel system is also comprised of $n$ units, and each unit of the system is independent. When all of its units are inactive, the parallel system will be inactive. The reliability mathematical model is shown as

$$R_s = 1 - \prod_{i=1}^{n} F_i = 1 - \prod_{i=1}^{n} (1 - R_i), \quad i = 1, 2, \ldots, n; \tag{2}$$

where $F_i$ and $R_i$ are the unreliability and reliability of unit $i$.

Using the voting and parallel structures, it not only has the voting structure's advantage of high safety, but also has the parallel structure's advantage of availability and maintainability. The reliability mathematics model of the system is shown as

$$R_s = 1 - \prod_{i=1}^{N} F_i = 1 - \prod_{i=1}^{N} \left[1 - \sum_{j=k}^{n} \binom{n}{j} R^j (1 - R)^{n-j}\right], \quad k \leq n, \; i = 1, 2, \ldots, N. \tag{3}$$

As shown in Eqs. (1)–(3), we can obtain the reliability function of the 2-out-of-2 system, the 2-out-of-3 system, and the double 2-out-of-2 system, which can be formulated as follows:

$$\begin{cases} R_{2-\mathrm{out-of}-2} = R, \\ R_{2-\mathrm{out-of}-3} = 3 \times R^2 - 2 \times R^3, \\ R_{\mathrm{double}\;2-\mathrm{out-of}-2} = 2 \times R^2 - R^4. \end{cases} \tag{4}$$

However, the above analysis shows that there is no disparity between the double 2-out-of-2 system and the 2-out-of-3 system. In fact, the 2-out-of-3 system is of the highest reliability. In order to provide full play to the advantages of four modules architecture and ensure the safety of the system, while improving the reliability and availability of the system, we designed a novel HAQVC system, based on the research on the framework and the mechanism of data interaction and redundant degeneration of the VC platform. At the same time, we draw the curves of reliability of the HAQVC system, the double 2-out-of-2 system, the 2-out-of-3 system, and the 2-out-of-2 system in the same figure as a contrast (Fig. 1). Obviously, the HAQVC system is of the highest reliability.

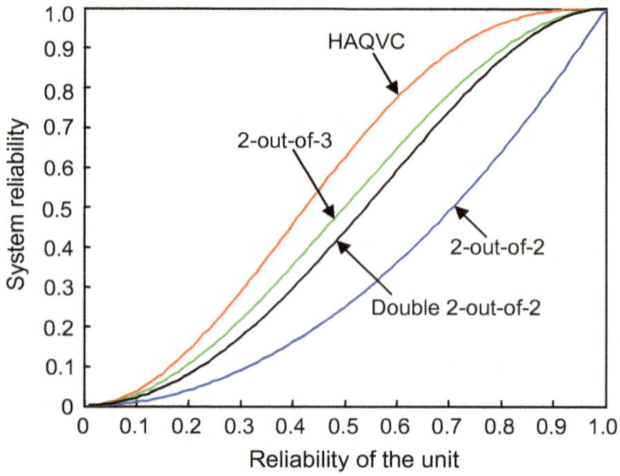


**Fig. 1** Reliability of each system

## 2 System Design

Compared with a general industrial control system, the onboard ATP system is essentially a special safety control system with high-speed trains as its controlled object. The interface units and the function modules of the onboard ATP system adopt the general modules except for the TIU, which mainly falls into two categories: multifunction vehicle bus (MVB) and relay interface. The function and scale of the system controller, types of input–output (IO) module, IO knot number, and types and number of communication module have been mostly determined. On the basis of meeting the requirements of the installation of rolling stock mechanical structure, electromagnetic compatibility, and convenient maintenance, a thorough study has been conducted for the key elements of the system, such as the requirements of safety, reliability, availability, maintainability, and real-timing and their restrictive relationships, so as to determine the scale of the control system, system architecture, network topology, hardware platform, and the mechanisms of communication scheduling and redundancy switching-over.

The VC module, input module, and output module of the HAQVC system are all of quadruple structure, linked by four redundancy safety buses (SBUS1, SBUS2, SBUS3, and SBUS4). The structure of the system is shown in Fig. 2. The system is double 2-out-of-2 when the four VC modules are operating correctly, in which one subsystem with 2-out-of-2 redundancy structure is composed of VC-A and VC-B, while the other subsystem is composed of VC-C and VC-D. We use four safety buses to achieve the interconnection, clock synchronization, and the communications scheduling between each subsystem. With the safety bus, the module can achieve the purpose of fault diagnosis, data synchronization, state information interaction, and safety data verification.

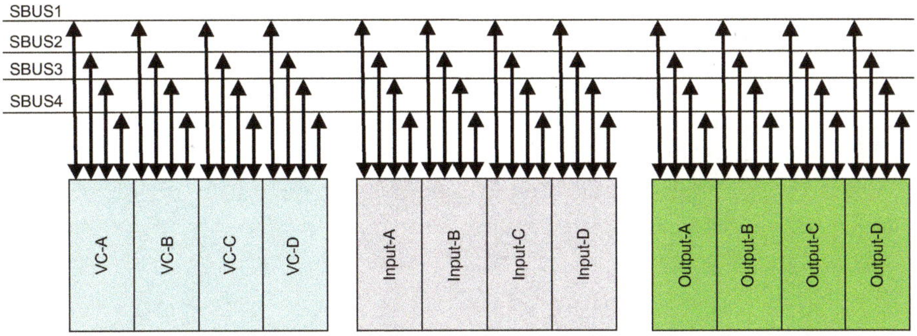


**Fig. 2** Architecture of the high availability quadruple vital computer (HAQVC) system

The traditional computer system of the double 2-out-of-2 includes two subsystems I and II, which are in a standby mode. Generally, only the master subsystem sends the results out, while the other subsystem is in the standby mode. Should the master subsystem break down, the slave subsystem will switch to the master subsystem and be run by the communication module. The way it works can significantly affect the availability and real-time attributes of the system. If some data are lost because of the handover process, it will lead to an emergency brake. The HAQVC system works in a parallel operation mode. The subsystems I and II are in output states; thus, there is no disturbance caused by the shutdown process in the HAQVC system.

In the traditional double 2-out-of-2 VC system, the system transforms into the 2-out-of-2 architecture when a failure is found in a certain module. Although there are two modules operating correctly, the system will be in the fail-safe state if subsystems I and II both have failures. However, if any module fails, the HAQVC system will degenerate to the 2-out-of-3 architecture. And if any two modules have faults, the system will transform into the 2-out-of-2 architecture. If more than three modules have faults, the HAQVC system will be in the fail-safe state. Table 1 lists the working state of the HAQVC system and the traditional double 2-out-of-2 system. States 2–5 mean the operating states of the system when only one module fails. The HAQVC system is operating with a 2-out-of-3 architecture, and the hardware fault tolerance is 1. In the same condition, the traditional double 2-out-of-2 system is operating with a 2-out-of-2 architecture and the hardware fault tolerance is 1. States 7–10 indicate that two modules of subsystems I and II have faults. The HAQVC will be operating with 2-out-of-2 architecture, while the double 2-out-of-2 system is in the fail-safe state. The HAQVC system has a distinct advantage over the traditional double 2-out-of-2 system.

In the system based on the HAQVC architecture, the key IO module and communication module use the similar architecture. And the interfaces of the ATP system, vehicle, and wayside equipment are more susceptible to the surge current and group impulse (Paul 2006). In the application and engineering, we have found

**Table 1** Working state of the high availability quadruple vital computer (HAQVC) system and the double 2-out-of-2 vital computer[a]

| Sequence no. | VC-A | VC-B | VC-C | VC-D | HAQVC | Double 2-out-of-2 vital computer |
|---|---|---|---|---|---|---|
| State 1 | ○ | ○ | ○ | ○ | Double 2-out-of-2 | Double 2-out-of-2 |
| State 2 | × | ○ | ○ | ○ | 2-out-of-3 | 2-out-of-2 |
| State 3 | ○ | × | ○ | ○ | 2-out-of-3 | 2-out-of-2 |
| State 4 | ○ | ○ | × | ○ | 2-out-of-3 | 2-out-of-2 |
| State 5 | ○ | ○ | ○ | × | 2-out-of-3 | 2-out-of-2 |
| State 6 | × | × | ○ | ○ | 2-out-of-2 | 2-out-of-2 |
| State 7 | × | ○ | × | ○ | 2-out-of-2 | Fail-safe |
| State 8 | × | ○ | ○ | × | 2-out-of-2 | Fail-safe |
| State 9 | ○ | × | × | ○ | 2-out-of-2 | Fail-safe |
| State 10 | ○ | × | ○ | × | 2-out-of-2 | Fail-safe |
| State 11 | ○ | ○ | × | × | 2-out-of-2 | 2-out-of-2 |
| State 12 | × | × | × | ○ | Fail-safe | Fail-safe |
| State 13 | × | × | ○ | × | Fail-safe | Fail-safe |
| State 14 | × | ○ | × | × | Fail-safe | Fail-safe |
| State 15 | ○ | × | × | × | Fail-safe | Fail-safe |
| State 16 | × | × | × | × | Fail-safe | Fail-safe |

[a]○ Normal; × Fault

that the maintenance ratio of those interfaces is high. Therefore, using the HAQVC architecture can ensure the high availability and safety of the system and enhance system reliability and maintainability.

## 3 Hardware and Embedded Safe Operation System (ES-OS)

In fault-tolerant design techniques, there are passive hardware redundancy, active hardware redundancy, and hybrid hardware redundancy. The HAQVC system is passive hardware redundancy, which has a fault masking and detection. If the HAQVC system has no more than two faults, the fault is masked and has no effect on the system operation before repair. The HAQVC system is designed on ARM7.

The hardware fault diagnosis is one of the core contents of the hardware design of the safety-related system. According to the safety requirement of SIL4, the system hardware should be designed with high diagnostic coverage (DC). The DC should be more than 99 and 90% for the systems whose hardware fault tolerances are 1 and 2, respectively (IEC 61508-2:2000). The DC of the HAQVC system is over 99% by self-diagnosis of single module and the diagnosis between the modules. The vital CPU module will test the hardware equipment to ensure that the

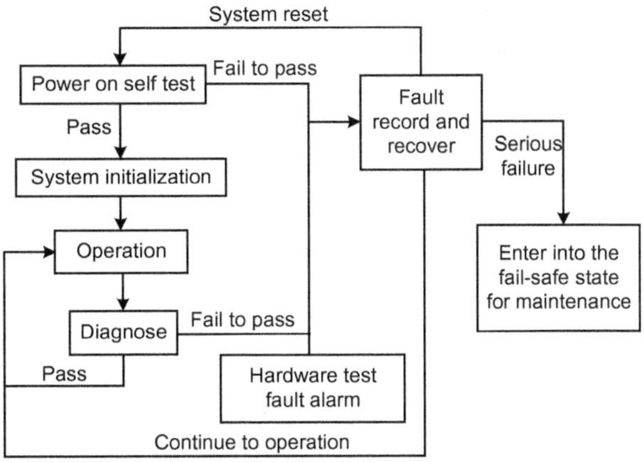


**Fig. 3** System test and diagnostic flow

hardware is normal. The detecting items include instruction set, register, RAM, FLASH, the stack pointer, program sequence, crystal oscillator frequency, and power In terms of function, the detection module falls into four categories: power-up detection sub-module, periodic check sub-module, the sub-module of interface of hardware detection circuit, and fault alarm sub-module. The system test and diagnostic flow are shown in Fig. 3.

The power-up detection sub-module will conduct a complete hardware detection of key hardware to ensure its normal operation when the system starts. The detecting objects include instruction set, register, RAM, and FLASH. The detection should not be complex such that the power-on time of equipment is not too long. However, the detecting range must include the whole target objects.

In order to find hardware faults, the function of periodic check sub-module is to detect each part of hardware in real time during equipment operation. The detecting objects include instruction set, register, RAM, the stack pointer, and the chip.

The sub-module of interface of hardware detection circuit will receive the failure warning signal from the hardware detection circuit, such as the detection of the sequencing of programs using watchdog circuit, the crystal failure, the power management chip, and fault diagnosis of power.

The function of fault alarm sub-module is to record and report the failure detected by the hardware and provide corresponding measures.

The vital CPU hardware has been developed for nearly five years (Fig. 4). Besides the technical part of the development, an internal organization that corresponds to the European Committee for Electrotechnical Standardization (CENELEC) standards has to be built (EN 50126:1999; EN 50128:2001; EN 50129:2003). During the development period, the general and complex CENELEC process was stripped to an easier-to-handle specific process for generic developments. Thus, the development process and the product itself fulfill the requirements of EN 50126:1999, EN 50128:2001, and EN 50129:2003.

According to the requirements of the safety system, a real-time embedded safe operation system (ES-OS) has been designed for the VC system. The ES-OS and

**Fig. 4** Picture of the hardware

application program are non-volatile stored in compact flashcards. Power-on programs will be transferred to synchronous dynamic random access memory (SDRAM) and executed from there. Execution of the ES-OS begins with short self-tests, and cyclic longtime testing is proceeded by the ES-OS in background. The interface between the ES-OS and application via the application interface (API) and application works with cyclic proceeded main loops.

Hardware-abstraction-layer (HAL) communicates with hardware on chip and on board. HAL functions include connector localization in rack, address-switch, on board/chip universal asynchronous receiver/transmitters (UARTs), test-signal reference, bus arbitration, power supervision, watchdog supervision, pushing buttons, reset, light-emitting diode (LED) display.

The ES-OS functional modules communicate with HAL and API, serving as an interface between HAL and API. The functions of the ES-OS modules include general control module, communication between HAL and API, output of massages to display, scanning and control of pushing buttons, self-tests, exception handling, system functions.

API communicates with application, serving as an interface between ES-OS and application. Based on practical application, six kinds of API are designed in the ES-OS. EA-input-API announces the state of the inputs to the application. EA-output-API controls the outputs by the application layer and reading back the state of the outputs to the application layer. Man-machine-interface-API (MMI-API) displays messages and reacts of pushing buttons. External-communication-API (E-COM-API) sends and receives data via the buses and local area network (LAN), building and analyzing the safeguarded telegrams. Internal-communication-API (I-COM-API) communicates with the other channel. Error-handling-API (Errhler-API) announces errors to the application and exception handling. System API is for special non-safe system functions such as timers.

# 4 Safety Bus and Deterministic Communication Schedule

The safety communication protocol has been adopted to develop the quadruple safety bus. By adding three bytes safe cyclic redundancy check (CRC) and extending a complete byte, a total of four bytes are transmitted to ensure the safety of the message. CRC adopts the Hamming distance $h$ which equals 7 to ensure the safety of the 24-byte data. In this study, the bit false rate (BFR), which is usually set to the value of $10^{-9}$ in good transmission equipment, is $10^{-4}$. The maximum transmission rate of the applied communication equipment is 500 frames/s. Each packet of the transmission frame format contains all the relevant safety performance. After the camouflage identification, the packets will be rejected and the message will be resent.

When the Hamming distance equals 7, 6-bit camouflage data can be identified. If 7-bit or more camouflage data emerges, there will be risks. Since the probability of over 7-bit camouflage data appearing is far lower than 7-bit camouflage data whose frame format is 216-bit, it can be neglected. Supposing that the number of bit is $n$, the binomial distribution would be: In a message frame format, the probability of the existence of $k$-bit camouflage data is

$$p(k) = \binom{n}{k} \times \text{BFR}^k \times (1 - \text{BFR})^{n-k}. \tag{5}$$

Because BFR <<1, and $p(k) \approx \binom{n}{k} \times \text{BFR}^k$. Based on BFR = $10^{-4}$, $n = 216$, and $k = 7$, the value of $p(7)$ can be obtained by $p(7) \approx \binom{216}{7} \times (10^{-4})^7 = 4 \times 10^{-16}$. In every 500 frames/s, the risk rate is shown as

$$\text{HR} = 4\times10^{-16}\times500\times3600\,\text{h}^{-1} = 7.2\times10^{-10}\,\text{h}^{-1}. \tag{6}$$

In the double-channel system, the risk rate when the packet data of channels A and B are camouflaged simultaneously is $5.2 \times 10^{-19}$ $\text{h}^{-1}$. Because of the low rate, it can be neglected during the actual calculation process, especially for the situation when only 20 addresses can be used for the system.

All the communications between modules and CPU rely on the safety communication bus. The HAQVC system adopts a fixed address coding technique. The first address of a set of modules ($n$) is a multiple of 4, $n = 4k$. The other three address codes of the group module are $4k + 1$, $4k + 2$, and $4k + 3$, and the following codes $4(k + 1)$, $4(k + 1) + 1$, $4(k + 1) + 2$, and $4(k + 1) + 3$ are for the next set of modules. The CPU modules of HAQVC occupy four communication buses. For the ATP system, the relationship among communication links of the modules is fixed, and the content of communication will change over time and circle. Although the communication among different VC modules occupies the bus, it will not take the CPU resources of other modules, because the irrelevant information will be blocked in the link layer of the bus interface chip. At a certain time, some emergency concerning driving happens, such as the track circuit's code sequence mutates, active balise information, and IO information on train safety status change. Even though the fixed communication slot has passed, the transmission of relevant messages will be through event-trigger communication so that the vital CPU module can take safety measures in time and not need to wait until the next cycle (IEC/PAS 62409:2005). The specific scheduling diagram is shown in Fig. 5.

The vital CPU module of HAQVC provides interactive information exchange between diagnostic message and synchronic information through the safety bus in the second and third time slots. In addition, through the inter-communication, vital CPU module receives real-time working status and calculation results of other CPU modules and thus guarantees the effectiveness and real-time attribute of safety output function based on the voting structure, false diagnosis and screening function, rapid regression function, and failure-oriented safety function. The control data can be obtained in the fourth time slot.

The certainty of communication scheduling of the whole system is based on precision clock synchronization. The measurement accuracy is shown in Fig. 6. The system, which combines hardware and software synchronization, achieves the precise synchronization among modules and ensures the safety of clock synchronization. Master clock sends synchronization-related messages within the first communication slot. The system supports the access of absolute clock of global positioning system (GPS) to realize global synchronization, and provides a guarantee for accident and error recording.

The device modules of the HAQVC system are connected with system bus through isolating communication interface modules so that the fault module can be cut off in case of failure and thus ensure the system bus safety when modules are disconnected from it.

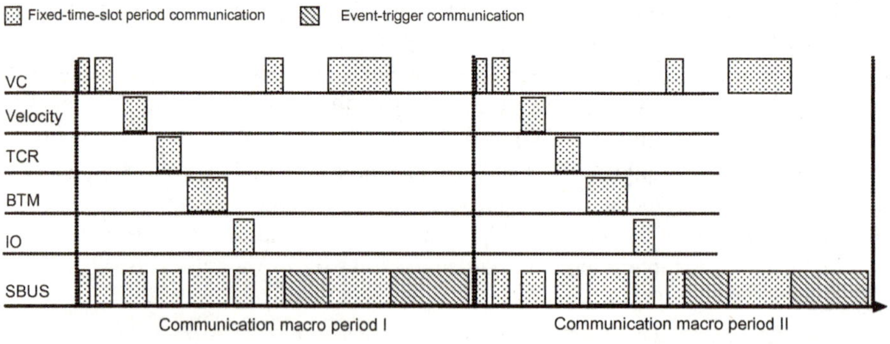


**Fig. 5** Communication scheduling sequence

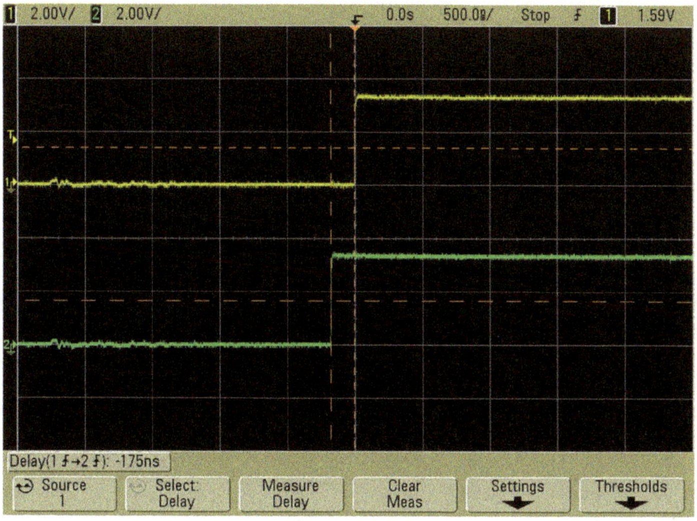


**Fig. 6** Diagram of accuracy of clock synchronization. *Green* and *yellow* represent pulses generated by any two different modules. The synchronization accuracy is represented by the time difference of the two pulse rising edge

## 5 System Modeling

The proposed HAQVC system structure is shown in Fig. 2. This system is comprised of four subsystems. As shown in Fig. 2, each CPU module receives voting data from four input modules and each output module receives voting data from four CPU modules. Thus, two CPU module, input module, or output module failures have no influence on the system. To start the modeling, we have adopted the following assumptions.

1. The system starts in the perfect operation when all of the system's modules are operating correctly.
2. Only one failure will occur at a time.
3. The error probabilities are the same for the same module of the system, which show symmetry.
4. Errors affecting different components of the same unit are statistically independent.

A Markov model of the HAQVC system is proposed in Fig. 7. The system is composed of 28 states. The PF state is a failure state and the rest state is operating. $\lambda_P$, $\lambda_{IN}$, and $\lambda_{OUT}$ are the failure rates of the CPU module, the input module, and the output module, respectively, and $r$ designates the system repair rate of the HAQVC system. For simplicity, we assume that the repair rate is a specific value for all states

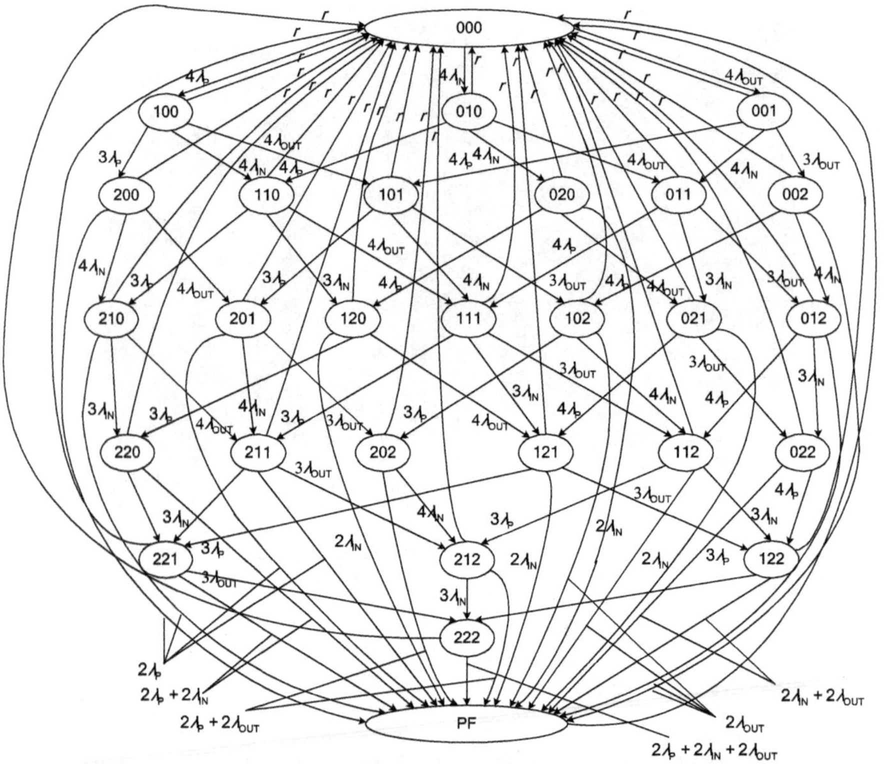


The number indicates the module state. From left to right, the first number is the CPU module state, the second one is the input module state, and the third is the output module state. For example, 200 means that only two CPU modules break down, 101 means that just one CPU module and one output module have failed, while the other modules are operating correctly.

**Fig. 7** Markov model of the high availability quadruple vital computer (HAQVC) system

of the system. The discrete system equation is highly complex, and thus, it is represented in a simple form as

$$\boldsymbol{P} = \begin{bmatrix} 1-\sum & S_{1,2} & \cdots & S_{1,27} & S_{1,28} \\ r & 1-\sum & \cdots & S_{2,27} & S_{2,28} \\ \vdots & \vdots & & \vdots & \vdots \\ r & S_{27,2} & \cdots & 1-\sum & S_{27,28} \\ r & S_{28,2} & \cdots & S_{28,27} & 1-\sum \end{bmatrix} \quad (7)$$

where $S_{i,j}$ is the state transition probability from state $i$ to state $j$, and $\Sigma$ is the state transition probability sum of row of the matrix.

## 6 Evaluation

In order to be sure that this design is meaningful, we have carried out experiments to compare the performance of the HAQVC system, the all voting triple modular redundancy (AVTMR) system (Kim et al. 2005), and the double 2-out-of-2 system (Wang et al. 2007). Keeping the failure rate and the repair rate unchanged, we draw the curves of reliability, availability, maintainability, and safety (RAMS) parameters of the three systems in the same figure as contrast using MATLAB.

### *6.1 Reliability*

As is well known, reliability is the ability of a system to perform its required functions under stated conditions for a specified period of time. As shown in Fig. 8a, initially, the HAQVC system is of the highest reliability until about 730,000 h, and from 730,000 h, the double 2-out-of-2 system is the same with the HAQVC system. Anyway, the AVTMR system is not of good reliability among the three systems for a lengthy time.

### *6.2 Availability*

Availability means the ability of a product to be in a state to perform a required function under given conditions at a given instant of time or over a given time interval assuming that the required external resources are provided. Availability of the HAQVC system, the AVTMR system, and the double 2-out-of-2 system is shown in Fig. 8b. For simplicity, the repair rate of each system is assumed to be 0.003 for simulation. The availability of each system is close to that shown in Fig. 8b. The HAQVC system has the highest availability, and the double 2-out-of-2 system is better than the AVTMR system.

### *6.3 Maintainability*

Maintainability is the probability that the failed system will be restored to an operational state within a specified period of time. As shown in Fig. 8c, the AVTMR and double 2-out-of-2 systems are of higher maintainability than HAQVC. Thus, this design can improve the availability of the system.

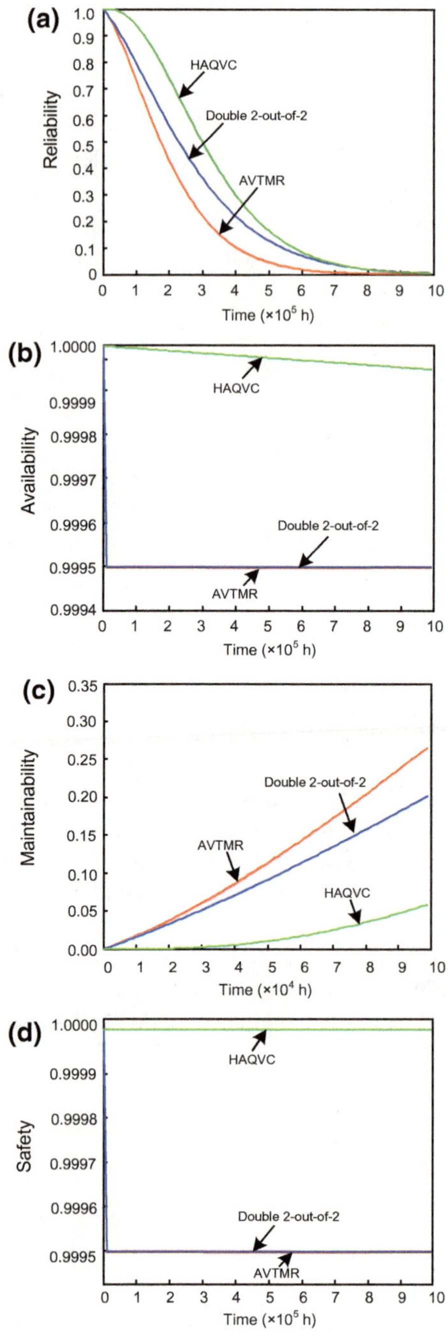


**Fig. 8** Reliability (**a**), availability (**b**), maintainability (**c**), and safety (**d**) of each system

### 6.4 *Safety*

Safety is the probability of systems where there is no dangerous failure. That is, safety is a state in which there is no danger. We take each system as a repairable system, and the repair rate is 0.003. As shown in Fig. 8d, the HAQVC system has the highest safety.

## 7 Conclusions

In this paper, based on the analysis of the architecture of the traditional double 2-out-of-2 system and the 2-out-of-3 system, we propose the HAQVC system, which is a novel fault-tolerant system with fire-new redundancy structure, and can significantly improve the reliability and safety. Its working process has been described and compared with the AVTMR system and the double 2-out-of-2 system in RAMS. Simulation results indicate that the HAQVC system has the best characteristic in RAMS and it is a better VC platform for a railway signal system.

## References

Dou, F. S., Cao, Z., Luo, L., & Long, Z. Q. (2007). Design and realization of safety computer systems based on double 2-vote-2 redundancy. Chinese Control Decision Conference, Wuxi, China (pp. 1059–1061, 1066).

EN 50126:1999. *Railway applications—The specification and demonstration of Reliability, Availability, Maintainability and Safety (RAMS)*. European Committee for Electrotechnical Standardization.

EN 50128:2001. *Railway applications—Communication, Signaling and Processing Systems-Software for Railway Control and Protection Systems*. European Committee for Electrotechnical Standardization.

EN 50129:2003. *Railway applications—Communication, Signaling and Processing Systems-Safety Related Electronic Systems for Signaling*. European Committee for Electrotechnical Standardization.

IEC 61508-2:2000. *Functional Safety of Electrical/Electronic/Programmable Electronic Safety-Related Systems—Part 2: Requirements for Electrical/Electronic/Programmable Electronic Safety-Related Systems*. International Electrotechnical Commission.

IEC 61508-6:2000. *Functional Safety of Electrical/Electronic/ Programmable Electronic Safety-Related Systems—Part 6: Guidelines on the Application of IEC 61508-2 and IEC 61508-3*. International Electrotechnical Commission.

IEC/PAS 62409:2005. *Real-time Ethernet for Plant Automation (EPA)*. International Electro Technical Commission.

Kim, H., Jeon, H. J., Lee, K., & Lee, H. (2002). The design and evaluation of all voting triple modular redundancy system. *Annual Reliability and Maintainability Symposium* (pp. 439–444). doi:10.1109/RAMS.2002.981682.

Kim, H., Lee, H., & Lee, K. (2005). The design and analysis of AVTMR (all voting triple modular redundancy) and dual-duplex system. *Reliability Engineering and System Safety, 88*(3), 291–300. doi:10.1016/j.ress.2004.08.012.

Paul, C. R. (2006). *Introduction to electromagnetic compatibility* (2nd ed.). Inc., Hoboken, NJ, USA. doi:10.1002/0471758159.Fmatter.

Qin, Q. N., Wei, X. Y., Yu, R. R., Han, L. (2010). Simplified design of embedded double 2-vote-2 computer system. In *3rd International Symposium on Test Automation and Instrumentation*, Xiamen, China (pp. 233–236).

Wang, S., Ji, Y. D., Dong, W., & Yang, S. Y. (2007). Design and RAMS analysis of a fault-tolerant computer control system. *Tsinghua Science and Technology, 12*(S1), 116–121. doi:10.1016/S1007-0214(07)70095-0.

## Author Biography

**Ping Tan** obtained his Ph.D. degree from the College of Control Science and Engineering at Zhejiang University, Hangzhou, China, in 2014. He has done research and development work on control systems in the SUPCON and INSIGMA Groups for more than ten years. He has been selected as one of the 151 Engineering Training Talents of Zhejiang province, one of the 131 Middle-aged and Young talents of Hangzhou. He is the specialist of the Hangzhou Industrial and Information Technology Commission. He has won the first prize of Zhejiang science and Technology Progress Award in 2006. He is the Senior Engineer of the Zhejiang University of Science and Technology. His research interests include distributed control systems, industrial Ethernet, fieldbus, vital computer and signaling systems for the rail transit system.

# Design and Analysis of the Hybrid Excitation Rail Eddy Brake System of High-Speed Trains

**Ji-en Ma, Bin Zhang, Xiao-yan Huang, You-tong Fang and Wen-ping Cao**

## Contents



## 1 Introduction

The braking system, which is one of the most important technologies of the high-speed trains, is very crucial to operation reliance and safety. The braking methods are mainly the regenerative braking system and disc braking system,

J. Ma · X. Huang · Y. Fang
College of Electrical Engineering, Zhejiang University, Hangzhou 310027, China

B. Zhang (✉)
State Key Laboratory of Fluid Power and Mechatronic Systems, Zhejiang University, Hangzhou 310027, China
e-mail: zbzju@163.com

W. Cao
Newcastle University, Newcastle upon Tyne NE1 7RU, UK



Y. Fang and Y.H. Zhang (eds.), *China's High-Speed Rail Technology*,
Advances in High-speed Rail Technology, https://doi.org/10.1007/978-981-10-5610-9_26

which are widely used in high-speed trains under 300 km/h (Zhang 2009). However, the velocities of present high-speed trains are over 350 km/h. Thus, it is very important to explore new braking methods that can be applied to these high-speed trains. The eddy brake can be used when the velocity of the train velocities is very high. It cannot only reduce energy consumption, but it also produces economic benefits and improves technical features (Dietrich et al. 2001).

The existing eddy brake system has some limitations. First of all, down pass system through pressure relief can ensure that the air cylinder acts when the air system is compromised. However, when the electric circuit is compromised, the loss of excitation can lead to loss of braking force. As a result, ICE3 (Inter City Express 3) installs a lot of batteries as a back-up electrical source, which adds to the weight of the train (Bottauscio et al. 2006; Guo et al. 2006). Second, if the electrical sources are switched off when the train stops, the eddy brake system cannot operate. In addition, the power of excitation and heat is too large (Graber 2003; Kunz 2005). This paper deals with a hybrid excitation rail eddy brake system, which can help to deal with the limitations of steering malfunction and parking brake.

## 2 Theory of Eddy Brake System

The structure of the hybrid excitation rail eddy brake system is shown in Fig. 1. This system is mainly composed of an airlift system, braking magnetic system, and braking auxiliaries. When the train is traveling at a high speed, the hybrid excitation rail eddy brake system causes an eddy brake through the excitation of the permanent magnet. At the same time, the exciting electricity in the exciting coils can be modified in order to assist the positive motivation, which can ensure that the train's

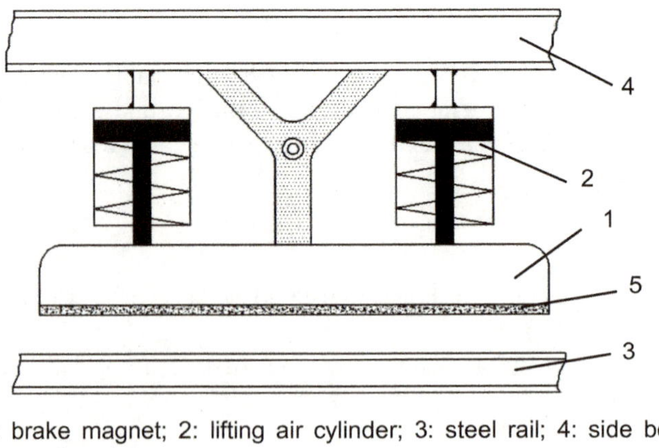


1: brake magnet; 2: lifting air cylinder; 3: steel rail; 4: side beam; 5: wearing plate

**Fig. 1** Structure of the hybrid excitation rail eddy brake system

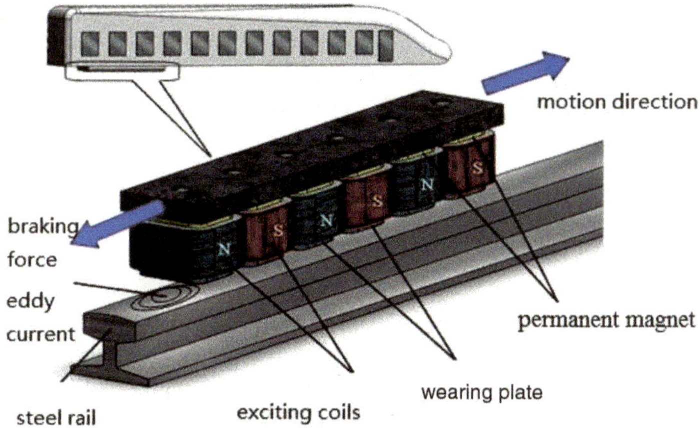


**Fig. 2** Principle diagram of the electronic magnetic system of the brake system

braking process can be dynamically adjusted. When the train's speed reaches the threshold, which means that the condition of the friction braking is satisfied, the electronic magnetic system will contact the rail. The exciting electricity is reduced or it disappears. The brake system presses the wearing plate on the rail through the attractive force of the permanent magnets. The energy of motion is changed into the energy of heat through the friction force.

The braking magnetic system includes the permanent magnets and the exciting coils (Fig. 2). When the high-speed train is braked, it produces the eddy brake through the permanent magnets. At the same time, we can input the auxiliary exciting electricity in the exciting coils so as to produce the positive excitation, and at the same time to keep the same negative acceleration, which allows the high-speed train to be controlled dynamically. When the train's speed reaches the threshold, which means that the condition of friction braking is satisfied, the auxiliary electricity in the exciting coils will be closed. Meanwhile, the attractive force of the permanent magnets and the wearing plates can help to produce the friction brake and then realize the energy-saving effect in the braking process. When the brake system resumes to the relief state, the auxiliary electricity in the exciting coils produces a negative excitation, which can offset the attractive force of the permanent magnets. As a result, we can lift the brake magnetic system through only a small force, and the relief status can resume more quickly.

The force of the electronic magnetic system is

$$F_k - mg - PS - F_{\mathrm{A}} - f = 0, \tag{1}$$

where $F_k$ is the elastic force of the built-in springs, $mg$ is the gravity of the magnetic system, $P$ is the total pressure of the pistons, $S$ is the cross-sectional area of the pistons, $F_{\mathrm{A}}$ is the attractive of the magnetic system, and $f$ is the friction of the pistons.

When the brake system is relieved, the electricity in the exciting coils produces reverse motivation, which can offset the attractive force of the permanent magnets. As a result, the lift system can use a very small force to hoist the magnetic system, which can help the recovery of the brake relief process. The two subsystems of the brake system are shown in detail.

## *2.1 Lift System of the Brake System*

The hybrid excitation rail eddy brake system controls the ups and downs of the brake exciting system through the air cylinder, which makes the brake exciting system and the wearing plate reach or keep off the rail, and thus, the working status of the brake exciting system is determined.

### 2.1.1 Status of Relief

When the brake system is relieved, the control valve is squeezed so that the up cavity of the lift air cylinder can be connected with the air. The lift air cylinder is hoisted via the built-in springs in order that the brake exciting system can maintain a certain distance off the rail. Figure 3 shows the lift system of relief.

### 2.1.2 Status of Brake

When the brake system is broken, the left cavity of the control valve releases the pressure, and the control valve moves left through the built-in springs. The up

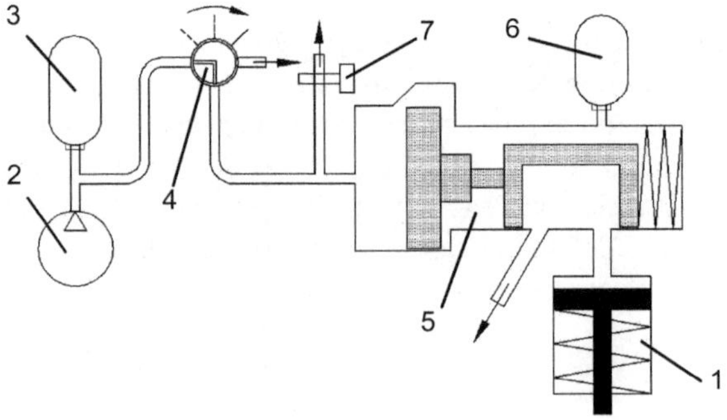


1: lifting air cylinder; 2: air compressor; 3: main air cylinder; 4: brake valve; 5: control valve; 6: auxiliary air cylinder; 7: urgent control valve

**Fig. 3** Lift system of relief

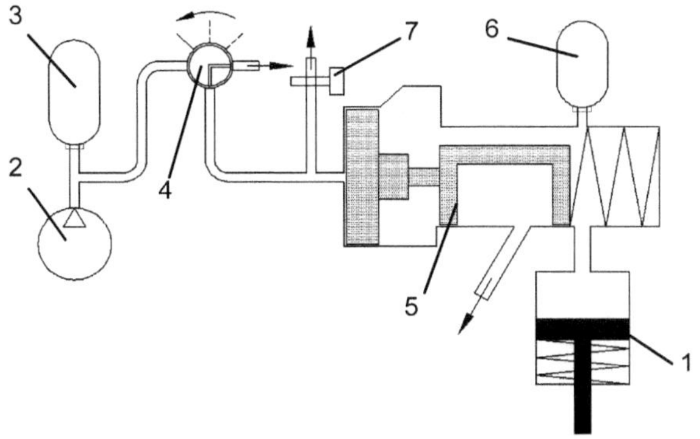


1: lifting air cylinder; 2: air compressor; 3: main air cylinder; 4: brake valve; 5: control valve; 6: auxiliary air cylinder; 7: urgent control valve

**Fig. 4** Lift system of brake

**Fig. 5** Brake exciting system

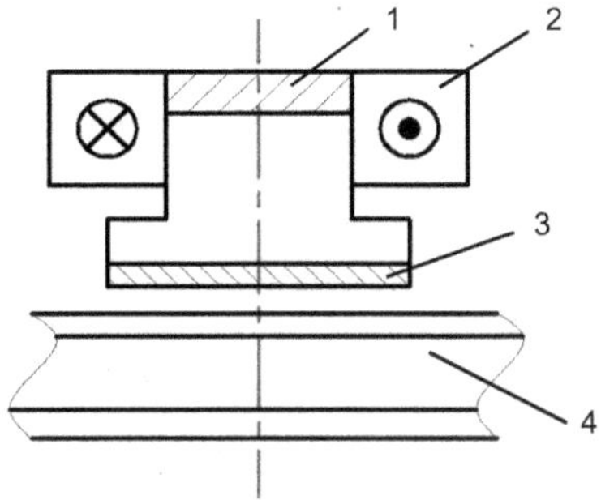


1: permanent magnet; 2: exciting coils; 3: wearing plate; 4: steel rail

cavity of the lift air cylinder is connected with the auxiliary air cylinder and moves down via the pressure of the air in the auxiliary air cylinder, which brings the brake exciting system close to the rail (Fig. 4).

## 2.2 *Brake Exciting System*

The magnetic pole of the brake system includes the permanent magnet and the exciting coils, whose structure is indicated in Fig. 5.

In principle, the hybrid excitation rail eddy brake system can be regarded as a kind of special induction linear motor whose magnetic field is non-sin and its primary is short. This motor includes the primary yoke, the primary exciting coils, the permanent magnets, the wearing plates, and the secondary steel rail. The

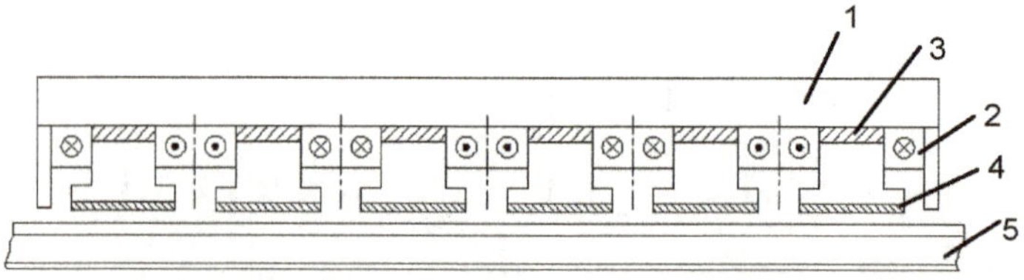


**Fig. 6** Whole diagram of the brake magnetic system

**Table 1** Parameters of hybrid excitation rail eddy brake system

| Parameter | Value |
|---|---|
| Pole number | 6 |
| Pole pitch (mm) | 45 |
| Measure of the permanent magnet (length × thickness) (mm) | 20 × 4 |
| Model of the permanent magnet | N35SH |
| Thickness of magnetic yoke (mm) | 32 |
| Length of the wearing plate (mm) | 35 |
| Thickness of the wearing plate (mm) | 3 |
| Turn number of primary exciting coils | 395 |

primary exciting coils use DC. Each of the permanent magnets is set between the magnetic pole and the magnetic yoke. The secondary is the steel rail. The whole diagram of the brake magnetic system is shown in Fig. 6.

The air magnetic field between the brake system and the steel rail is produced by both the permanent magnet and the exciting coils. When the train travels, the air magnetic field is mainly produced by the permanent magnet, and the DC exciting coils only supply a small part. As a result, it can be modified by the electricity, which helps to control the magnetic field and the braking force.

## 2.3 *Main Parameters of the Brake System*

In (Lu and Ye 2005), the structural design of the hybrid excitation linear motor provides a model of the eddy brake and simulates this model (Table 1).

In the hybrid excitation rail eddy brake system, the magnetic field that is produced by the permanent magnet is the main part of the air magnetic field. The air magnetic field is affected by both the eddy current magnetic field and the electrical exciting magnetic field. As a result, the largest degaussing working point must be paid attention to when the brake system is designed in order to prevent irreversible degaussing. Note that too strong electrical exciting magnetic field can lead to an over-saturation of the magnetic circuit (Graber 2003). The permanent magnet in this study is NdFeB of N35SH. If the working temperature is assumed to be 75 °C:

$$B_r = \left(1 + (t - 20)\frac{\alpha_{B_r}}{100}\right) B_{r20}, \tag{2}$$

where the residual flux density $B_r$ in the working temperature is 1.13 T. $B_{r20}$ is the residual flux density when the temperature is 20 °C, $t$ is the temperature, $\alpha_{Br}$ is the irreversible conversion coefficient of the residual flux density, and it is often $-0.0012\ K^{-1}$.

The coercive force of the permanent magnet in this temperature is often $H_c = 847{,}138$ A/m.

The relative permeability of the degaussing curve is

$$\mu = \frac{B_{r20}}{\mu_0 H_{c20}} = 1.062, \mu_0 = 4\pi \times 10^{-7}. \tag{3}$$

In the above model, the magnetization of the permanent magnet is along the $y$ axis, and its length is 4 mm. As a result, the calculating magneto motive force is

$$F_c = H_c h_{pm} = 3388.552\ \text{A}. \tag{4}$$

where $F_c$ is the magneto motive force, and $h_{pm}$ is the length of the permanent magnet along the $y$ axis.

Because too large exciting electricity can lead to the over-saturation of the magnetic field in the motor, the positive exciting electricity should not be too large. If the exciting electricity is reversed, the electrical exciting can cause degaussing of the permanent magnet. As a result, the absolute value of the electrical exciting magneto motive force should be smaller than the calculating magneto motive force of the permanent magnet. According to calculations, when the exciting electricity is lower than 4.2 A, the magneto motive force produced by the exciting coils is always smaller than the calculated magneto motive force of the permanent magnet. As a result, the electrical exciting coils will not generate irreversible degaussing.

# 3 Simulation

## *3.1 Creating Finite Element Method (FEM) Model*

The hypotheses are as follows (Gay and Ehsani 2006):

1. In the studied electronic magnetic field, the magnetic field only has two elements, one is in the $x$ direction, and the other is in the $y$ direction. In addition,

both the vectors of the electricity density and magnetic potential only have element in the $z$ direction.

2. Neglect the effect of the temperature on the iron's conductivity; that is to say, its conductivity is equal to zero.
3. Neglect the magnetic field outside the motor's shell. As a result, the outside surface of both the primary and secondary can be regarded as the equimagnetic potential surface with the zero vector.
4. The electricity density inside the conductors is uniformly distributed, and there is no free charge inside the conductors.

According to the abovementioned hypotheses and the brake system model, we can create a 2D FEM model using the ANSYS software (Fig. 7).

In this model, the brake system has no relative motion to the steel rail, and the gap between them is 2 mm. In addition, there is no electricity in the exciting coils. We can change this model into grids (Fig. 8).

When we have loaded the model, we can use the solution to obtain the magnetic field intensity distribution graph, and then use the postprocessor to gain the magnetic induction intensity, which is indicated in Fig. 9. In this graph, the magnetic induction intensity is larger where the color is deeper.

In the same way, we can obtain the distribution graph of the magnetic lines of force (Fig. 10).

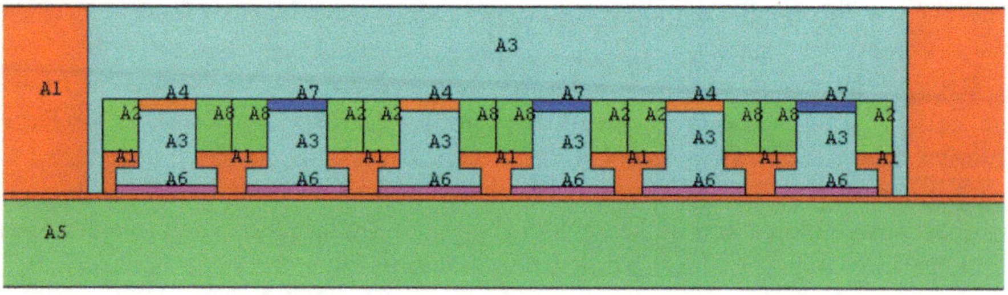


**Fig. 7** FEM analyzing model of the brake system

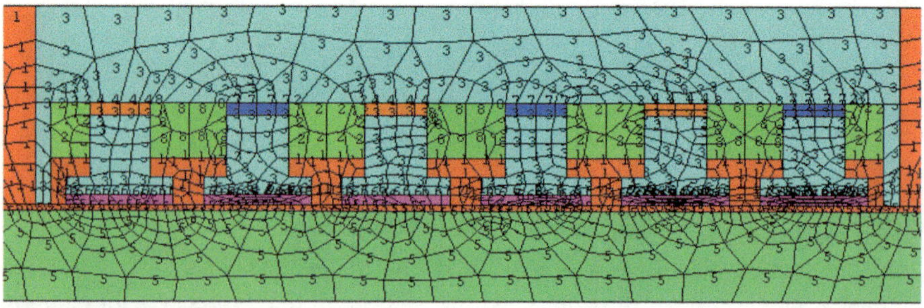

**Fig. 8** Grids of the FEM model

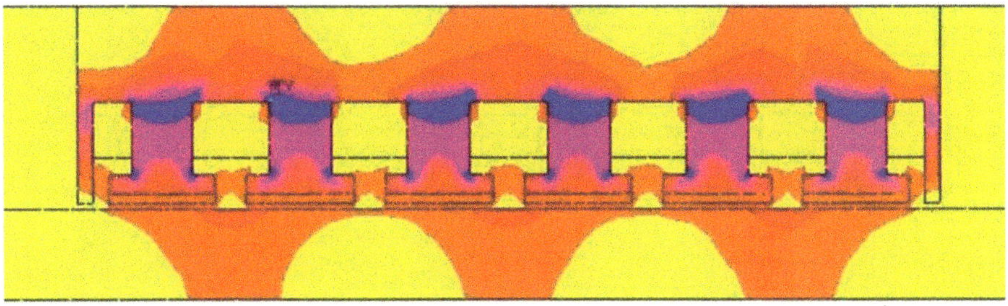

**Fig. 9** Magnetic induction intensity diagram when the speed is zero and the gap is 2 mm

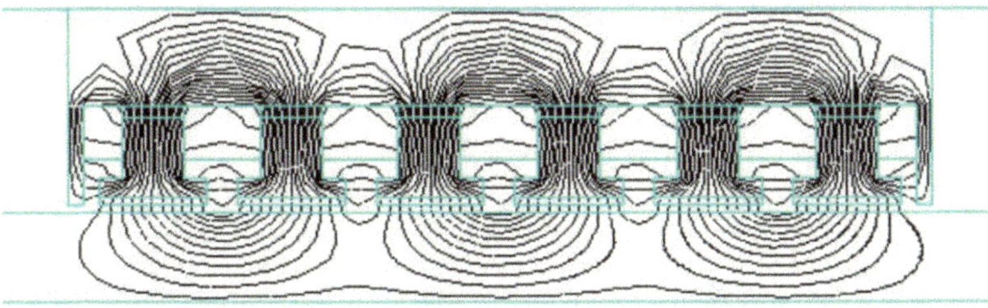

**Fig. 10** Distribution of the magnetic lines of force when the speed is zero and the gap is 2 mm

The abovementioned is on condition that the bake system has no relative motion to the steel rail. However, what this paper deals with is the magnetic field when the brake system is static, but the train is traveling at varying speed. The direction of the speed is horizontal-right and the gap is 2 mm.

From Fig. 11, we can safely come to the conclusion that the faster the speed is, the more seriously the gap magnetic field is distorted.

## 3.2 *Effects of Gap*

The factors such as air gap and exciting electricity both can affect the braking force of the brake system when independent excitation is used (Tang and Ye 2006; Cai et al. 2007). To analyze the function of the hybrid excitation rail eddy brake system, one constructs the model, and then analyzes it for both, when there is exciting and when there is no exciting.

### 3.2.1 When There is no Exciting

When there is no exciting, the hybrid excitation rail eddy brake system is changed into the eddy brake system when only the permanent magnet can excite. Simulating

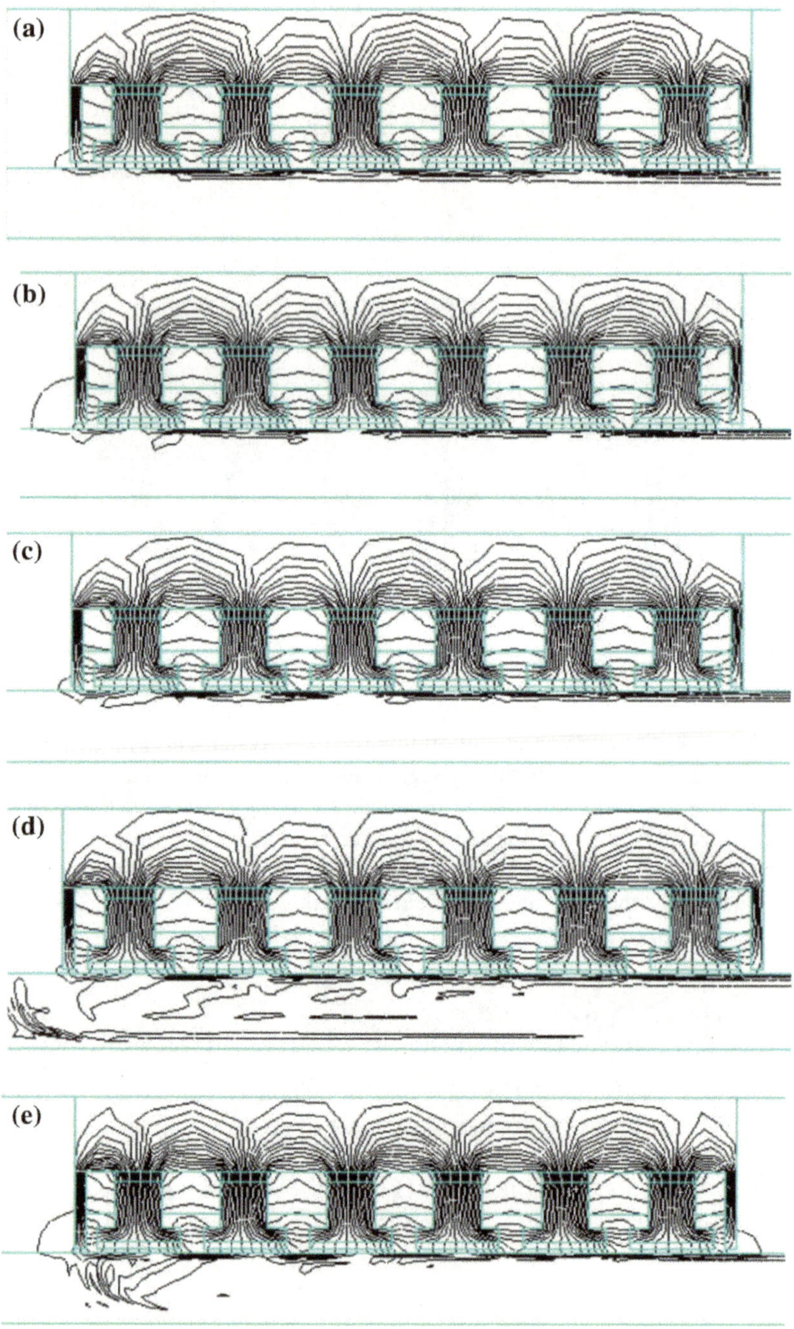


**Fig. 11** Magnetic lines of force when the speed is: **a** 100 km/h; **b** 200 km/h; **c** 300 km/h; **d** 400 km/h; and **e** 500 km/h

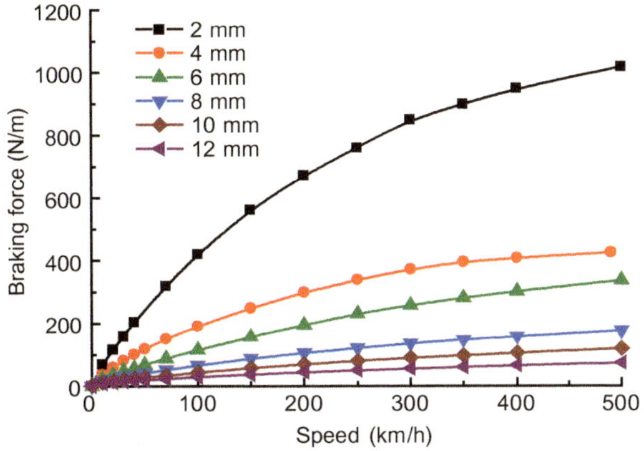


**Fig. 12** Effects of different air gaps on the braking force

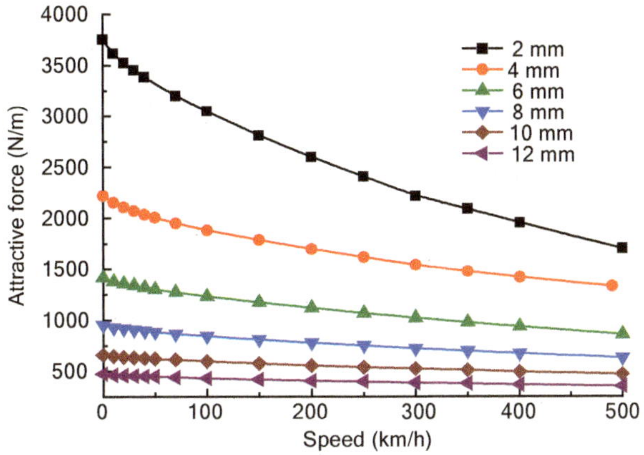


**Fig. 13** Effects of different air gaps on the attractive force

the eddy brake system when the value of the air gap varies, one can obtain the curve of the braking force with the velocity (Fig. 12). We can see that when the air gap is smaller, the braking force is larger, and the variation of the braking force is larger. However, when the velocity magnifies, this variation is smaller, which shows that the brake system can supply a steady force whose variation is rather small.

Because the attractive force has an effect on the lift system, the simulation can obtain the relation of the attractive force and the velocity (Fig. 13). When the air gap is thicker than 6 mm, both the braking force and the attractive force have little variation when the velocity varies. When the air gap is thicker, this kind of variation is smaller and the force is constant.

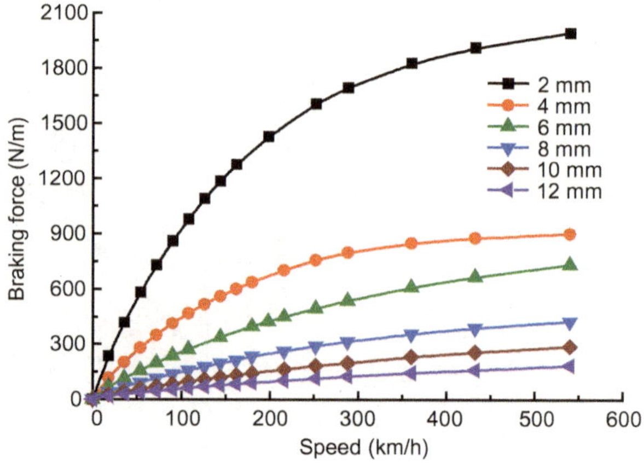


**Fig. 14** Effects of different air gaps on the braking force in hybrid exciting

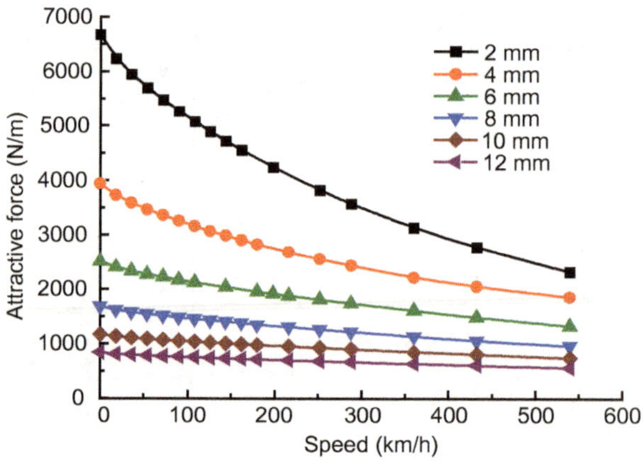


**Fig. 15** Effects of different air gaps on the attractive force in hybrid exciting

### 3.2.2 When There is Exciting

The DC exciting coils of the eddy brake system have 395 turns. When the electricity is 2 A, the relationship between the braking force and the velocity with different air gaps is as shown in Fig. 14.

In the case of hybrid exciting, the variation range of the braking force is much larger than that when only the permanent magnets excite, which leads to the braking force being raised so that the brake distance can be reduced.

Of course, the variation range of the attractive force in hybrid exciting also magnifies a lot, which makes it easier for the lift system to raise or to put down the brake system more easily. In addition, the braking force or the attractive force can be easily controlled through the regulation of the exiting electricity, such that the braking force is dynamically controlled and the braking force is kept constant. Figure 15 shows the relationship of the attractive force and the velocity in hybrid exciting.

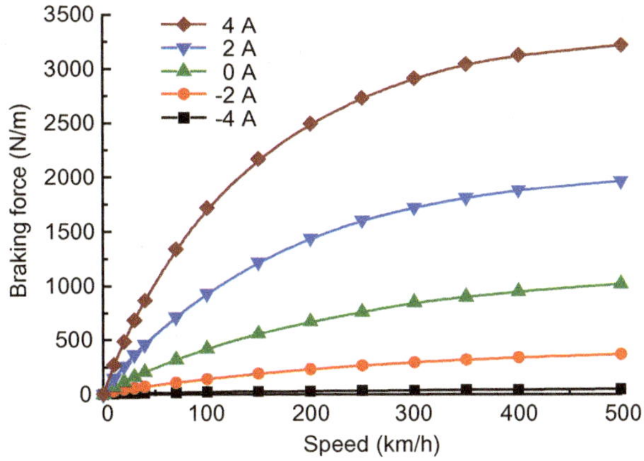


**Fig. 16** Effects of different electricity on the braking force with the gap of 2 mm

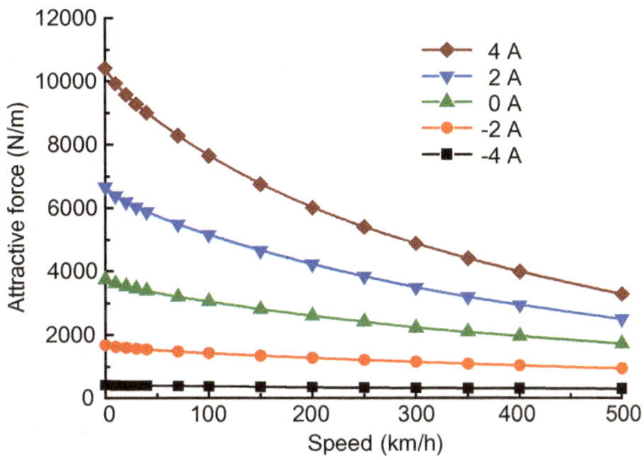


**Fig. 17** Effects of different electricity on the attractive force with the gap of 2 mm

## *3.3 Effects of Electricity*

We modify the exciting electricity so that we can control the value of the attractive force or braking force, and dynamically control the braking force or keep it constant. When the exciting electricity is positive, it will enhance the magnetic field; but when it is negative, it will decrease the magnetic field, so that we can lift the electronic magnetic system with a smaller force. As a result, we can examine how the braking force and attractive force of the brake system vary with the speed in different electricity. We suppose that the gap is 2 mm.

From Fig. 16, we can see that the smaller the electricity, the smaller the braking force. When the electricity is negative, the braking force reduces very quickly at the same speed, while the varying amplitude of the braking force is smaller and smaller when at different speeds. The reason is that when the electricity is negative, it will produce a negative magnetic field that is opposite to the permanent magnetic field.

As a result, the total gap magnetic field decreases greatly, and then the braking force reduces very quickly. Likewise, the attractive force is smaller when the electricity is smaller.

From Fig. 17, we can see that when the electricity is negative, the attractive force decreases quite quickly. When the electricity is −4 A, it does not bring irreversible demagnetization of the permanent magnets, and it reduces the magnetic field, decreasing the attractive force considerably. As a result, the air lifting system uses little force to lift the magnetic system through the built-in springs.

From the above analyses, we conclude that both the gap and the electricity have obvious effects on the performance of the braking system and that the smaller the gap, the larger the braking force and the attractive force. In addition, the larger the electricity, the bigger the braking force and attractive force. When the speed increases, the braking force also increases, but the attractive force decreases. The larger the speed, the smaller the variations of the attractive force and baking force.

## 4 Optimization

The braking force is related to the magnetic induction density and its distribution. As a result, to choose the rational magnetic materials and to optimize the structure of the magnetic road and the shape of the permanent magnets both can improve the energy density of the braking system, such that the braking performances can be dynamically modified.

Form the above simulation analyses, we can see that the magnetic road of the original magnetic system needs further optimization. Because the magnetic leakage

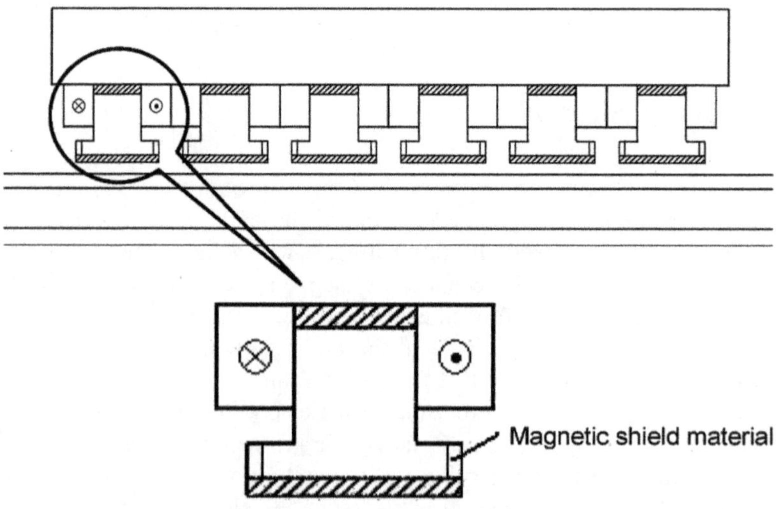


**Fig. 18** Structure of the brake system after being optimized

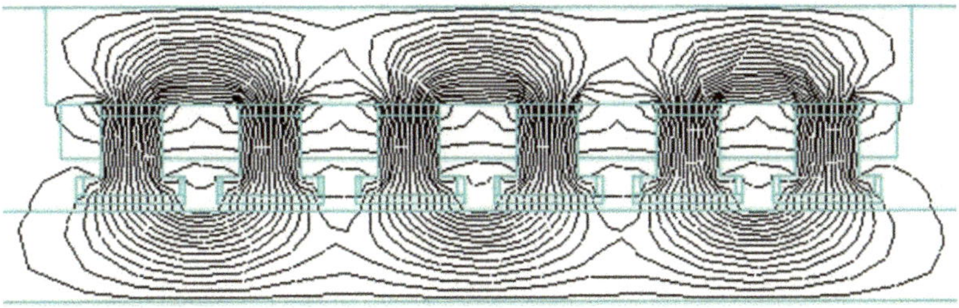

**Fig. 19** Magnetic force lines distribution of the brake system after optimization

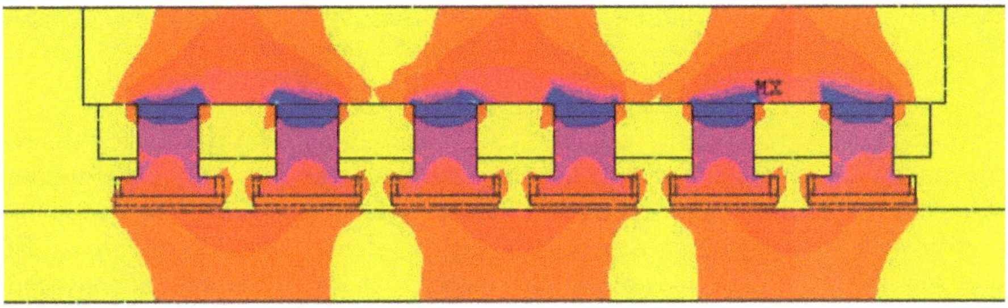

**Fig. 20** Magnetic induction density of the brake system after optimization

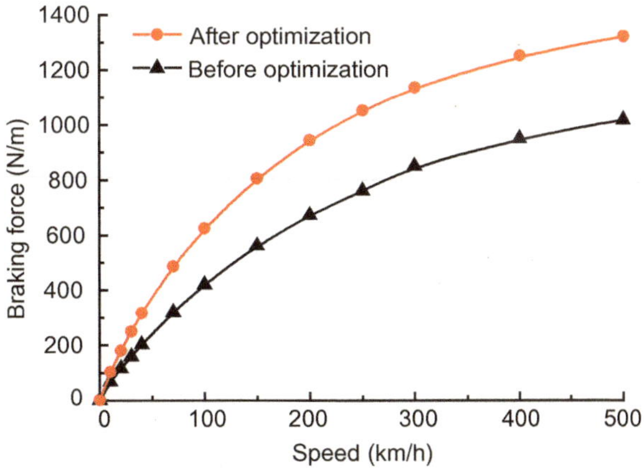


**Fig. 21** Comparison between the braking force after and before optimization

in the teeth of the magnetic poles is quite large, and that in the partition besides the two ends of the brake system is also large, the exciting magnetic field is coupled with the rail eddy current magnetic field. We try to add a magnetic shield material and remove the partitions besides the two ends of the brake system, the optimized structure of the brake system is shown in Fig. 18.

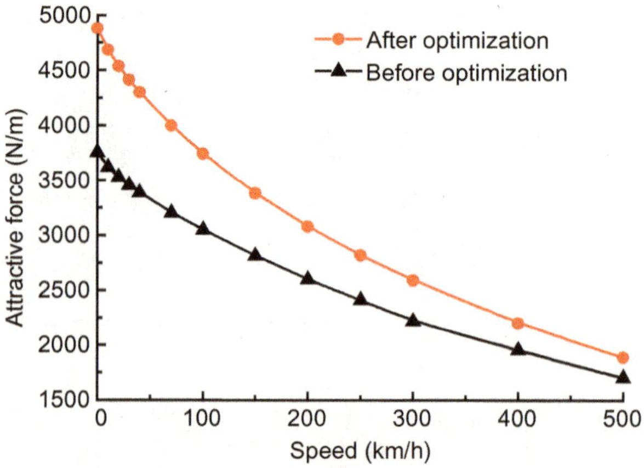


**Fig. 22** Comparison between the attractive force after and before optimization

When there is no electricity in the exciting coils and the gap is 2 mm, we simulate the brake system and then obtain the magnetic force lines distribution graph (Fig. 19). After optimization, the magnetic leakage is reduced, and at the same time the coupling between the magnetic system and the steel rail is also improved, such that the forces of the brake system and the steel rail are enhanced. As a result, it is beneficial to the braking process of the high-speed train.

Likewise, the magnetic induction density of the brake system changes is shown in Fig. 20.

From Figs. 19 and 20, we can conclude that, after structure optimization of the brake system, both of the two kinds of forces will be affected. Supposing that the gap is 2 mm, and there is no electricity, we simulate the brake system.

It is obvious that the optimization of the structure improves the gap magnetic field between the brake system and the steel rail; that is to say, the coupling of the exciting magnetic field with the eddy current magnetic field is enhanced, which leads to the addition of the braking force. As a result, we can say that the optimization of the magnetic road reaches its expected effects. Likewise, we can compare the variation of the attractive force. From Figs. 21 and 22, we can see that the optimization of the magnetic road also reaches its expected effects.

## 5 Conclusions

The hybrid excitation rail eddy brake system is a kind of non-adhesion brake system. As a result, it is not limited by the adhesion conditions and compared to the adhesion system. It has the advantages such as non-abrasion and better brake effects. In addition, compared to the single exciting eddy brake system, on one hand, it can be controlled well, because it can modify the exciting current to control the variation of the braking force. On the other hand, it does not consume a lot of electricity energy.

This system not only realizes the safety of the malfunction steering through the airlift system, but also uses the exciting electricity and reduces the attractive force of the permanent magnet to make the wearing plate touch the steel rail easily. As a result, it integrates the advantages of both the rail eddy brake and the magnetic rail brake. It not only saves a lot of energy, but also reduces the loss of the motive energy.

## References


Bottauscio, O., Chiampi, M., & Manzin, A. (2006). Element-free Galerkin method in eddy current problems with ferromagnetic media. *IEEE Transactions on Magnetics, 42*(5), 1577–1584. doi:10.1109/TMAG.2005.863932.

Cai, J. L., Liu, Z., & Zhang, Z. C. (2007). Analysis and design of electrical magnetic eddy brake. *Electromechanical Engineering, 28*(8), 84–86. (in Chinese).

Dietrich, A. B., Chabu, I. E., & Cardoso, J. R. (2001). Eddy-current brake analysis using analytic and FEM calculations. *IEEE Transaction on Magnetics, 37*(5), 454–461.

Gay, S. E., & Ehsani, M. (2006). Parametric analysis of eddy current brake performance by 3-D finite element analysis. *IEEE Transactions on Magnetics, 42*(2), 319–328. doi:10.1109/TMAG.2005.860782.

Graber, J. (2003). The linear eddy current brake system of ICE3 trains. *Technology of Trains Abroad, 5,* 1–6. (in Chinese).

Guo, Q. Y., Hu, J. T., & Hu, X. Y. (2006). Features research of linear eddy brake in high-speed trains. *Journal of Tongji University (Natural Science), 34*(6), 804–807. (in Chinese).

Kunz, M. (2005). The linear eddy brake of ICE3-technical views and operating experience. *Current Transformer and Electrical Haul, 2,* 4–8. (in Chinese).

Lu, Q. F., & Ye, Y. Y. (2005). Magnetic field and thrust force of the hybrid exciting linear synchronous machine. *China Society for Electrical Engineering, 25*(10), 127–130. (in Chinese).

Tang, Y. C., & Ye, Y. Y. (2006). FEM analysis and design of permanent magnetic eddy brake. *Micromotor, 36*(3), 34–36. (in Chinese).

Zhang, S. G. (2009). *The Research of the Design Methods of High-Speed Trains*. Beijing (in Chinese): China Rail Publishing House.


### Author Biographies

**Ji-en Ma** PhD, is an associate professor in Zhejiang University, Hangzhou, China. She received a PhD degree in mechatronics from Zhejiang University in 2009. Then she did postdoctoral work in the College of Electrical Engineering, Zhejiang University, China. Her recent work is on electrical machines and drives. Her research interests include PM machines and drives for traction applications, and mechatronics machines such as the magneto fluid bearing.

**Bin Zhang** PhD, is a Research Associate in Mechanical Engineering at Zhejiang University, Hangzhou, China. He received a Doctor degree from Zhejiang University in 2009 in mechanical engineering. From 2015 to 2016, His research interests are Intelligent machine and Biomanufacture.

**You-tong Fang** is a professor at Zhejiang University. He is chairman of the High Speed Rail Research Center in Zhejiang University, deputy director of the National Intelligent Train Research Center, on the committee of the China High Speed Rail Innovation Plan, and expert of the National 863 Program (China High Technology Development Program) in modern transportation and advanced carrying technology. He is also director of three projects of the National Natural Science Foundation of China (NSFC) and more than ten National 863 projects and Technology Support Programs. Recent work has been on electrical machines and drives. His research interests include permanent magnet (PM) machines and drives for traction applications.

**Wen-ping Cao** PhD, is the Chair Professor in Electrical Power Engineering and the Head of Power Electronics, Machines and Power System Group at Aston University, UK. In 2015 he was a Marie Curie Fellow at the Massachusetts Institute of Technology, MA, USA. Prof. Cao is presently a Royal Society Wolfson Research Merit Award holder in the UK. He was a semi-finalist at the Annual MIT-China Innovation and Entrepreneurship Forum (MIT-CHIEF) Business Plan Contest, USA, in 2015, the Dragon's Den Competition Award winner from Queen's University Belfast, UK, in 2014, and the Innovator of the Year Award winner from Newcastle University, UK, in 2013.

# Simulation Software for CRH2 and CRH3 Traction Driver Systems Based on SIMULINK and VC

**Qin-fen Lu, Bin Wang, Xiao-yan Huang, Ji-en Ma, You-tong Fang, Jin Yu and Wen-ping Cao**

**Contents**



## 1 Introduction

In recent years, the construction of high-speed trains has developed very quickly in China, especially the China Railway High-Speed 2 (CRH2) and China Railway High-Speed 3 (CRH3) models. Along with this development, the basic theories relative to CRH have been given more attention. Research on traction driver systems is particularly important because such systems transform the energy between the electric network and the train. They should not only supply enough energy to the train but also have high efficiency and high quality. To improve their perfor-

Q. Lu · B. Wang · X. Huang (✉) · J. Ma · Y. Fang
College of Electrical Engineering, Zhejiang University, Hangzhou 310027, China
e-mail: eezxh@zju.edu.cn

J. Yu
National Engineering Laboratory for System Integration of High-Speed Train (South), CSR Qingdao Sifang Co., Ltd, Qingdao 266111, China

W. Cao
Newcastle University, Newcastle upon Type NE1 7RU, UK



Y. Fang and Y.H. Zhang (eds.), *China's High-Speed Rail Technology*,
Advances in High-speed Rail Technology, https://doi.org/10.1007/978-981-10-5610-9_27

mance, detailed simulation results are required. Thus, a suitable simulation method needs to be investigated (Song 2009).

Normally, a simulation model of a traction driver system is developed using SIMULINK software. Due to strong computer power and the abundant components of SIMULINK, a model is easily established. But this kind of model cannot run without SIMULINK and also lacks a friendly interface to input parameters and output results. Moreover, users need to master the SIMULINK software and know the model well if they want to use it. These disadvantages limit its wide application. To overcome these disadvantages, a combined programming method using SIMULINK and VC++ is adopted because VC++ is popular object oriented software. A strong simulation program with a friendly interface can then be obtained. Moreover, it can run on any computer and without SIMULINK (Zhang and Wang 2008).

## 2 Simulation Model

The CRH2 and CRH3 are typical types of high-speed train and are becoming increasingly popular in China. Their traction driver systems consist of a traction transformer, a traction converter, and traction motors. The function of the traction transformer is to change the 25 kV from the electric network to 1500 V. This single-phase AC power is supplied to the traction converter, which adopts a popular AC-DC-AC transmission method. It includes a single-phase pulse rectifier, a traction inverter, and an intermediate DC link. Using this inverter, the 1500 V single-phase AC voltage is rectified to 2600 V DC voltage, and then the DC power is inverted to three-phase AC power with variable frequency and variable voltage, and supplied to four traction induction motors (Song 2009).

For the CRH2, the topology of both the single-phase pulse rectifier and the traction inverter is three-level, and has a controlled bridge for energy conversion (Celanovic 2001). For the CRH3, the topology of both the single-phase pulse rectifier and the traction inverter is two-level, with a controlled bridge. The controlled bridge operation principle can be illustrated based on their equivalent circuit (Carter et al. 1997; Lin and Lu 2000). For example, Fig. 1 shows the equivalent circuit of a single-phase pulse rectifier. Its voltage equation is

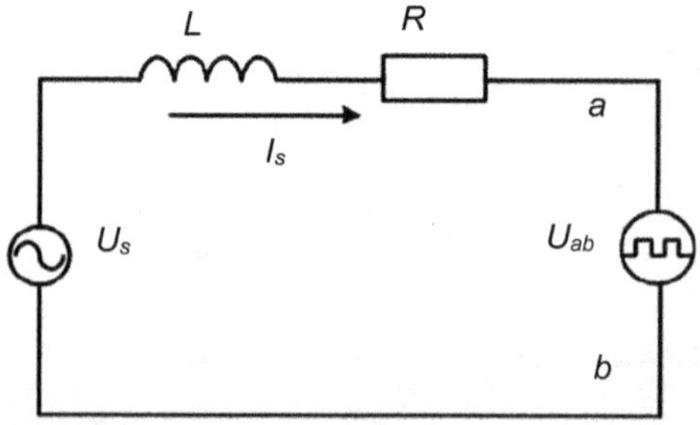


**Fig. 1** Equivalent circuit of a single-phase pulse rectifier

$$\dot{U}_s = \mathrm{j}\omega L\dot{I}_s + R\dot{I}_s + \dot{U}_{ab}, \quad (1)$$

where $\dot{U}_s$ is the input voltage, $\omega$ is the angular speed, $R$ is the equivalent resistance, $L$ is the equivalent inductance, $\dot{U}_{ab}$ is the output voltage of the single-phase pulse rectifier, and $\dot{I}_s$ is the input current.

The traction motor adopts indirect rotor magnetic field orientation, vector control strategy, and space vector pulse width modulation (SVPWM) method (Tolbert and Habetler 1999). The position of the rotor flux need not be measured or directly calculated, but can be indirectly deduced from the slip frequency. The estimated value is taken as the position angle of stator current coordinate transformation. Comparing the measured motor speed with the given speed, the speed regulator obtains the given torque component of the stator current. The given rotor flux is deduced from the given curve of the rotor flux and speed. According to this flux, the excitation component of the stator current is calculated. By comparing these two given components of stator current to the corresponding current measured by the proportional-integral (PI) regulator, the given d$q$ components of the voltage are obtained. These are then transformed in $\alpha$–$\beta$ coordinates with the estimated position of the rotor flux (Depenbrock 1988; Lascu et al. 2000). Finally, the two voltage components are supplied to the SVPWM unit and output three-phase variable voltage and variable frequency (VVVF) voltage.

Based on the CRH2 traction driver system, the simulation model is constructed based on SIMULINK (Fig. 2).

Based on the CRH3 traction driver system, the simulation model is constructed based on SIMULINK (Fig. 3). Compared to CRH2, there are two secondary

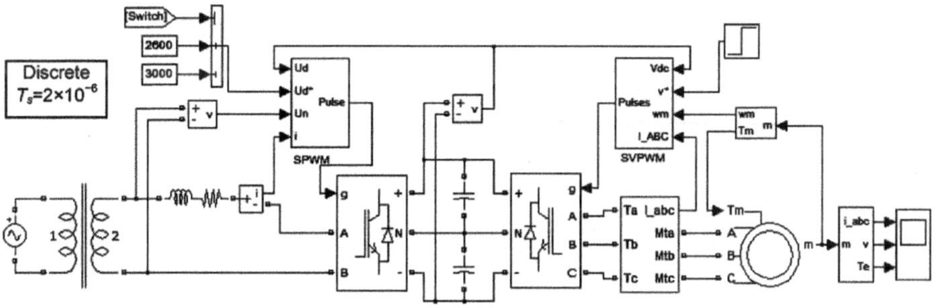


**Fig. 2** Simulated model of the traction driver system of CRH2

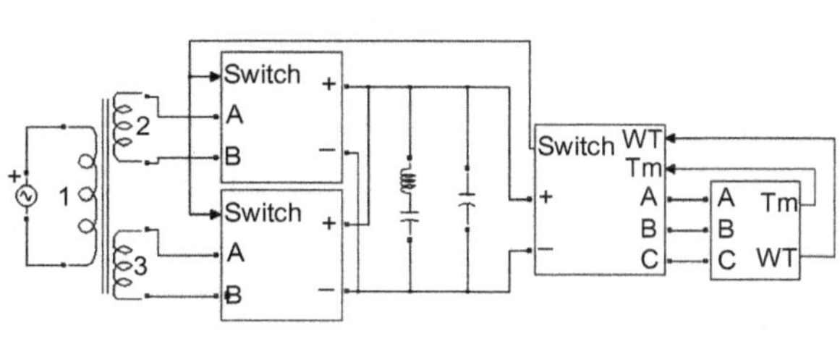


**Fig. 3** Simulated model of the traction driver system of CRH3

windings and two single-phase pulse rectifiers. The two output DC powers are in parallel connection, and supply an inverter. The inverter is easier to control due to the two-level topology, but the LC filter is needed in the DC link. In addition, the power of the traction system is much greater than that of CRH2.

## 3 Developed Simulation Software

The simulation software of the CRH2 traction driver system consists of a main program, an input and output interface program, and a simulation core program. All programs are developed using VC++. The main program supplies a friendly interface to accept the simulation parameters and shows the output results. The input and output interface program can not only exchange more data between the main program and the simulation model, but also deals with the input parameters and output results as required. The diagram of this simulation software is shown in Fig. 4.

The simulation core program of VC++ is changed from that model shown in Fig. 2 by real-time workshop (RTW), which includes the main code file (*.cpp), data file (*-data.cpp), and three basic files (Grt_main. c, rt_logging. c, and rt_sim. c). The main code file describes the simulation model in C code, initiates the program, sets up the sample time, and outputs the results. The input and output interface function is also added to this file. The data file gives values to every module such as the motor, inverter, and transformer. Because the other basic three files are the same in all models, they do not have any detailed information within this simulation structure. Although it can not add any detailed code, the adding input and output interface function should declare for call. Figure 5 shows the simulation software.

For the CRH2 and CRH3, each has two simulation models, such as a two-level one motor model, a two-level four motor model or a three-level four motor model. To every model, the parameters of the motor, power supply, and given train speed can be input. The simulation results are read in real time and displayed dynamically. At the same time, the results are analyzed and calculated, and then the fundamental component, harmonic component, efficiency, power factor, and heating power of every link are obtained.

**Fig. 4** Diagram of simulation software

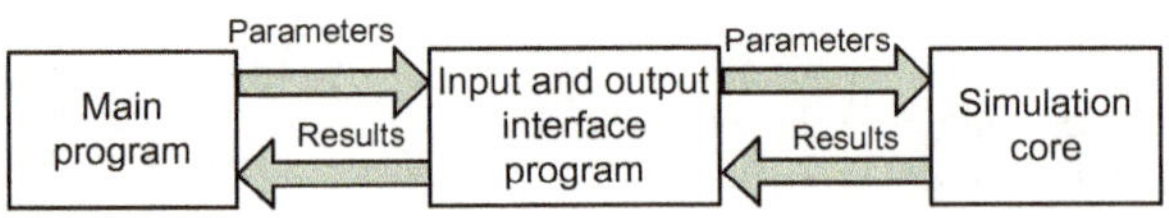

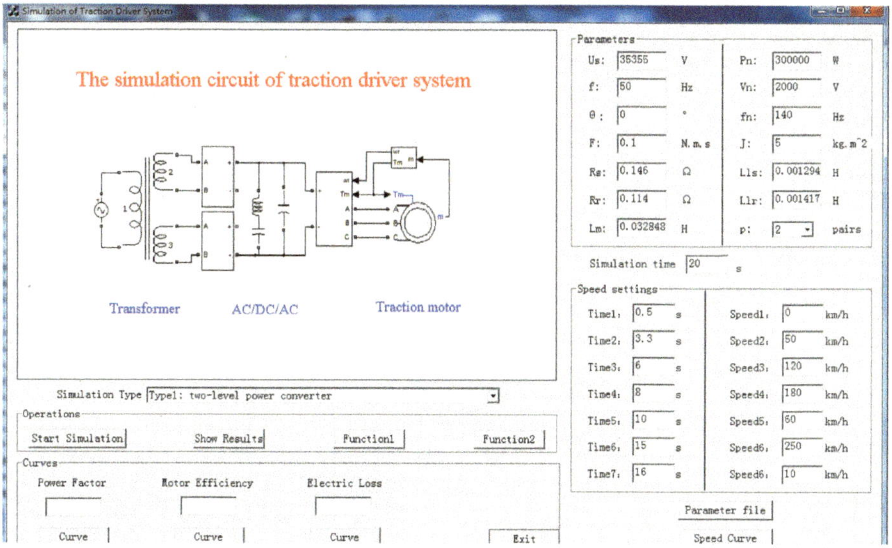


**Fig. 5** Simulation software

# 4 Simulation Results

In the developed simulation software, the traction performance can be simulated. The parameters were as follows: power supply voltage $U_s$ = 25,000 V, transformer ratio $k$ = 25,000/1500, carrier frequency of pulse rectifier $f_c$ = 1250 Hz, DC voltage $U_{DC}$ = 2600 V (traction)/3000 V(brake), DC capacitance $C_1 = C_2$ = 16 mF, motor rated power $P_N$ = 300 kW, rated voltage $U_N$ = 2000 V, rated frequency $f_N$ = 140 Hz, stator resistance $R_s$ = 0.114 Ω, stator leakage inductance $L_s$ = 1.417 mH, rotor resistance $R_r$ = 0.146 Ω, rotor leakage inductance $L_r$ = 1.294 mH, mutual inductance $L_{rm}$ = 32.8 mH, and pole pair $p$ = 2 (Venkataraman et al. 1980).

## *4.1 Traction and Brake Performance of CRH2*

The simulation assumes the train accelerates to 200 km/h during 2.2 s, keeps this speed for 1.3 s, and decreases speed to zero during 1.3 s. Figure 6 shows the simulation results of traction and brake conditions.

The simulated value of the train speed coincides with the given value. The traction driver system has a large torque only when the train accelerates or decelerates. The voltage of the DC link remains at 2600 V in traction condition and at

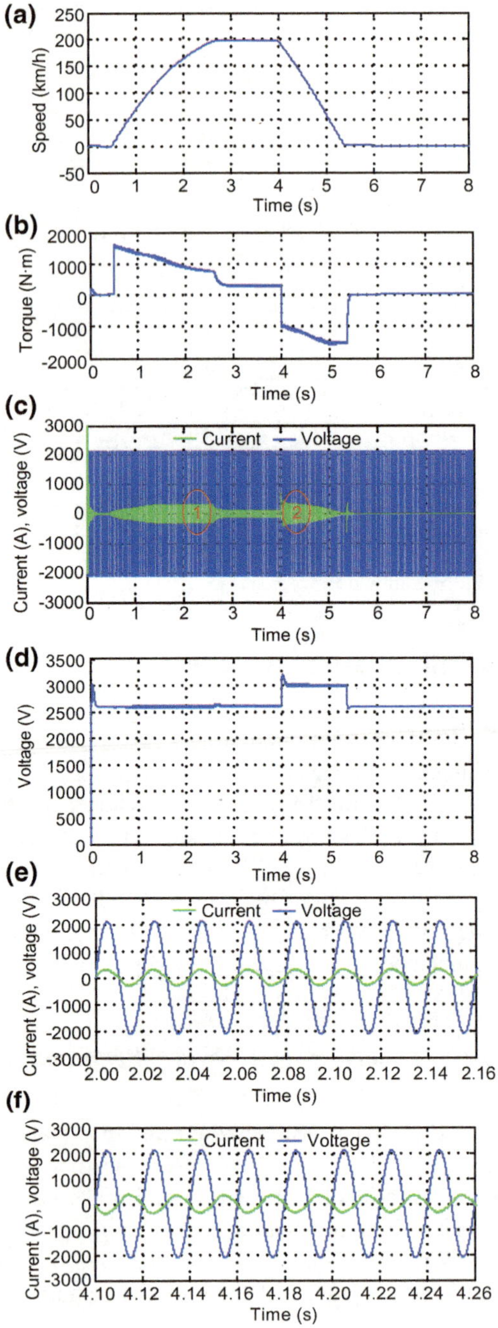


**Fig. 6** Simulated traction and brake performance results of CRH2. **a** simulated value of train speed; **b** electromagnetic torque; **c** output of transformer in the secondary winding; **d** DC voltage; **e** traction condition (*circle 1* in **c**); **f** brake condition (*circle 2* in **c**)

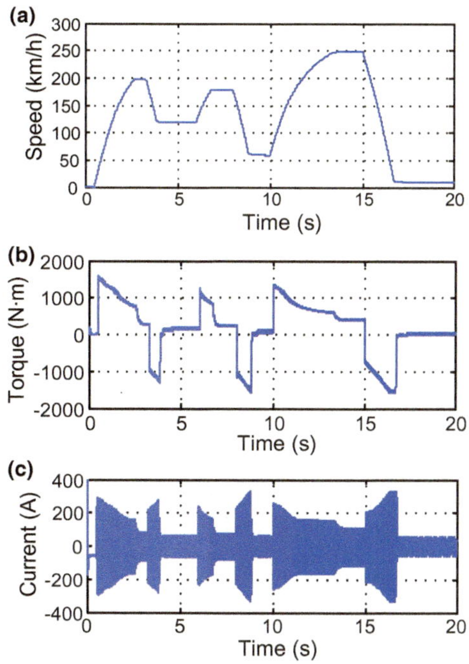


**Fig. 7** Simulated speed regulation performance results of CRH2. **a** simulated value of train speed; **b** electromagnetic torque; **c** motor current

3000 V in brake condition, which meets the requirements of CRH2. In traction condition, the current and voltage of the transformer secondary winding are almost in the same phase, i.e., the power factor is almost one. In brake condition, the current and voltage of the transformer secondary winding are in positive phase.

As all simulation results met the actual work conditions, it can be concluded that the developed simulation software is useful and of benefit to system optimization.

## 4.2 Speed Regulation Performance Simulation

This simulation assumed the train runs at 200, 120, 180, 60, and 250 km/h, in turn (Fig. 7). The electromagnetic torque and motor current not only increase quickly as the train accelerates or decelerates, but also increase with the train speed. But the effect of acceleration is much greater than that of speed. Thus, the speed regulation performance of this traction system is fine.

## 4.3 Transient Performance of Fault Condition

Apart from normal operation, the simulation software can also be used to simulate the transient performance of a fault condition such as the pantograph temporarily

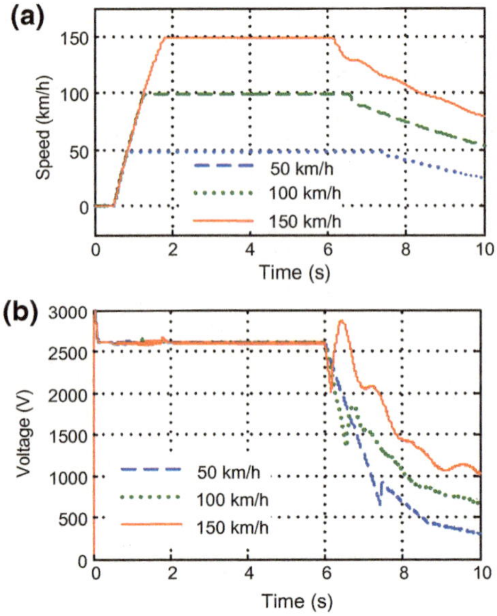


**Fig. 8** Transient performance of a fault condition of CRH2. **a** transient train speed; **b** voltage of DC links

disconnecting to the grid, four wheels with different radii, or motors with small differences in parameters. Figure 8a shows the transient train speed when the pantograph temporarily disconnects from the grid at speeds of 50, 100, and 150 km/h. The time for which the speed is maintained is 1.2, 0.5, and 0.13 s, respectively. Thus, the time decreases rapidly as the speed increases. Figure 8b shows the voltage of the DC link. It decreases firstly due to the lost power supply, and then increases after a short time when the motor acts as a generator. But the increase is very short-lived. Both the value and duration of the increase depend on the train speed.

## 5 Conclusions

This paper describes the development of simulation software in VC++ which can simulate operation performance and fault diagnosis. After the simulation, the cores of the CRH2 and CRH3 traction driver system differ from the corresponding models established by SIMULINK. The developed simulation software combines a friendly interface and data processing. In the software, the traction motor adopts transient current control and an indirect rotor magnetic field orientation vector control strategy, and the traction converter uses SPWM and SVPWM methods. On this basis, the typical operation performance and fault condition of a traction converter were simulated and analyzed at different train speeds. The simulation software has proved useful and can provide a reference for the actual design and production of traction converters.

## References


Carter, J., Goodman, C. J., & Zelaya, H. (1997). Analysis of the single-phase four-quadrant PWM converter resulting in steady-state and small-signal dynamic models. *IEE Proceedings Electric Power Applications, 144*(4), 241–247. doi:10.1049/ip-epa:19971058.

Celanovic, N. (2001). A fast space-vector modulation algorithm for multilevel three-phase converters. *IEEE Transactions on Industry Applications, 37*(2), 637–641. doi:10.1109/28.913731.

Depenbrock, M. (1988). Direct self-control (DSC) of inverter fed induction machine. *IEEE Transaction on Power Electronics, 3*(4), 420–429. doi:10.1109/63.17963.

Lascu, C., Boldea, I., & Blaabjerg, F. (2000). A modified direct torque control for induction motor sensorless drive. *IEEE Transaction on Industry Applications, 36*(1), 122–130. doi:10.1109/28.821806.

Lin, B. R., & Lu, H. H. (2000). A novel PWM scheme for single-phase three-level power-factor-correction circuit. *IEEE Transactions on Industrial Electronics, 47*(2), 245–252. doi:10.1109/41.836339.

Song, L. M. (2009). *The transmission and control of EMU* (pp 134–217). Beijing, China: The Chinese Railway Press. (in Chinese).

Tolbert, L. M., & Habetler, T. G. (1999). Novel multilevel inverter carrier-based PWM methods. *IEEE Transactions on Industry Applications, 35*(5), 1098–1107. doi:10.1109/28.793371.

Venkataraman, R., Ramaswami, B., Hotlz, J. (1980). Electronic analog slip calculator for induction motor driver. *IEEE Transaction on Industrial Electronics and Control Instrumentation,* ***IECI-27***(2),110–116. doi:10.1109/TIECI.1980.351637.

Zhang, L., & Wang, J. Y. (2008). *Combining Programming of MATLAB and C/C++* (pp 44–88). Beijing, China: The People's Posts and Telecommunications Press. (in Chinese).


### Author Biographies

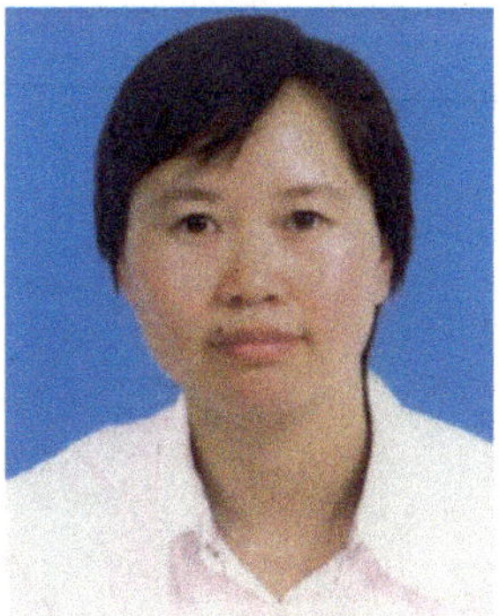

**Qin-fen Lu** Ph.D., is a professor of the College of Electrical Engineering, Zhejiang University, China. She is an IET Fellow, IEEE Senior Member, Vice Chairman and Secretary-General of the Linear Motor Committee, China Electrotechnical Society.

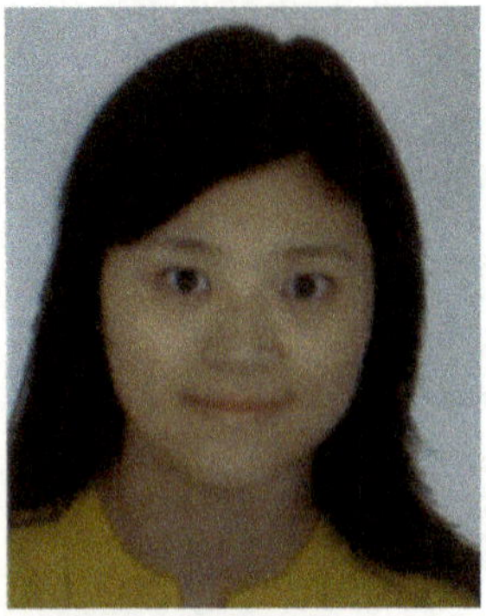

**Xiao-yan Huang** Ph.D., is a professor in Electrical Engineering at Zhejiang University, Hangzhou, China. She received a BE degree from Zhejiang University in 2003, and received a PhD degree in electrical machines and drives from the University of Nottingham, Nottingham, UK, in 2008. From 2008 to 2009, she was a Research Fellow with the University of Nottingham. Her research interests are PM machines and drives for aerospace and traction applications, and generator systems for urban networks.

**Ji-en Ma** Ph.D., is an associate professor in Zhejiang University, Hangzhou, China. She received a PhD degree in mechatronics from Zhejiang University in 2009. Then she did postdoctoral work in the College of Electrical Engineering, Zhejiang University, China. Her recent work is on electrical machines and drives. Her research interests include PM machines and drives for traction applications, and mechatronics machines such as the magneto fluid bearing.

**You-tong Fang** is a professor at Zhejiang University. He is chairman of the High-Speed Rail Research Center in Zhejiang University, deputy director of the National Intelligent Train Research Center, on the committee of the China High-Speed Rail Innovation Plan, and expert of the National 863 Program (China High Technology Development Program) in modern transportation and advanced carrying technology. He is also director of three projects of the National Natural Science Foundation of China (NSFC) and more than ten National 863 projects and Technology Support Programs. Recent work has been on electrical machines and drives. His research interests include permanent magnet (PM) machines and drives for traction applications.

**Wen-ping Cao** Ph.D., is the Chair Professor in Electrical Power Engineering and the Head of Power Electronics, Machines and Power System Group at Aston University, UK. In 2015, he was a Marie Curie Fellow at the *Massachusetts Institute of Technology, MA, USA*. Prof. Cao is presently a *Royal Society Wolfson* Research *Merit Award holder in the UK*. He was a semi-finalist at the Annual MIT-China Innovation and Entrepreneurship Forum (MIT-CHIEF) Business Plan Contest, USA, in 2015, the Dragon's Den Competition Award winner from Queen's University Belfast, UK, in 2014, and the Innovator of the Year Award winner from Newcastle University, UK, in 2013.

# Electromagnetic Environment Around a High-Speed Railway Using Analytical Technique

**Yong-jian Zhi, Bin Zhang, Kai Li, Xiao-yan Huang, You-tong Fang and Wen-ping Cao**

## Contents



Y. Zhi · X. Huang · Y. Fang
College of Electrical Engineering, Zhejiang University, Hangzhou 310027, China

B. Zhang (✉)
State Key Laboratory of Fluid Power and Mechatronic Systems,
Zhejiang University, Hangzhou 310027, China
e-mail: zbzju@163.com

K. Li
Department of Information Science and Electronic Engineering,
Zhejiang University, Hangzhou 310027, China

W. Cao
Newcastle University, Newcastle upon Type NE1 7RU, UK

 
Y. Fang and Y.H. Zhang (eds.), *China's High-Speed Rail Technology*,
Advances in High-speed Rail Technology, https://doi.org/10.1007/978-981-10-5610-9_28

## 1 Introduction

The Beijing-Shanghai High-Speed Railway officially started operating on June 30, 2011. It is a milestone in the development of China's railway. A mass of high-powered electronic devices used in the Beijing-Shanghai High-Speed Railway System has led to the possibility of interference to the facilities in the environment, such as telecommunications lines and wireless systems, and also the possibility of affecting human health. So it is necessary to evaluate the electromagnetic environment around the railway lines. That is why the electromagnetic compatibility (EMC) standard (GB/T 24338-2009 2009), translated from CENELEC Standard EN 50121 (2006), has to be established. Where the electromagnetic interference (EMI) source is and how it propagates to the observer are the primary problems to be solved. The main high frequency EMI radiates electromagnetic energy from train converters. A part of the EMI is injected to the contact line, and three patterns are considered for the EMI affecting other systems: (1) as a horizontal wire antenna radiating electromagnetic energy; (2) as a multiconductor line propagating the EMI signals away from the source point; (3) as a multiconductor line leading to the crosstalk between the lines (Cozza and Demoulin 2008).

The typical configuration of an electric railway line on the Beijing-Shanghai High-Speed Railway is sketched in Fig. 1, which depicts the electric locomotive, power supply line, and autotransformer (AT) used in the power supply system. Electric energy mainly supplied to traction electric machine is regulated by a switched-mode unit, which generates a strong conducted EMI due to the fast switch on and off. The conducted EMI goes back to the overhead supply line through a pantograph, thus imposing external interference through radiation and propagation along the supply line. The interference mentioned above is the secondary radiation, and there is also primary radiation from the high-powered electronic devices themselves (Cozza and Demoulin 2008).

An actual railway system is regarded as an assembly of uniform multiconductor transmission lines, which consists of catenary wire, contact wire, ground wire, negative feeder wire, and two rails. Multiconductor transmission lines structures are generally solved by the transmission line theory (Paul 2008). In his classical book, all aspects about multiconductor transmission lines are considered. The drawback of transmission line theory is that only quasi-transverse electromagnetic (TEM) mode is considered. However, when the frequency becomes higher, other modes' influence will be enhanced. The mode analysis of an infinitely long line above the earth was first solved by Carson (1926). In his theory, the distributed parameters of a quasi-TEM transmission line were analyzed and calculated. Since then, many investigators have revisited the problem, and numerous studies have been carried out. Using Hertz potentials in three media, Sunde (1968) derived the results for a wider frequency range including the displacement currents in the soil, which was neglected in Carson (1926)'s model. Sunde (1968)'s model is suitable over wider frequency range. Excepting the quasi-TEM mode, other models that proved to be very important have been overlooked. Wait (1972) proposed a modal equation

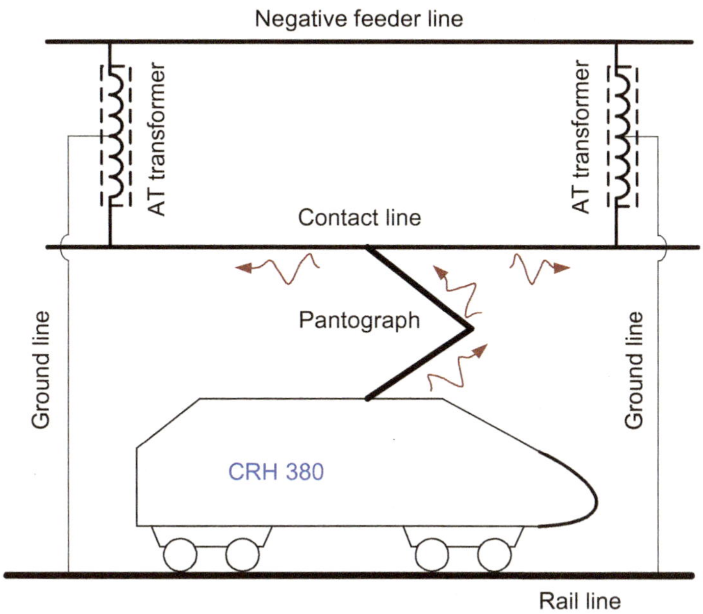


**Fig. 1** Typical configuration used on the Beijing-Shanghai high-speed railway, China

through a full-wave approach. The only approximations were thin-wire and complex exponential current distribution. An extension of Wait (1972)'s model was provided by D'Amore and Starto (1996a, b), pointing out that the transmission-line mode is dominant even when the quasi-TEM approximation does not hold. Kuester et al. (1978) and Olsen et al. (1978) proposed modal equation solutions including quasi-TEM mode, surface-attached mode, radiation mode, and surface wave. A good summary of this research can be found in Olsen et al. (2000). The line is assumed continuous in all the cases above.

But in fact the contact line is disconnected by an AT every 25 km, and a model considering the lumped-circuit is discussed in (Mazloom et al. 2009). He composited the Finite Difference Time Domain (FDTD) routine and Alternative Transients Program (ATP), and proved that it was effective in dealing with the lumped components, such as boost transformers (BTs), ATs, track circuits, and line interconnections.

Under some condition, the disconnected wire can be seen as infinitely long (Cozza and Demoulin 2008). Moreover, it is obvious that the overhead lines in a railway system are above the stratified earth, specially the rails above the sleeper. Thus, stratified earth has been investigated. The most influential research findings were summarized by Wait (1970) and Li (2009).

# 2 Magnetic Field Formulae

## 2.1 Geometry of Railway System

The typical configuration of the multiconductor transmission lines system under consideration is shown in Fig. 2. R1 and R2 are the rails, whose cross-section is not presented realistically. The distance between them is 1435 mm, and track circuits are connected with them to ensure safety. R3 and R4 are, respectively, a catenary wire and contact wire 6 m above the rails, connected by droppers in order to assure equipotential along the overhead line. R5 is a negative feeder wire connected to one end of the AT. R6 is a ground wire, one end of which is connected to the middle of the AT, and the other connected to the rail. Regions 0, 1, and 2 are the air, the middle layer, and the earth, respectively, characterized by $\mu_j$, $\varepsilon_j$, $\sigma_j$, $j = 0, 1, 2$, where $\mu_0$ and $\varepsilon_0$ are respectively permeability and permittivity of vacuum, and $\sigma_j$ is the conductivity of media $i$.

## 2.2 Integrated Formulae

The wave numbers in the three-layered region are

$$k_0 = \omega\sqrt{\mu_0 \varepsilon_0}, \tag{1}$$

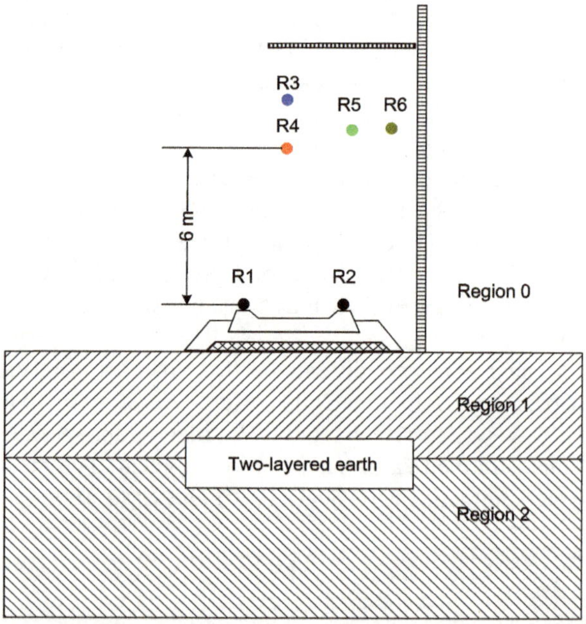


**Fig. 2** Cross-section of the multiconductor lines under consideration

$$k_j = \omega\sqrt{\mu_0\left(\varepsilon_0\varepsilon_{rj} + \mathrm{i}\sigma_j/\omega\right)}, \quad j = 1, 2, \tag{2}$$

where $\omega$ is the angular speed, $\varepsilon_{rj}$ is the relative dielectric constant of media $j$, and $\mathrm{e}^{-\mathrm{i}\omega t}$ is used, the integrated formulae of the magnetic field in the air radiated by a horizontal infinitely long wire over a two-layered earth are (Wait 1970)

$$B_{0x}(x, y, z) = \frac{I\mu_0}{4\pi}\left(B_{0x}^1 + B_{0x}^2 + B_{0x}^3\right), \tag{3}$$

$$B_{0z}(x, y, z) = \frac{I\mu_0}{4\pi}\left(B_{0z}^1 + B_{0z}^2 + B_{0z}^3\right), \tag{4}$$

where $B^1$ and $B^2$ are direct wave and ideal reflected wave,

$$\begin{aligned} B_{0x}^1 &= \int_{-\infty}^{+\infty} \mathrm{e}^{\pm \mathrm{i}\gamma_0|z-d|}\mathrm{e}^{\mathrm{i}\lambda x}\mathrm{d}\lambda \\ &= \frac{2\mathrm{i}k_0(z-d)}{\sqrt{x^2+(z-d)^2}}K_1\left[\mathrm{i}k_0\sqrt{x^2+(z-d)^2}\right], \end{aligned} \tag{5}$$

$$\begin{aligned} B_{0x}^2 &= \int_{-\infty}^{+\infty} \mathrm{e}^{\mathrm{i}\gamma_0(z+d)}\mathrm{e}^{\mathrm{i}\lambda x}\mathrm{d}\lambda \\ &= \frac{2\mathrm{i}k_0(z+d)}{\sqrt{x^2+(z+d)^2}}K_1\left[\mathrm{i}k_0\sqrt{x^2+(z+d)^2}\right], \end{aligned} \tag{6}$$

$$\begin{aligned} B_{0z}^1 &= -\int_{-\infty}^{+\infty} \mathrm{e}^{\pm \mathrm{i}\gamma_0|z-d|}\mathrm{e}^{\mathrm{i}\lambda x}\frac{\lambda}{\gamma_0}\mathrm{d}\lambda \\ &= \frac{2\mathrm{i}k_0 x}{\sqrt{x^2+(z-d)^2}}K_1\left[\mathrm{i}k_0\sqrt{x^2+(z-d)^2}\right], \end{aligned} \tag{7}$$

$$\begin{aligned} B_{0z}^2 &= \int_{-\infty}^{+\infty} \mathrm{e}^{\pm \mathrm{i}\gamma_0(z+d)}\mathrm{e}^{\mathrm{i}\lambda x}\frac{\lambda}{\gamma_0}\mathrm{d}\lambda \\ &= -\frac{2\mathrm{i}k_0 x}{\sqrt{x^2+(z+d)^2}}K_1\left[\mathrm{i}k_0\sqrt{x^2+(z+d)^2}\right], \end{aligned} \tag{8}$$

where $d$ is the height of line source to the ground surface, and $K_0$ and $K_1$ are the modified Bessel functions.

$$B_{0x}^{3} = \int\limits_{-\infty}^{+\infty} e^{i\gamma_0(z+d)} e^{i\lambda x}(Q-1)d\lambda, \tag{9}$$

$$B_{0z}^{3} = \int\limits_{-\infty}^{+\infty} \frac{\lambda}{\gamma_0} e^{i\gamma_0(z+d)} e^{i\lambda x}(Q-1)d\lambda, \tag{10}$$

where

$$Q-1 = 2\frac{-\gamma_0\gamma_1 + i\gamma_0\gamma_2\tan(\gamma_1 l)}{\gamma_0\gamma_1 + \gamma_1\gamma_2 - i(\gamma_1^2 + \gamma_0\gamma_2)\tan(\gamma_1 l)}, \tag{11}$$

$$\gamma_j = \sqrt{k_j^2 - \lambda^2},\ \ j = 0, 1, 2, \tag{12}$$

where $l$ is the thickness of middle layer.

The main aim is to evaluate the integrals using analytical techniques. Contour integral technique is applied to the above integrals. The contribution of residue to the integrals is called trapped surface wave, and the contribution of the integral along the branch lines is called the lateral wave. The complete wave contributions are provided next.

## 2.3 Trapped Surface Wave

From the above derivations, we know that the trapped surface wave is from the residue. The two factors of the residue are poles and expressions. First, let us examine the pole equation:

$$\gamma_0\gamma_1 + \gamma_1\gamma_2 - i(\gamma_1^2 + \gamma_0\gamma_2)\tan(\gamma_1 l) = 0. \tag{13}$$

The roots are obtained using Newton's iteration method from the pole equation (Zhi et al. 2012).

Note that when the roots of the pole equation are solved, the residue is obtained easily. The terms of the trapped surface wave can be achieved as

$$\begin{aligned} B_{0x}^{S} = {} & -i\mu_0 I \sum_j e^{ix\lambda_j^* + i\gamma_0(\lambda_j^*)(z+d)} \\ & \times \frac{-\gamma_0(\lambda_j^*)\gamma_1(\lambda_j^*) + i\gamma_0(\lambda_j^*)\gamma_2(\lambda_j^*)\tan\gamma_1(\lambda_j^*)l}{\gamma_1(\lambda_j^*)q'(\lambda_j^*)}, \end{aligned} \tag{14}$$

$$B_{0z}^{S} = i\mu_0 I \sum_j \lambda_j^* e^{ix\lambda_j^* + i\gamma_0(\lambda_j^*)(z+d)} \times \frac{-\gamma_0(\lambda_j^*)\gamma_1(\lambda_j^*) + i\gamma_0(\lambda_j^*)\gamma_2(\lambda_j^*)\tan\gamma_1(\lambda_j^*)l}{\gamma_0(\lambda_j^*)\gamma_1(\lambda_j^*)q'(\lambda_j^*)}, \tag{15}$$

where

$$q'(\lambda) = -\lambda\left(\frac{\gamma_0}{\gamma_1} + \frac{\gamma_1}{\gamma_0} + \frac{\gamma_2}{\gamma_1} + \frac{\gamma_1}{\gamma_2}\right) + i\lambda\tan(\gamma_1 l)\left(\frac{\gamma_0}{\gamma_2} + \frac{\gamma_2}{\gamma_0} + 2\right) + i\sec^2(\gamma_1 l)\left(\gamma_1^2 + \gamma_0\gamma_2\right)\frac{l\lambda}{\gamma_1}. \tag{16}$$

## 2.4 *Lateral Wave*

From the above derivations, we know that the lateral wave is from the integral along the branch lines. Figure 3 shows that there are three branch lines; thus, lateral wave consists of three compositions. The evaluation of the integral along the branch cut $\Gamma_1$ is zero, and the integral along the branch cut $\Gamma_2$ can be neglected (Li 2009). Thus, the lateral wave is mainly the contribution of branch cut $\Gamma_0$, which is presented in Eq. (A2) in the Appendix.

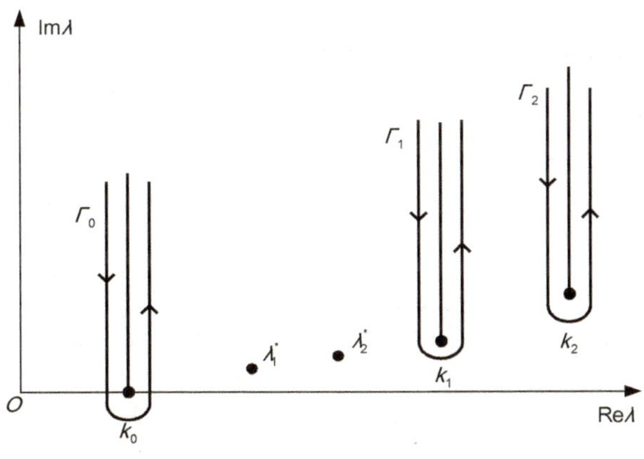


**Fig. 3** Configuration of poles and branch lines

### 2.5 Final Formulae for the Magnetic Field

Based on the above results, the final formulae of the magnetic fields $B_{0x}$ and $B_{0z}$ are shown in the Appendix, where

$$\begin{aligned} A &= -1 + \mathrm{i}\frac{\gamma_0}{\gamma_1}\tan(\gamma_1 l) \\ &\approx -1 + \mathrm{i}\frac{\sqrt{k_2^2 - k_0^2}}{\sqrt{k_1^2 - k_0^2}}\tan\left(\sqrt{k_1^2 - k_0^2 l}\right), \end{aligned} \tag{17}$$

$$\begin{aligned} B &= \gamma_2 - \mathrm{i}\gamma_1 \tan(\gamma_1 l) \\ &\approx -\sqrt{k_2^2 - k_0^2} + \mathrm{i}\sqrt{k_1^2 - k_0^2}\tan\left(\sqrt{k_1^2 - k_0^2 l}\right), \end{aligned} \tag{18}$$

$$\Delta = \frac{1}{\sqrt{2}}\left(\frac{B}{Ak_0} - \mathrm{i}\frac{\mathrm{z} + \mathrm{d}}{\mathrm{x}}\right), \tag{19}$$

$$p^* = k_0 x \Delta^2. \tag{20}$$

It is seen that the total field is composed of the direct wave, the ideal reflected wave or image wave, the trapped surface wave, and the lateral wave.

## 3 Computation Results and Comparison

With the measuring point on the interface $z$ = 0 m and the height of the source line $d$ = 6 m, $\varepsilon_1$ = 2.65, $\varepsilon_2$ = 8, $\sigma_2$ = 0.4 S/m, and $f$ = 10, 50 and 500 MHz according to (CENELEC Standard EN50121, 2006) are considered for contact line. Magnitudes of the total field, the trapped surface wave, and DRL wave, which is composed of the direct wave, the ideal reflected wave, and the lateral wave, are computed at $k_1 l = 2.97\pi$ as shown in Fig. 4a–c.

With the measuring point on the interface $z$ = 0 m and the height of the source line $d$ = 0.2 m, $\varepsilon_1$ = 2.65, $\varepsilon_2$ = 8, $\sigma_2$ = 0.4 S/m, and $f$ = 10, 50 and 500 MHz are considered for rail line. Magnitudes of the total field, the trapped surface wave, and the DRL waves are computed at $k_1 l = 2.97\pi$ as shown in Fig. 5a–c.

Compared with trapped surface wave by a dipole in (Li 2009), the amplitude of the trapped surface wave by a horizontal infinitely long wire has no dispersion losses, due to the wavefront of the trapped surface wave radiated over the two-layered earth is always a plane. This characteristic is different from that of a dipole source in the presence of a three-layered region. In the case of a dipole source, the wavefront is enlarged as the dispersion of the trapped surface, the

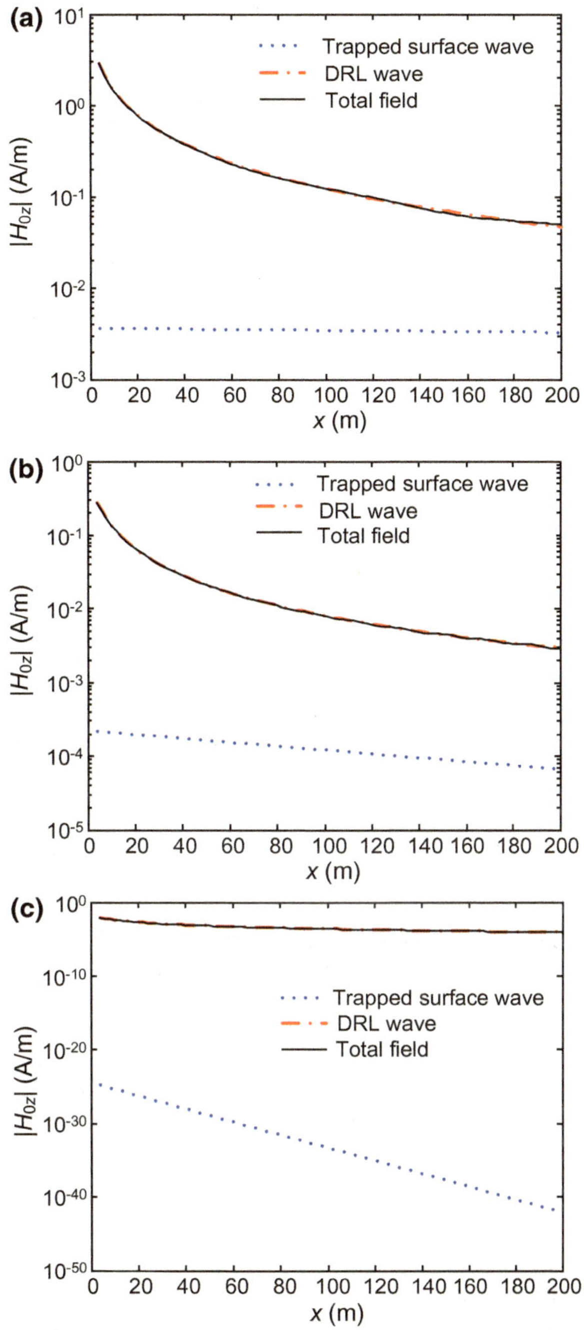


**Fig. 4** Magnitude of $H_{0z}$ versus propagating distances at $z = 0$, $d = 6$ m with $\varepsilon_1 = 2.65$, $\varepsilon_2 = 8$, $\sigma_2 = 0.4$ S/m and $k_1 l = 2.97\pi$. **a** $f = 10$ MHz; **b** $f = 50$ MHz; **c** $f = 500$ MHz

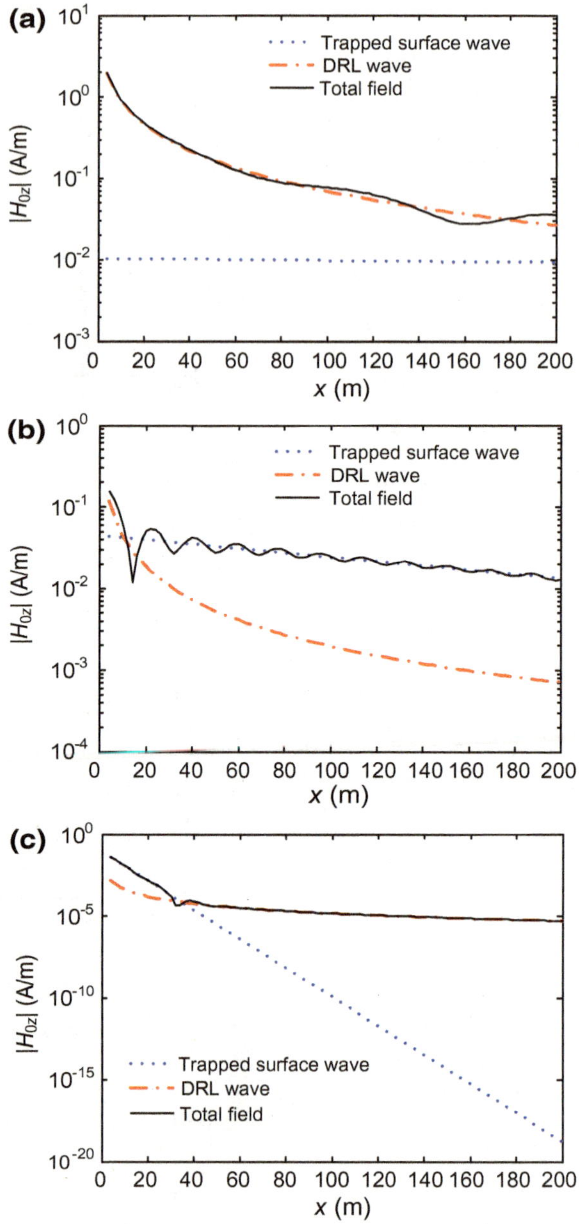


**Fig. 5** Magnitude of $H_{0z}$ versus propagating distances at $z = 0$, $d = 0.2$ m with $\varepsilon_1 = 2.65$, $\varepsilon_2 = 8$, $\sigma_2 = 0.4$ S/m and $k_1 l = 2.97\pi$. **a** $f = 10$ MHz; **b** $f = 50$ MHz; **c** $f = 500$ MHz

amplitude of the trapped surface wave is attenuated as $\rho^{-1/2}$ in the $\rho$ direction. The attenuation of trapped surface wave in all the figures is mainly due to the dielectric losses. And the nearer the observer is to the interface, the more important the strapped surface wave is when considering the thickness of the middle layer.

# 4 Conclusions

Overhead lines in high-speed railway consist of catenary wire, contact wire, negative feeder wire, ground wire, and rails, which constitute multiconductor transmission lines. These lines are divided into two groups: contact lines and rail lines. Magnetic field around the lines is discussed as single line for contact lines and rail lines.

The total magnetic field consists of the direct wave, the ideal reflected wave or image wave, the trapped surface wave, and the lateral wave. The trend of the field and wave is attenuated along the $x$ direction.

When the frequency becomes higher and the source is near the interface, the trapped surface wave is dominant. But when the source is away from the interface, the total field is dominated by DRL waves.

# Appendix

$$\begin{aligned} B_{0x} = & -\frac{\mathrm{i}I\mu_0 k_0(z-d)}{2\pi\sqrt{x^2+(z-d)^2}} K_1\left[\mathrm{i}k_0\sqrt{x^2+(z-d)^2}\right] \\ & + \frac{\mathrm{i}I\mu_0 k_0(z+d)}{2\pi\sqrt{x^2+(z+d)^2}} K_1\left[\mathrm{i}k_0\sqrt{x^2+(z+d)^2}\right] \\ & - \mathrm{i}\mu_0 I\sum_j \mathrm{e}^{\mathrm{i}x\lambda_j^* + \mathrm{i}\gamma_0(\lambda_j^*)(z+d)} \frac{-\gamma_0(\lambda_j^*)\gamma_1(\lambda_j^*) + \mathrm{i}\gamma_0(\lambda_j^*)\gamma_2(\lambda_j^*)\tan\gamma_1(\lambda_j^*)l}{\gamma_1(\lambda_j^*)q'(\lambda_j^*)} \\ & + \sqrt{2}\frac{I\mu_0}{4\pi}\left[\frac{B}{A}\mathrm{e}^{\mathrm{i}0.75\pi + \mathrm{i}k_0x - \mathrm{i}\frac{k_0(z+d)^2}{2x}}\sqrt{\frac{\pi}{k_0x}} - i\pi\frac{B^2}{A^2k_0}\mathrm{e}^{\mathrm{i}k_0x - \mathrm{i}0.75\pi - \mathrm{i}\frac{k_0(z+d)^2}{2x} - \mathrm{i}k_0x\Delta^2}F(p^*)\right], \end{aligned} \tag{A1}$$

$$\begin{aligned} B_{0z} = & +\frac{\mathrm{i}I\mu_0 k_0(z-d)}{2\pi\sqrt{x^2+(z-d)^2}} K_1\left[\mathrm{i}k_0\sqrt{x^2+(z-d)^2}\right] \\ & - \frac{\mathrm{i}I\mu_0 k_0(z+d)}{2\pi\sqrt{x^2+(z+d)^2}} K_1\left[\mathrm{i}k_0\sqrt{x^2+(z+d)^2}\right] \\ & + \mathrm{i}\mu_0 I\sum_j \mathrm{e}^{\mathrm{i}x\lambda_j^* + \mathrm{i}\gamma_0(\lambda_j^*)(z+d)} .\lambda_j^* \frac{-\gamma_0(\lambda_j^*)\gamma_1(\lambda_j^*) + \mathrm{i}\gamma_0(\lambda_j^*)\gamma_2(\lambda_j^*)\tan\gamma_1(\lambda_j^*)l}{\gamma_0(\lambda_j^*)\gamma_1(\lambda_j^*)q'(\lambda_j^*)} \\ & + \sqrt{2}\frac{I\mu_0}{4\pi}\left[k_0\mathrm{e}^{\mathrm{i}0.75\pi + \mathrm{i}k_0x - \mathrm{i}\frac{k_0(z+d)^2}{2x}}\sqrt{\frac{\pi}{k_0.x}} - i\pi\frac{B}{A}\mathrm{e}^{\mathrm{i}k_0x - \mathrm{i}0.75\pi - \mathrm{i}\frac{k_0(z+d)^2}{2x} - \mathrm{i}k_0x\Delta^2}F(p^*)\right]. \end{aligned} \tag{A2}$$

## References

Carson, J. R. (1926). Wave propagation in overhead wires with ground return. *Bell System Technology Journal*, 539–554.

CENELEC Standard EN 50121. (2006). Railway applications-electromagnetic compatibility. European Committee for Electrotechnical Standardization.

Cozza, A., & Demoulin, B. (2008). On the modeling of electric railway lines for the assessment of infrastructure impact in radiated emission tests of rolling stock. *IEEE Transactions on Electromagnetic Compatibility, 50*(3), 566–576. doi:10.1109/TEMC.2008.924387.

D'Amore, M., & Starto, M. S. (1996a). Simulation models of a dissipative transmission line above a lossy ground for a wide frequency range-Part I: single conductor configuration. *IEEE Transactions on Electromagnetic Compatibility, 38*(2), 127–138. doi:10.1109/15.494615.

D'Amore, M., & Starto, M. S. (1996b). Simulation models of a dissipative transmission line above a lossy ground for a wide frequency range-Part II: multiconductor configuration. *IEEE Transactions on Electromagnetic Compatibility, 38*(2), 139–149. doi:10.1109/15.494616.

GB/T 24338-2009. (2009). *Railway applications-electromagnetic compatibility-Part 1: general*. CSR Zhuzhou Times Electric Co., Ltd., CSR Qingdao Sifang Co., Ltd., and China Railway Electrification Survey Design & Research Institute Co. Ltd.

Kuester, E. F., Chang, D. C., & Olsen, R. G. (1978). Modal theory of long horizontal wire structures above the earth, 1, excitation. *Radio Science, 13*(4), 605–613. doi:10.1029/RS013i004p00605.

Li, K. (2009). *Electromagnetic fields in stratified media*. Zhejiang University Press and Springer.

Mazloom, Z., Theethayi, N., & Thottappillil, R. (2009). A method for interfacing lumped-circuit models and transmission-line system models with application to railways. *IEEE Transactions on Electromagnetic Compatibility, 51*(3), 833–841. doi:10.1109/TEMC.2009.2023112.

Olsen, R. G., Kuester, E. F., & Chang, D. C. (1978). Modal theory of long horizontal wire structures above the earth, 2, properties of discrete modes. *Radio Science, 13*(4), 615–623. doi:10.1029/RS013i004p00615.

Olsen, R. G., Young, J. L., & Chang, D. C. (2000). Electromagnetic wave propagation on a thin wire above earth. *IEEE Transactions on Antennas and Propagation, 48*(9), 1413–1419. doi:10.1109/8.898775.

Paul, C. R. (2008). *Analysis of multiconductor transmission lines*. New York, Chichester, Brisbane, Toronto, Singapore: Wiley.

Sunde, E. D. (1968). *Earth conduction effects in transmission systems*. New York: Dover.

Wait, J. R. (1970). *Electromagnetic waves in stratified media* (2nd ed.). Oxford, London, New York, Paris: Pergamon Press.

Wait, J. R. (1972). Theory of wave propagation along a thin wire parallel to an interface. *Radio Science, 7*(6), 675–679. doi:10.1029/RS007i006p00675.

Zhi, Y. J., Li, K., & Fang, Y. T. (2012). Electromagnetic field of a horizontal infinitely long wire over the dielectric-coated earth. *IEEE Transactions on Antennas and Propagation, 60*(1), 360–366. doi:10.1109/TAP.2011.2167917.

## Author Biographies

**Bin Zhang** Ph.D., is a Research Associate in Mechanical Engineering at Zhejiang University, Hangzhou, China. He received a Doctor degree from Zhejiang University in 2009 in mechanical engineering. From 2015 to 2016, His research interests are Intelligent machine and Biomanufacture.

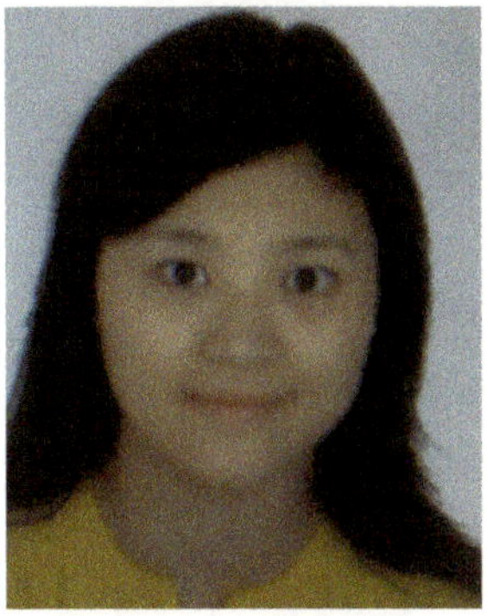

**Xiao-Yan Huang** Ph.D., is a professor in Electrical Engineering at Zhejiang University, Hangzhou, China. She received a BE degree from Zhejiang University in 2003, and received a Ph.D. degree in electrical machines and drives from the University of Nottingham, Nottingham, UK, in 2008. From 2008 to 2009, she was a Research Fellow with the University of Nottingham. Her research interests are PM machines and drives for aerospace and traction applications, and generator systems for urban networks.

**You-Tong Fang** is a professor at Zhejiang University. He is chairman of the High Speed Rail Research Center in Zhejiang University, deputy director of the National Intelligent Train Research Center, on the committee of the China High Speed Rail Innovation Plan, and expert of the National 863 Program (China High Technology Development Program) in modern transportation and advanced carrying technology. He is also director of three projects of the National Natural Science Foundation of China (NSFC) and more than ten National 863 projects and Technology Support Programs. Recent work has been on electrical machines and drives. His research interests include permanent magnet (PM) machines and drives for traction applications.

**Wen-Ping Cao** Ph.D., is the Chair Professor in Electrical Power Engineering and the Head of Power Electronics, Machines and Power System Group at Aston University, UK. In 2015, he was a Marie Curie Fellow at the Massachusetts Institute of Technology, MA, USA. Prof. Cao is presently a Royal Society Wolfson Research Merit Award holder in the UK. He was a semi-finalist at the Annual MIT-China Innovation and Entrepreneurship Forum (MIT-CHIEF) Business Plan Contest, USA, in 2015, the Dragon's Den Competition Award winner from Queen's University Belfast, UK, in 2014, and the Innovator of the Year Award winner from Newcastle University, UK, in 2013.

# Optimal Condition-Based Maintenance Strategy Under Periodic Inspections for Traction Motor Insulations

**Jian Zhang, Ji-en Ma, Xiao-yan Huang, You-tong Fang and He Zhang**

## Contents




J. Zhang · X. Huang · Y. Fang
College of Electrical Engineering, Zhejiang University,
Hangzhou 310027, China

J. Ma (✉)
China Academy of West Region Development, Zhejiang University,
Hangzhou 310027, China
e-mail: jienma@126.com

H. Zhang
Department of Electrical and Electronic Engineering,
The University of Nottingham, Nottingham NG7 2RD, UK




Y. Fang and Y.H. Zhang (eds.), *China's High-Speed Rail Technology*,
Advances in High-speed Rail Technology, https://doi.org/10.1007/978-981-10-5610-9_29

# 1 Introduction

Traction motors are key parts of high-speed trains (Fang 2011), and insulation failure is a key failure mode. There are about 37% of traction motor failures resulting from insulation failures (Zhou et al. 2006). The reliability of insulation has a direct effect on the safety and performance of high-speed trains.

At the rated rotating speed, the winding insulation of traction motors is subject to a continuous high-voltage square wave pulse instead of a sinusoidal voltage. Compared with a sinusoidal voltage, a high-voltage square wave may apply a higher voltage with a high frequency. The high frequency also intensifies the aging effect of dielectric loss, partial discharge, and space charges on the insulation, which may provoke premature failure.

Fatigue and shock play a primary role in traction motor insulation failure. Fatigue is associated with electrical, thermal, and mechanical aging, and shocks arise from the interventions of extremely high-voltage pulses of regional over-voltage, over switched voltage, ring effect, and so on. Therefore, a traction motor insulation can be treated as a deteriorating system subject to the combination of fatigue load and shock load.

Timely maintenance can free traction motor insulations from sudden catastrophic breakdown. Maintenance planning is of vital importance to guarantee safety and availability. Time-based maintenance (TBM) and condition-based maintenance (CBM) are two mainstream strategies of preventive maintenance (PM) (Liu et al. 2007; Asadzadeh and Azadeh 2014; Gao et al. 2014). Up to now, TBM based on historical life information is the main method for insulation maintenance, where maintenance is scheduled based on a pre-specified age threshold. It has obvious drawbacks, such as either excessive or insufficient maintenance (Ma 2008). With the development of sophisticated inspection technology on key performance parameters, such as the quantity of partial discharge, the polarization index, and winding insulation resistance, CBM shows great potential to substitute for TBM. However, as far as we are aware no CBM policy specific for motor insulations has been proposed in previous studies.

The issue of CBM policies for deteriorating systems has attracted a lot of attention. Most CBM models rely on the accessibility to degradation process information (Lu and Liu 2014; Shi and Zeng 2014). Gebraeel et al. (2005) presented a maintenance model combining the estimation of residual life distribution based on the Bayes theory and PM, in which the degrading function was assumed to be linear. Yang and Klutke (2000) and van Noortwijk (2009) described the degrading characteristics of deteriorating systems by a Gamma process and a Levy process.

Life data collected by the manufactures of insulation materials and the accelerating test results before delivery both make the life distribution accessible, and the integration of CBM and TBM is both possible and a significant development in engineering practice. Up to now research on TBM and CBM has remained separate. Data sharing between these two maintenance modes is a problem and worthy of in-depth research (Ahmad and Kamaruddin 2012). By sharing the statistical process data, some researchers attempted to establish the joint optimal models of CBM and TBM for manufacturing systems (Panagiotidou and Tagaras 2008, 2010, 2012; Yin et al. 2015). This model is innovative by maximally using the life information and state information and can be extended to deteriorating system maintenance with modification. In addition, Tang et al. (2015) provided an autoregressive model to describe system degradation.

For shock models, Tang and Lam (2006) and Lam (2009) proposed a maintenance model for deteriorating systems under $\delta$-shock, which worked out the maintenance plan on the basis of the interval between two shocks. Hu (1995) and Montoro-Cazorla and Pérez-Ocón (2014) established the model considering the shock effect by adopting a Markovian process. All the above-mentioned shock models assumed that each shock may lead to a discount of residual life. The accumulation of life loss is used as a reference to set up maintenance models, but these do not match the extreme shock situation of traction motor insulations. Some researchers also extended the degradation-threshold-shock model of optimal maintenance, which integrated degradation and shock effects and described them with stochastic process models (Wang et al. 2011; Castro 2013; Lin et al. 2014, 2015; Caballé et al. 2015; Castro et al. 2015).

In summary, for the application of CBM in insulation maintenance, it is a big challenge to establish an integrated model of TBM and CBM, one which explicitly takes extremely high-voltage shock effect into account. In this paper, we propose a maintenance policy in which periodic inspections are conducted to evaluate insulation performance, and the preventive maintenance actions are initialized by exceeding the critical age or exceeding a pre-specified inspection threshold. An optimization model is set up, which takes inspection interval and the critical age as the optimization variables, and minimal maintenance cost per time unit as the optimization objective.

In this paper, we first describe the problem setting and the basic assumption of the maintenance strategy with details. Then, we focus on the optimization model and establish the optimization model ignoring the shock effect as a preparation for future analysis. At last, the proposed model is validated through a numerical example.

## 2 Problem Statement and Description

The proposed maintenance strategy is described as follows.

1. The failure modes of traction motor insulations can be divided into two categories (Yao et al. 2013):

Failure *a*: The first type of failure, the situation that functional failure occurs while inspection data does not reach or exceed the threshold value. A corrective maintenance (CM) should be executed.

Failure *b*: The second type of failure, the situation that inspection data reaches or exceeds the threshold value with no occurrence of functional failure. A PM should be executed.

Failure *a* is directly observable and self-announcing, while Failure *b* can only be detected through offline inspections. Figure 1 describes the operating and failure modes of traction motor insulations.

2. Traction motor insulations can work in two states, normal state and latent failure state. Latent failure state is a transition state between normal state and functional failure state. During this period, insulations have shifted from a stable period to an accelerating aging period, and their failure rate increases acutely, that is to say the probability that functional failure occurs in this state is higher than that in normal state. $s = \{0, 1, 2\}$ is the state set of traction motor insulations, and states 0, 1, and 2 represent normal state, latent failure state, and functional failure state, respectively. When insulation is under state 1, the operation cost of traction motors will increase due to the decrease of motor efficiency. Under either state 0 or state 1, insulation can jump to state 2. Note that the probability of the jump from state 1 to state 2 is higher than that from state 0 to state 2. Figure 2 clarifies the state transition mechanism, in which $P_{Si}$ is the failure probability at state $i$ ($i = 0, 1$) after each shock, $f(t)$ is the probability density function (PDF) of the time from normal state to latent failure state, $\varphi_0(t)$ is the PDF of the time to functional failure state in normal state, and $\varphi_1(t)$ is the PDF of the time to functional failure state in latent failure state. The difference between Fig. 2 and the one proposed by Panagiotidou and Tagaras (2010) is that shocks are integrated into the state transition process.
3. The offline inspections on the performance indicator are performed periodically (assuming inspection is perfect). Generally, the inspection times should be an arbitrary value independent of each other. However, the optimization model itself is already extremely complicated, and if we set the inspection times as

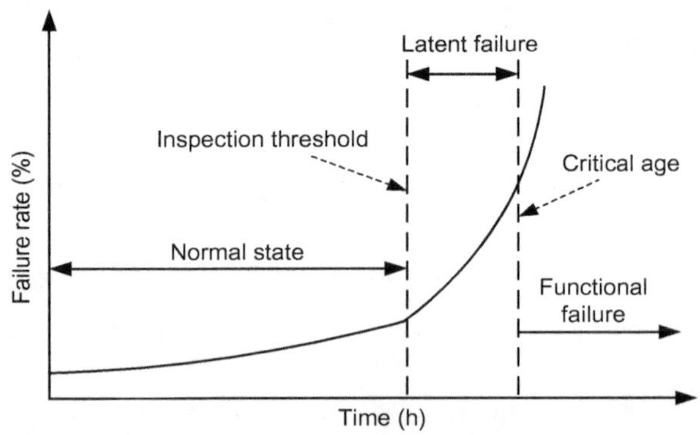


**Fig. 1** Operating and failure modes of insulations

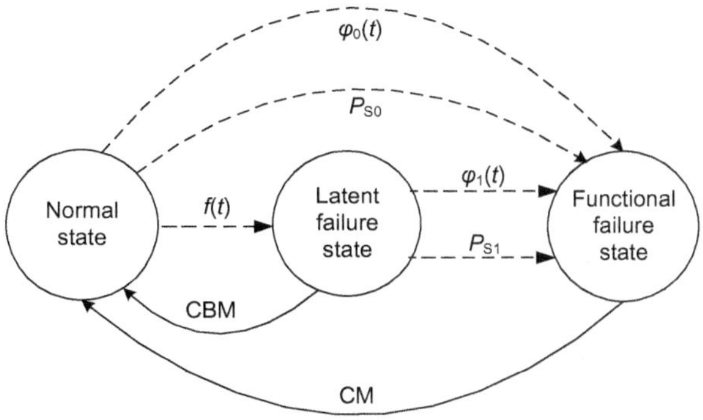


**Fig. 2** State transition mechanism

arbitrary values, which may add more complexity to the optimization model, it is too hard to obtain a convergent solution in practice. So for practical feasibility, the inspection will be taken with a constant interval or the interval should obey a simple rule. In this study, a constant inspection interval is adopted to simplify the optimization model. When the performance of insulations is lower than the national standards, Failure *b* happens, and PM is called for to restore insulations to the normal state. For simplicity, we define the duration of *N* inspection intervals as the critical age, beyond which the failure rate of insulations will be unacceptably high. If insulations can survive *N* inspection intervals without any state transition or failure, a PM is scheduled to be carried out. Once Failure *a* happens, the motor has to stop without any delay and an emergency control mechanism should be launched to let other traction motors share its traction load. In this situation, CM is indispensable. Once a maintenance action is triggered, either PM or CM, an operation cycle ends, and a replacement action is performed due to the high safety requirement of high-speed trains without considering the delivery time of spare parts. Insulations will restore to an "as good as new" state after maintenance, which also indicates the start of a new operation cycle.

4. Insulation is subject to external random extremely high-voltage shock which occurs according to a Poisson process with $\lambda$ as the intensity parameter. The failure probability after each shock for state 0 is smaller than that for state 1 ($P_{S0} < P_{S1}$). Here, we ignore the influence of extreme shocks on insulation residual life (Chen and Li 2008).
5. $A(t_i) = \{a_0, a_1, a_2\}$ is the maintenance action set at time $t_i$, where $a_0$ represents no maintenance action, and $a_1$ and $a_2$ are PM and CM, respectively.
6. $E = \{E(T_0), E(T_1), P_{PM}, P_{CM}, T_C, T_{IN}\}$ is the operation state parameter set. $E(T_i)$ is the expected time at state $i$ ($i = 0, 1$) in an operation cycle; $P_{PM}$ and $P_{CM}$ are the probabilities to carry out PM and CM within one cycle, respectively; $T_C$ is the pre-defined critical age, and $T_{IN}$ is the interval between two continuous inspections. Note that $T_{IN}$ is a fixed value and $T_C = NT_{IN}$.

7. $C = \{C_0, C_1, C_{PM}, C_{CM}, C_{IN}, \text{CPT}\}$ is the cost set. $C_i$ is the operation cost at state $i$ ($i = 0, 1$) and $C_1 > C_0$; $C_{PM}$, $C_{CM}$, and $C_{IN}$ are the costs of one PM, one CM, and one inspection, respectively; and CPT is the operation cost per unit time within one cycle.

## 3 Mathematical Model Development

In this section, we will establish an optimization model for traction motor insulation maintenance, in which the minimal CPT in an operation cycle is regarded as the optimization objective. Since CPT is a function of the variables of $N$ and $T_{IN}$, they are defined to be the decision variables with the constraints set as follows: $\{T_C > T_{IN} > 0; C_1 > C_0 > 0; C_{CM} > C_{PM} > 0; P_{S1} > P_{S0}; N \geq 2\}$.

The expected operation time within one operation cycle comprises the expected operation time at state 0 and the expected operation time at state 1. The expected cost within one cycle includes the expected costs of operation at state 0, operation at state 1, inspections, PM, and CM.

Then, the objective function can be expressed as

$$\text{CPT} = \frac{E(C)}{E(T)}, \tag{1}$$

where $E(C)$ and $E(T)$ are the expected cost and expected operation time of one cycle, respectively, and

$$\begin{aligned} E(C) &= E(T_0)C_0 + E(T_1)C_1 + P_{PM}C_{PM} \\ &\quad + P_{CM}C_{CM} + E(\text{IN})C_{IN}, \\ E(T) &= E(T_0) + E(T_1), \end{aligned}$$

where $E(\text{IN})$ is the expected number of inspections within one cycle.

In the rest of this section, we focus on the derivation of the variables in this model.

### 3.1 Failure Rate of Insulations Under Shocks

Suppose $\Psi(t, i)$ is the survival cumulative distribution function (CDF) of insulations under high-voltage shocks in the duration of $t$ when operating in state $i$. According to Poisson process theory, we can obtain

$$\Psi(t,\ i) = \sum_{k=0}^{+\infty}\left[\frac{(\lambda t)^k}{k!}\mathrm{e}^{-\lambda t}(1 - P_{\mathrm{S}i})^k\right] = \exp(-\lambda t P_{\mathrm{S}i}). \tag{2}$$

The failure CDF under shocks within period $t$ at state $i$ ($i$ = 0, 1) is

$$\bar{\Psi}(t,\ i) = 1 - \Psi(t,\ i),$$

and the corresponding PDF is

$$\phi(t,\ i) = \lambda P_{\mathrm{S}i}\exp(-\lambda t P_{\mathrm{S}i}).$$

## 3.2 *State Transition Probability During a Single Inspection Interval*

Suppose $t_{i-1}$ ($i \geq 1$) is the time when the ($i - 1$)th inspection is taken, in which $t_0$ means the initial time of a new operation cycle and at $t_0$ insulations are as good as new. Suppose $s(t_i)$ is the state at time $t_i$, and if $s(t_{i-1}) = 0$, traction motor insulation will continue to work. In the duration of $t_{i-1}$–$t_i$, four scenarios may occur:

(1) Scenario $a$: During the interval of $t_{i-1}$–$t_i$, no state transition occurs, that is to say $s(t_i) = 0$. Then we can obtain the condition probability of occurrence of Scenario $a$ as

$$\begin{aligned} P_a^{t_i} &= P\{s(t_i) = 0|s(t_{i-1}) = 0 \\ &\quad \cup (\forall t \in (t_{i-1},\ t_i],\ s(t) \neq 1 \cup s(t) \neq 2)\} \\ &= \frac{1 - \Phi_0(t_i)}{1 - \Phi_0(t_{i-1})}\frac{1 - F(t_i)}{1 - F(t_{i-1})}\frac{\Psi(t_i, 0)}{\Psi(t_{i-1}, 0)} \\ &= \frac{\bar{\Phi}_0(t_i)}{\bar{\Phi}_0(t_{i-1})}\frac{\bar{F}(t_i)}{\bar{F}(t_{i-1})}\Psi(T_{\mathrm{IN}}, 0), \end{aligned} \tag{3}$$

where $F(t)$ is the CDF of the time from normal state to latent failure state, $\Phi_0(t)$ is CDF of the time to functional failure state in normal state, and $\Phi_1(t)$ is the CDF of the time to functional failure state in latent failure state, and

$$\begin{aligned} \bar{\Phi}_0(t) &= 1 - \Phi_0(t), \\ \bar{F}(t_i) &= 1 - F(t_i). \end{aligned}$$

(2) Scenario $b$: During the interval of $t_{i-1}$–$t_i$, the operation state starts at state 0 but ends at state 1, which means that one state transition from state 0 to state 1

occurs without the occurrence of functional failures. The condition probability of occurrence of Scenario $b$ can be expressed by

$$\begin{aligned} P_b^{t_i} &= P\{s(t_i) = 1|s(t_{i-1}) = 0 \cup (\forall t \in (t_{i-1}, t_i], s(t) \neq 2)\} \\ &= \int\limits_{t_{i-1}}^{t_i} \frac{f(t)}{\bar{F}(t_{i-1})} \frac{\bar{\Phi}_0(t)}{\bar{\Phi}_0(t_{i-1})} \frac{\bar{\Phi}_1(t_i)}{\bar{\Phi}_1(t)} \\ &\quad \cdot \Psi(t - t_{i-1}, 0)\, \Psi(T_{\mathrm{IN}} - t + t_{i-1}, 1)\mathrm{d}t, \end{aligned} \tag{4}$$

where

$$\bar{\Phi}_1(t) = 1 - \Phi_1(t).$$

(3) Scenario $c$: At some time between the interval of $t_{i-1}$–$t_i$, one functional failure happens without state transition from state 0 to state 1. The operation cycle is interrupted. Due to the fact that the probability that functional failure resulting from both the internal factors and shocks simultaneously is quite slim, it is ignored to simplify calculation. The condition probability of occurrence of Scenario $c$ is given by

$$\begin{aligned} P_c^{t_i} &= P(s(t_i) = 2|s(t_{i-1}) = 0 \cup (\forall t \in (t_{i-1}, t_i], s(t) \neq 1) \\ &= \int\limits_{t_{i-1}}^{t_i} \frac{\bar{F}(t)}{\bar{F}(t_{i-1})} \left[ \frac{\varphi_0(t)}{\bar{\Phi}_0(t_{i-1})} \Psi(t - t_{i-1}, 0) \right. \\ &\quad \left. + \frac{\bar{\Phi}_0(t)}{\bar{\Phi}_0(t_{i-1})} \phi(t - t_{i-1}, 0) \right] \mathrm{d}t. \end{aligned} \tag{5}$$

(4) Scenario $d$: At some time between the interval of $t_{i-1}$–$t_i$, one state transition from state 0 to state 1 happens and then one state transition from state 1 to state 2 takes place in order, when the operation cycle is interrupted. The condition probability of occurrence of Scenario $d$ is given by

$$\begin{aligned} P_d^{t_i} &= P\{s(t_i) = 2|s(t_{i-1}) = 0 \\ &\quad \cup (\forall t \in (t_{i-1}, t'], s(t) = 1) \\ &\quad \cup (\forall t \in (t', t_i], s(t) = 2\} \\ &= \int\limits_{t_{i-1}}^{t_i} \Delta_1 \int\limits_{t}^{t_i} \Delta_2 \mathrm{d}t' \mathrm{d}t, \end{aligned} \tag{6}$$

where

$$\Delta_1 = \frac{f(t)}{\bar{F}(t_{i-1})} \frac{\bar{\Phi}_0(t)}{\bar{\Phi}_0(t_{i-1})} \Psi(t - t_{i-1}, 0),$$

$$\Delta_2 = \left[\frac{\varphi_1(t')}{\bar{\Phi}_1(t)} \Psi(t' - t, 1) + \frac{\bar{\Phi}_1(t')}{\bar{\Phi}_1(t)} \phi(t' - t_{i-1}, 1)\right].$$

## 3.3 *Mathematical Models for the Variables in the Optimization Model*

### 3.3.1 Expected Time at State 0 within a Cycle

The expected time at state 0 for traction motor insulations in a cycle includes the following three parts:

(1) No state transition happens in a cycle, and the expected time at state 0 is represented by $T_0^0$;
(2) One state transition from state 0 to state 1 happens, while no state transition from state 0 to state 2 happens, and the expected time at state 0 is represented by $T_0^1$;
(3) One state transition from state 0 to state 2 happens, while no state transition from state 0 to state 1 happens, and the expected time at state 0 is represented by $T_0^2$.

Then the expected time at state 0 yields the following expression:

$$E(T_0) = T_0^0 + T_0^1 + T_0^2, \tag{7}$$

where

$$T_0^0 = T_C \bar{F}(T_C) \bar{\Phi}_0(T_C) \Psi(T_C, 0),$$

$$T_0^1 = \int_0^{T_C} t f(t) \bar{\Phi}_0(t) \Psi(t, 0) \mathrm{d}t,$$

$$T_0^2 = \int_0^{T_C} t \bar{F}(t) [\varphi_0(t) \Psi(t, 0) + \bar{\Phi}_0(t) \phi(t, 0)] \mathrm{d}t,$$

and after simplification, we can obtain

$$E(T_0) = \int_0^{T_C} \bar{\Phi}_0(t)\bar{F}(t)\Psi(t,0)\mathrm{d}t. \tag{8}$$

### 3.3.2 Expected Time at State 1 in a Cycle

If the inspection at time $t_{i-1}$ is completed with a result that insulation is at state 0, we obtain the expression of the probability of its occurrence

$$\begin{aligned} P(S_0^{t_{i-1}}) &= P_a^{t_{i-1}}\bar{F}(t_{i-2})\bar{\Phi}_0(t_{i-2})\Psi(t_{i-2},0) \\ &= \bar{F}(t_{i-1})\bar{\Phi}_0(t_{i-1})\Psi(t_{i-1},0). \end{aligned}$$

If and only if Scenario $b$ or Scenario $d$ happens during $t_{i-1}$–$t_i$, there exists time that insulation is at state 1. The expression of the expected time at state 1 during $t_{i-1}$–$t_i$ is given by

$$E(T_{1i}) = P(S_0^{t_{i-1}}) \int_{t_{i-1}}^{t_i} \left[ P_b^{t_i}(t_i - t) + \Delta_1 \int_t^{t_i} \Delta_2(t' - t)\mathrm{d}t' \right] \mathrm{d}t. \tag{9}$$

For the whole cycle, the expression of the expected time at state 1 is given by

$$E(T_1) = \sum_{i=1}^{N} E(T_{1i}). \tag{10}$$

### 3.3.3 Probability of PM and CM in a Cycle

PM is called for in the following two situations:

Situation $a$: insulation is detected to be at state 1 through the periodic inspection.

Situation $b$: after $N$ inspection intervals, no state transition happens, but insulation is unacceptable. Then a PM is processed.

Through the above analysis, it can be seen that the probability that Situation $a$ happens is equal to the summation of Scenario $b$ happens in each inspection interval, and then it can be derived as

$$P_{\mathrm{PM}}^a = \sum_{i=1}^{N} P_b^{t_i}. \tag{11}$$

Then we can obtain the probability that PM is called for in a cycle as follows:

$$P_{\mathrm{PM}} = \sum_{i=1}^{N} P_b^{t_i} + P(S_0^{t_N}), \tag{12}$$

where $P(S_0^{t_i})$ is the probability of being at state 0 at time $t_i$.

Because a cycle will end when either a PM or a CM is launched, it is clear that

$$P_{\mathrm{CM}} = 1 - P_{\mathrm{PM}}. \tag{13}$$

### 3.3.4 Expected Number of Inspections in a Cycle

When the inspection at time $t_{i-1}$ is completed, if and only if Scenario $a$ or Scenario $b$ happens, the inspection at time $t_i$ is called for, whose probability can be represented by

$$P_{t_i}^{\mathrm{IN}} = P(S_0^{t_{i-1}})(P_a^{t_i} + P_b^{t_i}).$$

So the expected number of inspections within one cycle is

$$E(\mathrm{IN}) = \sum_{i=1}^{m} P(S_0^{t_{i-1}})(P_a^{t_i} + P_b^{t_i}), \quad m \geq 2. \tag{14}$$

## 4 Special Case

The above section describes the optimal maintenance model combining TBM with CBM, in which the shock effect is taken into account. In this part, we will establish the model of merely integrating TBM and CBM without the shock effect as a special case for future comparison of their efficiency and analyzing the effect of the shock parameter variations on optimization solutions. In the following section, we will put a ^ above the variables to differentiate from the ones used in previous sections. For example, $\hat{P}_{\mathrm{PM}}$ stands for the probability to carry out PM within one cycle in the model without considering the shock effect.

The mathematical model under such circumstance is shown as follows.

### 4.1 State Transition Probability Within One Inspection Interval

Scenario *a*:

$$\hat{P}_a^{t_i} = \frac{\bar{\Phi}_0(t_i)}{\bar{\Phi}_0(t_{i-1})}\frac{\bar{F}(t_i)}{\bar{F}(t_{i-1})}.$$

Scenario *b*:

$$\hat{P}_b^{t_i} = \int_{t_{i-1}}^{t_i} \frac{f(t)}{\bar{F}(t_{i-1})}\frac{\bar{\Phi}_0(t)}{\bar{\Phi}_0(t_{i-1})}\frac{\bar{\Phi}_1(t_i)}{\bar{\Phi}_1(t)}\mathrm{d}t.$$

Scenario *c*:

$$\hat{P}_c^{t_i} = \int_{t_{i-1}}^{t_i} \frac{\bar{F}(t)}{\bar{F}(t_{i-1})}\frac{\varphi_0(t)}{\bar{\Phi}_0(t_{i-1})}\mathrm{d}t.$$

Scenario *d*:

$$\hat{P}_d^{t_i} = \int_{t_{i-1}}^{t_i} \hat{\Delta}_1 \int_{t}^{t_i} \hat{\Delta}_2 \mathrm{d}t' \mathrm{d}t,$$

where

$$\hat{\Delta}_1 = \frac{f(t)}{\bar{F}(t_{i-1})}\frac{\bar{\Phi}_0(t)}{\bar{\Phi}_0(t_{i-1})},$$

$$\hat{\Delta}_2 = \frac{\varphi_1(t')}{\bar{\Phi}_1(t)}.$$

### 4.2 Mathematical Models for the Variables in Optimization Model

The expected time of one cycle is

$$E(\hat{T}) = E(\hat{T}_0) + E(\hat{T}_1),$$

where

$$E(\hat{T}_0) = \int_0^{T_C} \bar{\Phi}_0(t)\bar{F}(t)\mathrm{d}t,$$

$$E(\hat{T}_1) = \sum_{i=1}^{N} E(\hat{T}_{1i}),$$

$$E(\hat{T}_{1i}) = \hat{P}(S_0^{t_{i-1}})\left[\int_{t_{i-1}}^{t_i} \hat{P}_b^{t_i}(t_i - t)\mathrm{d}t + \int_{t_{i-1}}^{t_i} \hat{\Delta}_1 \int_t^{t_i} \hat{\Delta}_2(t' - t)\mathrm{d}t'\mathrm{d}t\right].$$

### 4.3 *Probability of PM, CM, and the Expected Number of Inspections in One Cycle*

The probability of PM and CM, and the expected number of inspections in one cycle are

$$\hat{P}_{\mathrm{PM}}^{a} = \sum_{i=1}^{N} \hat{P}_b^{t_i},$$

$$\hat{P}_{\mathrm{CM}} = 1 - \hat{P}_{\mathrm{PM}},$$

$$E(\mathrm{IN}) = \sum_{i=1}^{m-1} \hat{P}(\hat{S}_0^{t_{i-1}})(\hat{P}_a^{t_i} + \hat{P}_b^{t_i}), \quad m \geq 2.$$

## 5 Numerical Investigation and Discussion

Our numerical example used the data from a specific kind of insulation material for traction motors. A primary assumption is that the times for state transitions all follow a Weibull distribution. The time unit is 10,000 h. The shape parameters and scale parameters are (2.198, 96.542) for state 0 to state 1, (2.653, 41.737) for state 1 to state 2, and (2.207, 89.246) for state 0 to state 2. The failure rates under extreme shock are 0.0021 and 0.0057 at state 0 and state 1, respectively.

The basic objective of maintenance optimization is to avoid both excessive and insufficient maintenance and finally reach the best compromise between maintenance

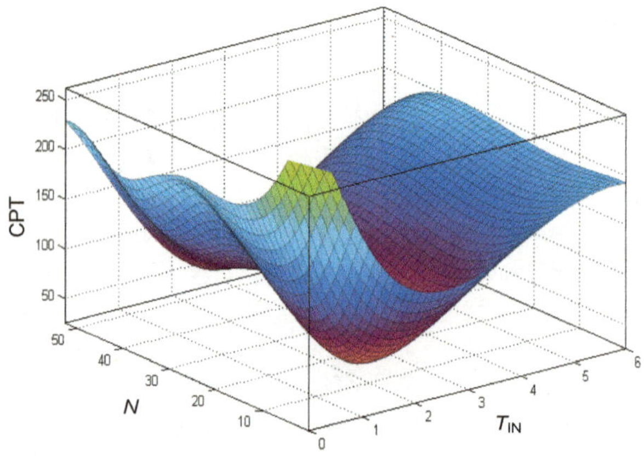


**Fig. 3** Mapping relationship among CPT, $T_{IN}$, and $N$

**Table 1** Optimization solutions

| No. | $\lambda$ | $P_{S0}$ | $P_{S1}$ | $N$ | $T_{IN}$ | $T_C$ | $C$ (%) |
|---|---|---|---|---|---|---|---|
| 1 | 0.01 | 0.0004 | 0.0007 | 14 | 1.60 | 22.40 | 0.74 |
| 2 | 0.02 | 0.0017 | 0.0153 | 16 | 1.21 | 19.36 | 3.84 |
| 3 | 0.02 | 0.0021 | 0.0029 | 13 | 1.58 | 20.54 | 4.03 |
| 4 | 0.02 | 0.0150 | 0.0200 | 12 | 1.39 | 16.68 | 5.76 |
| 5 | 0.05 | 0.0021 | 0.0057 | 12 | 1.48 | 17.76 | 5.38 |
| 6 | 0.20 | 0.0004 | 0.0007 | 13 | 1.57 | 20.41 | 1.65 |
| 7 | 0.20 | 0.0170 | 0.0650 | 9 | 1.38 | 12.42 | 13.70 |
| 8 | 1.00 | 0.1000 | 0.1200 | 4 | 1.31 | 5.24 | 40.70 |

cost and reliability. Using the above data, we can obtain the mapping relationship among CPT, $T_{IN}$, and $N$ as shown in Fig. 3. It can be observed that:

1. The convexity of the surface indicates that there exists an optimal set of $N$ and $T_{IN}$.
2. The CPT of the proposed model is much higher with a small $N$, which results from the poor availability of traction motor insulations and excessively frequent PM. In this case, excessive maintenance occurs.
3. For very large $N$, the maintenance strategy turns into a total CBM model. Furthermore, if $T_{IN}$ is large simultaneously, both the life information and the condition monitoring information are given up. This maintenance transfers to a pure CM strategy and is inevitably costly.
4. At the optimal point, operation cost and failure risk reach the best trade-off. It can be observed that from this point, with the increase of $T_{IN}$ and $N$, the risk of sudden functional failure increases greatly because of insufficient maintenance, which will finally lead to expensive CM.

The optimal solutions for $T_{IN}$ and $N$ are 1.48 and 12, respectively, and the critical age is 17.76. In this case, the optimization solution $\hat{T}_{IN}$ and $\hat{N}$ for the special case are 1.65 and 14, respectively, and the critical age is 23.1.

The influence of shock parameters on the maintenance strategy is another crucial concern. By varying the values of $\lambda$ and $P_{Si}$, a group of solutions, listed in Table 1, is obtained. $C\%$ indicates the cost decrease percent between the CPT of the proposed model and that of the optimal solution of the special case. We can observe that:

1. According to the 1st set of data in Table 1, for small values of shock frequency as well as small failure rate after shocks, the proposed model is very close to the special case. However, for high frequency of shocks as well as high failure rates after shocks, such as the No. 8 dataset, the pre-defined critical age is too short to fulfill the availability requirements, and it is too expensive to operate under such a situation. It can be concluded that the traction motors are inadequate for working in this case.
2. $T_{\mathrm{IN}}$ is more susceptible to the difference between $P_{S0}$ and $P_{S1}$. With an increasing difference, more frequent inspections are demanded. In state 1, shocks are more inclined to lead to functional failure and more frequent inspections should be performed to avoid the operation in state 1.
3. With the increase of the frequency of shocks and failure rate, the optimal critical age decreases. More shocks may lead to the increase of the probability of high-cost CM, and therefore, the critical age should be reduced to lower the probability of CM provoked by time-related internal factors.
4. Comparing with the optimization solution of the special case, both $T_{\mathrm{IN}}$ and $T_{\mathrm{C}}$ are diminished. That means the system will operate under insufficient maintenance if the maintenance strategy obtained from the model of the special case is selected. This may bring about a higher risk of sudden breakdown. With the increase of the shock parameters, the impact of shocks on maintenance strategies becomes more remarkable, and the strategy may even eventually mislead the maintenance actions.

## 6 Conclusions

We developed an optimization model of PM strategy for traction motor insulations, which combines TBM and CBM as well as taking random shocks into account.

The primary contribution of this study is to work out a feasible and practical CBM strategy for traction motor insulations based on their operating characteristics. Through periodic inspections, an operating mechanism comprising two failure modes and three operating states is proposed, which can make the best use of historic life information and inspection information. Moreover, it can overcome the drawbacks of excessive and insufficient maintenance of traditional time-based maintenance schemes.

Another main contribution is to take extreme shock effect into account in the maintenance model. Traction motor insulation is different from normal deteriorating systems because of suffering random extreme shock load during operation. In the proposed model, extreme shocks following a Poisson process are integrated into the maintenance model, which ultimately increases the accuracy of the maintenance model.

Numerical investigation validates that by taking shock effect into account, insulations can operate at a lower risk of catastrophic breakdown, and the operating cost is reduced considerably.

## References

Ahmad, R., & Kamaruddin, S. (2012). An overview of time-based and condition-based maintenance in industrial application. *Computer and Industrial Engineering, 63*(1), 135–149. doi:10.1016/j.cie.2012.02.002.

Asadzadeh, S. M., & Azadeh, A. (2014). An integrated systemic model for optimization of condition-based maintenance with human error. *Reliability Engineering & System Safety, 124,* 117–131. doi:10.1016/j.ress.2013.11.008.

Caballé, N. C., Castro, I. T., Pézrez, C. J., et al. (2015). A condition-based maintenance of a dependent degradation-threshold-shock model in a system with multiple degradation processes. *Reliability Engineering & System Safety, 134,* 98–109. doi:10.1016/j.ress.2014.09.024.

Castro, I. T. (2013). An age-based maintenance strategy for a degradation-threshold-shock-model for a system subjected to multiple defects. *Asia-Pacific Journal of Operational Research, 30* (05), 1350016–1350029. doi:10.1142/S0217595913500164.

Castro, I. T., Caballé, N. C., & Pérez, N. C. (2015). A condition-based maintenance for a system subject to multiple degradation processes and external shocks. *International Journal of Systems Science, 46*(9), 1692–1704. doi:10.1080/00207721.2013.828796.

Chen, J. Y., & Li, Z. H. (2008). An extended extreme shock maintenance model for a deteriorating system. *Reliability Engineering & System Safety, 93*(8), 1123–1129. doi:10.1016/j.ress.2007.09.008.

Fang, Y. T. (2011). On China's high-speed railway technology. *Journal of Zhejiang University-SCIENCE A (Applied Physics & Engineering), 12*(12), 883–884. doi:10.1631/jzus.A11GT000.

Gao, Y. C., Feng, Y. X., Tan, J. R. (2014). Multi-principle preventive maintenance: a design-oriented scheduling study for mechanical systems. *Journal of Zhejiang University-SCIENCE A (Applied Physics & Engineering), 15*(11), 862–872. doi:10.1631/jzus.A1400102.

Gebraeel, N. Z., Lawley, M. A., Li, R., et al. (2005). Residual life distributions from component degradation signals: a bayesian approach. *IIE Transactions, 37*(6), 543–557. doi:10.1080/07408170590929018.

Hu, Q. Y. (1995). The optimal replacement of Markov deteriorative under stochastic shocks. *Microelectronics Reliability, 35*(1), 27–31. doi:10.1016/0026-2714(94)00074-X.

Lam, Y. (2009). A geometric process $\delta$-shock maintenance model. *IEEE Transactions on Reliability, 58*(2), 389–396. doi:10.1109/TR.2009.2020261.

Lin, Y. H., Li, Y. F., & Zio, E. (2014). Multi-state physics model for the reliability assessment of a component under degradation processes and random shocks. In ESREL, Amsterdam, the Netherlands pp. 1–7.

Lin, Y. H., Li, Y. F., & Zio, E. (2015). Integrating random shocks into multi-state physics models of degradation processes for component reliability assessment. *IEEE Transactions on Reliability, 64*(1), 154–166. doi:10.1109/TR.2014.2354874.

Liu, B. Y., Fang, Y. T., Wei, J. X., et al. (2007). Inspection-replacement policy of system under predictive maintenance. *Journal of Zhejiang University-SCIENCE A, 8*(3), 495–500. doi:10.1631/jzus.2007.A0495.

Lu, X. F., & Liu, M. (2014). Hazard rate function in dynamic environment. *Reliability Engineering & System Safety, 130,* 50–60. doi:10.1016/j.ress.2014.04.020.

Ma, H. Z., 2008. *Motor state monitoring and fault diagnosis* (pp. 354–370 ). Beijing, China: China Machine Press (in Chinese).

Montoro-Cazorla, D., & Pérez-Ocón, R. (2014). A reliability system under different types of shock governed by a Markovian process and maintenance policy *K. European Journal of Operational Research, 235*(3), 636–642. doi:10.1016/j.ejor.2014.01.021.

Panagiotidou, S., & Tagaras, G. (2008). Evalution of maintenance policies for equipment subject to quality shifts and failures. *International Journal of Production Research, 46*(20), 5761–5779. doi:10.1080/00207540601182260.

Panagiotidou, S., & Tagaras, G. (2010). Statistical process control and condition based maintenance: a meaningful relationship through data sharing. *Production and Operations Management, 19*(2), 156–171. doi:10.1111/j.1937-5956.2009.01073.x.

Panagiotidou, S., & Tagaras, G. (2012). Optimal integrated process control and maintenance under general deterioration. *Reliability Engineering & System Safety, 104,* 58–70. doi:10.1016/j.ress.2012.03.019.

Shi, H., & Zeng, J. C. (2014). Preventive maintenance strategy based on life prediction. *Computer Integrated Manufacturing Systems, 20*(5), 1133–1140. (in Chinese).

Tang, D. Y., Makis, V., Jafari, L., et al. (2015). Optimal maintenance policy and residual life estimation for a slowly degrading system subject to condition monitoring. *Reliability Engineering & System Safety, 134,* 198–207. doi:10.1016/j.ress.2014.10.015.

Tang, Y. Y., & Lam, Y. (2006). A $\delta$-shock maintenance model for deteriorating system. *European Journal of Operational Research, 168*(2), 541–556. doi:10.1016/j.ejor.2004.05.006.

van Noortwijk, J. M. (2009). Survey of the application of gamma processes in maintenance. *Reliability Engineering & System Safety, 94*(1), 2–21. doi:10.1016/j.ress.2007.03.019.

Wang, Z., Huang, H. Z., Li, Y., et al. (2011). An approach to reliability assessment under degradation and shock process. *IEEE Transaction of Reliability, 60*(4), 852–863. doi:10.1109/TR.2011.2170254.

Yang, Y., & Klutke, G. A. (2000). Lifetime-characteristics and inspection schemes for levy processes. *IEEE Transactions on Reliability, 49*(4), 377–382. doi:10.1109/24.922490.

Yao, Y. Z., Meng, C., Wang, C., et al. (2013). Optimal preventive maintenance policies for equipment under condition monitoring. *Computer Integrated Manufacturing Systems, 19*(12), 2968–2975. (in Chinese).

Yin, H., Zhang, G. J., Zhu, H. P., et al. (2015). An integrated model of statistical process control and maintenance based on the delayed monitoring. *Reliability Engineering & System Safety, 133,* 323–333. doi:10.1016/j.ress.2014.09.020.

Zhou, K., Wu, G. N., Deng, T., et al. (2006). Aging time effect on PD and space charge behavior in magnet wire under high PWM voltage. In *IEEE International Symposium on Electrical Insulation*, Toronto, Canada pp. 159–162. doi:10.1109/ELINSL.2006.1665281.

## Author Biographies

**Jian Zhang** Ph.D., is an assistant researcher of Electrical Engineering at Zhejiang University, China. He received his Ph.D. in mechanical engineering in 2010 from Zhejiang University. In 2015 he was an Academic Visitor at the University of Nottingham, UK. His main research interests are on reliability engineering and motor design.

**Ji-en Ma** Ph.D., is an associate professor in Zhejiang University, Hangzhou, China. She received a Ph.D. degree in mechatronics from Zhejiang University in 2009. Then she did postdoctoral work in the College of Electrical Engineering, Zhejiang University, China. Her recent work is on electrical machines and drives. Her research interests include PM machines and drives for traction applications, and mechatronics machines such as the magneto fluid bearing.

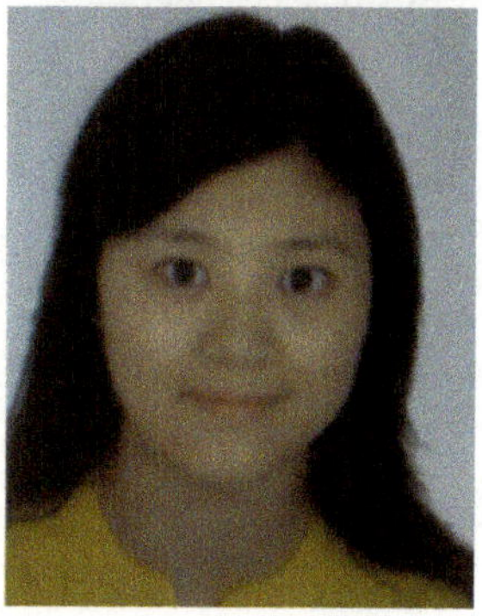

**Xiao-yan Huang** Ph.D., is a professor in Electrical Engineering at Zhejiang University, Hangzhou, China. She received a BE degree from Zhejiang University in 2003, and received a Ph.D. degree in electrical machines and drives from the University of Nottingham, Nottingham, UK, in 2008. From 2008 to 2009, she was a Research Fellow with the University of Nottingham. Her research interests are PM machines and drives for aerospace and traction applications, and generator systems for urban networks.

**You-tong Fang** is a professor at Zhejiang University. He is chairman of the High Speed Rail Research Center in Zhejiang University, deputy director of the National Intelligent Train Research Center, on the committee of the China High Speed Rail Innovation Plan, and expert of the National 863 Program (China High Technology Development Program) in modern transportation and advanced carrying technology. He is also director of three projects of the National Natural Science Foundation of China (NSFC) and more than ten National 863 projects and Technology Support Programs. Recent work has been on electrical machines and drives. His research interests include permanent magnet (PM) machines and drives for traction applications.

# A Combined Simulation of High-Speed Train Permanent Magnet Traction System Using Dynamic Reluctance Mesh Model and Simulink

**Xiao-yan Huang, Jian-cheng Zhang, Chuan-ming Sun, Zhang-wen Huang, Qin-fen Lu, You-tong Fang and Li Yao**

## Contents



X. Huang · J. Zhang · Z. Huang · Q. Lu (✉) · Y. Fang
College of Electrical Engineering, Zhejiang University, Hangzhou 310027, China
e-mail: luqinfen@zju.edu.cn

C. Sun
CSR Qingdao Sifang Co. Ltd., Qingdao 266111, China

L. Yao
Danfoss (Tianjin) Ltd., Tianjin 301700, China



Y. Fang and Y.H. Zhang (eds.), *China's High-Speed Rail Technology*, Advances in High-speed Rail Technology, https://doi.org/10.1007/978-981-10-5610-9_30

# 1 Introduction

Permanent magnet (PM) motor traction systems for high-speed railways have been highlighted recently due to their inherent advantages in terms of compact structure, high efficiency, and high-power factor (Matsuoka 2007). Some prototype trains have been independently developed by Alstom, Bombardier, and Siemens. Although large-scale commercial application is not in prospect at the moment, the PM motor traction system is considered as one of the most advanced high-speed train technologies for the future (Mermet-Guyennet 2010; Lee 2012).

A common way to evaluate the performance of the PM motor traction system is a dynamic simulation using the MATLAB Simulink models (Lu et al. 2011). However, the PM motor Simulink model is based on constant machine parameters such as resistance, *d*-axis inductance ($L_d$), *q*-axis inductance ($L_q$) but, in practice, the parameters such as $L_d$ and $L_q$ vary during operation. Therefore, the PM motor Simulink model cannot represent the real situation.

The finite element PM motor model can be used together with the Simulink power electronics model for accurate and complete solutions for detailed behavior (Ugalde et al. 2009). However, a substantial amount of computational resources and much longer computational time are needed, especially for the traction motor which is normally rated at a few hundred kilowatts.

The dynamic mesh modeling (DRM) method was developed by Carpenter (1968), Ostovic (1986, 1988, 1989), Sewell et al. (1999), and Yao (2006). It is based on the reluctance mesh. The magnetic field behavior is mapped onto an equivalent lumped equivalent circuit. This DRM method can save computation time without compromising accuracy (Dogan et al. 2013; Araujo et al. 2014; Nguyen-Xuan et al. 2014). Therefore, in this paper, a dynamic mesh PM motor model is built and combined with Simulink power electronics models for better accuracy and higher simulation efficiency.

# 2 Dynamic Reluctance Mesh Model

## *2.1 Principle of Basic Reluctance Mesh (RM) Model*

The DRM method is based on the concept of equivalent magnetic circuits. Many laws and analysis methods for magnetic circuits can be analogous to those of electric circuit theory. In electric circuits, the fundamental parameters are voltage ($V$), current ($I$), and resistance ($R_e$), and their behavior is described by network constraints, such as Kirchhoff's voltage and current laws, and their constitutive relationships are described by, for example, Ohm's law. In magnetic circuits, the corresponding fundamental parameters are magnetomotive (MMF), flux ($\Phi$), and reluctance ($R_m$) as shown in Table 1. A simplified RM motor model consists of nine

**Table 1** Fundamental parameters in magnetic and electric circuits

| Magnetic circuit | Reluctance | MMF | Flux |
|---|---|---|---|
| | $R_m$ | $MMF = \Phi \cdot R_m$ | $\Phi = MMF/R_m$ |
| Electric circuit | Resistance | Voltage | Current |
| | $R_e$ | $V = IR_e$ | $I = V/R_e$ |

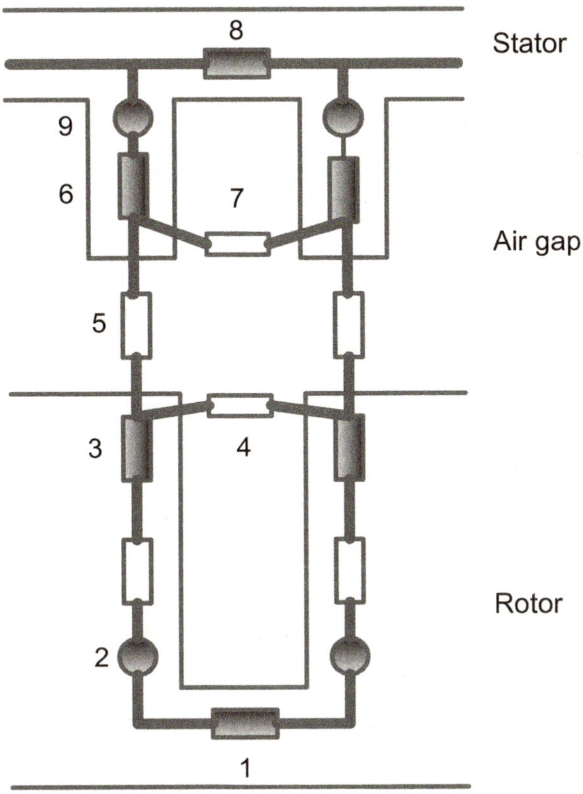


**Fig. 1** Simplified RM model. *1* rotor yoke; *2* magnetic source in rotor (*permanent magnet*); *3* rotor pole; *4* leakage between rotor poles; *5* air gap between stator and rotor; *6* stator teeth; *7* air gap between stator teeth; *8* stator yoke; *9* magnetic source in stator

components as shown in Fig. 1, where the square components represent the reluctance, while the circle component represents the MMF.

The magnetic circuit laws are analogous to those of electric circuits as follows: Kirchhoff's current and voltage law for magnetic circuits, for each node,

$$\sum_k \phi_k = 0, \tag{1}$$

$$\sum_k F_k = 0, \tag{2}$$

where $\phi_k$ is the flux, and $F_k$ is the MMF.

The reluctance of each component can be calculated by

$$R_{\mathrm{m}} = \frac{l}{\mu A}, \tag{3}$$

where $l$ is the length, $A$ is the cross section area of each component, and $\mu$ is the permeability.

Based upon the above equations, the performance of the machine can be investigated.

## 2.2 *Dynamic Reluctance Mesh Model*

The air gap plays a very important role in performance of the PM motor, because its reluctance is large compared to that of the other parts. Therefore, the air gap model is a key factor in simulating electrical machine performance accurately. However, air gap reluctances change as the rotor rotates, because under different rotor positions, the coupling between stator and rotor teeth is different.

In this study, the air gap reluctance is recalculated at every instantaneous rotor position. This is also the reason why the reluctance mesh is "dynamic." The torque can be calculated from the derivatives of the energy (co-energy) versus the displacement of the rotor. Since the air gap is the only parameter that varies with the rotor position, the torque can be written as (Yao 2006)

$$T_{\mathrm{e}} = \sum_{i=0}^{N-1} \left( \frac{1}{2} \frac{B_i^2 d_i l_i}{\mu_0} \frac{\partial w_i}{\partial \theta} \right), \tag{4}$$

where $B_i$ is the flux density in the $i$th air gap reluctance, $\mu_0$ is the permeability of the vacuum, $\theta$ is the angle, $N$ is the turn number of the winding, and $d_i$, $l_i$, $w_i$ are the depth, length, width of the $i$th air gap reluctance, respectively.

## 2.3 *DRM Model of the PM Traction Motor*

A quarter model of the PM traction motor analyzed in this study is shown in Fig. 2. The interior PM rotor structure is used to improve the torque performance of the traction motor. However, due to the complex rotor structure, the magnetic calculation will be more challenging, especially in the calculation of $L_d$ and $L_q$. Based on the analysis above, the reluctance network model per pole pair is built as shown in Fig. 3. In addition, the equivalent magnetic network model of the air gap at a certain rotor position is shown in Fig. 4, which should be changed as the rotor rotates as discussed above.

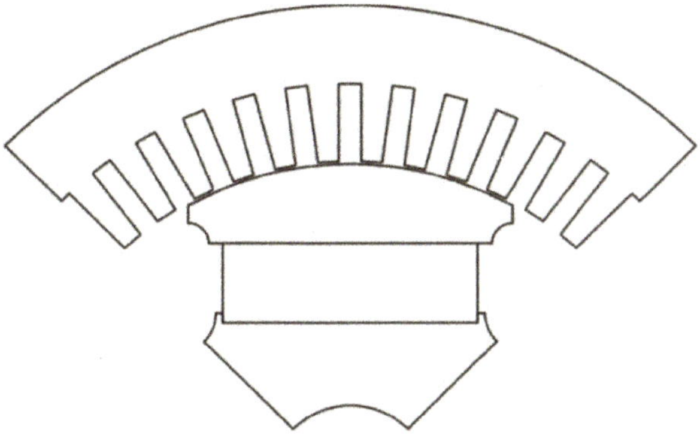

**Fig. 2** Quarter model of the PM traction motor

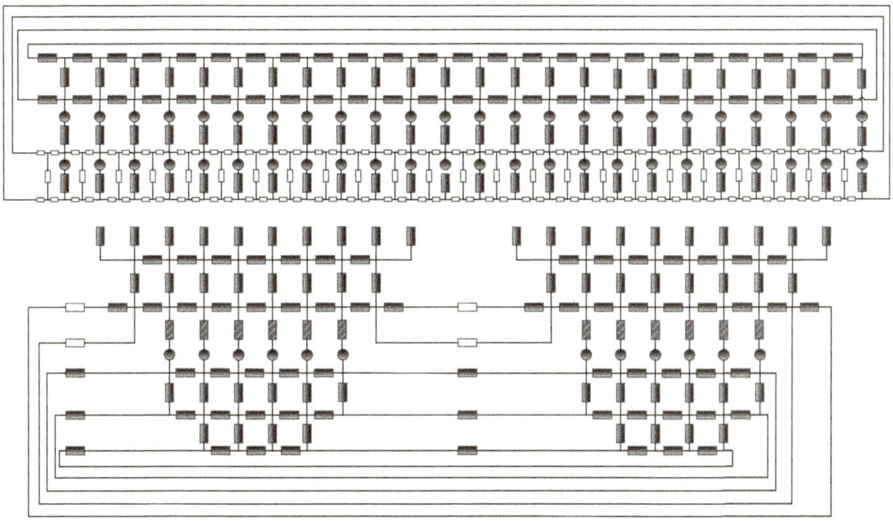

**Fig. 3** Reluctance network model of the PM motor

## 2.4 Simulation Results of DRM Model

Using the DRM model, the main motor parameters such as electromotive force (EMF) at no-load conditions, $L_d$ and $L_q$ can be accurately calculated in less time.

The back EMF at the no-load condition is calculated by Eq. (5). Based on the winding distribution, the phase-$A$ winding is across a number $k$ of stator teeth, so the flux linkage of phase-$A$ winding can be expressed as Eq. (6). The comparison results for the back EMF separately calculated by DRM and finite element method (FEM) are presented in Fig. 5. It can be seen that the result calculated by DRM has a good agreement with that calculated by FEM.

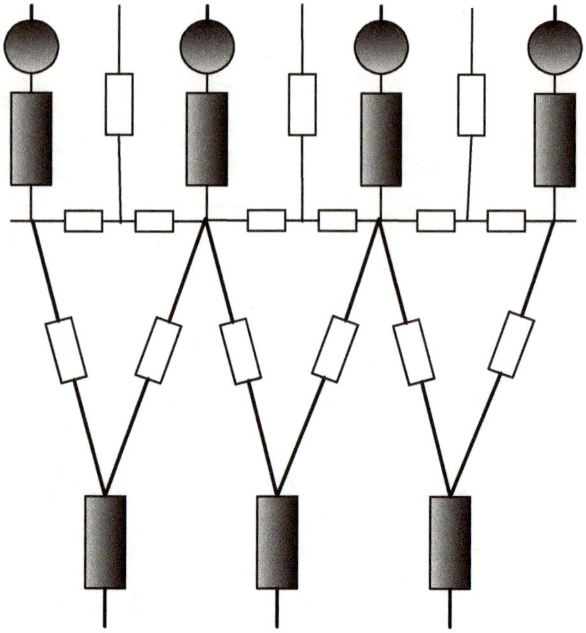

**Fig. 4** Equivalent magnetic network model of the air gap

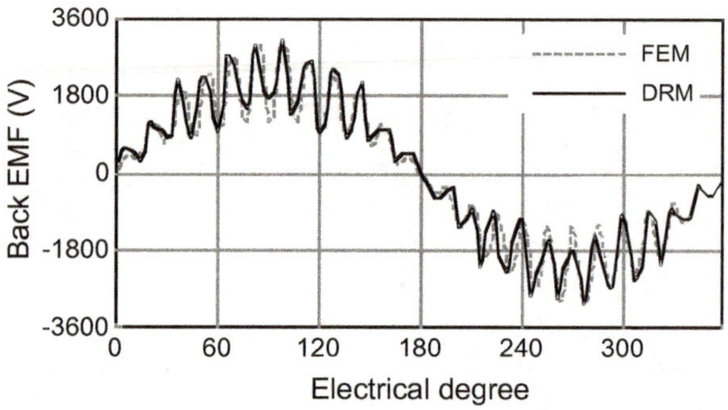


**Fig. 5** Back EMF results calculated by DRM and FEM

$$E_A = -\frac{d\Phi_A}{dt}, \tag{5}$$

$$\Phi_A = \sum_{i=1}^{k} N \cdot \varphi_i, \tag{6}$$

where $E_A$ is the back EMF at the no-load condition of phase-$A$ winding, $\Phi_A$ is the flux linkage of phase-$A$ winding, and $\varphi_i$ is the flux through stator tooth $i$.

As for calculation of $L_d$ and $L_q$, an advanced method called frozen permeability is used to consider the effect of saturation caused by magnets and current. The frozen permeability method calculates the inductance through two calculations. In

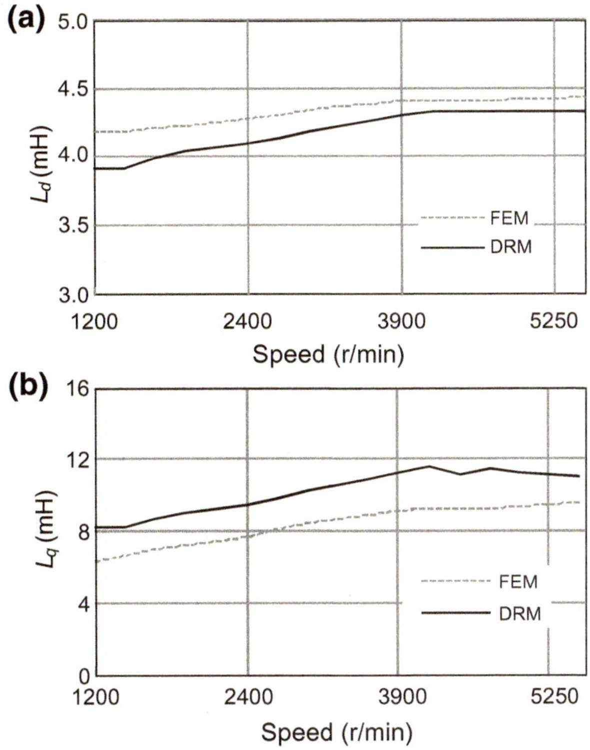


**Fig. 6** $L_d$ (**a**) and $L_q$ (**b**) calculated by DRM and FEM

the first calculation, excitation sources of both the winding current and the magnet are active. The resulting permeability distribution from the first calculation is used in the second calculation. In addition, in the second calculation, the magnet is set to inactive, so only the winding current is involved. Thus, the inductance can be calculated through Eq. (7). The inductance calculation results are presented in Fig. 6 and show good agreement with the FEM calculation.

$$L = \frac{d\psi}{dI}. \tag{7}$$

where $L$ is the inductance, and $\psi$ is the flux linkage.

## 3 Combined Model Using DRM and Simulink

The basic concept of the combined Simulink-DRM simulation is using the DRM model as the motor model, while using the control system model provided by Simulink. There are two common ways to implement the combined Simulink-DRM simulation:

1. Replacing the DRM of PM motors for the constant parameter model (CPM) of PM motors provided by Simulink. DRM and Simulink will be running simultaneously. The dynamic parameters will be provided by DRM. However, for every time-stepping, the DRM and Simulink models need iteration calculation, which makes the simulation more complicated and time-consuming.
2. A dynamic parameters database is generated using DRM simulation. Then the dynamic parameters are put in the CPM in Simulink using the database index or lookup table. In this case, the constant parameters in the traditional Simulink model will be modified instantaneously according to the motor operating condition. Compared to the first method, this method is more efficient without sacrifice of accuracy. Therefore, it will be adopted in this study.

The combined simulation carried out in this paper is based on the following assumptions:

1. Ignoring the effect of rotor skew on the stator back-EMF harmonics;
2. Keeping the excitation from the PM constant;
3. Ignoring the saturation of stator end winding flux leakage.

## 3.1 Dynamic Modeling and Interface

The dynamic parameter permanent magnet synchronous motor (PMSM) model was built using the Simulink S-function block, to replace the traditional CPM. The following equations are used with the S-function block to build the *d–q* coordinated PMSM model (Wang 2011):

$$\begin{cases} u_d = R_1 i_d - \omega\psi_d + \dfrac{\mathrm{d}\psi_d}{\mathrm{d}t}, \\ u_q = R_1 i_q - \omega\psi_q + \dfrac{\mathrm{d}\psi_q}{\mathrm{d}t}, \end{cases} \tag{8}$$

$$\begin{cases} \psi_d = L_d i_d + \psi_{\mathrm{m}}, \\ \psi_q = L_q i_q, \end{cases} \tag{9}$$

$$\frac{\mathrm{d}\omega}{\mathrm{d}t} = \frac{1}{J}(T_{\mathrm{e}} - T_{\mathrm{f}} - F\omega - T_{\mathrm{m}}), \tag{10}$$

$$T_{\mathrm{e}} = \frac{3}{2}p[\psi_{\mathrm{m}} i_q + (L_d - L_q) i_d i_q], \tag{11}$$

where $u_d$, $u_q$ are the voltage of the *d*-, *q*-axis; $i_d$, $i_q$ are the current of the *d*-, *q*-axis; $\psi_d$, $\psi_q$, $\psi_{\mathrm{m}}$ are the flux linkage of the *d*-, *q*-axis, and permanent magnet flux linkage; $T_{\mathrm{e}}$, $T_{\mathrm{f}}$, $T_{\mathrm{m}}$ are the electric torque, friction torque, and load torque; *J*, *F* are the inertia, friction; $L_d$, $L_q$ are the inductance of the $d$-, $q$-axis, respectively, and $\omega$ is the angular speed. In dynamic parameter models, $L_d$, $L_q$ vary with the rotor position.

## 3.2 Vector Control System

The basic concept of PMSM vector control is to control the PMSM like a separately excited DC motor. The stator current can be defined in the $d$–$q$ coordinate system. The field flux linkage component of the current is normally aligned with the $d$-axis, while the torque component of the current is aligned along the $q$-axis. By controlling $i_d$ and $i_q$, the field and torque can be controlled respectively. In this study, the traction system behavior under $i_d = 0$ and maximum torque per ampere control (MTPA) will be investigated.

1. PMSM with $i_d = 0$ control

According to Eq. (11), when $i_d = 0$, the torque will be determined by controlling $i_q$. In the case of $i_d = 0$, there is no reluctance torque, which means it is more suitable for a non-saliency motor. In this study, the salient PMSM will be used. Thus, this method is not the best choice and is only be used here to provide comparison with other methods.

2. MTPA

For the MTPA control, the following equations are used (Tang 1997):

$$\begin{cases} \dfrac{\partial(T_e/i_s)}{\partial i_d} = 0, \\ \dfrac{\partial(T_e/i_s)}{\partial i_q} = 0. \end{cases} \tag{12}$$

Keeping stator current $i_s$ constant, and substituting $i_s = \sqrt{i_d^2 + i_q^2}$ into Eq. (12), $i_d$ can be written as

$$i_d = \frac{-\psi_m + \sqrt{\psi_m^2 + 4(L_d - L_q)^2 i_q^2}}{2(L_d - L_q)}, \tag{13}$$

and then the torque can be rewritten as

$$T_e = \frac{3}{4} p i_q \left( \sqrt{\psi_m^2 + 4(L_d - L_q)^2 i_q^2} + \psi_m \right). \tag{14}$$

For any given torque, the minimum $i_d$, $i_q$ can be found.

### 3.3 Combine DRM and Simulink Model

1. Traction system model with $i_d = 0$

Figure 7 shows the PMSM traction system diagram with $i_d = 0$. The $i_d$ reference is set to zero. $i_q$ will be controlled by the proportional integral (PI) controlled according to the speed error. The PMSM model is built using an S-function, and $L_d$, $L_q$ vary with the rotor position.

2. Traction system model with MTPA

PMSM traction system with MTPA control is similar to that with the $i_d = 0$ control. The only difference is that $i_d$, $i_q$ will be calculated according to the torque demand using Eq. (14) as shown in Fig. 8. It can be seen from the equation that the torque and current relationship cannot be solved directly. In this study, an MTPA control block using the S-function will be built to determine $i_q$ using a bi-section method for the numeric solution. Then $i_d$ can be found.

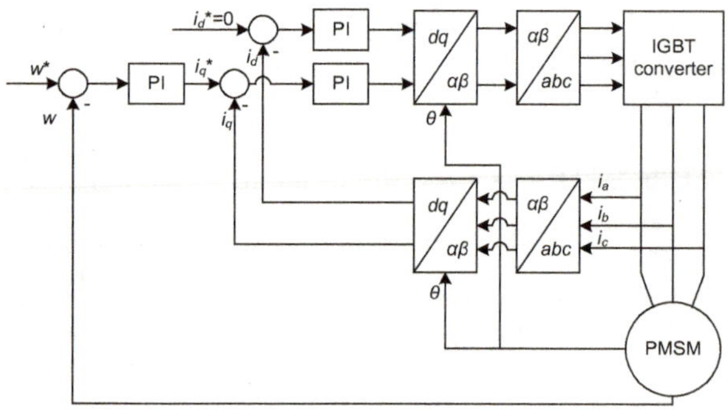


**Fig. 7** PMSM traction system diagram with $i_d = 0$

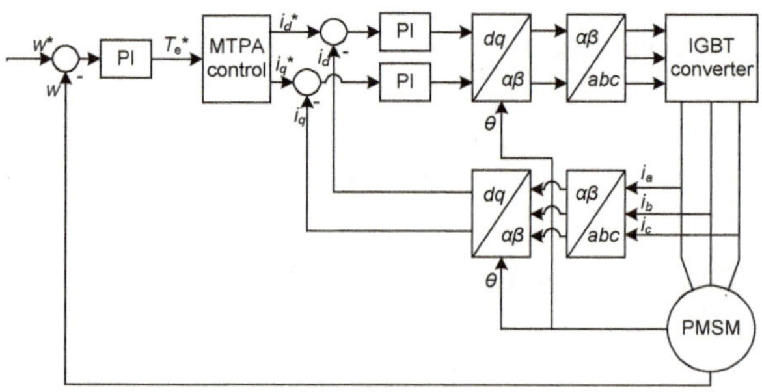


**Fig. 8** PMSM traction system diagram with MTPA

## 4 Simulation Results

A 600 kW PMSM traction system model for a high-speed train was built. The required starting torque is 3500 N m. In the constant torque area, the rated torque is 1350 N m, and the speed is 3000 r/min. In the brake mode, the speed will reduce to zero. The output current and torque were observed.

### *4.1 Constant Parameter PMSM Traction System Simulink Models*

The first model had $i_d = 0$ using Simulink built in blocks. Both $i_d = 0$ and MTPA control were adopted. The motor dynamic speed response, torque, $i_d$, $i_q$, and power factor were compared under both control strategies, as shown in Fig. 9.

The speed response is faster and smoother under the MTPA control as shown in Fig. 9a. It also can be seen from Fig. 9b, under both control methods, that the output torque can be controlled precisely, providing starting and braking torque of 3500 N m. However, with $i_d = 0$ control, the torque pulsation was greater than that under MTPA control. It means that the system has better transient performance under MTPA control. For the phase-*A* current, as shown in Fig. 9c, for the same torque, with MTPA control the current is much less than with $i_d = 0$ control, which will lead to high efficiency of the PMSM. $i_d$ could be less than zero to realize field weakening control under MTPA control as shown in Fig. 9d, e. In this case, the operation speed range can be extended. The power factor with MTPA control was better as shown in Fig. 9f.

Based upon the above simulation results, it can be concluded that although the MTPA is more complicated, they have advantages in terms of fast transient behavior, high efficiency, and high-power factor.

### *4.2 Combined Simulation Using DRM and Simulink*

The key feature of the combined simulation using DRM and Simulink is that $L_d$ and $L_q$ vary with the rotor position. However, the changes of $L_d$ and $L_q$ will not have much effect under $i_d = 0$ control. Thus, the system performance under $i_d = 0$ control will not be investigated in this study. The transient behavior using CPM and combined DPM under MTPA control will be compared. The simulations were carried out. The results are shown in Fig. 10.

The transient behavior of the system under each model is similar except for the $L_d$ and $L_q$ as shown in Fig. 10. However, during the acceleration of the train, the system was operating in the constant torque area, and the output torque waveform is as shown in Fig. 11. Because of the variation of $i_d$ and $i_q$, the degree of motor

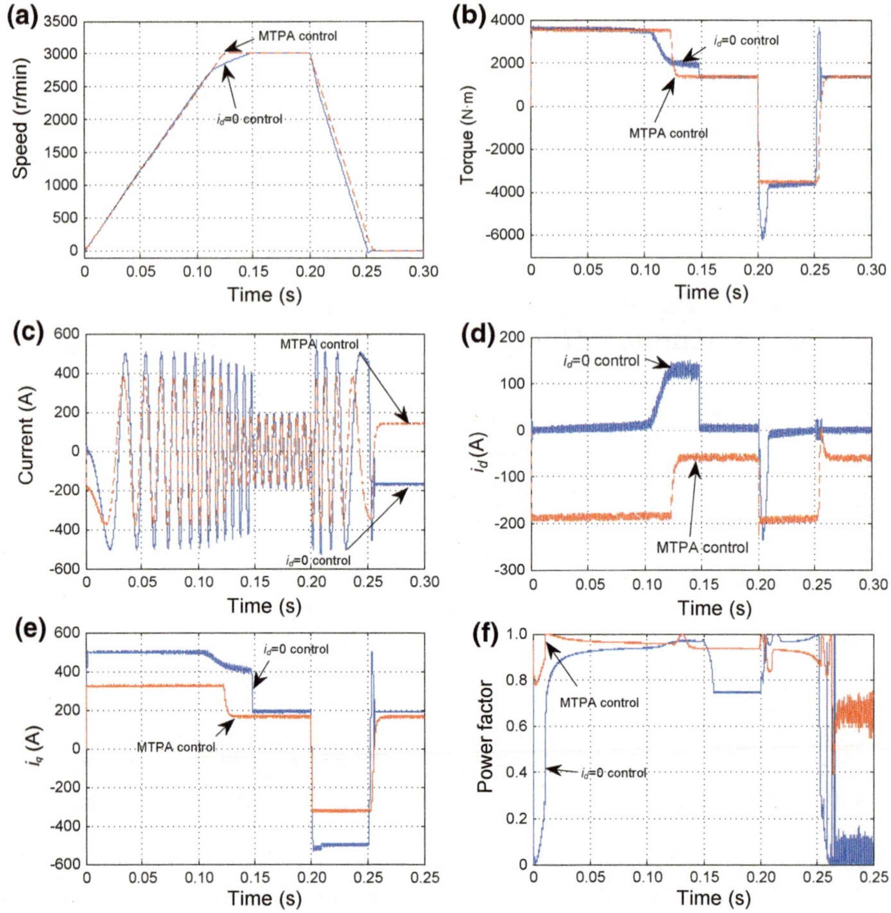


**Fig. 9** Comparison of system behavior under $i_d = 0$ and MTPA control using Simulink model. **a** speed response; **b** torque; **c** phase-*A* current; **d** $i_d$; **e** $i_q$; **f** power factor

saturation varied and therefore $L_d$ and $L_q$ varied. The output torque varied accordingly, especially when the speed was further increased, and the torque pulsation increased correspondingly.

During the braking operation, the output torque waveforms using the CPM and DPM are shown in Fig. 12. The system was in the field weakening operation; therefore, $L_d$ and $L_q$ varied dramatically due to saturation, which leads to a large torque pulsation.

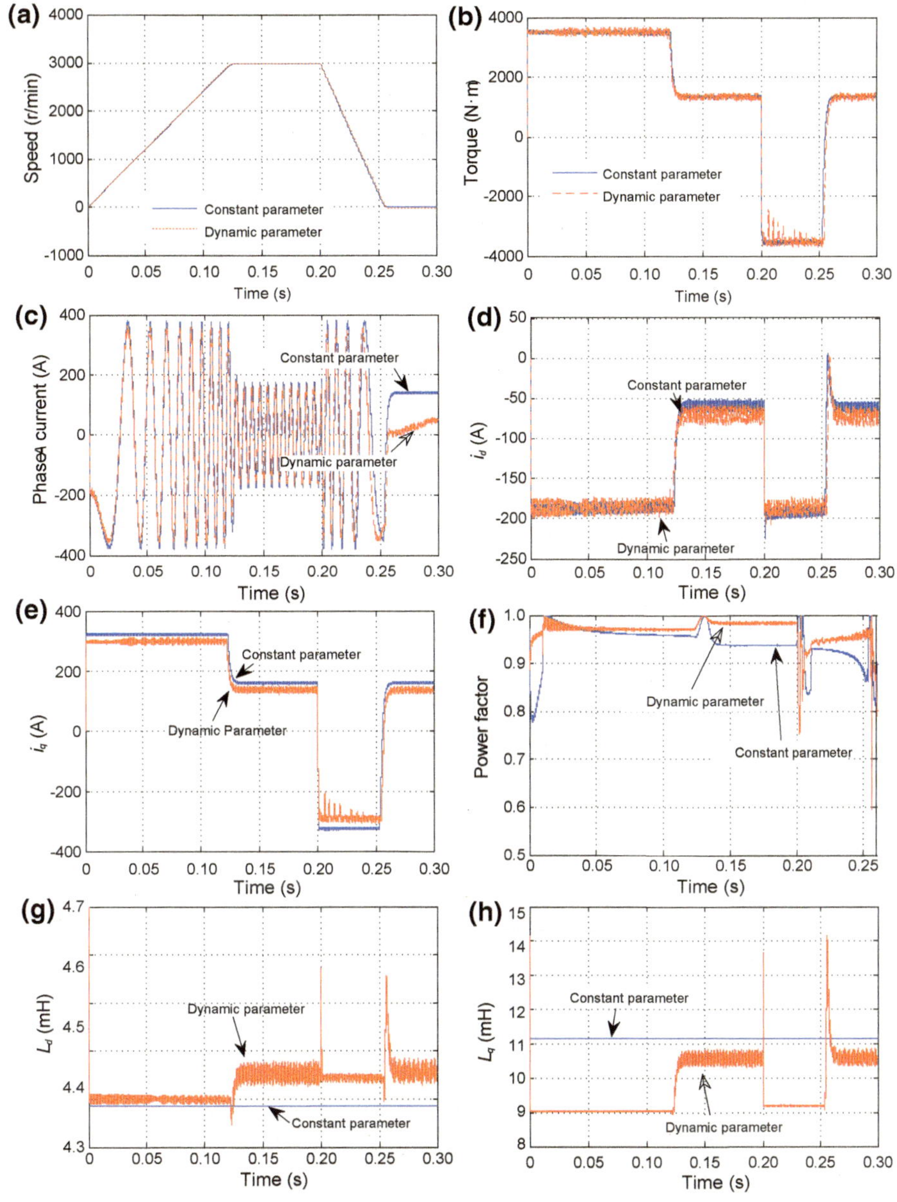


**Fig. 10** Comparison of simulations using constant parameter model and dynamic parameters model under MTPA control. **a** speed response; **b** torque; **c** phase current; **d** $i_d$; **e** $i_q$; **f** power factor; **g** $L_d$; **h** $L_q$

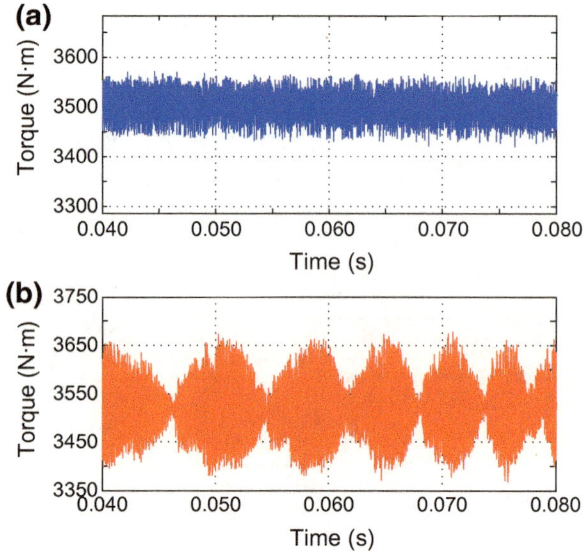


**Fig. 11** Output torque under speed accelerating operation: **a** CPM; **b** DPM

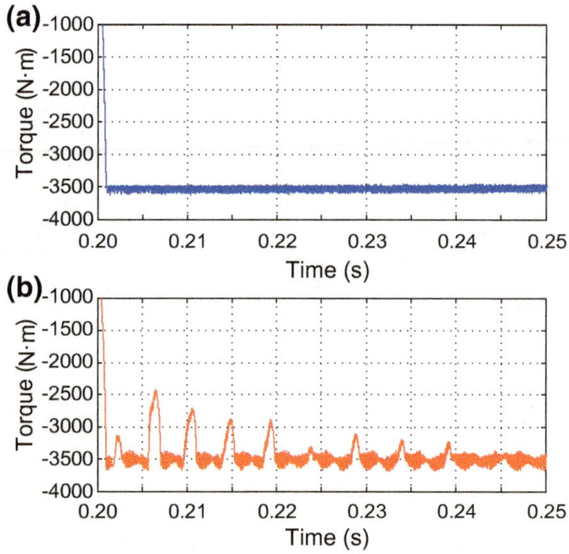


**Fig. 12** Comparison of output torque during the braking operation: **a** CPM; **b** DPM

## 5 Conclusions

This paper presented a combined dynamic parameter PMSM traction system model using the DRM and MATLAB Simulink. Models under $i_d = 0$ and MTPA control were built. Simulations were carried out under real high-speed train operation including acceleration, constant speed, and braking. The results proved that the system with MTPA control has advantages in terms of fast transient response, and high-power factor and efficiency. Simulations using both CPM and DPM models were carried out. The results confirm that the DPM model provides higher accuracy without much sacrifice of time consumption or computation resource. The combined DPM model was shown to be useful and able to provide a reference for the actual design and production of a high-speed train traction system.

## References

Araujo, D. M., Coulomb, J. L., Chadebec, O., et al. (2014). A hybrid boundary element method-reluctance network method for open boundary 3-d nonlinear problems. *IEEE Transactions on Magnetics, 50*(2), 77–80. doi:10.1109/TMAG.2013.2281759.

Carpenter, C. J. (1968). Magnetic equivalent circuits. *Proceedings of the Institution of Electrical Engineers, 115*(10), 1503–1511.

Dogan, H., Garbuio, L., Nguyen-Xuan, H., et al. (2013). Multistatic reluctance network modeling for the design of permanent-magnet synchronous machines. *IEEE Transactions on Magnetics, 49*(5), 2347–2350. doi:10.1109/TMAG.2013.2243426.

Lee, K. K. (2012). *The evolution and outlook of core technologies for high-speed railway in China*. 1st international workshop on high-speed and intercity railways, **2**:495–507. doi:10.1007/978-3-642-27963-8_45.

Lu, Q. F., Wang, B., Huang, X. Y., et al. (2011). Simulation software for CRH2 and CRH3 traction driver systems based on Simulink and VC. *Journal of Zhejiang University-SCIENCE A (Applied Physics & Engineering), 12*(12), 945–949. doi:10.1631/jzus.A11GT006.

Matsuoka, K. (2007). Development trend of the permanent magnet synchronous motor for railway traction. *IEEJ Transactions on Electrical and Electronic Engineering, 2*(2), 154–161. doi:10.1002/tee.20121.

Mermet-Guyennet, M. (2010). *New power technologies for traction drives*. International symposium on power electronics, electrical drives, automation and motion, Pisa, Italy, pp. 719–723.

Nguyen-Xuan, H., Dogan, H., Perez, S., et al. (2014). Efficient reluctance network formulation for electrical machine design using optimization. *IEEE Transactions on Magnetics, 50*(2), 869–872. doi:10.1109/TMAG.2013.2282407.

Ostovic, V. (1986). A method for evaluation of transient and steady state performance in saturated squirrel cage induction machines. *IEEE Transactions on Energy Conversion, 1*(3), 190–197.

Ostovic, V. (1988). A simplified approach to magnetic equivalent-circuit modeling of induction machines. *IEEE Transactions on Industry Applications, 24*(2), 308–316.

Ostovic, V. (1989). A novel method for evaluation of transient states in saturated electric machines. *IEEE Transactions on Industry Applications, 25*(1), 96–100.

Sewell, P., Bradley, K. J., Clare, J. C., et al. (1999). Efficient dynamic models for induction machines. *International Journal of Numerical Modelling: Electronic Networks, Devices and Fields, 12*(6), 449–464. doi:10.1002/(sici)1099-1204(199911/12)12:6<449:aid-jnm365>3.0.co;2-w.

Tang, R. Y. (1997). *Modern permanent magnet machines: theory and design* (p. 473). Beijing, China: China Machine Pres. (in Chinese).

Ugalde, G., Almandoz, G., Poza, J., et al. (2009). Computation of iron losses in permanent magnet machines by multi-domain simulations. In: *13th European Conference on Power Electronics and Applications*, Barcelona (pp. 1–10).

Wang, X. H. (2011). *Permanent magnet machines* (p. 327). Beijing, China: China Electric Power Press. (in Chinese).

Yao, L. (2006). Magnetic field modelling of machine and multiple machine systems using dynamic reluctance mesh modelling. PhD Thesis, University of Nottingham, Nottingham, UK.

## Author Biographies

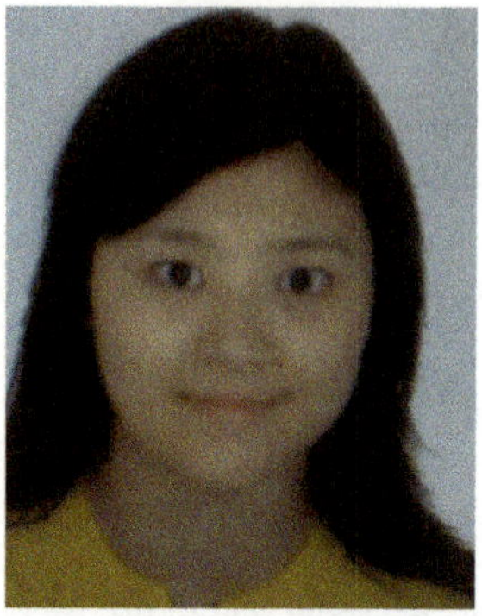

**Xiao-yan Huang** Ph.D., is a professor in Electrical Engineering at Zhejiang University, Hangzhou, China. She received a BE degree from Zhejiang University in 2003, and received a Ph.D. degree in electrical machines and drives from the University of Nottingham, Nottingham, UK, in 2008. From 2008 to 2009, she was a Research Fellow with the University of Nottingham. Her research interests are PM machines and drives for aerospace and traction applications, and generator systems for urban networks.

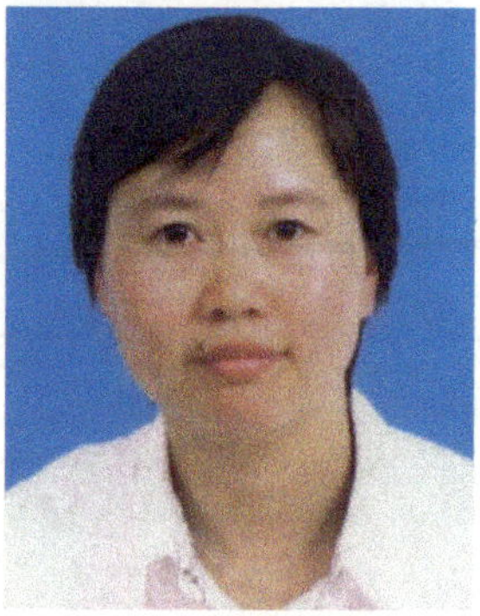

**Qin-fen Lu** Ph.D., is a professor of the College of Electrical Engineering, Zhejiang University, China. She is an IET Fellow, IEEE Senior Member, Vice Chairman, and Secretary-General of the Linear Motor Committee, China Electrotechnical Society.

**You-tong Fang** is a professor at Zhejiang University. He is chairman of the High Speed Rail Research Center in Zhejiang University, deputy director of the National Intelligent Train Research Center, on the committee of the China High Speed Rail Innovation Plan, and expert of the National 863 Program (China High Technology Development Program) in modern transportation and advanced carrying technology. He is also director of three projects of the National Natural Science Foundation of China (NSFC) and more than ten National 863 projects and Technology Support Programs. Recent work has been on electrical machines and drives. His research interests include permanent magnet (PM) machines and drives for traction applications.

# 3D Thermal Analysis of a Permanent Magnet Motor with Cooling Fans

**Zheng Tan, Xue-guan Song, Bing Ji, Zheng Liu, Ji-en Ma and Wen-ping Cao**

## Contents



B. Ji · Z. Liu
School of Electrical and Electronic Engineering,
Newcastle University, Newcastle upon Tyne NE1 7RU, UK

X. Song
School of Mechanical Engineering, Dalian University of Technology, Dalian 116024, China

J. Ma
College of Electrical Engineering, Zhejiang University, Hangzhou 310027, China

J. Ma
Department of Mechanical Engineering, University of California at Berkeley,
Berkeley, CA 94720-1740, USA

W. Cao (✉)
School of Electronics, Electrical Engineering and Computer Science,
Queen's University Belfast, Belfast BT9 5AH, UK
e-mail: w.cao@qub.ac.uk

W. Cao
Department of Electrical Engineering and Computer Science,
Massachusetts Institute of Technology, Cambridge, MA 02139, USA

Z. Tan
Power Grid Technology Center, State Grid Jibei Electric Power Co. Ltd
Research Institute, Beijing 100045, China

 
Y. Fang and Y.H. Zhang (eds.), *China's High-Speed Rail Technology*,
Advances in High-speed Rail Technology, https://doi.org/10.1007/978-981-10-5610-9_31

## 1 Introduction

In recent years, permanent magnet (PM) machines have attracted much attention in various industries, such as electric vehicles, wind generators, electric ships, and high-speed trains, due to their high-torque density and high efficiency in compact packages. With the ever-growing demand for high-torque density and drive power, the thermal problems of PM machines have become increasingly studied, and the efficient analysis and design of cooling techniques for PM machines have become very important and necessary.

It is widely known that the performance of PM machines in many applications, including high-speed trains, is affected by overheating, which can easily cause magnet demagnetization and degradation of the isolation materials and of motor efficiency. Therefore, it is very important to develop an effective heat dissipation technique for PM machines. In this paper, a novel and special cooling system design for an axial flux permanent magnet (AFPM) machine are carried out as a case study. In the literature, there has been much research on the electromagnetic aspects of AFPM machines, but their thermal performance has not been fully investigated (Mellor et al. 1991; Hendershot and Miller 1994; Lee et al. 2000; El-Refaie et al. 2004; Boglietti et al. 2005; Mezani et al. 2005; Dorrrell et al. 2006; Trigeol et al. 2006; Yu et al. 2010; Wrobel et al. 2013; Chong et al. 2014; Dong et al. 2014). For other types of electrical machines, there are four methods generally used for thermal analysis. These are the lumped parameter network method, the finite element method (FEM), the computational fluid dynamics (CFD) method, and the experimental method. The lumped parameter network method is widely used due to its fast prediction and reasonable simplification. However, to achieve accurate results, the model parameters such as convective heat transfer coefficients, thermal contact resistances, and the properties of components have to be defined or assumed before the analysis, and some of these properties are difficult or impractical to measure in reality. In comparison, the FEM needs less experimental or empirical values for an accurate analysis, and it can provide more details about localized temperature distribution. However, the FEM is usually used for conduction heat transfer problem rather than convection heat transfer and conjugate heat transfer problems and that restricts its wide application for AFPM machines with air-forced cooling. The experimental method is the most effective method, but the test facilities required can be costly and time-consuming. Therefore, for convection heat transfer analysis of AFPM machines with air-forced cooling devices, the CFD method is chosen here because it can consider both conduction and convection phenomena, even though it usually takes a little more time and effort than the lumped parameter network and FEM.

This paper presents a 3D thermal study of an AFPM machine using CFD analysis. The heat source is obtained and defined based on physical experiments. Two kinds of fans for forced convection cooling are designed and installed on the AFPM machine without adding any cooling equipment. The results show that CFD is very efficient for thermal analysis of the AFPM machine at the design stage.

The design of cooling fans is helpful in lowering the temperature and avoiding overheating issues in the machine.

## 2 AFPM Machine

The AFPM used in this study for thermal analysis is shown in Fig. 1. It is used for traction drive applications. There are three main components of the machine, of which the first is the rotor disc with 30 embedded permanent magnets, the second is the stator with copper wires wound around a steel core, and the third is the casing supporting the rotor disc and the stator. The copper windings are bundles of individual wires of 0.5 mm diameter coated with a layer of insulation. Figure 2 shows a simplified model in 3D, which considers the bundles as one non-individual body. The air enters through the cutout on the casing around the shaft,and flows out from the cutout on the casing in the axial direction.

## 3 CFD Modeling

The approach used for thermal analysis in this study is based on steady CFD analysis. As mentioned above, it involves both heat conduction and convection transfer. To solve such a conjugate heat transfer problem with a high accuracy, 3D steady Reynolds-averaged Navier-Stokes equations and the energy equation are developed and numerically solved.

**Fig. 1** Photo of the AFPM

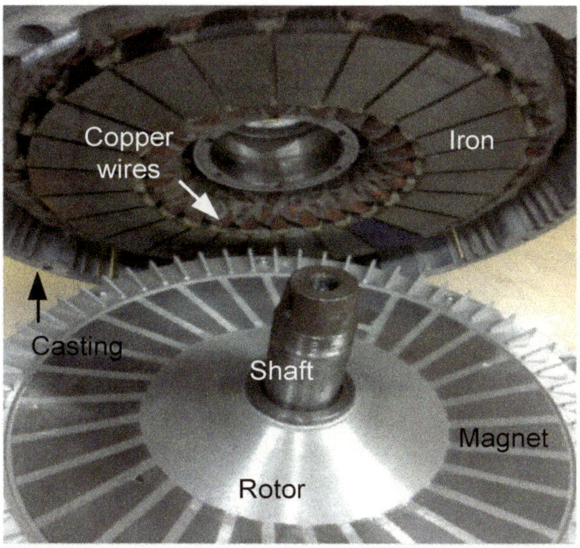

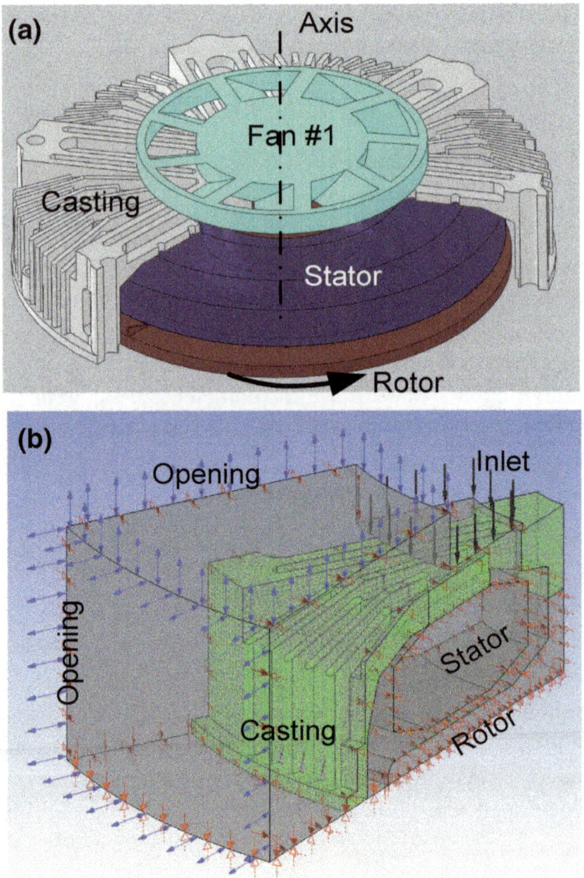


**Fig. 2** Numerical model of the AFPM. **a** CAD model; **b** CFD model

Some assumptions are required for solving the heat transfer problem. These are as follows: (1) The flow is turbulent and incompressible, (2) The flow is steady state, (3) Buoyancy and radiation heat transfer are neglected, and (4) the thermo-physical properties of the fluid are temperature independent. Based on these assumptions, the governing equations (continuity, momentum, and energy equations) are established as follows:

$$\frac{\partial(\rho u_i)}{\partial x_i} = 0, \tag{1}$$

$$\frac{\partial(\rho u_i u_j)}{\partial x_j} = -\frac{\partial p}{\partial x_i} + \frac{\partial}{\partial x_j}\left[\mu\left(\frac{\partial u_i}{\partial x_j} + \frac{\partial u_j}{\partial x_i} - \frac{2}{3}\frac{\partial u_k}{\partial x_k}\delta_{ij}\right)\right] + \frac{\partial}{\partial x_j}\left(-\rho\overline{u_i' u_j'}\right), \tag{2}$$

$$\frac{\partial[u_i(\rho E + p)]}{\partial x_j} = \frac{\partial}{\partial x_j}\left[\lambda_{\text{eff}}\frac{\partial T}{\partial x_j} + u_i(\tau_{ij})_{\text{eff}}\right] + S_E, \tag{3}$$

where $\rho$ is the fluid density, $u_i$ and $u_j$ are the velocity vectors in two perpendicular directions $i$ and $j$, respectively, $p$ is the static pressure, $\mu$ is the molecular viscosity, also referred to as the dynamic viscosity, $\delta_{ij}$ is the Kronecker delta function, $E$ is the total energy per unit mass, $S_E$ is the energy generation rate per unit volume, $T$ is the temperature, $u'_i$ and $u'_j$ represent the velocity fluctuations, $\overline{\rho u'_i u'_j}$ is the Reynolds stress, $\lambda_{\text{eff}}$ is the effective thermal conductivity, and $(\tau_{ij})_{\text{eff}}$ is the deviatoric stress tensor. Since this is also a conjugate heat transfer problem, which involves heat conduction, the governing equation of the solid is also required:

$$\frac{\partial}{\partial x_j}\left(\lambda_s\frac{\partial T}{\partial x_j}\right) + S_E = 0, \tag{4}$$

where $\lambda_s$ is the thermal conductivity of the solid.

A 3D model is built to contain all the rig components and the cooling fan. Figure 2a shows the sectional structure model with reasonable simplification and with addition of the cooling fan which does not exist in the original model shown in Fig. 1. Figure 2b depicts the CFD model used for the thermal analysis of the AFPM based on the simplified structural model in ANSYS CFX 14.0. The CFD model is verified in terms of grid size, turbulence model selection, and discretization schemes, and it encloses all components in an air cabinet by using solid–solid interfaces and fluidsolid interfaces. A total of four parts including the flow field (outside and inside) and structure model are built and connected together. Taking advantage of symmetrical geometry, it is only necessary to develop a 1/24 model (30° in the symmetrical model) to reduce the computational burden, in which symmetrical boundary conditions instead of periodic boundary conditions are used in the cut planes, since the planar smooth surfaces of the rotor have less effect on heat transfer even during high-speed rotation. The effect of small fins installed along the diameter edge is ignored, because the air gap between the rotor and stator is very small and its precise implementation in the calculation is difficult. From a basic electromagnetic simulation, the total heat generated in the full model is 780 W at 2600 r/min (including winding copper loss and iron loss), so an approximate estimation of heat generated at the stator and rotor can be made as shown in Table 1.

To optimize the cooling design, a comparative study is conducted for the AFPM machine under three conditions: no fan, fan #1, and fan #2. Fan #1 is an empirical design based on engineering experience, which is shown at the upper left corner of

**Table 1** Heat generated in the 1/24 model

| Component | Volume ($m^3$) | Heat (W) | Source ($W/m^3$) |
|---|---|---|---|
| Stator | $3.74 \times 10^{-5}$ | 16.25 | $4.34 \times 10^5$ |
| Rotor | $1.29 \times 10^{-4}$ | 16.25 | $1.26 \times 10^5$ |

Fig. 3. Fan #2 with curved surfaces is optimized by aerodynamic study and is shown at the lower right corner of Fig. 3. The two have identical outer diameters and the numbers of blades. For the model without a cooling fan, the inlet region in Fig. 2b is set as an opening boundary. In the case of the models with cooling fans, the opening boundary condition at the inlet region is replaced by a constant inlet air mass flow, which is calculated in terms of CFD simulation as shown in Fig. 3; Table 2. This gives a straightforward way of estimating the mass flow condition for the thermal performance of the AFPM machine, and it can be easily used for other similar CFD analyzes without prior knowledge of the mass flow condition.

Grid quality and independence have been examined to minimize their influence, where the grid of flow field has 462,117 nodes, the casting model has 55,574 nodes, and the rotor and stator models have 8577 and 1593 nodes, respectively. In addition, the standard $k$-$\varepsilon$ turbulence model is used, and a scalable wall function is implemented over the solid walls. Thermal energy instead of total energy is used to model the transport of enthalpy through the fluid domain due to the low speed flows through and around the machine. Automatic time step control is used and the convergence criteria are assumed to be satisfied if the root mean square (RMS) residual is lower than $1 \times 10^{-5}$.

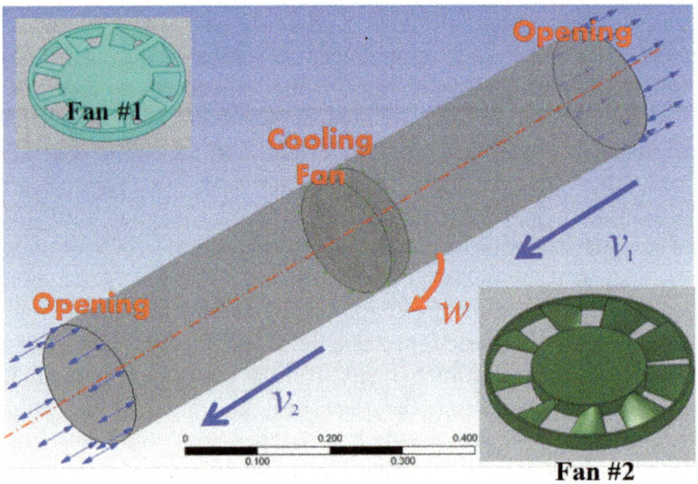


**Fig. 3** Simple cooling fan model for mass flow calculations

**Table 2** Mass flow rate at three cooling conditions

| Fan type | Mass flow rate (kg/s) | | |
|---|---|---|---|
| | 2600 r/min | 4600 r/min | 6600 r/min |
| No fan | No forced ventilation | | |
| #1 | 0.060 | 0.106 | 0.152 |
| #2 | 0.128 | 0.227 | 0.327 |

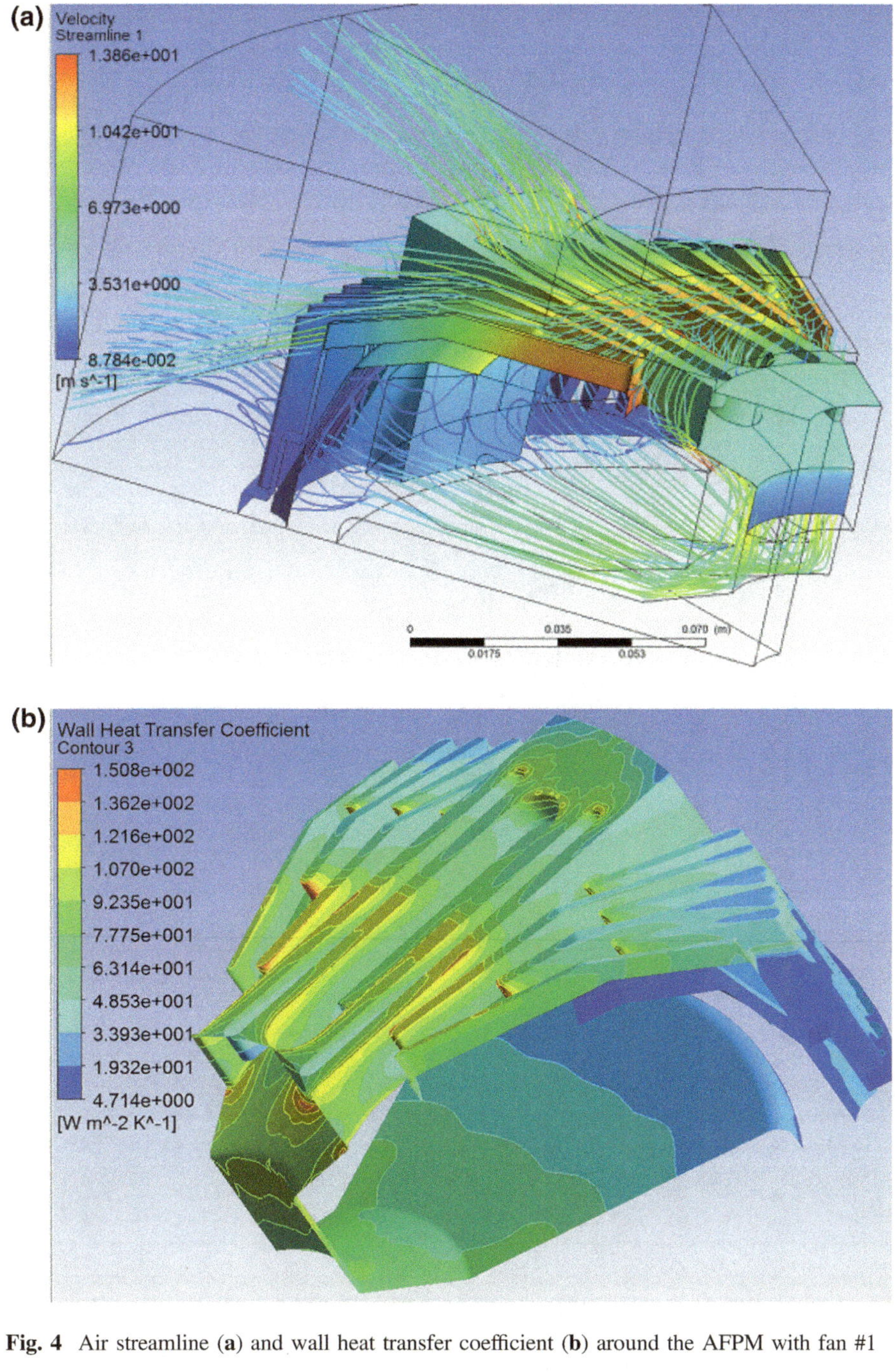


**Fig. 4** Air streamline (**a**) and wall heat transfer coefficient (**b**) around the AFPM with fan #1

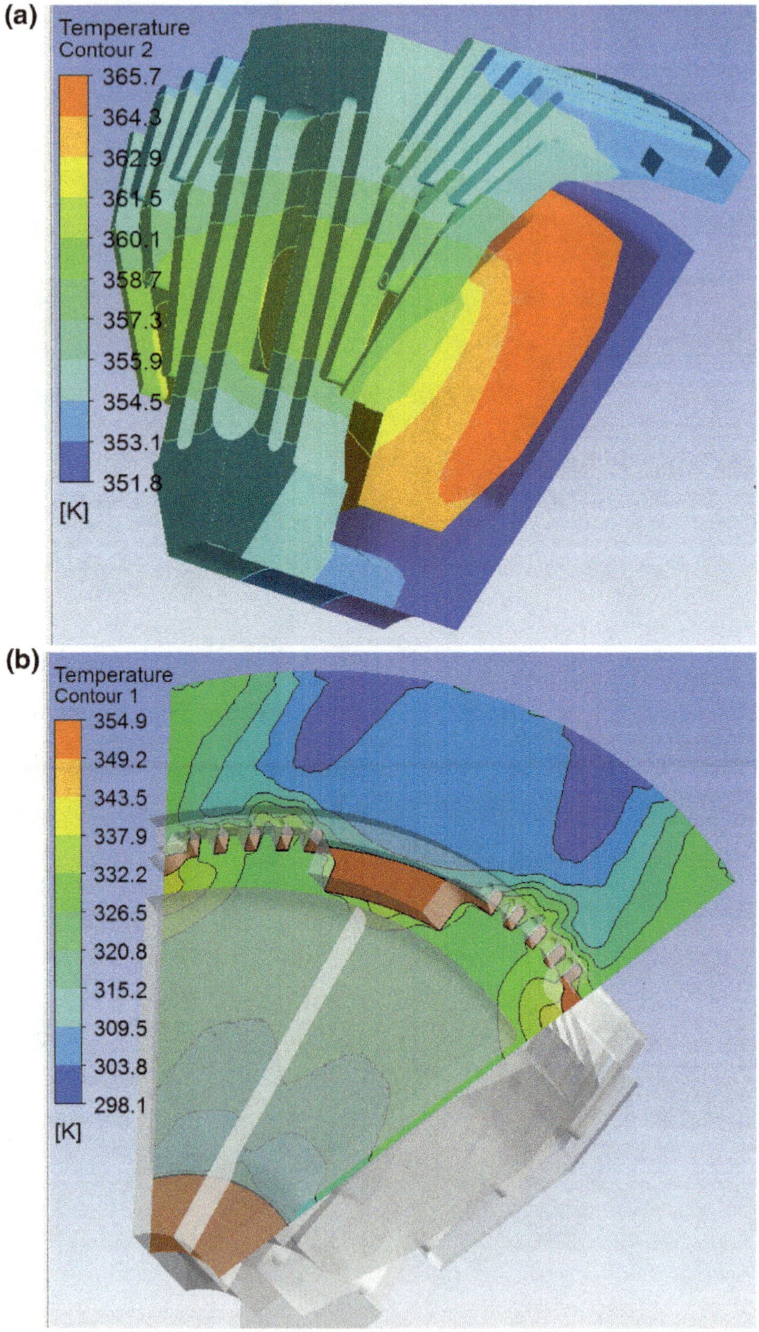


**Fig. 5** Temperature distribution of the machine (**a**) and the middle plane at the air gap (**b**)

**Table 3** Component temperature at three ventilation conditions

| Fan type | Temperature (°C) | | | | | |
|---|---|---|---|---|---|---|
| | Stator | | Rotor | | Casting | |
| | Min | Max | Min | Max | Min | Max |
| No fan | 237.5 | 247.2 | 192.1 | 215.4 | 213.6 | 242.5 |
| #1 | 86.8 | 92.5 | 78.6 | 80.5 | 80.5 | 87.6 |
| #2 | 72.1 | 77.7 | 65.4 | 66.7 | 66.0 | 72.9 |

## 4 Analysis of CFD Results

The thermal analysis is carried out on a 64-bit workstation, and each analysis takes about 3 h to converge. Figure 4a plots the streamline through and around the AFPM with the cooling fan #1. It is found that 74.8% of the total mass flow pumped by the cooling fan flows outside the machine and the remaining 25.2% of the total mass flow ventilates the inside space through the cutout on the casting. It is evident that both the external air flow and the internal flow play a significant role in heat dissipation. However, the wall heat transfer coefficient plot in Fig. 4b indicates that the external air flow provides the majority of the total heat dissipation as the external flow is much greater than the inside one on the disc surface. To enhance the total cooling performance, it is desired to design a better cutout on the casing, which would have less pressure loss and thus allow more air flow through it to directly cool the inside surfaces.

Figure 5 plots the temperature distribution on the three main components of the machine and the middle plane between the stator and rotor discs, where the maximum temperature can be found at the stator at the side approaching the rotor. The contact surfaces between the casting and stator play a significant part in the heat conduction, which leads to a temperature reduction in the stator. The temperature distribution on the middle plane illustrates the flow temperature and shows that, as expected, much heat is taken away by means of the cutout on the casting.

Table 3 lists the minimum and maximum temperatures in each component with and without cooling fans. Fans #1 and #2 are efficient in decreasing the temperature. In particular, the temperatures of the stator, rotor, and casting with fan #1 decrease by approximately 150, 110, and 150 °C, respectively, as compared to the machine without a cooling fan. The design with fan #2 has an even better cooling performance, resulting in a further 15 °C reduction compared to fan #1.

## 5 Conclusions

In this study, the CFD method has been used to investigate heat transfer in an axial flux permanent magnet machine and to optimize a cooling fan design. Some simplifications and assumptions are made to facilitate a numerical analysis, based on the steady CFD method, to predict the coolant air mass flow, velocity, and pressures outside and inside of the machine and the temperature distribution of the machine. The comparison results of the AFPM machine without a cooling fan, and with two designs of cooling fans reveal that cooling fans are helpful in enhancing heat transfer and reducing the temperature of the machine, and a good design of cooling fans based on aerodynamics is efficient and essential to avoid overheating problems. Further, work will be set out to conduct extensive experimental tests on these prototypes.

## References

Boglietti, A., Cavagnino, A., & Staton, D. A. (2005). TEFC induction motors thermal models: a parameter sensitivity analysis. *IEEE Transactions on Industry Applications, 41*(3), 756–763. doi:10.1109/TIA.2005.847311.

Chong, Y. C., Echenique-Subiabre, E. J. P., Mueller, M. A., et al. (2014). The ventilation effect on stator convective heat transfer of an axial flux permanent magnet machine. *IEEE Transactions on Industrial Electronics, 61*(8), 4392–4403. doi:10.1109/TIE.2013.2284151.

Dong, J., Huang, Y., Jin, L., et al. (2014). Thermal optimization of a high-speed permanent magnet motor. *IEEE Transactions on Magnetics, 50*(2), 749–752. doi:10.1109/TMAG.2013.2285017.

Dorrell, D.G., Staton, D.A., Hahout, J., et al. (2006). Linked electromagnetic and thermal modelling of a permanent magnet motor. In *3rd IET International Conference on Power Electronics, Machines and Drives.* (pp. 536–540). Dublin, Ireland. doi:10.1049/cp:20060166.

El-Refaie, A. M., Harris, N. C., Jahns, T. M., et al. (2004). Thermal analysis of multibarrier interior PM synchronous machine using lumped parameter model. *IEEE Transactions on Energy Conversion, 19*(2), 303–309. doi:10.1109/TEC.2004.827011.

Hendershot, J. R., & Miller, T. J. E. (1994). *Design of Brushless Permanent Magnet Motors.* Oxford, UK: Magna Physics and Oxford Science Publications.

Lee, Y. S., Hahn, S., & Kauh, S. K. (2000). Thermal analysis of induction motor with forced cooling channels. *IEEE Transactions on Magnetics, 36*(4), 1398–1402. doi:10.1109/20.877700.

Mellor, P. H., Roberts, D., & Turner, D. R. (1991). Lumped parameter thermal model for electrical machines of TEFC design. *IEE Proceedings B (Electric Power Applications), 138*(5), 205–218. doi:10.1049/ip-b.1991.0025.

Mezani, S., Talorabet, N., & Laporte, B. (2005). A combined electromagnetic and thermal analysis of induction motors. *IEEE Transactions on Magnetics, 41*(5), 1572–1575. doi:10.1109/TMAG.2005.845044.

Trigeol, J. F., Bertin, Y., & Lagonotte, P. (2006). Thermal modeling of an induction machine through the association of two numerical approaches. *IEEE Transactions on Energy Conversion, 21*(2), 314–323. doi:10.1109/TEC.2005.859964.

Wrobel, R., Vainel, G., Copeland, C., et al. (2013). Investigation of mechanical loss and heat transfer in an axial-flux PM machine. In *IEEE Energy Conversion Congress and Exposition.* (pp. 4372–4379). Denver, USA. doi:10.1109/ECCE.2013.6647285.

Yu, Q., Laudensack, C., & Gerling, D. (2010). Improved lumped parameter thermal model and sensitivity analysis for SR drives. In *XIX International Conference On Electrical Machines.* (pp. 1–6). Rome, Italy. doi:10.1109/ICELMACH.2010.5607746.

## Author Biography

**Zheng Tan** received the B.Eng. degree in electrical engineering and automation from the Inner Mongolia University of Science and Technology, Baotou, China, in 2010; and the M.Sc. and Ph.D. degree in electrical power from Newcastle University, Newcastle upon Tyne, U.K., in 2011 and 2016 respectively. He is currently a research staff with State Grid Jibei Electric Power Co. Ltd Research Institute, Beijing, China.

# Index





Y. Fang and Y.H. Zhang (eds.), *China's High-Speed Rail Technology*,
Advances in High-speed Rail Technology, https://doi.org/10.1007/978-981-10-5610-9